optimal statistical decisions

optimal
statistical
decisions

Morris H. DeGroot

Professor of Mathematical Statistics
and
Head, Department of Statistics
Carnegie-Mellon University

Wiley Classics Library Edition Published 2004

WILEY-
INTERSCIENCE

A JOHN WILEY & SONS, INC., PUBLICATION

Published by John Wiley & Sons, Inc., Hoboken, New Jersey.
Published simultaneously in Canada.

For general information on our other products and services please contact our Customer Care Department within the U.S. at 877-762-2974, outside the U.S. at 317-572-3993 or fax 317-572-4002.

Wiley also publishes its books in a variety of electronic formats. Some content that appears in print, however, may not be available in electronic format.

Library of Congress Cataloging-in-Publication Data is available.

ISBN 978-0-471-68029-1

10 9 8 7 6 5 4 3

To Dolores

*I'm afraid we have been a little too sure that
the Stoat turns white to make itself invisible
against its background of snow, in accordance with
the theory of protective coloration. That explanation is very pretty and snug, but it looks silly
sometimes, especially when there is no snow within
miles. Perhaps we need more data on whether the
Stoat was expecting snow when it turned white.*

foreword to the
Wiley Classics Libary edition

Optimal Statistical Decisions is a landmark book. It is still about the clearest introduction to Bayesian Statistical Decision Theory available. Written in the late 1960's, it does not cover the computational advances that have become so popular and well-used. But with the rapid growth of interest in computational methods, it is all too easy to neglect what purposes are served by those computations. DeGroot's book, with its clear exposition of Bayesian principles, will be useful to help keep those purposes in mind.

The book is divided into four parts. The first, a review of probability theory, is useful as a reference and for clarifying notation. The second, a derivation of the principle of maximizing expected utility is original in that it takes two primitives: "is more likely than" for probability, and "is preferred to" for utility. This resolves the uncomfortable fact that using only "is preferred to" among gambles identifies the product of probability and utility, but does not clarify how to disentangle them. The third part, on statistical decisions, rehearses the facts about fixed-sample posterior decisions. Chapter 11 compares Bayesian answers to classical ones concerning the linear model (and goes further in compromising Bayesian ideas than I find comfortable now).

The strongest part of the book, in my mind, is the last part, on sequential decisions. This is the subject that most engaged DeGroot as a researcher, and here we see the full power of his intellect.

To appreciate a book, it is useful to understand the author. Even a reader who never knew DeGroot will appreciate what a wonderful writer he was. While some may disagree with what he writes, it is hard to imagine that anyone will be in doubt as to what he means. His papers are as clear as his books. But this is only one aspect of an extraordinary scholar and person. He was an institutional builder, as founder of the Statistics Department at Carnegie Mellon University and as first Executive Editor of Statistical Science. He was a wonderful colleague and friend, always ready for a chat about principles, a research problem, a departmental problem, a reference or personal advice.

It is a great pleasure to me to play a role in reintroducing a great book and a great man to a new generation of Bayesian statisticians.

JOSEPH B. ("JAY") KADANE

Pittsburgh, Pa.
Nov. 22, 2003

preface to the original edition

Statistical decision theory and methods of Bayesian statistical inference have been both intensively and extensively developed during the past twenty years. A unified theory has been constructed during this period, and the concepts and methods have been widely applied to problems in the areas of engineering and communications, economics and management, psychology and behavioral science, and systems and operations research. Because of these developments, interest in decision theory and its applications has greatly increased at all mathematical levels. The purpose of this book is to provide, at an advanced undergraduate or beginning graduate level, a thorough course in the theory and methodology of optimal statistical decisions.

The book is intended for students in the areas of application mentioned above, as well as for students in statistics and mathematics. Throughout the book, expository discussions are presented to ease the reader's path through the technical material. Complete proofs and derivations are given for almost all theoretical results, but these results are usually introduced or followed by explanations and examples.

A mathematical background of calculus and elementary matrix theory is assumed in this book. In addition, a one-year introductory course in probability and mathematical statistics is highly desirable.

Such a course is not strictly necessary, since the probability concepts that will be used in the book are surveyed in the early chapters. Nevertheless, it is strongly recommended that the reader have had at least an introduction to probability, random variables, and distributions. Since the book contains both rigorous derivations and expository discussions, it is hoped that each reader with this minimum background will be able to proceed at a mathematical level appropriate to his training.

There is enough material for a one-year course in decision theory, but the book can also easily be used for a one-semester course or a one-quarter course. Chapters 2 and 3 provide a survey of probability theory and related topics. It is recommended that these chapters be skimmed lightly or be skipped entirely in a first reading. The reader can refer to them later if necessary. Chapter 4 contains a catalog of univariate distributions for later reference.

Throughout the book, a decision problem which involves the multivariate normal distribution, or some other multivariate distribution, is usually considered in a separate section from that in which the corresponding problem involving a univariate distribution is considered. Hence, discussions involving multivariate distributions can easily be skipped by less advanced readers. For this reason, the multivariate distributions described in Chapter 5 may be studied in varying amounts of detail by different readers, and the chapter may even be omitted entirely by some readers.

In Chapters 6 and 7, the fundamental concepts of subjective probability and utility are developed and discussed. In a short course, it is recommended that the expository sections of these chapters be included and the axiomatic derivations be omitted.

Chapters 8 to 11 contain a comprehensive development of Bayesian methods in statistical decision problems. These chapters can serve as the basis of a one-semester course, and it is recommended that they be studied carefully.

In the final part of the book, Chapters 12 to 14, problems of sequential decision making and dynamic programming are introduced and studied. Since all books must come to an end, problems in which time is regarded as a continuous parameter are not considered, and only sequential processes which evolve in discrete stages are included. Much of the material presented here appears in a textbook for the first time. In the final chapter, applications are described in detail in inventory theory, control theory, search problems, gambling systems, and information theory.

Many exercises are presented at the end of each chapter, and the correct answers are included with almost all of them. A few exercises, which require the development of the reader's own probabilities and utilities, are of a psychological nature and have no single correct answer.

More than 400 references are cited in the text. These references are listed at the end of the book, together with a supplementary bibliography containing about 150 additional items.

The comments of many colleagues, students, and friends have been extremely helpful to me. I am indebted to all of them and regret that they cannot be mentioned individually. Versions of the manuscript at various stages of its development were typed by Lee Martin, Dolores DeGroot, Harriet Merrill, and JoEllen Luncher. Each of them has been of great assistance, and each has made an essential contribution to this book. The excellent and comprehensive editorial work of my father, Archibald DeGroot, brought the final draft of this manuscript much closer to my goal of a text that is both technical and readable. Finally, I am indebted to Dean Richard M. Cyert of Carnegie-Mellon University. His encouragement and support have enabled me to devote significant portions of my time to the preparation of this book.

Morris H. DeGroot

contents

part two: subjective probability and utility

part three: statistical decision problems

optimal statistical decisions

survey of probability theory

introduction

The science of statistics deals with the development of theories and techniques which are appropriate for making inferences under the conditions of uncertainty and partial ignorance that necessarily exist in a wide range of activities. Current statistical practice depends mainly on the formulation of probability models of various physical systems, on methods of collecting and analyzing numerical data, and on the design of efficient and informative experiments. The term *decision theory* refers to the class of statistical problems in which the statistician must gain information about certain critical parameters in order to be able to make effective decisions in situations where the consequences of his decisions will depend on the values of these parameters. The theory of optimal statistical decisions that will be developed in this book can be of value in many types of activities and has been successfully applied in a wide variety of problems. This theory is commonly called subjective statistical decision theory or Bayesian statistical decision theory.

In most of the book, the decision-making aspects of each problem can be formalized through an explicit specification covering the available decisions, the items of cost or gain involved in making these decisions, and the relevant probability distributions. The theory and techniques on which this specification is based will be discussed thoroughly. The discussions will include mathematical methods of characterizing the statistician's information and uncertainty relating to the values of the parameters and also methods of changing this characterization as additional information about the parameters is acquired.

Subjective, or Bayesian, statistical decision theory is applicable to those problems in which the information and uncertainty about the parameters can, at any time, be summarized by a probability distribution of their possible values. Therefore, this book will deal only with those statistical decision problems which meet the following two requirements: (1) The conditions can be formulated in terms of a manageable number of parameters. (2) Although the values of these parameters are not known exactly, any uncertainty about the values can be represented by a suitable probability distribution. In connection with these requirements, the following two points should be kept in mind. First, the number of parameters that would be manageable in a given situation is, to a large extent, a function of the current state of computer science and technology (obviously, this number is larger now than ever before). Second, there is considerable controversy among statisticians and other persons who study the foundations of probability as to whether or not uncertainty about the value of a particular parameter can be represented by a probability distribution. Some authorities believe that such a distribution is appropriate only when values of the parameter clearly have relative frequencies. Another group of authorities maintains that probability is a logical concept which can be applied to parameters in a much wider class of problems. Furthermore, these authorities believe that in each such problem there is a uniquely defined distribution which is appropriate for a particular parameter and must necessarily be assigned to that parameter. Authorities in a third group believe that essentially all probability distributions are subjective and that whenever anyone carries out a statistical investigation involving a parameter, he can represent his uncertainty pertaining to the values of that parameter by a suitable probability distribution. There are also other distinctive opinions and many modifications of the three here mentioned. Authors of some of the well-known books on these matters are, in alphabetical order, Carnap (1962), Fisher (1956), Good (1950), Jeffreys (1961), Keynes (1921), Nagel (1937), Reichenbach (1949), Savage (1954), and von Mises (1957).

It will be assumed in all problems in this book that each parameter can be assigned a particular probability distribution. Reasoning of the type given in Chap. 6 should be relevant in a broad class of decision problems, and this reasoning suggests how suitable distributions might be assigned in such problems. Even though there is a great deal of controversy in regard to certain aspects of the foundations of probability and statistics, the differences of opinion do not extend to the propriety of assigning probability distributions to the parameters in many of the specific problems that will be studied here. Moreover, in problems where the parameters do have appropriate probability distributions, including many important problems of common occurrence, it is generally agreed that the theory and techniques of statistical decision making to be presented in this book are applicable, correct, and useful.

Since the modern theory of mathematical probability is the basic tool in the development of statistical methods, the next four chapters are devoted to a survey of the portions of that theory which will be used later in the book. In order that this survey may be somewhat condensed and still be essentially complete with respect to the results which will be needed, proofs and motivation will generally be omitted. Readers for whom this brief survey is insufficient should consult any of the standard texts on probability and statistics. Some recent texts at an introductory level are Brunk (1965), Feller (1957), Freeman (1963), Harris (1966), Hogg and Craig (1965), Lindgren (1962), McCord and Moroney (1964), Mood and Graybill (1963), Papoulis (1965), Parzen (1960), Pfeiffer (1965), and Tucker (1962). Some others at a more advanced level are Cramér (1946), Feller (1966), Fisz (1963), Gnedenko (1962), Krickeberg (1965), Loève (1963), Rao (1965), and Wilks (1962).

experiments, sample spaces, and probability

2.1 EXPERIMENTS AND SAMPLE SPACES

Decisions of the type to be studied in this book pertain to the design, performance, or analysis of experiments. Throughout the book, the person or committee making these decisions will be called the *statistician*. The word "experiment" is used here in a very general sense to describe virtually any process of which all possible outcomes can be specified in advance and of which the actual outcome will be one of those specified. The outcome of an experiment may be random or nonrandom, in the usual senses of those terms. (Indeed, the problem of distinguishing between experiments with random outcomes and those with nonrandom outcomes is at the heart of the controversy described in Chap. 1.) For the purposes of this book, the interesting feature of an experiment is simply that its outcome is not definitely known by the statistician beforehand. Some examples of experiments are:

1. Ascertaining the year of birth of a specified person famous in history
2. Determining whether a certain baby, who was born this morning, will ever have at least one grandchild
3. Selecting a sample of 300 items at random from a very large

manufactured lot of similar items and counting the number of these selected items that are defective

4. Observing the air temperature at a certain location at specified intervals of time during a period of several years

5. Determining the average lifetime of a large number of light bulbs produced by some manufacturing process

6. Tossing a coin repeatedly until three heads have been obtained and observing the total number of tosses required

The set of all possible outcomes of an experiment is called the *sample space* S of the experiment. For a given experiment, the number of outcomes and the names of the outcomes in S may be chosen differently by different statisticians. The basic requirements are that S must include *all* possible outcomes and that each outcome must be described with all essential detail. Since the statistician must assign probabilities to various sets of outcomes, he usually establishes the specification for the sample space so as to facilitate this assignment. For this reason, in the specification for S, the outcomes will often be described in much greater detail than would be normal for the type of decision envisaged. For instance, suppose that in a certain experiment the statistician is interested in the sum of the numbers that appear when five dice are tossed at one time. It is usually simplest for him to compute the relevant probabilities when each outcome is regarded as an ordered sequence of five numbers, because all the outcomes of S can then be assigned the same probability.

2.2 SET THEORY

The sample space S of any experiment is conveniently regarded as a set of points or elements, each element being a possible outcome of the experiment. In this section we shall review briefly the standard notation, terminology, and fundamentals of set theory that will be used throughout the book.

The statement that a point x belongs to a set A, or is a member of a set A, is written $x \in A$. The statement that the set A is contained in the set B is written $A \subset B$ or $B \supset A$; this statement means that each point that belongs to A also belongs to B. If $A \subset B$, then A is said to be a *subset* of B.

Let S be the sample space of some experiment. For any set $A \subset S$, the complement of A, denoted by A^c, is the set of points in S that do not belong to A. The set $\emptyset = S^c$ is the empty set.

Let α be an arbitrary collection of sets, each of which is a subset of S. The *union* of the sets in α is defined to be the set that contains the points in S that belong to at least one of the sets in α. The *intersection* of the sets in $\mathcal{C}$ is defined to be the set that contains the points in S that are common to all of the sets in $\mathcal{C}$. The union of any n sets $A_1, \ldots, A_n$ is written as $\bigcup_{i=1}^{n} A_i$ or as $A_1 \cup \cdots \cup A_n$. The intersection of $A_1, \ldots, A_n$ is written as $\bigcap_{i=1}^{n} A_i$ or as $A_1 \cap \cdots \cap A_n$, or simply as $A_1 A_2 \cdots A_n$. Similar notation is used for unions and intersections of an infinite sequence of sets.

Two sets A and B are said to be *disjoint* if they have no points in common, i.e., if $AB = \emptyset$. A finite or infinite sequence of disjoint sets is a sequence in which all pairs of sets are disjoint.

Consider a certain property Π with the feature that any particular point $s \in S$ either has property Π or does not have property Π. The set A containing all points that have property Π will be denoted by the relation

$$A = \{s: s \text{ has property } \Pi\}.$$

Some basic results derived from the preceding principles are presented as Exercises 2 and 3 at the end of this chapter.

Product Spaces

Suppose that $S_1, \ldots, S_n$ are the sample spaces of n experiments. The cartesian product, or product space, $S_1 \times \cdots \times S_n$ is defined to be the set of all ordered n-tuples $(s_1, \ldots, s_n)$, where $s_i \in S_i$ for $i = 1, \ldots, n$. The product space $S_1 \times \cdots \times S_n$ will be the sample space of a composite experiment in which all n individual experiments with sample spaces $S_1, \ldots, S_n$ are performed. Similarly, if $S_1, S_2, \ldots$ is an infinite sequence of sample spaces, the infinite cartesian product $S_1 \times S_2 \times \cdots$ is defined to be the set of all infinite sequences $(s_1, s_2, \ldots)$, where $s_i \in S_i$ for $i = 1, 2, \ldots$.

In many problems, the sample spaces of the n individual experiments are identical. If S is the common sample space of each of the n experiments, the n-fold product space $S \times \cdots \times S$ will be written as S^n. In other words, S^n is the set of all ordered n-tuples $(s_1, \ldots, s_n)$, where $s_i \in S$ for $i = 1, \ldots, n$. Similarly, S^∞ denotes the set of all infinite sequences $(s_1, s_2, \ldots)$, where $s_i \in S$ for $i = 1, 2, \ldots$.

The Real Line and R^n

The outcomes of most of the experiments that will be considered in this book will be real numbers or n-tuples of real numbers. The set of all real numbers, i.e., the real line, will be denoted by R^1. Furthermore, the n-dimensional Euclidean space of all ordered n-tuples of real num-

bers will be denoted by R^n. It follows from these definitions that
$R^n = R^1 \times \cdots \times R^1$.

2.3 EVENTS AND PROBABILITY

If A is a subset of the sample space S of some experiment, the probability
of A (to be precisely defined shortly) is a numerical measure of the chances
that the actual outcome of the experiment will be an element of A. Much
of the elementary theory of probability deals with techniques for computing the probabilities of certain relatively complex sets that are of special interest from the probabilities of certain simpler sets. However, some
particular subset of S may be so irregular in structure that it is not possible to assign a probability to this subset. For example, suppose that
a projectile is aimed at a two-dimensional target. It is often reasonable
to assume that the probability of the projectile hitting in any given subset
of the target is proportional to the area of the subset. However, certain
subsets of the plane are so irregular that no meaningful definition of area
can be applied to them. The reader is referred to any standard text on
measure theory, such as Halmos (1950) or Kolmogorov and Fomin (1961),
for examples of such nonmeasurable subsets, not only in the plane but on
the real line as well. Therefore, for certain experiments, probabilities are
assigned only to those subsets of S that are of special interest and are
sufficiently regular, rather than to all subsets of S. In a practical experiment, a limitation of this kind is not particularly restrictive because
virtually every subset of interest is sufficiently regular to have a probability. The actual computation of the probabilities of some subsets may
be extremely difficult, but this difficulty is not a consideration in the
present discussion.

A family α of sets, each of which is a subset of S, is called a σ-field
if the family meets the following three requirements:

1. $S \in \alpha$.
2. If $A \in \alpha$, then $A^c \in \alpha$.
3. If $A_1, A_2, \ldots$ is an infinite sequence of sets from α, then
   $\cup_{i=1}^{\infty} A_i \in \alpha$.

It is not difficult to show that the union of a finite number of sets from α
will again belong to α, as will the intersection of a finite or countable
number of sets from α (see Exercise 4).

For any experiment with a sample space S, probabilities will be
defined for all sets in an appropriately chosen σ-field α. The sets in α
are called *events*. In this book, as in most work in probability, if S contains only a finite or countable number of outcomes, then α will include

all subsets of S. If S is the n-dimensional space R^n, then $\mathcal{C}$ will be the σ-field of *Borel sets*, i.e., the smallest σ-field containing all n-dimensional intervals.

For a given sample space S and σ-field $\mathcal{C}$, a *probability distribution* P on $(S, \mathcal{C})$ is a nonnegative function the value of which is defined for each event in $\mathcal{C}$ and which has the following two properties:

1. $P(S) = 1$.
2. If $A_1, A_2, \ldots$ is a sequence of disjoint events, then

$$P(\bigcup_{i=1}^{\infty} A_i) = \sum_{i=1}^{\infty} P(A_i).$$

The triple $(S, \mathcal{C}, P)$ is called a *probability space*. Some basic properties of probability distributions are presented in Exercise 5.

For any event A, the symbol $\Pr(A)$ will be used regularly in this book to denote the probability of A, regardless of any other symbol that might also be used in a special context to indicate the particular probability distribution under consideration.

Independent Events

Two events A and B are said to be independent if $\Pr(AB) = \Pr(A)\Pr(B)$. The events in a family $\mathcal{B}$ of events are *independent* if, for every finite sequence $A_1, \ldots, A_k$ of events in $\mathcal{B}$,

$$\Pr(\bigcap_{i=1}^{k} A_i) = \prod_{i=1}^{k} \Pr(A_i). \tag{1}$$

This concept is very important in assigning probabilities to events. For instance, if two or more events are regarded as being physically independent, in the sense that the occurrence or nonoccurrence of some of them has no influence on the occurrence or nonoccurrence of the others, then this condition is translated into mathematical terms through the assignment of probabilities satisfying Eq. (1).

Measurable Functions

Consider a given sample space S and σ-field $\mathcal{C}$. Let g be a real-valued function defined at all points of S, and for each real number x, let A_x be the subset of S defined by the relation

$$A_x = \{s: g(s) \le x\}. \tag{2}$$

The function g is said to be *measurable with respect to* $\mathcal{C}$, or $\mathcal{C}$-*measurable*, if the set A_x belongs to the σ-field $\mathcal{C}$ for every number $x \in R^1$.

If g is $\mathcal{Q}$-measurable and B is any Borel set on the real line, then the subset $g^{-1}(B)$ of S, defined by the relation

$$g^{-1}(B) = \{s: g(s) \in B\}, \tag{3}$$

also belongs to $\mathcal{Q}$.

Although the concept of a measurable function is vital to the general theory of measure and integration, this concept will not be emphasized in this book, and technical problems concerning the measurability of functions will typically not be discussed. Nevertheless, the measurability of certain functions of special interest will be explicitly demonstrated for the benefit of more advanced readers.

2.4 CONDITIONAL PROBABILITY

In any given experiment, it is often necessary to consider the probability of the occurrence of an event A when additional information about the outcome of the experiment has been obtained from the occurrence of some other event B. This is called the *conditional probability* of A when B is given. If $\Pr(B) > 0$, then the conditional probability $\Pr(A|B)$ of A, given B, is defined to be

$$\Pr(A|B) = \frac{\Pr(AB)}{\Pr(B)}. \tag{1}$$

The definition of the conditional probability $\Pr(A|B)$ when $\Pr(B) = 0$ is discussed in the next chapter.

The concept of conditional probability as defined by Eq. (1) is useful when probabilities must be assigned to events. In some problems, unconditional probabilities are most easily obtained by first assigning certain conditional probabilities directly and then calculating from them the desired unconditional probabilities. Exercise 9 illustrates how this procedure is often used. Of course, in a certain sense, all probabilities are conditional probabilities because they depend on such factors as the statistician's knowledge and the particular mathematical model he chooses to represent an experiment. In this book, we shall avoid the philosophical complexities of this comment by simply considering as given and fixed the information available to the statistician at the time a specific problem is being analyzed and by regarding those probabilities which do not depend on any further information or change as unconditional probabilities.

Although the following theorem, which is commonly known as *Bayes' theorem*, is elementary, it is used extensively and will be applied in various guises throughout this book. The proof is presented here in detail.

Bayes' theorem   *Let A_1, A_2, . . . be an infinite sequence of disjoint events with $\bigcup_{i=1}^{\infty} A_i = S$ and $\Pr(A_i) > 0$ for $i = 1$, 2, ...   Also, let B be another event such that $\Pr(B) > 0$.   Then*

$$\Pr(A_i|B) = \frac{\Pr(B|A_i)\Pr(A_i)}{\sum_{j=1}^{\infty} \Pr(B|A_j)\Pr(A_j)} \qquad i = 1, 2, \ldots \qquad (2)$$

A similar result applies to a finite sequence of disjoint events A_1, . . . , A_n satisfying the above conditions.

Proof   It follows from Eq. (1) that for any fixed value of i ($i = 1, 2, \ldots$), $\Pr(A_i|B) = \Pr(BA_i)/\Pr(B)$.   But again by Eq. (1),

$$\Pr(BA_i) = \Pr(B|A_i)\Pr(A_i).$$

It follows from the hypothesis about the sequence A_1, A_2, . . . that $B = \bigcup_{j=1}^{\infty} BA_j$.   Since the events BA_1, BA_2, . . . are disjoint, it follows from property 2 in the definition of a probability distribution given in Sec. 2.3 that

$$\Pr(B) = \sum_{j=1}^{\infty} \Pr(BA_j) = \sum_{j=1}^{\infty} \Pr(B|A_j)\Pr(A_j). \blacksquare$$

2.5 BINOMIAL COEFFICIENTS

Binomial coefficients can be used to simplify the computation of probabilities of events in a wide variety of experiments.   They are especially helpful in combinatorial problems in which the sample space S has a finite or countable number of outcomes and the computation of the probability of a particular event requires counting the number of outcomes in S that satisfy certain specified conditions.   For any real number a and any positive integer x, the binomial coefficient $\binom{a}{x}$ is defined by the relation

$$\binom{a}{x} = \frac{\Pi_{i=0}^{x-1}(a - i)}{x!}. \qquad (1)$$

Furthermore, by definition, $\binom{a}{0} = 1$ for any real number a.

By these relations, the binomial coefficient $\binom{a}{x}$ is defined for any nonnegative integer x, and only for these values of x.   If a is a positive integer with $a \geq x$, then it follows from Eq. (1) that

$$\binom{a}{x} = \frac{a!}{x!(a - x)!},$$

where $0! = 1$. Also, if a is a positive integer with $a < x$, then it follows from Eq. (1) that $\binom{a}{x} = 0$.

EXERCISES

1. Give at least two different specifications for the sample space in each of the six examples of experiments mentioned in Sec. 2.1.

2. Show that for any sets A, B, and D,

$$A(B \cup D) = (AB) \cup (AD).$$

Also show that

$$(A \cup B)^c = A^c B^c \qquad \text{and} \qquad (AB)^c = A^c \cup B^c.$$

3. Suppose that A_1, A_2, $\ldots$ is a given sequence of sets. Let $B_1 = A_1$ and

$$B_n = (A_1 \cup \cdots \cup A_{n-1})^c A_n \qquad n = 2, 3, \ldots.$$

Show that B_1, B_2, $\ldots$ is a sequence of disjoint sets such that

$$\bigcup_{j=1}^{n} A_j = \bigcup_{j=1}^{n} B_j \qquad n = 1, 2, \ldots$$

and

$$\bigcup_{j=1}^{\infty} A_j = \bigcup_{j=1}^{\infty} B_j.$$

4. Suppose that a family $\mathcal{Q}$ of sets is a σ-field. Assuming that A_1, A_2, $\ldots$ is an infinite sequence of sets from $\mathcal{Q}$, prove that $\bigcap_{i=1}^{\infty} A_i \in \mathcal{Q}$. Also, prove that $\bigcup_{i=1}^{n} A_i \in \mathcal{Q}$ and $\bigcap_{i=1}^{n} A_i \in \mathcal{Q}$ for $n = 2, 3, \ldots$.

5. Prove that any probability distribution P has the following properties:

(a) If A_1, $\ldots$, A_n are disjoint events, then $P(\bigcup_{i=1}^{n} A_i) = \sum_{i=1}^{n} P(A_i)$ and $P(\emptyset) = 0$.

(b) $P(A^c) = 1 - P(A)$.

(c) If A and B are events such that $A \subset B$, then $P(A) \leq P(B)$.

(d) If A_1, A_2, $\ldots$ is any sequence of events, then

$$P\left(\bigcup_{i=1}^{\infty} A_i\right) \leq \sum_{i=1}^{\infty} P(A_i).$$

(e) If A_1, A_2, and A_3 are any three events, then

$$P(A_1 \cup A_2) = P(A_1) + P(A_2) - P(A_1 A_2)$$

and

$$P(A_1 \cup A_2 \cup A_3) = P(A_1) + P(A_2) + P(A_3) - P(A_1 A_2) - P(A_1 A_3)$$
$$- P(A_2 A_3) + P(A_1 A_2 A_3).$$

Prove by induction the analogous result for n events A_1, $\ldots$, A_n.

(f) If $A_1 \subset A_2 \subset \cdots$ is a nondecreasing sequence of events, then

$$P\left(\bigcup_{i=1}^{\infty} A_i\right) = \lim_{n \to \infty} P(A_n).$$

(g) If $A_1 \supset A_2 \supset \cdots$ is a nonincreasing sequence of events, then

$$P(\bigcap_{i=1}^{\infty} A_i) = \lim_{n \to \infty} P(A_n).$$

6. A certain system has n components which operate independently of each other. If each component has probability 0.01 of failing, what is the probability that at least one of the components will fail?

7. A secretary types five letters and the corresponding five envelopes and then puts the letters into the envelopes at random. Find the probability that no letter is put into its proper envelope.

8. Suppose that $A_1, \ldots, A_k$ are independent events. Let $B_1, \ldots, B_k$ be another sequence of k events such that for each value of i ($i = 1, \ldots, k$), either $B_i = A_i$ or $B_i = A_i{}^c$. Prove that $B_1, \ldots, B_k$ are independent events. *Hint:* Use an induction argument on the number of indices i such that $B_i = A_i{}^c$.

9. If $A_1, \ldots, A_k$ are events such that $P(\bigcap_{i=1}^{j} A_i) > 0$ for $j = 1, 2, \ldots,$ $k - 1$, prove that

$$P(\bigcap_{i=1}^{k} A_i) = P(A_1)P(A_2|A_1)P(A_3|A_1A_2) \cdots P(A_k|A_1A_2 \cdots A_{k-1}).$$

Also, if $P(\bigcap_{i=1}^{j} A_i) = 0$ for some value of j, prove that $P(\bigcap_{i=1}^{k} A_i) = 0$.

10. (a) A gambler has in his pocket a fair coin (with a head and a tail) and a two-headed coin. He selects one of the coins at random, and when he flips it, it shows heads. What is the probability that it is the fair coin? (b) Suppose that he flips the same coin a second time and again it shows heads. What is now the probability that it is the fair coin? (c) Suppose that he flips the same coin a third time and it shows tails. What is now the probability that it is the fair coin?

11. Three different machines were used for producing a large lot of similar manufactured items. Suppose that 20 percent of the items were produced by machine A, 30 percent by machine B, and 50 percent by machine C. Suppose further that 4 percent of the items produced by machine A are defective, that 3 percent of the items produced by machine B are defective, and that 1 percent of the items produced by machine C are defective. (a) If one item selected at random from the entire lot is found to be defective, by which machine is it most likely to have been produced? (b) If 100 items are selected at random from the entire lot and exactly two of them are found to be defective, what is the probability that both were produced by machine A?

12. There are three coins in a box. One is a two-headed coin, another is a two-tailed coin, and the third is a fair coin. When one of the three coins is selected at random and flipped, it shows heads. What is the probability that it is the two-headed coin?

13. A box contains five slips of paper numbered, respectively, 1, 2, 3, 4, and 5. Suppose that two independent drawings are made from the box with replacement; in other words, the first slip is returned to the box before the second slip is drawn. If it is known that the number on the second slip is at least as large as the number on the first slip, what is the probability that the number on the first slip is 2?

14. Show that for any nonnegative integer r,

$$\binom{-1}{r} = (-1)^r \quad \text{and} \quad \binom{-2}{r} = (-1)^r(r + 1).$$

15. Show that for any positive integer n,

$$\binom{n}{0} + \binom{n}{1} + \binom{n}{2} + \cdots + \binom{n}{n} = 2^n.$$

and

$$\binom{n}{0} - \binom{n}{1} + \binom{n}{2} - \cdots \pm \binom{n}{n} = 0.$$

16. (a) Show that for any number x and any positive integer r,

$$\binom{x}{r} + \binom{x}{r-1} = \binom{x+1}{r}.$$

(b) Show that for any positive integers r and k,

$$\sum_{n=0}^{r} \binom{n+k-1}{k-1} = \binom{r+k}{k}.$$

random variables, random vectors, and distribution functions

3.1 RANDOM VARIABLES AND THEIR DISTRIBUTIONS

Let $(S, \mathcal{Q}, P)$ define a given probability space. A *random variable* X is a real-valued, $\mathcal{Q}$-measurable function which has a specified value $X(s)$ at each point $s \in S$. Every random variable X induces a probability distribution P_X on the real line R^1. For any Borel set $B \subset R^1$, the probability $P_X(B)$ is defined by the relation

$$P_X(B) = \Pr(X \in B) = P\{s \colon X(s) \in B\}. \tag{1}$$

The *distribution function* (abbreviated d.f.) of a random variable X is defined to be the function F whose value for any real number t is specified by the relation

$$F(t) = \Pr(X \leq t). \tag{2}$$

Every d.f. F is nondecreasing. That is, if x and y are any two numbers such that $x \leq y$, then $F(x) \leq F(y)$ (see Exercise 1). Furthermore, for any number $x \in R^1$, let $F(x^+)$ denote the limit of the values $F(y)$ as y converges to x through values greater than x. Then F is continuous from the right in the sense that $F(x^+) = F(x)$ for any value $x \in R^1$. Also,

$$\lim_{t \to -\infty} F(t) = 0 \quad \text{and} \quad \lim_{t \to \infty} F(t) = 1.$$

Conversely, any function F with these properties can be considered to be a d.f., and it defines a probability distribution on the Borel sets of the real line.

In practical applications, almost every distribution that occurs is of one of the following three types.

Discrete Distributions

A random variable X is said to have a *discrete* distribution if X can take only a countable (perhaps finite) number of distinct values $x_1, x_2, \ldots$. The *probability function* (abbreviated p.f.) of X is the function f defined by the relation

$$f(x) = \Pr(X = x) \qquad x = x_1, x_2, \ldots \tag{3}$$

Hence, for any Borel set $B \subset R^1$,

$$P_X(B) = \sum_{\{i:\ x_i \,\epsilon\, B\}} f(x_i).$$

Absolutely Continuous Distributions

A random variable X has an *absolutely continuous* distribution if there exists a nonnegative *probability density function* (abbreviated p.d.f.) f such that for any Borel set $B \subset R^1$,

$$P_X(B) = \int_B f(x)\, dx. \tag{4}$$

It should be noted that on a countable set of points (or, more generally, on a set of points having Lebesgue measure 0), the values of the p.d.f. f can be changed without affecting any of the probabilities given by Eq. (4). If X has an absolutely continuous distribution, its d.f. can be differentiated almost everywhere, and at any continuity point x of the p.d.f. f, $F'(x) = f(x)$.

Mixed Distributions

The distribution of a random variable may have both a discrete part and an absolutely continuous part. That is, there may be a countable set of points $x_1, x_2, \ldots$ having total probability p $(0 < p < 1)$, while the remaining probability $(1 - p)$ is spread over the real line in accordance with some p.d.f.

3.2 MULTIVARIATE DISTRIBUTIONS

The preceding section dealt with *univariate distributions*, i.e., with distributions of a single random variable. The joint distribution of two random variables is called a *bivariate distribution*, and more generally, the joint distribution of n random variables $X_1, \ldots, X_n$ is called a *multivariate distribution*. The *joint distribution function* of $X_1, \ldots, X_n$ is the function F whose value is specified at each point $(x_1, \ldots, x_n) \in R^n$ by the relation

$$F(x_1, \ldots, x_n) = \Pr(X_1 \leq x_1, \ldots, X_n \leq x_n). \tag{1}$$

Every multivariate d.f. as defined by Eq. (1) satisfies conditions analogous to those given in Sec. 3.1 for the d.f. of a single random variable. In practical applications, almost every distribution that occurs is of one of the following three types.

Discrete Distributions

The joint distribution of $X_1, \ldots, X_n$ is *discrete* if the distribution is concentrated on a countable (perhaps finite) set $T \subset R^n$. Any discrete distribution is specified by the joint probability function f defined at each point $(x_1, \ldots, x_n) \in T$ by the relation

$$f(x_1, \ldots, x_n) = \Pr(X_1 = x_1, \ldots, X_n = x_n). \tag{2}$$

Absolutely Continuous Distributions

The joint distribution of $X_1, \ldots, X_n$ is *absolutely continuous* if there exists a nonnegative joint probability density function such that for every Borel set $B \subset R^n$,

$$\Pr[(X_1, \ldots, X_n) \in B] = \int \cdots \int_B f(x_1, \ldots, x_n) \, dx_1 \cdots dx_n. \tag{3}$$

Mixed Distributions

It is possible that some of the random variables $X_1, \ldots, X_n$ have discrete distributions while the others have absolutely continuous distributions. It is still convenient to represent the joint distribution of $X_1, \ldots, X_n$ by a function f which might be called the joint p.f.–p.d.f. It is, of course, essential to keep in mind that probabilities of various events in R^n are computed from any joint p.f.–p.d.f. by integrating for some components and summing for the others.   More generally, the marginal distribution of one of the random variables $X_1, \ldots, X_n$ might itself be mixed and have both a discrete part and an absolutely continuous part.

3.3 SUMS AND INTEGRALS

It is convenient to have terminology and notation that will make it possible to discuss discrete distributions and absolutely continuous distributions simultaneously.   Accordingly, to designate a function f that may be
either a p.d.f. or a p.f., we shall use the term *generalized probability density
function* (abbreviated g.p.d.f.).   Suppose that X is a random variable with
g.p.d.f. f.   Let g be any suitable function which has a domain of definition in R^1 that includes all possible values of the random variable X.   Let
B be any Borel set in R^1.   In accordance with the abstract theory of
measure and integration, either the sum

$$\sum_{x \, \epsilon \, B} g(x)f(x)$$

or the integral

$$\int_B g(x)f(x) \, dx$$

will be described by the notation

$$\int_B g(x)f(x) \, d\mu(x). \tag{1}$$

The integral (1) is said to exist if, and only if,

$$\int_B |g(x)|f(x) \, d\mu(x) < \infty. \tag{2}$$

If the integral in the relation (2) is finite when $B = R^1$, then the function
g is said to be an *integrable* function of X.

We shall also make use of the notation of Lebesgue-Stieltjes integrals to simplify the presentation.   Hence, if F is the d.f. of X, then
instead of writing the integral (1), we shall sometimes write

$$\int_B g(x) \, dF(x). \tag{3}$$

Similarly, when referring to the joint distribution of n random variables $X_1, \ldots, X_n$, we shall use the term *joint g.p.d.f.* to describe a
function f that may be either their joint p.f., their joint p.d.f., or their
joint p.f.–p.d.f.   For any suitable function g and any Borel set $B \subset R^n$,
the multivariate analogs of the integrals (1) and (3) will be written in the
form

$$\int \cdots \int_B g(x_1, \ldots, x_n)f(x_1, \ldots, x_n) \, d\mu(x_1) \cdots d\mu(x_n) \tag{4}$$

and

$$\int \cdots \int_B g(x_1, \ldots, x_n) \, dF(x_1, \ldots, x_n). \tag{5}$$

Remember that in the integrals (4) and (5), some components may be discrete while the others are absolutely continuous.

3.4 MARGINAL DISTRIBUTIONS AND INDEPENDENCE

When a statistician is dealing with a problem involving n random variables $X_1, \ldots, X_n$, he often finds it advisable to consider the joint distribution of just a few of those variables.   In such a case, the joint distribution of a subset of the random variables $X_1, \ldots, X_n$ is called a *marginal distribution*. Thus, the marginal joint d.f. F_k of $X_1, \ldots, X_k$ $(1 \leq k < n)$ is determined from the joint d.f. F_n of $X_1, \ldots, X_n$ by the relation

$$
\begin{aligned}
F_k(x_1, \ldots, x_k) &= \Pr(X_1 \leq x_1, \ldots, X_k \leq x_k) \\
&= \lim F_n(x_1, \ldots, x_n) \\
&\qquad \text{as } x_j \to \infty \ (j = k+1, \ldots, n). \quad (1)
\end{aligned}
$$

Similarly, the marginal joint g.p.d.f. f_k of $X_1, \ldots, X_k$ is determined from the joint g.p.d.f. f_n of $X_1, \ldots, X_n$ by the relation

$$
f_k(x_1, \ldots, x_k) = \int \cdots \int_{R^{n-k}} f_n(x_1, \ldots, x_n) \, d\mu(x_{k+1}) \cdots d\mu(x_n).
$$
$$(2)$$

For $i = 1, \ldots, n$, let G_i denote the marginal univariate d.f. of the random variable X_i.   The random variables $X_1, \ldots, X_n$ are said to be *independent* if, and only if, their joint d.f. F_n can be factored at every point $(x_1, \ldots, x_n) \in R^n$ as follows:

$$
F_n(x_1, \ldots, x_n) = G_1(x_1) \cdots G_n(x_n). \quad (3)
$$

It follows that the random variables $X_1, \ldots, X_n$ are independent if, and only if, their joint g.p.d.f. f_n and the marginal univariate g.p.d.f.'s $g_1, \ldots, g_n$ can be chosen so that at every point $(x_1, \ldots, x_n) \in R^n$,

$$
f_n(x_1, \ldots, x_n) = g_1(x_1) \cdots g_n(x_n). \quad (4)
$$

The random variables $X_1, \ldots, X_n$ are said to be a *random sample from the g.p.d.f. f* if they are independent and identically distributed and if each of them has the marginal g.p.d.f. f.

In other words, the joint g.p.d.f. f_n of the random variables $X_1, \ldots, X_n$ must be specified by the equation

$$
f_n(x_1, \ldots, x_n) = f(x_1) \cdots f(x_n) \qquad (x_1, \ldots, x_n) \in R^n. \quad (5)
$$

Suppose that $X_1, \ldots, X_n$ is a random sample from some g.p.d.f., and suppose that $Y_1 \leq Y_2 \leq \cdots \leq Y_n$ are the values of the random variables $X_1, \ldots, X_n$ rearranged in order of increasing magnitude.   The random variables $Y_1, \ldots, Y_n$ are called the *order statistics* of the random sample.

3.5 VECTORS AND MATRICES

The basic concepts of the theory of vectors and matrices will be used freely throughout this book. Any reader who is not familiar with these concepts as they are reviewed here should consult an introductory text on matrix theory or linear algebra. This section is intended simply to remind the reader of the concepts that will be needed, and some definitions, such as those of the determinant and the rank of a matrix, are omitted here.

The convention will be adopted throughout this book that unless explicitly defined otherwise, every vector $\mathbf{x}$ is a column vector. Thus, a k-dimensional vector $\mathbf{x}$ is a column of k real numbers of the form

$$\mathbf{x} = \begin{bmatrix} x_1 \\ \cdot \\ \cdot \\ \cdot \\ x_k \end{bmatrix}.$$

Clearly, every k-dimensional vector can be regarded as a point in R^k.

For ease of printing and reading, it is natural in certain contexts to discuss a vector $\mathbf{x}$ in terms of its transpose $\mathbf{x}'$. Thus, $\mathbf{x}' = (x_1, \ldots, x_k)$ or $\mathbf{x} = (x_1, \ldots, x_k)'$. In general, if $\mathbf{A}$ is any $k \times m$ matrix defined by the equation

$$\mathbf{A} = \begin{bmatrix} a_{11} & \cdots & a_{1m} \\ \cdot & \cdots & \cdot \\ a_{k1} & \cdots & a_{km} \end{bmatrix},$$

then its transpose $\mathbf{A}'$ is the $m \times k$ matrix defined by the equation

$$\mathbf{A}' = \begin{bmatrix} a_{11} & \cdots & a_{k1} \\ \cdot & \cdots & \cdot \\ a_{1m} & \cdots & a_{km} \end{bmatrix}.$$

The k-dimensional vector $\mathbf{0}$ is the vector each of whose components is 0. The $k \times k$ identity matrix $\mathbf{I}$ is the matrix each of whose k diagonal elements is 1 and each of whose other elements is 0.

A *square* matrix is a matrix having the same number of rows and columns. If $\mathbf{A}$ is a square matrix, the *determinant* of $\mathbf{A}$ is represented by the notation $|\mathbf{A}|$.

A square matrix $\mathbf{A}$ is *nonsingular* if $|\mathbf{A}| \neq 0$, and $\mathbf{A}$ is *singular* if $|\mathbf{A}| = 0$. A nonsingular $k \times k$ matrix has *rank* k; a singular $k \times k$

matrix has rank less than k.   If a matrix $\mathbf{A}$ is nonsingular, then there is a unique *inverse matrix* $\mathbf{A}^{-1}$ having the property that $\mathbf{AA}^{-1} = \mathbf{A}^{-1}\mathbf{A} = \mathbf{I}$.

The *trace* tr $(\mathbf{A})$ of any $k \times k$ matrix $\mathbf{A}$ is defined to be the sum of the k diagonal elements of $\mathbf{A}$.   Two basic properties of the trace follow (see Exercise 4):

1.   If $\mathbf{A}$ is a $k \times m$ matrix and $\mathbf{B}$ is an $m \times k$ matrix, for any positive integers k and m, then

$$\text{tr } (\mathbf{AB}) = \text{tr } (\mathbf{BA}). \tag{1}$$

2.   If each of the matrices $\mathbf{A}_1, \ldots, \mathbf{A}_n$ is a $k \times k$ matrix, then

$$\text{tr} \left(\sum_{i=1}^{n} \mathbf{A}_i \right) = \sum_{i=1}^{n} \text{tr } (\mathbf{A}_i). \tag{2}$$

A square matrix $\mathbf{A}$ is called *symmetric* if $\mathbf{A} = \mathbf{A}'$.   A symmetric $k \times k$ matrix $\mathbf{A}$ is defined to be *positive definite* if, for every k-dimensional vector $\mathbf{x} \neq \mathbf{0}$,

$$\mathbf{x}'\mathbf{Ax} > 0. \tag{3}$$

A symmetric positive definite matrix is necessarily nonsingular, and its inverse must also be a symmetric positive definite matrix.

A symmetric $k \times k$ matrix $\mathbf{A}$ is defined to be *nonnegative definite* if, for every k-dimensional vector $\mathbf{x}$,

$$\mathbf{x}'\mathbf{Ax} \geq 0. \tag{4}$$

In some books, a nonnegative definite matrix is called positive semidefinite.

If $\mathbf{A}$ is a symmetric $k \times k$ nonnegative definite matrix, then there exists a $k \times k$ matrix $\mathbf{B}$ with the property that $\mathbf{A} = \mathbf{BB}'$.   If, in fact, $\mathbf{A}$ is positive definite, then $\mathbf{B}$ is nonsingular.   Moreover, then $\mathbf{B}$ itself can be chosen to be a symmetric positive definite matrix.

Random Vectors

An n-dimensional random vector $\mathbf{X}$ is simply a sequence of n random variables $\mathbf{X} = (X_1, \ldots, X_n)'$.   The random vector $\mathbf{X}$ takes values in R^n, and the distribution of $\mathbf{X}$ is the same as the joint distribution of the component random variables $X_1, \ldots, X_n$.   Thus, the d.f. and g.p.d.f. of the random vector $\mathbf{X}$ are the joint d.f. and joint g.p.d.f. of the random variables $X_1, \ldots, X_n$.

Suppose that f is the g.p.d.f. and F is the d.f. of the random vector $\mathbf{X} = (X_1, \ldots, X_n)'$.   If $\mathbf{x} = (x_1, \ldots, x_n)'$, then, in terms of vectors, the integrals (4) and (5) in Sec. 3.3 are written in the form

$$\int_B g(\mathbf{x}) f(\mathbf{x}) \, d\mu(\mathbf{x}) \tag{5}$$

and

$$\int_B g(\mathbf{x}) \, dF(\mathbf{x}). \tag{6}$$

Again, a function g is said to be an integrable function of $\mathbf{X}$ if

$$\int_{R^n} |g(\mathbf{x})| \, dF(\mathbf{x}) < \infty. \tag{7}$$

The definition of the independence of k random vectors $\mathbf{X}_1, \ldots, \mathbf{X}_k$ is similar to the definition given in Sec. 3.4 for random variables. Also, as in Sec. 3.4, the n-dimensional random vectors $\mathbf{X}_1, \ldots, \mathbf{X}_k$ are said to be a random sample from an n-dimensional g.p.d.f. f if the vectors $\mathbf{X}_1, \ldots, \mathbf{X}_k$ are independent and the marginal g.p.d.f. of each of them is f.

3.6 EXPECTATIONS, MOMENTS, AND CHARACTERISTIC FUNCTIONS

The *expectation* $E(X)$ of any random variable X with d.f. F is defined by the equation

$$E(X) = \int_{R^1} x \, dF(x). \tag{1}$$

The expectation $E(X)$ is said to exist if, and only if, the integral in Eq. (1) exists. The expectation $E(X)$ of any random variable X is also called the *mean* of X or the *expected value* of X.

More generally, suppose that $\mathbf{X} = (X_1, \ldots, X_n)'$ is a random vector with d.f. F, and suppose that g is an integrable function of $\mathbf{X}$. The expectation $E[g(\mathbf{X})]$ is defined by the equation

$$E[g(\mathbf{X})] = \int_{R^n} g(\mathbf{x}) \, dF(\mathbf{x}). \tag{2}$$

This definition of expectation cannot be ambiguous because of the following fact: If Y is the random variable defined by the equation $Y = g(\mathbf{X})$ and if F_Y denotes the d.f. of Y, then the integral in Eq. (2) must have the same value as the integral

$$E(Y) = \int_{R^1} y \, dF_Y(y). \tag{3}$$

For any random variable X and any positive integer r, the expectation $E(X^r)$ is called the rth *moment* of X. The moments $E\{[X - E(X)]^r\}$ with respect to the mean are called the *central moments* of X. Of course, for certain random variables, some of these expectations may not exist.

In particular, the *variance* $\mathrm{Var}(X)$ of X is the second central moment, as defined by the equation

$$\text{Var}(X) = E\{[X - E(X)]^2\} = E(X^2) - E^2(X). \tag{4}$$

The *standard deviation* of X is the square root of $\text{Var}(X)$. If $E(X) \neq 0$, the *coefficient of variation* of X is the ratio

$$\frac{[\text{Var}(X)]^{\frac{1}{2}}}{E(X)}. \tag{5}$$

If X and Y denote any two random variables, the *covariance* $\text{Cov}(X, Y)$ of X and Y is defined as follows:

$$\text{Cov}(X, Y) = E\{[X - E(X)][Y - E(Y)]\} = E(XY) - E(X)E(Y). \tag{6}$$

In particular, $\text{Cov}(X, X) = \text{Var}(X)$. The *correlation* of X and Y is the ratio

$$\frac{\text{Cov}(X, Y)}{[\text{Var}(X)\text{Var}(Y)]^{\frac{1}{2}}}. \tag{7}$$

If the correlation (7) is 0, then X and Y are said to be *uncorrelated*.

Consider a matrix

$$\mathbf{T} = \begin{bmatrix} T_{11} & \cdots & T_{1m} \\ \cdot & \cdots & \cdot \\ T_{k1} & \cdots & T_{km} \end{bmatrix} \tag{8}$$

in which each component T_{ij} $(i = 1, \ldots, k; j = 1, \ldots, m)$ is a random variable. The expectation $E(\mathbf{T})$ of $\mathbf{T}$ is defined to be the matrix in which each component is the expectation of the corresponding component of $\mathbf{T}$. Hence,

$$E(\mathbf{T}) = \begin{bmatrix} E(T_{11}) & \cdots & E(T_{1m}) \\ \cdot & \cdots & \cdot \\ E(T_{k1}) & \cdots & E(T_{km}) \end{bmatrix}. \tag{9}$$

In particular, for any random vector $\mathbf{X} = (X_1, \ldots, X_n)'$, the *mean vector* $E(\mathbf{X})$ of $\mathbf{X}$ is defined by the equation

$$E(\mathbf{X}) = [E(X), \ldots, E(X_n)]'. \tag{10}$$

The *covariance matrix* $\text{Cov}(\mathbf{X})$ of $\mathbf{X}$ is defined to be the $n \times n$ symmetric matrix whose (i, j)th component is $\text{Cov}(X_i, X_j)$, for $i = 1, \ldots, n$ and $j = 1, \ldots, n$. Hence,

$$\text{Cov}(\mathbf{X}) = \begin{bmatrix} \text{Var}(X_1) & \text{Cov}(X_1, X_2) & \cdots & \text{Cov}(X_1, X_n) \\ \text{Cov}(X_2, X_1) & \text{Var}(X_2) & \cdots & \text{Cov}(X_2, X_n) \\ \cdot & \cdots & & \cdot \\ \text{Cov}(X_n, X_1) & \text{Cov}(X_n, X_2) & \cdots & \text{Var}(X_n) \end{bmatrix}. \tag{11}$$

Suppose that the random vector $\mathbf{X} = (X_1, \ldots, X_n)'$ has mean vector $\boldsymbol{\mu}$ and covariance matrix $\boldsymbol{\Sigma}$. Also, let the random vector $\mathbf{Y} = (Y_1, \ldots, Y_m)'$ be defined by the transformation $\mathbf{Y} = \mathbf{AX} + \mathbf{b}$, where $\mathbf{A}$ is an $m \times n$ matrix of constants and $\mathbf{b}$ is an m-dimensional vector of constants. Then $E(\mathbf{Y}) = \mathbf{A\mu} + \mathbf{b}$ and $\mathrm{Cov}(\mathbf{Y}) = \mathbf{A\Sigma A'}$. In particular, the variance of any linear combination $\mathbf{a'X}$ is $\mathrm{Var}(\mathbf{a'X}) = \mathbf{a'\Sigma a} \geq 0$. Thus, every covariance matrix $\boldsymbol{\Sigma}$ is nonnegative definite. In fact, $\boldsymbol{\Sigma}$ is positive definite unless there is a linear combination $\mathbf{a'X}$ (where $\mathbf{a} \neq \mathbf{0}$) that has variance 0.

Two other important properties of expectations are the following: For any set of random variables $X_1, \ldots, X_n$ whose expectations exist,

$$E\left(\sum_{i=1}^{n} a_i X_i\right) = \sum_{i=1}^{n} a_i E(X_i). \tag{12}$$

Moreover, if the variables $X_1, \ldots, X_n$ are independent, then

$$E\left(\prod_{i=1}^{n} X_i\right) = \prod_{i=1}^{n} E(X_i). \tag{13}$$

Characteristic Functions and Moment-generating Functions

The *characteristic function* (abbreviated c.f.) of a random variable X is the complex-valued function ζ defined at each number t $(-\infty < t < \infty)$ by the equation

$$\zeta(t) = E(e^{itX}) = E(\cos tX) + iE(\sin tX). \tag{14}$$

The c.f. of any random variable always exists, and no two different distributions yield the same c.f. Thus, there is a one-to-one correspondence between d.f.'s and c.f.'s.

Some of the useful properties of c.f.'s may be stated as follows: If $X_1, \ldots, X_k$ are k independent random variables with c.f.'s $\zeta_1, \ldots, \zeta_k$, respectively, and if the random variable Y is defined as

$$Y = X_1 + \cdots + X_k,$$

then the c.f. ζ of Y is given by the equation

$$\zeta(t) = \prod_{j=1}^{k} \zeta_j(t) \qquad -\infty < t < \infty. \tag{15}$$

If, for a given random variable X with c.f. ζ, the expectation $E(X^r)$ exists for some positive integer r, then this expectation can be found from the relation

$$E(X^r) = i^{-r}\zeta^{(r)}(0), \tag{16}$$

where $\zeta^{(r)}(0)$ denotes the derivative of ζ of order r evaluated at $t = 0$.

The c.f. of an n-dimensional random vector $\mathbf{X} = (X_1, \ldots, X_n)'$ is the complex-valued function ζ defined at each point $\mathbf{t} = (t_1, \ldots, t_n)'$ ($\mathbf{t} \in R^n$) by the equation

$$\zeta(\mathbf{t}) = E(e^{i\mathbf{t}'\mathbf{x}}). \tag{17}$$

The properties of the c.f. of an n-dimensional random vector are similar to the properties presented above for a random variable. In particular, suppose that for a given random vector $\mathbf{X} = (X_1, \ldots, X_n)'$ which has c.f. ζ, the expectation $E(\Pi_{j=1}^n X_j^{r_j})$ exists for some set of non-negative integers $r_1, \ldots, r_n$. Then this expectation can be found from the relation

$$E\left(\prod_{j=1}^n X_j^{r_j}\right) = \frac{1}{i^{r_1 + \cdots + r_n}} \left[\frac{\partial^{r_1 + \cdots + r_n}\zeta(\mathbf{t})}{\partial t_1^{r_1} \cdots \partial t_n^{r_n}}\right]_{\mathbf{t}=0}. \tag{18}$$

Another useful property of c.f.'s may be stated as follows: Let $\mathbf{X} = (X_1, \ldots, X_n)'$ be a random vector which has c.f. ζ. Let $\mathbf{A}$ be an $m \times n$ matrix of constants, let $\mathbf{b}$ be an m-dimensional vector of constants, and let $\mathbf{Y} = (Y_1, \ldots, Y_m)'$ be the random vector defined by the equation $\mathbf{Y} = \mathbf{AX} + \mathbf{b}$. Then the c.f. ζ_Y of $\mathbf{Y}$ can be determined at any point $\mathbf{t} \in R^m$ by the relation

$$\zeta_Y(\mathbf{t}) = e^{i\mathbf{t}'\mathbf{b}}\zeta(\mathbf{A}'\mathbf{t}). \tag{19}$$

The *moment-generating function* of a random variable X is the real-valued function ψ defined by the equation

$$\psi(t) = E(e^{tX}). \tag{20}$$

It follows that the value of the moment-generating function is specified at any real number t such that the expectation in Eq. (20) is finite. It is often advantageous to consider moment-generating functions rather than characteristic functions because they are real-valued rather than complex-valued. Their disadvantage is that for some random variables the moment-generating function is finite only for certain limited values of t. However, if the moment-generating function of a random variable X is finite for all values of t in an interval around the value $t = 0$, then all the moments $E(X^r)$ exist ($r = 1, 2, \ldots$). Furthermore, their values can be found by differentiating the moment-generating function at $t = 0$.

Multivariate moment-generating functions are defined similarly, and equations analogous to Eqs. (15) and (18) can be derived.

3.7 TRANSFORMATIONS OF RANDOM VARIABLES

Suppose that the joint distribution of n random variables $X_1, \ldots, X_n$ is absolutely continuous, and let f denote their joint p.d.f. Suppose that we are interested in finding the joint p.d.f. of n other random variables $Y_1, \ldots, Y_n$, each of which is a function of the variables $X_1, \ldots, X_n$. A standard result of this type will be presented in this section [see, e.g., Rao (1965)].

Let $\mathbf{X} = (X_1, \ldots, X_n)'$ and $\mathbf{Y} = (Y_1, \ldots, Y_n)'$, and consider a transformation from the random vector $\mathbf{X}$ to the random vector $\mathbf{Y}$ defined by the equations

$$Y_1 = g_1(\mathbf{X}), \ldots, Y_n = g_n(\mathbf{X}). \tag{1}$$

We shall assume that some subset $R^0 \subset R^n$, for which $\Pr(\mathbf{X} \in R^0) = 1$, has the following property:

R^0 can be partitioned into a finite number of disjoint sets $S_1, \ldots, S_k$ $(k \geq 1)$ such that in each set S_i, no two distinct values $\mathbf{x}_1 \in S_i$ and $\mathbf{x}_2 \in S_i$ of the vector $\mathbf{X}$ yield the same value of the vector $\mathbf{Y}$. In other words, we assume that on each set S_i of the partition, the transformation defined by Eq. (1) is a *one-to-one* transformation. Of course, if this transformation is a one-to-one transformation on the set R^0 itself, then there is no need to partition R^0 further.

For $i = 1, \ldots, k$, let the set $T_i \subset R^n$ be the image of the set S_i under the transformation defined by Eq. (1). (Note that the sets $T_1, \ldots, T_k$ need not be disjoint.) Also, for $i = 1, \ldots, k$, let the inverse transformation from the set T_i to the set S_i be defined at any point $(y_1, \ldots, y_n)' \in T_i$ by the equations

$$\begin{aligned} x_1 &= h_{i1}(y_1, \ldots, y_n), \\ & \cdot \cdot \cdot \cdot \cdot \cdot \cdot \cdot \cdot \cdot \cdot \cdot \\ x_n &= h_{in}(y_1, \ldots, y_n). \end{aligned} \tag{2}$$

It will be assumed that the n first-order partial derivatives of each function h_{ij} are continuous. By definition, the Jacobian $J_i(y_1, \ldots, y_n)$ of the transformation (2) from T_i to S_i is the determinant

$$J_i(y_1, \ldots, y_n) = \begin{vmatrix} \dfrac{\partial h_{i1}}{\partial y_1} & \cdots & \dfrac{\partial h_{i1}}{\partial y_n} \\ \cdot & \cdots & \cdot \\ \dfrac{\partial h_{in}}{\partial y_1} & \cdots & \dfrac{\partial h_{in}}{\partial y_n} \end{vmatrix}. \tag{3}$$

If it is assumed that the Jacobian $J_i(y_1, \ldots, y_n)$ does not vanish at any point $(y_1, \ldots, y_n)' \in T_i$, the following conclusion results: At any point $\mathbf{y} = (y_1, \ldots, y_n)' \in \cup_{i=1}^k T_i$, the value of the p.d.f. ϕ of $\mathbf{Y}$ is given by the equation

$$\phi(\mathbf{y}) = \sum_i f[h_{i1}(\mathbf{y}), \ldots, h_{in}(\mathbf{y})]|J_i(\mathbf{y})|. \tag{4}$$

Here, $|J_i(\mathbf{y})|$ denotes the absolute value of the Jacobian, and the summation is taken over all values of i $(i = 1, \ldots, k)$ such that $\mathbf{y} \in T_i$.

Moreover, $\phi(\mathbf{y}) = 0$ at any point $\mathbf{y} \in (\cup_{i=1}^{k} T_i)^c$. This fact, together with Eq. (4), specifies the value of the p.d.f. ϕ at every point $\mathbf{y} \in R^n$.

3.8 CONDITIONAL DISTRIBUTIONS

Let $\mathbf{X} = (X_1, \ldots, X_m)'$ and $\mathbf{Y} = (Y_1, \ldots, Y_n)'$ be two random vectors, possibly of different dimensions, and let $f(\mathbf{x}, \mathbf{y})$ be the value of their joint g.p.d.f. at any points $\mathbf{x} \in R^m$ and $\mathbf{y} \in R^n$. Also, let g denote the marginal g.p.d.f. of $\mathbf{Y}$, and consider any point $\mathbf{y} \in R^n$ such that $g(\mathbf{y}) > 0$. Then the *conditional g.p.d.f.* $h(\cdot \,|\mathbf{y})$ of $\mathbf{X}$, when the value $\mathbf{Y} = \mathbf{y}$ is given, is defined at any point $\mathbf{x} \in R^m$ as follows:

$$h(\mathbf{x}|\mathbf{y}) = \frac{f(\mathbf{x}, \mathbf{y})}{g(\mathbf{y})}. \tag{1}$$

The definition of the conditional g.p.d.f. $h(\cdot \,|\mathbf{y})$ is irrelevant for any point $\mathbf{y} \in R^n$ such that $g(\mathbf{y}) = 0$, since these points form a set having probability 0.

Note that the event $\{\mathbf{Y} = \mathbf{y}\}$, on which the distribution in Eq. (1) is conditioned, has probability 0 if some of the components of $\mathbf{Y}$ have absolutely continuous distributions. Hence, Eq. (1) represents an extension of the definition of conditional probability given in Sec. 2.4. The next result is the corresponding extension of Bayes' theorem as given in Sec. 2.4. It represents the conditional g.p.d.f. $h_2(\cdot \,|\mathbf{x})$ of $\mathbf{Y}$, when the value $\mathbf{X} = \mathbf{x}$ is given, in terms of the conditional g.p.d.f. $h_1(\cdot \,|\mathbf{y})$ of $\mathbf{X}$, given $\mathbf{Y}$, and the marginal g.p.d.f. g_2 of $\mathbf{Y}$. Also, let g_1 denote the marginal g.p.d.f. of $\mathbf{X}$.

Bayes' theorem   *Let* $\mathbf{X} = (X_1, \ldots, X_m)'$ *and* $\mathbf{Y} = (Y_1, \ldots, Y_n)'$ *be two random vectors. Let the marginal g.p.d.f.'s* g_1 *and* g_2 *and the conditional g.p.d.f.'s* $h_1(\cdot \,|\mathbf{y})$ *and* $h_2(\cdot \,|\mathbf{x})$ *be as defined above. Then for any point* $\mathbf{x} \in R^m$ *such that* $g_1(\mathbf{x}) > 0$ *and any point* $\mathbf{y} \in R^n$,

$$h_2(\mathbf{y}|\mathbf{x}) = \frac{h_1(\mathbf{x}|\mathbf{y})g_2(\mathbf{y})}{\displaystyle\int_{R^n} h_1(\mathbf{x}|\mathbf{t})g_2(\mathbf{t}) \, d\mu(\mathbf{t})}. \tag{2}$$

Proof   The proof is immediate because the numerator in Eq. (2) is the value of the joint g.p.d.f. $f(\mathbf{x}, \mathbf{y})$ and the denominator is $g_1(\mathbf{x})$. ∎

Conditional Expectations

Let $\mathbf{X} = (X_1, \ldots, X_m)'$ and $\mathbf{Y} = (Y_1, \ldots, Y_n)'$ be two random vectors, and let $\phi(\mathbf{X}, \mathbf{Y})$ be an integrable function of $\mathbf{X}$ and $\mathbf{Y}$. The condi-

tional expectation $E[\phi(\mathbf{X}, \mathbf{Y})|\mathbf{Y}]$ of the random variable $\phi(\mathbf{X}, \mathbf{Y})$, when $\mathbf{Y}$ is given, is defined as a function of the random vector $\mathbf{Y}$ whose value $E[\phi(\mathbf{X}, \mathbf{Y})|\mathbf{y}]$ when $\mathbf{Y} = \mathbf{y}$ is specified as follows:

$$E[\phi(\mathbf{X}, \mathbf{Y})|\mathbf{y}] = \int_{R^m} \phi(\mathbf{x}, \mathbf{y}) h_1(\mathbf{x}|\mathbf{y}) \, d\mu(\mathbf{x}). \tag{3}$$

In other words, Eq. (3) gives the expectation of the random variable $\phi(\mathbf{X}, \mathbf{y})$ under the conditional distribution of $\mathbf{X}$, given the value $\mathbf{Y} = \mathbf{y}$. A very important fact, which can be easily verified, is that

$$E\{E[\phi(\mathbf{X}, \mathbf{Y})|\mathbf{Y}]\} = E[\phi(\mathbf{X}, \mathbf{Y})]. \tag{4}$$

A useful identity involving the means and variances of conditional distributions is the following. Suppose that X and Y are two random variables, and suppose that $\mathrm{Var}(X) < \infty$. Let $\mathrm{Var}(X|Y)$ denote the function of Y whose value when $Y = y$ is the variance of X when its distribution is specified by the conditional g.p.d.f. $h_1(\cdot \, |y)$. Then (see Exercise 23)

$$\mathrm{Var}(X) = E[\mathrm{Var}(X|Y)] + \mathrm{Var}[E(X|Y)]. \tag{5}$$

Random Samples

Throughout this book we shall be considering random variables $X_1, \ldots,$ X_n and W whose joint distribution has the following properties. Suppose that W takes values in the space Ω, and let ξ denote the g.p.d.f. of W. Suppose also that conditionally on any given value $W = w$, the random variables $X_1, \ldots, X_n$ are independent and identically distributed and that the common g.p.d.f. of each of these random variables is $f(\cdot \, |w)$. Hence, the conditional joint g.p.d.f. of $X_1, \ldots, X_n$, given the value $W = w$, is the product $f(x_1|w) \cdots f(x_n|w)$, and the marginal joint g.p.d.f. f_n of $X_1, \ldots, X_n$ is given by the relation

$$f_n(x_1, \ldots, x_n) = \int_\Omega f(x_1|w) \cdots f(x_n|w) \xi(w) \, d\mu(w). \tag{6}$$

It follows from Bayes' theorem that the conditional g.p.d.f. $\xi(\cdot \, |x_1, \ldots, x_n)$ of W, given any values $X_1 = x_1, \ldots, X_n = x_n$ such that Eq. (6) is positive, can be written in the following form:

$$\xi(w|x_1, \ldots, x_n) = \frac{f(x_1|w) \cdots f(x_n|w) \xi(w)}{f_n(x_1, \ldots, x_n)}. \tag{7}$$

Since the family of g.p.d.f.'s $\{f(\cdot \, |w), w \, \epsilon \, \Omega\}$ is indexed by the possible values of the random variable W, this variable is called a *parameter* of the family. The conditional joint distribution of $X_1, \ldots, X_n$, for any given value w of the parameter W, is the same as the joint distribution of a random sample from the g.p.d.f. $f(\cdot \, |w)$.

EXERCISES

1. Prove that every d.f. F is nondecreasing and continuous from the right. Also, prove that

$$\lim_{t \to -\infty} F(t) = 0 \quad \text{and} \quad \lim_{t \to \infty} F(t) = 1.$$

2. Prove that if the d.f. F of a random variable X is continuous at the point t, then $\Pr(X = t) = 0$. As a more general result, show that for any value t,

$$\Pr(X = t) = F(t) - F(t^-).$$

[Here, $F(t^-)$ is the limit of the values $F(y)$ as y converges to t through values smaller than t.]

3. Mr. A and Mr. B agree to meet at a certain place between 1 and 2 o'clock. Suppose that they arrive at the meeting place independently and randomly during the hour. Find the distribution of the length of time for which Mr. A has to wait for Mr. B.

4. Prove Eqs. (1) and (2) in Sec. 3.5.

5. Suppose that X_1 and X_2 are two random variables whose joint p.d.f. is

$$f(x_1, x_2) = \begin{cases} c(x_1 + x_2) & \text{for } 0 \le x_1 \le x_2 \le 1, \\ 0 & \text{otherwise.} \end{cases}$$

Find c, the marginal p.d.f.'s of X_1 and X_2, $E(X_1)$, and the p.d.f. of the random variable $X_1 + X_2$. Are X_1 and X_2 independent?

6. Suppose that X_1 and X_2 are two random variables whose joint p.f. is

$$f(x_1, x_2) = \begin{cases} cx_1x_2 & \text{for } x_1 = 1, 2; x_2 = 1, 2, \\ 0 & \text{otherwise.} \end{cases}$$

Find c, the marginal p.f. of X_1, $E(X_1X_2)$, the p.f. of $X_1 + X_2$, and the c.f. of X_1. Are X_1 and X_2 independent?

7. Suppose that X_1, X_2, and X_3 are three random variables whose joint p.d.f. is

$$f(x_1, x_2, x_3) = \begin{cases} c & \text{for } 0 \le x_1 \le 1, -2 \le x_2 \le 3, 4 \le x_3 \le 5, \\ 0 & \text{otherwise.} \end{cases}$$

Find c, the marginal p.d.f. of X_1, and the joint c.f. of X_1, X_2, and X_3. Are X_1, X_2, and X_3 independent?

8. Suppose that X_1, X_2, and X_3 are three random variables whose joint p.d.f. is

$$f(x_1, x_2, x_3) = \begin{cases} c & \text{for } x_1^2 + x_2^2 + x_3^2 \le 1, \\ 0 & \text{otherwise.} \end{cases}$$

Find c and the marginal p.d.f. of X_1. Are X_1, X_2, and X_3 independent?

9. Let X_1 and X_2 be independent random variables such that $0 < \text{Var}(X_i) < \infty$ for $i = 1, 2$. Show that X_1 and X_2 are uncorrelated.

10. Suppose that X is a random variable whose p.d.f. is

$$f(x) = \begin{cases} \frac{1}{2} & \text{for } -1 \le x \le 1, \\ 0 & \text{otherwise.} \end{cases}$$

If the random variable Y is defined by the equation $Y = X^2$, show that X and Y are uncorrelated but not independent. (Note that X and Y are obviously not independent since Y is a specific function of X.)

11. Assuming that X is a random variable whose p.d.f. is f, derive an expression for the p.d.f. of the random variable $Y = aX + b$.

12. Assuming that X is a random variable whose p.d.f. is f, derive an expression for the p.d.f. of the random variable $Y = X^2$.

13. Suppose that X_1 and X_2 are two random variables whose joint p.d.f. is f. Derive expressions for the univariate p.d.f. of each of the following random variables: $X_1 + X_2$, $X_1 X_2$, and X_1/X_2.

14. Suppose that X_1 and X_2 are random variables whose joint p.d.f. is f. Let $Y_1 = X_1^2$ and $Y_2 = X_2^2$. Derive an expression for the joint p.d.f. of Y_1 and Y_2.

15. Let $X_1, \ldots, X_n$ be a random sample from the p.d.f. f, and let $Y_1 \leq Y_2 \leq \cdots \leq Y_n$ be the order statistics of this sample. Show that the joint p.d.f. g of the random variables $Y_1, \ldots, Y_n$ is

$$g(y_1, \ldots, y_n) = \begin{cases} n! f(y_1) \cdots f(y_n) & \text{for } y_1 < y_2 < \cdots < y_n, \\ 0 & \text{otherwise.} \end{cases}$$

16. Let $X_1, \ldots, X_n$ be a random sample from the p.d.f. f which, for fixed positive constants a and b, is defined by the equation

$$f(x) = \begin{cases} \dfrac{ax^{a-1}}{b^a} & \text{for } 0 < x < b, \\ 0 & \text{otherwise.} \end{cases}$$

If $Y_1 \leq \cdots \leq Y_n$ are the order statistics of the sample, prove that the n random variables $Y_1/Y_2, Y_2/Y_3, \ldots, Y_{n-1}/Y_n$, and Y_n are independent.

17. (a) Suppose that a random variable X has a continuous d.f. F. Show that the random variable $Y = F(X)$ has a uniform distribution on the interval $(0, 1)$. (This transformation from X to Y is called the *probability integral transformation*.) (b) Suppose that X has a continuous d.f. F, and let G be any other continuous d.f. on the real line. Find a transformation $Z = \phi(X)$ such that the d.f. of Z is G. (*Caution:* Neither F nor G need be strictly increasing.)

18. Suppose that k random variables $X_1, \ldots, X_k$ have a joint distribution such that the correlation between any two of them is ρ. Show that $\rho \geq -1/(k - 1)$.

19. Suppose that X is a random variable which can take only the values 0, 1, 2, ... Prove that

$$E(X) = \sum_{n=1}^{\infty} \Pr(X \geq n)$$

and that $E(X)$ exists if, and only if, this infinite series converges.

20. Suppose that X_1 and X_2 are random variables whose joint p.d.f. f is

$$f(x_1, x_2) = \begin{cases} c & \text{for } x_1^2 + x_2^2 \leq 1, \\ 0 & \text{otherwise.} \end{cases}$$

Find the conditional p.d.f. of X_1 when X_2 is given, and find the conditional expectation $E(X_1|X_2)$.

21. Either of two instruments might be used for making a certain measurement. Instrument 1 yields a measurement Y_1 whose p.d.f. f_1 is

$$f_1(t) = \begin{cases} 1 & \text{for } 0 \leq t \leq 1, \\ 0 & \text{otherwise.} \end{cases}$$

Instrument 2 yields a measurement Y_2 whose p.d.f. f_2 is

$$f_2(t) = \begin{cases} 2t & \text{for } 0 \le t \le 1, \\ 0 & \text{otherwise.} \end{cases}$$

Suppose that one of these instruments, designated as X, is chosen at random and that a measurement Y is made with it. (a) Find the joint p.f.–p.d.f. of X and Y, and find the marginal p.d.f. of Y. (b) If the observed value for the measurement Y is found to be $Y = y$ $(0 \le y \le 1)$, what is the probability that instrument 2 was used for making the measurement?

22. Suppose that a point X_1 is first chosen at random (i.e., with a constant p.d.f.) from the interval $(0, 1)$, that a point X_2 is next chosen at random from the interval $(X_1, 1)$, that a point X_3 is then chosen at random from the interval $(X_2, 1)$, and that the process is continued for n stages $(n = 2, 3, \ldots)$. (a) Derive an expression for $E(X_n)$. (b) Derive an expression for the marginal p.d.f. of X_2. (c) Find the conditional p.d.f. of X_1 when X_2 is given, and find the conditional expectation $E(X_1|X_2)$.

23. Prove Eq. (5) in Sec. 3.8.

24. Suppose that W is a random variable whose p.d.f. ξ is defined as follows:

$$\xi(w) = \begin{cases} \dfrac{1}{n!}\, w^n e^{-w} & \text{for } w > 0, \\ 0 & \text{otherwise.} \end{cases}$$

Suppose also that conditionally on any given value $W = w (w > 0)$, the random variables $X_1, \ldots, X_n$ are independent and that each of them has p.d.f. $f(\cdot\,|w)$ defined as follows:

$$f(x|w) = \begin{cases} \dfrac{1}{w} & \text{for } 0 < x < w, \\ 0 & \text{otherwise.} \end{cases}$$

Find the joint marginal p.d.f. of $X_1, \ldots, X_n$ and also the conditional p.d.f. of W, given any set of values $X_1 = x_1, \ldots, X_n = x_n$.

some special univariate distributions

4.1 INTRODUCTION

The purpose of this chapter is to present for review and for later reference a relatively concise catalog of several of the special but important univariate distributions that are widely used in practical applications. In the next five sections we shall list some discrete univariate distributions, and in the remaining seven sections of the chapter we shall list some univariate distributions that are absolutely continuous. The procedure for each distribution is simply to record for later use the p.f. or the p.d.f., together with a few of its properties. The derivations of the few properties that are presented here are usually omitted in the text. However, together with some related results, these derivations serve as exercises at the end of the chapter.

Most readers will be familiar with at least the elementary properties of the distributions presented here, as well as with some standard examples for which probability models based on these distributions are appropriate. Consequently, we shall not attempt to present detailed motivation for the study of these particular distributions but shall include at most a few words and a few references. For derivations and additional discussion of the properties of these distributions, as well as for examples of their occurrence in probability and statistical contexts, the reader is referred to the standard texts listed at the end of Chap. 1. Further illustrations of the role of these distributions in statistical decision making will be brought out in this book when they are used later.

4.2 THE BERNOULLI DISTRIBUTION

A random variable X has a *Bernoulli distribution with parameter p* $(0 < p < 1)$ if X takes only the values 0 and 1 and

$$\Pr(X = 1) = p = 1 - \Pr(X = 0).$$

The p.f. $f(\cdot \,|p)$ of X can be written in the form

$$f(x|p) = \begin{cases} p^x q^{1-x} & \text{for } x = 0, 1, \\ 0 & \text{otherwise.} \end{cases} \tag{1}$$

Here $q = 1 - p$. Clearly, the Bernoulli distribution is of fundamental importance since it is the appropriate model for any experiment where the outcome must belong to one of two mutually exclusive classes. Familiar examples are those where the outcome must be either success or failure, or either defective or nondefective.

If a random variable X has the Bernoulli distribution given by Eq. (1), then it is easy to show that (see Exercise 1)

$$E(X) = p \quad \text{and} \quad \text{Var}(X) = pq. \tag{2}$$

When all the random variables in a finite or infinite sequence are independent and have the same Bernoulli distribution, the sequence is called a *sequence of Bernoulli trials*.

4.3 THE BINOMIAL DISTRIBUTION

A random variable X has a *binomial distribution with parameters n and p* $(n = 1, 2, \ldots ; 0 < p < 1)$ if X has a discrete distribution whose p.f. $f(\cdot \,|n, p)$ is

$$f(x|n, p) = \begin{cases} \binom{n}{x} p^x q^{n-x} & \text{for } x = 0, 1, \ldots, n, \\ 0 & \text{otherwise.} \end{cases} \tag{1}$$

Here, also, $q = 1 - p$. If $X_1, \ldots, X_n$ is a sequence of Bernoulli trials with parameter p, then the sum $X_1 + \cdots + X_n$ has a binomial distribution with parameters n and p. In other words, if an experiment consists of n independent trials and if, for each trial, the probability of the occurrence of a certain event is p, then the distribution of the total number of occurrences of that event in the experiment will be a binomial distribution with parameters n and p. The following results are easily derived from this fact.

If a random variable X has the binomial distribution given by Eq. (1), then (see Exercise 2)

$$E(X) = np \quad \text{and} \quad \text{Var}(X) = npq. \tag{2}$$

If $X_1, \ldots, X_k$ are independent random variables and X_i has a binomial distribution with parameters n_i and p $(i = 1, \ldots, k)$, then the sum $X_1 + \cdots + X_k$ has a binomial distribution with parameters $n_1 + \cdots + n_k$ and p (see Exercise 3).

4.4 THE POISSON DISTRIBUTION

A random variable X has a *Poisson distribution with mean* $\lambda(\lambda > 0)$ if X has a discrete distribution whose p.f. $f(\cdot | \lambda)$ is

$$f(x|\lambda) = \begin{cases} \dfrac{e^{-\lambda}\lambda^x}{x!} & \text{for } x = 0, 1, 2, \ldots, \\ 0 & \text{otherwise.} \end{cases} \tag{1}$$

A standard derivation shows that under certain mathematical conditions, the number of occurrences of a certain phenomenon in a fixed period of time, or in a fixed region of space, follows a Poisson distribution. It has also been observed that under certain physical conditions, the Poisson distribution provides an appropriate model for such applications as the number of particles emitted in a specified direction within a fixed period of time by some radioactive substance, the number of incoming telephone calls received at a switchboard during a given period, and the number of defects in a fixed length of cloth or tape. Finally, the Poisson distribution provides a useful approximation to the binomial distribution for large values of n and small values of p (see Exercise 4).

If a random variable X has the Poisson distribution given by Eq. (1), then it can be verified that (see Exercise 5)

$$E(X) = \lambda \quad \text{and} \quad \text{Var}(X) = \lambda. \tag{2}$$

If $X_1, \ldots, X_k$ are independent random variables and X_i has a Poisson distribution with mean λ_i $(i = 1, \ldots, k)$, then the sum $X_1 + \cdots + X_k$ has a Poisson distribution with mean $\lambda_1 + \cdots + \lambda_k$ (see Exercise 6).

4.5 THE NEGATIVE BINOMIAL DISTRIBUTION

A random variable X has a *negative binomial distribution with parameters r and p* $(r > 0; 0 < p < 1)$ if X has a discrete distribution whose p.f.

$f(\cdot |r, p)$ is

$$f(x|r, p) = \begin{cases} \dbinom{r + x - 1}{x} p^r q^x & \text{for } x = 0, 1, 2, \ldots, \\ 0 & \text{otherwise.} \end{cases} \tag{1}$$

Here again, $q = 1 - p$.

For any positive integer r, the negative binomial distribution arises in the following way: In a sequence of Bernoulli trials with parameter p, let Y denote the total number of trials that are required in order to obtain the value 1 exactly r times. If X denotes the number of trials on which the value 0 was obtained before the value 1 was obtained r times, then $X = Y - r$ and X has the negative binomial distribution given by Eq. (1). This distribution also arises in certain stochastic processes [see, e.g., Parzen (1962)] and has been used by Mosteller and Wallace (1964) to represent word frequencies. When $r = 1$, the negative binomial distribution is often called the *geometric distribution*.

If a random variable X has the negative binomial distribution given by Eq. (1), then (see Exercise 9)

$$E(X) = \frac{rq}{p} \quad \text{and} \quad \text{Var}(X) = \frac{rq}{p^2}. \tag{2}$$

If $X_1, \ldots, X_k$ are independent random variables and X_i has a negative binomial distribution with parameters r_i and p $(i = 1, \ldots, k)$, then the sum $X_1 + \cdots + X_k$ has a negative binomial distribution with parameters $r_1 + \cdots + r_k$ and p (see Exercise 10).

4.6 THE HYPERGEOMETRIC DISTRIBUTION

A random variable X has a *hypergeometric distribution with parameters A, B, and n* (where A, B, and n are positive integers such that $n \leq A + B$) if X has a discrete distribution whose p.f. $f(\cdot |A, B, n)$ is

$$f(x|A, B, n) = \begin{cases} \dfrac{\dbinom{A}{x} \dbinom{B}{n - x}}{\dbinom{A + B}{n}} & \text{for } x = 0, 1, \ldots, n, \\ 0 & \text{otherwise.} \end{cases} \tag{1}$$

The hypergeometric distribution arises in the following way: Consider a finite population of $A + B$ items in which there are A items of type 1 and B items of type 2. Suppose that n items are selected at random from the population, without being replaced, and let X denote the number of these items that are of type 1. Then X has the hypergeo metric distribution given by Eq. (1). It follows from this discussion and from the definition of binomial coefficients that $f(x|A, B, n) = 0$ for $x > A$ or $n - x > B$. Thus, the p.f. given by Eq. (1) is, in fact, non-zero only when x is an integer in the following interval:

$$\max \{0, n - B\} \leq x \leq \min \{n, A\}. \tag{2}$$

If a random variable X has the hypergeometric distribution given by Eq. (1), then (see Exercise 13)

$$E(X) = \frac{nA}{A + B} \quad \text{and} \quad \text{Var}(X) = \frac{nAB}{(A + B)^2} \cdot \frac{A + B - n}{A + B - 1}. \tag{3}$$

4.7 THE NORMAL DISTRIBUTION

A random variable X has a *normal distribution with mean μ and variance σ^2* ($-\infty < \mu < \infty$; $\sigma > 0$) if X has an absolutely continuous distribution whose p.d.f. $f(\cdot | \mu, \sigma^2)$ is specified at any point x ($-\infty < x < \infty$) by the equation

$$f(x|\mu, \sigma^2) = (2\pi\sigma^2)^{-\frac{1}{2}} \exp \left[\frac{-(x - \mu)^2}{(2\sigma^2)} \right]. \tag{1}$$

Here we are using the exponential notation whereby, for any real or complex number v,

$$\exp (v) = e^v. \tag{2}$$

It is shown in a very famous theorem, which is one of a class known as *central limit theorems*, that if $X_1, X_2, \ldots$ is a sequence of independent and identically distributed random variables which have positive variances, then the limiting distribution of the sum $X_1 + \cdots + X_n$, when suitably normalized, will be a normal distribution. Moreover, it is known from other central limit theorems that, under certain conditions, the limiting distribution of a sum like this may still be normal even though the random variables $X_1, X_2, \ldots$ are neither independent nor identically distributed. Perhaps as a reflection of these theorems, it has been found that large classes of empirical distributions in virtually all fields of study are approximately normal. For these reasons, it is often assumed when constructing the probability model for a given problem, that the distribution of each observation is approximately normal. Therefore, statistical methods for handling normally distributed random variables are of great importance.

If a random variable X is normally distributed and has the p.d.f.

given by Eq. (1), then (see Exercise 15)

$$E(X) = \mu \quad \text{and} \quad \text{Var}(X) = \sigma^2. \tag{3}$$

It can be shown that any linear combination of normally distributed random variables will itself be normally distributed. In particular, suppose that $X_1, \ldots, X_k$ are independent random variables and that X_i is normally distributed with mean μ_i and variance σ_i^2 $(i = 1, \ldots, k)$. If $a_1, \ldots, a_k$ and b are constants such that $a_i \neq 0$ for at least one value of i, then the random variable $a_1 X_1 + \cdots + a_k X_k + b$ is normally distributed with mean $a_1 \mu_1 + \cdots + a_k \mu_k + b$ and variance $a_1^2 \sigma_1^2 + \cdots + a_k^2 \sigma_k^2$ (see Exercise 16).

The normal distribution for which the mean is 0 and the variance is 1 is called the *standard normal distribution*. The p.d.f. and d.f. of this distribution are usually denoted by φ and Φ, respectively. Thus, for any number x $(-\infty < x < \infty)$,

$$\varphi(x) = (2\pi)^{-\frac{1}{2}} \exp\left(\frac{-x^2}{2}\right) \tag{4}$$

and

$$\Phi(x) = \int_{-\infty}^{x} \varphi(u) \, du. \tag{5}$$

The *precision* τ of a normal distribution is defined to be the reciprocal of the variance; that is,

$$\tau = \frac{1}{\sigma^2}. \tag{6}$$

In much of the work that comes later, it will be more convenient to specify a normal distribution by its mean μ and its precision τ rather than by its mean and variance. Thus, if a random variable X has a normal distribution with mean μ and precision τ, the p.d.f. $g(\cdot | \mu, \tau)$ is specified at any number x $(-\infty < x < \infty)$ by the equation

$$g(x | \mu, \tau) = \left(\frac{\tau}{2\pi}\right)^{\frac{1}{2}} \exp\left[\frac{-\tau(x - \mu)^2}{2}\right]. \tag{7}$$

As is well known, the variance of a normal distribution is a measure of its dispersion. Therefore, the precision of such a distribution is a measure of the concentration of the distribution around its mean. Obviously, all results presented in this section can be reformulated in terms of these new parameters. For example, if a random variable X is normally distributed and its precision is τ, then for any constants a and b, with $a \neq 0$, the random variable $aX + b$ is normally distributed and its precision is τ/a^2.

4.8 THE GAMMA DISTRIBUTION

A random variable X has a *gamma distribution with parameters α and β* ($\alpha > 0, \beta > 0$) if X has an absolutely continuous distribution whose p.d.f. $f(\cdot | \alpha, \beta)$ is

$$f(x | \alpha, \beta) = \begin{cases} \dfrac{\beta^\alpha}{\Gamma(\alpha)} x^{\alpha-1} e^{-\beta x} & \text{for } x > 0, \\ 0 & \text{otherwise.} \end{cases} \qquad (1)$$

The gamma function Γ that appears in Eq. (1) is defined as

$$\Gamma(\alpha) = \int_0^\infty u^{\alpha-1} e^{-u} \, du \qquad \alpha > 0. \qquad (2)$$

This function has the following property (see Exercise 18):

$$\Gamma(\alpha) = (\alpha - 1)\Gamma(\alpha - 1) \qquad \alpha > 1. \qquad (3)$$

Since $\Gamma(1) = 1$, as can be readily computed from Eq. (2), it follows from Eq. (3) that $\Gamma(n) = (n - 1)!$ for any positive integer n. In many statistical applications, it is necessary to consider gamma functions whose arguments are either integers or integers divided by 2. It can be shown that $\Gamma(\frac{1}{2}) = \pi^{\frac{1}{2}}$ (see Exercise 18). Hence, for any integer n, $\Gamma(n/2)$ is easily evaluated from Eq. (3).

If a random variable X has a gamma distribution whose p.d.f. is given by Eq. (1), then (see Exercise 19)

$$E(X) = \frac{\alpha}{\beta} \qquad \text{and} \qquad \text{Var}(X) = \frac{\alpha}{\beta^2}. \qquad (4)$$

Let $X_1, \ldots, X_k$ be independent random variables, and assume that X_i has a gamma distribution with parameters α_i and β ($i = 1, \ldots, k$). Then, for any constant $c > 0$, the random variable $c(X_1 + \cdots + X_k)$ has a gamma distribution with parameters $\alpha_1 + \cdots + \alpha_k$ and β/c (see Exercise 20).

The distribution defined by Eq. (1) when $\alpha = 1$ and $\beta > 0$ is called the *exponential distribution with parameter β*.

If n is any positive integer, the gamma distribution for which $\alpha = n/2$ and $\beta = \frac{1}{2}$ is called the *chi-square (χ^2) distribution with n degrees of freedom*. If n random variables $X_1, \ldots, X_n$ are independent and if each of these variables has a standard normal distribution, then the random variable $X_1^2 + \cdots + X_n^2$ has a χ^2 distribution with n degrees of freedom (see Exercise 21).

Another important result is the following: Let $X_1, \ldots, X_n$ be a random sample from a normal distribution with mean μ and variance σ^2, and let

$$\bar{X} = \frac{1}{n} \sum_{i=1}^{n} X_i \quad \text{and} \quad S^2 = \frac{1}{\sigma^2} \sum_{i=1}^{n} (X_i - \bar{X})^2.$$

Then S^2 has a χ^2 distribution with $n - 1$ degrees of freedom, and the random variables S^2 and $\bar{X}$ are independent.

4.9 THE BETA DISTRIBUTION

A random variable X has a *beta distribution with parameters α and β* ($\alpha > 0, \beta > 0$) if X has an absolutely continuous distribution whose p.d.f. $f(\cdot | \alpha, \beta)$ is

$$f(x|\alpha, \beta) = \begin{cases} \dfrac{\Gamma(\alpha + \beta)}{\Gamma(\alpha)\Gamma(\beta)} x^{\alpha-1}(1 - x)^{\beta-1} & \text{for } 0 < x < 1, \\ 0 & \text{otherwise.} \end{cases} \tag{1}$$

If X has a beta distribution whose p.d.f. is given by Eq. (1), then (see Exercise 26)

$$E(X) = \frac{\alpha}{\alpha + \beta} \tag{2}$$

and

$$\text{Var}(X) = \frac{\alpha\beta}{(\alpha + \beta)^2(\alpha + \beta + 1)}. \tag{3}$$

The beta distribution for which $\alpha = \beta = 1$ is called the *uniform distribution on the interval* $(0, 1)$. This distribution is also a member of the family to be considered next.

4.10 THE UNIFORM DISTRIBUTION

A random variable X has a *uniform distribution on the interval* (α, β), where $-\infty < \alpha < \beta < \infty$, if X has an absolutely continuous distribution whose p.d.f. $f(\cdot | \alpha, \beta)$ is

$$f(x|\alpha, \beta) = \begin{cases} \dfrac{1}{\beta - \alpha} & \text{for } \alpha < x < \beta, \\ 0 & \text{otherwise.} \end{cases} \tag{1}$$

If X has a uniform distribution whose p.d.f. is given by Eq. (1), then (see Exercise 28)

$$E(X) = \frac{\alpha + \beta}{2} \quad \text{and} \quad \text{Var}(X) = \frac{(\beta - \alpha)^2}{12}. \tag{2}$$

4.11 THE PARETO DISTRIBUTION

A random variable X has a *Pareto distribution with parameters x_0 and α* $(x_0 > 0, \alpha > 0)$ if X has an absolutely continuous distribution whose p.d.f. $f(\cdot \,|x_0, \alpha)$ is

$$f(x|x_0, \alpha) = \begin{cases} \dfrac{\alpha x_0^\alpha}{x^{\alpha+1}} & \text{for } x > x_0, \\ 0 & \text{otherwise.} \end{cases} \tag{1}$$

The Pareto distribution has been used, but not without some controversy, to represent a wide variety of empirical observations such as the distribution of income, the size of cities, and word frequencies. See Simon (1955; 1960; 1961a, b) and Mandelbrot (1959; 1961a, b; 1965) for an interesting discussion of these issues.

If X has a Pareto distribution whose p.d.f. is given by Eq. (1) and if $\alpha > 2$, then (see Exercise 30)

$$E(X) = \frac{\alpha x_0}{\alpha - 1} \tag{2}$$

and

$$\mathrm{Var}(X) = \frac{\alpha x_0^2}{(\alpha - 1)^2(\alpha - 2)}. \tag{3}$$

4.12 THE *t* DISTRIBUTION

A random variable X has a *t distribution with α degrees of freedom* $(\alpha > 0)$ if X has an absolutely continuous distribution whose p.d.f. $f(\cdot \,|\alpha)$ is specified at any point x $(-\infty < x < \infty)$ by the equation

$$f(x|\alpha) = \frac{\Gamma[(\alpha + 1)/2]}{(\alpha\pi)^{\frac{1}{2}}\Gamma(\alpha/2)} \left(1 + \frac{x^2}{\alpha}\right)^{-(\alpha+1)/2} \tag{1}$$

In many statistical applications, the parameter α is a positive integer. When $\alpha = 1$, the distribution defined by Eq. (1) is called the *Cauchy distribution*.

The *t* distribution is important in statistical work for the following reason: If X and Y are independent random variables, if X has a standard normal distribution, and if Y has a χ^2 distribution with n degrees of freedom, then the random variable

$$\frac{X}{(Y/n)^{\frac{1}{2}}} \tag{2}$$

has a *t* distribution with n degrees of freedom (see Exercise 32). It follows from this property and the comments made in Sec. 4.8 that if

$X_1, \ldots, X_n$ is a random sample from a normal distribution for which the mean is μ and the variance is σ^2, then the random variable

$$\frac{n^{\frac{1}{2}}(\bar{X} - \mu)}{\{[1/(n-1)]\Sigma_{i=1}^n (X_i - \bar{X})^2\}^{\frac{1}{2}}} \tag{3}$$

has a t distribution with $n - 1$ degrees of freedom.

If a random variable X has a t distribution with α degrees of freedom and if $\alpha > 2$, then (see Exercise 33)

$$E(X) = 0 \quad \text{and} \quad \text{Var}(X) = \frac{\alpha}{\alpha - 2}. \tag{4}$$

It is convenient to enlarge the family of t distributions so that it includes every p.d.f. which can be obtained from a p.d.f. of the form given in Eq. (1) through an arbitrary translation and change of scale. Specifically, a random variable Y is said to have a t *distribution with α degrees of freedom, location parameter μ, and precision τ* ($\alpha > 0$, $-\infty < \mu < \infty$, and $\tau > 0$) if the p.d.f. of the random variable $\tau^{\frac{1}{2}}(Y - \mu)$ is given by Eq. (1). Obviously, the p.d.f. $g(\cdot \,|\alpha, \mu, \tau)$ of Y is specified at any point y ($-\infty < y < \infty$) by the equation

$$g(y|\alpha, \mu, \tau) = \frac{\tau^{\frac{1}{2}}\Gamma[(\alpha + 1)/2]}{(\alpha\pi)^{\frac{1}{2}}\Gamma(\alpha/2)}\left[1 + \frac{\tau}{\alpha}(y - \mu)^2\right]^{-(\alpha+1)/2} \tag{5}$$

It should be noted that the parameter τ, which is here called the precision of the t distribution, is not the reciprocal of the variance of the distribution. Indeed, as pointed out above, the variance of the t distribution is not finite if $\alpha \leq 2$.

In this enlarged family of t distributions, the distribution whose p.d.f. is given by Eq. (1), for which $\mu = 0$ and $\tau = 1$, is sometimes called the *standardized t* distribution with α degrees of freedom.

4.13 THE F DISTRIBUTION

A random variable X has an F *distribution with α and β degrees of freedom* ($\alpha > 0, \beta > 0$) if X has an absolutely continuous distribution whose p.d.f. $f(\cdot \,|\alpha, \beta)$ is specified at any point $x > 0$ by the equation

$$f(x|\alpha, \beta) = \frac{\Gamma[(\alpha + \beta)/2]\alpha^{\alpha/2}\beta^{\beta/2}}{\Gamma(\alpha/2)\Gamma(\beta/2)} \cdot \frac{x^{(\alpha/2)-1}}{(\beta + \alpha x)^{(\alpha+\beta)/2}}. \tag{1}$$

Furthermore, $f(x|\alpha, \beta) = 0$ at any point $x \leq 0$.

The F distribution is important in statistical work for the following reason: If X and Y are independent random variables, if X has a χ^2 distribution with α degrees of freedom, and if Y has a χ^2 distribution with β

degrees of freedom, then the random variable

$$\frac{X/\alpha}{Y/\beta} \tag{2}$$

has an F distribution with α and β degrees of freedom (see Exercise 36).

One consequence of this fact is that the F distribution will represent the distribution of the ratio of certain independent quadratic forms which can be constructed from random samples from normal distributions. One of the simplest examples is the following: Let $X_1, \ldots, X_m$ ($m \geq 2$) be a random sample from a normal distribution for which the mean is μ_1 and the variance is σ^2, and let $Y_1, \ldots, Y_n$ ($n \geq 2$) be an independent random sample from a normal distribution for which the mean is μ_2 and the variance is again σ^2. Then the ratio

$$\frac{\sum_{i=1}^{m}(X_i - \bar{X})^2/(m-1)}{\sum_{i=1}^{n}(Y_i - \bar{Y})^2/(n-1)} \tag{3}$$

has an F distribution with $m-1$ and $n-1$ degrees of freedom.

If a random variable X has an F distribution with α and β degrees of freedom and if $\beta > 4$, then (see Exercise 39)

$$E(X) = \frac{\beta}{\beta - 2} \tag{4}$$

and

$$\text{Var}(X) = \frac{2\beta^2(\alpha + \beta - 2)}{\grave{\alpha}(\beta - 4)(\beta - 2)^2}. \tag{5}$$

EXERCISES

1. If X has a Bernoulli distribution with parameter p, show that $E(X)$ and $\text{Var}(X)$ are as given in Eq. (2) of Sec. 4.2. Also, show that the characteristic function ζ of X is $\zeta(t) = pe^{it} + q$.

2. If X has a binomial distribution with parameters n and p, show that $E(X)$ and $\text{Var}(X)$ are as given in Eq. (2) of Sec. 4.3. Also, show that the characteristic function ζ of X is $\zeta(t) = (pe^{it} + q)^n$.

3. If $X_1, \ldots, X_k$ are independent random variables and X_i has a binomial distribution with parameters n_i and p ($i = 1, \ldots, k$), show that the sum $X_1 + \cdots + X_k$ has a binomial distribution with parameters $n_1 + \cdots + n_k$ and p.

4. For any fixed nonnegative integer x, let $f(x|n, p)$ be the value of the binomial p.f. given by Eq. (1) of Sec. 4.3 and let $g(x|\lambda)$ be the value of the Poisson p.f. given by Eq. (1) of Sec. 4.4. If $n \to \infty$ and $p \to 0$ in such a way that $np \to \lambda$, prove that $f(x|n, p) \to g(x|\lambda)$.

5. If X has a Poisson distribution with mean λ, show that its characteristic function ζ is

$$\zeta(t) = \exp[\lambda(e^{it} - 1)].$$

Also, show that $E(X)$ and $\text{Var}(X)$ are as given in Eq. (2) of Sec. 4.4.

6. If $X_1, \ldots, X_k$ are independent random variables and X_i has a Poisson distribution for which the mean is λ_i $(i = 1, \ldots, k)$, show that the sum $X_1 + \cdots + X_k$ has a Poisson distribution with mean $\lambda_1 + \cdots + \lambda_k$.

7. Suppose that X_1 and X_2 are independent random variables and that X_i has a Poisson distribution for which the mean is λ_i $(i = 1, 2)$. If it is known that $X_1 + X_2 = n$, where n is a given positive integer, prove that the conditional distribution of X_1 is a binomial distribution with parameters n and $\lambda_1/(\lambda_1 + \lambda_2)$.

8. If X has a negative binomial distribution with parameters r and p, prove that its p.f. $f(\cdot \,|r, p)$ can be written in the form

$$
f(x|r, p) = \begin{cases} \binom{-r}{x} p^r(-q)^x & \text{for } x = 0, 1, 2, \ldots, \\ \\ 0 & \text{otherwise.} \end{cases}
$$

9. If X has a negative binomial distribution with parameters r and p, show that its characteristic function ζ is

$$
\zeta(t) = \left(\frac{p}{1 - qe^{it}} \right)^r.
$$

Also, show that $E(X)$ and $\text{Var}(X)$ are as given in Eq. (2) of Sec. 4.5.

10. Show that if $X_1, \ldots, X_k$ are independent random variables and X_i has a negative binomial distribution with parameters r_i and p $(i = 1, \ldots, k)$, then the sum $X_1 + \cdots + X_k$ has a negative binomial distribution with parameters $r_1 + \cdots + r_k$ and p.

11. For any fixed nonnegative integer x, let $f(x|r, p)$ be the value of the negative binomial p.f. given by Eq. (1) of Sec. 4.5 and let $g(x|\lambda)$ be the value of the Poisson p.f. given by Eq. (1) of Sec. 4.4. If $r \to \infty$ and $q \to 0$ in such a way that $rq \to \lambda$, prove that $f(x|r, p) \to g(x|\lambda)$.

12. Let $X_1, X_2, \ldots$ be a sequence of independent random variables all of which have the same discrete p.f. f defined as follows:

$$
f(x) = \begin{cases} \dfrac{1}{\log (1/p)} \dfrac{q^x}{x} & \text{for } x = 1, 2, \ldots, \\ \\ 0 & \text{otherwise.} \end{cases}
$$

Here $0 < p < 1$ and $q = 1 - p$. Let Y be a random variable that is independent of the sequence $X_1, X_2, \ldots$, and suppose that Y has a Poisson distribution with mean $r \log (1/p)$, where $r > 0$. Finally, define the random variable Z to be the sum of a random number Y of random variables in the sequence $X_1, X_2, \ldots$; that is, $Z = X_1 + X_2 + \cdots + X_Y$. Assume that if $Y = 0$, then $Z = 0$. Show that Z has a negative binomial distribution with parameters r and p.

13. If a random variable X has a hypergeometric distribution with parameters A, B, and n, prove that $E(X)$ and $\text{Var}(X)$ are as given in Eq. (3) of Sec. 4.6.

14. Prove that the normal p.d.f. as given by Eq. (1) of Sec. 4.7 is properly normalized by showing that

$$
\int_{-\infty}^{\infty} \exp \left(\frac{-x^2}{2} \right) dx = (2\pi)^{\frac{1}{2}}.
$$

Hint: Write the product of two integrals of this form as a double integral and change to polar coordinates.

15. If X has a normal distribution whose p.d.f. is given by Eq. (1) of Sec. 4.7,

show that its characteristic function ζ is

$$\zeta(t) = \exp\left(it\mu - \frac{t^2\sigma^2}{2}\right).$$

Also, show that $E(X) = \mu$ and $\mathrm{Var}(X) = \sigma^2$.

16. Let $X_1, \ldots, X_k$ be independent random variables such that X_i is normally distributed with mean μ_i and variance σ_i^2 $(i = 1, \ldots, k)$. Also, let $a_1, \ldots, a_k$ and b be constants such that $a_i \neq 0$ for at least one value of i. Show that the random variable $a_1X_1 + \cdots + a_kX_k + b$ is normally distributed with mean $a_1\mu_1 + \cdots + a_k\mu_k + b$ and variance $a_1^2\sigma_1^2 + \cdots + a_k^2\sigma_k^2$.

17. If X_1 and X_2 are independent and identically distributed random variables with the same normal distribution, show that the random variables $X_1 + X_2$ and $X_1 - X_2$ are independent.

18. Prove that the gamma function, as defined by Eq. (2) of Sec. 4.8, satisfies Eq. (3) of that section. Also, prove that $\Gamma(\frac{1}{2}) = \pi^{\frac{1}{2}}$. *Hint:* To prove the first result, use integration by parts. To prove the second result, change variables so that the integrand becomes a normal p.d.f.

19. If X has a gamma distribution with parameters α and β, show that its characteristic function ζ is

$$\zeta(t) = \left(1 - \frac{it}{\beta}\right)^{-\alpha}.$$

Also, show that $E(X)$ and $\mathrm{Var}(X)$ are as given in Eq. (4) of Sec. 4.8.

20. If $X_1, \ldots, X_k$ are independent random variables and X_i has a gamma distribution with parameters α_i and β $(i = 1, \ldots, k)$, show that for any constant c $(c > 0)$, the random variable $c(X_1 + \cdots + X_k)$ has a gamma distribution with parameters $\alpha_1 + \cdots + \alpha_k$ and β/c.

21. If $X_1, \ldots, X_n$ is a random sample from a standard normal distribution, prove that $X_1^2 + \cdots + X_n^2$ has a χ^2 distribution with n degrees of freedom.

22. If $X_1, \ldots, X_k$ are independent random variables and X_i has an exponential distribution with parameter β_i $(i = 1, \ldots, k)$, prove that the random variable min $\{X_1, \ldots, X_k\}$ has an exponential distribution with parameter $\beta_1 + \cdots + \beta_k$.

23. Let $X_1, \ldots, X_n$ be a random sample from an exponential distribution with parameter β, and let $Y_1 < Y_2 < \cdots < Y_n$ be the order statistics of the sample. Let n random variables $Z_1, \ldots, Z_n$ be defined by the equations

$$Z_1 = nY_1,$$
$$Z_2 = (n - 1)(Y_2 - Y_1),$$
$$Z_3 = (n - 2)(Y_3 - Y_2),$$
$$\cdots \cdots \cdots \cdots \cdots \cdots$$
$$Z_n = Y_n - Y_{n-1}.$$

Prove that the joint distribution of $Z_1, \ldots, Z_n$ is the same as the joint distribution of $X_1, \ldots, X_n$.

24. Let X and Y be two random variables whose joint distribution satisfies the following two conditions: (1) For any value y $(y > 0)$, the conditional distribution of X, given $Y = y$, is a Poisson distribution with mean y. (2) The marginal distribution of Y is a gamma distribution with parameters α and β. Show that the marginal distribution of X is a negative binomial distribution with parameters α and $\beta/(\beta + 1)$.

25. Prove that the beta p.d.f. as given by Eq. (1) of Sec. 4.9 is properly

normalized by showing that

$$\Gamma(\alpha)\Gamma(\beta) = \Gamma(\alpha + \beta) \int_0^1 x^{\alpha-1}(1 - x)^{\beta-1}\, dx.$$

Hint: Write the product $\Gamma(\alpha)\Gamma(\beta)$ as an iterated integral with the variables of integration designated as r and s. Then change the variables to r and $t = s/r$. After integrating over r, change the variable t to $u = 1/(1 + t)$.

26. Let X have a beta distribution with parameters α and β. For any positive integers m and n, find $E[X^m(1 - X)^n]$, and show that $E(X)$ and $\mathrm{Var}(X)$ are as given in Eqs. (2) and (3) of Sec. 4.9.

27. Suppose that X_1 and X_2 are independent random variables such that X_1 has a gamma distribution for which the parameters are α_1 and β and X_2 has a gamma distribution for which the parameters are α_2 and the same value of β. If $R = X_1/(X_1 + X_2)$ and $S = X_1 + X_2$, show that the random variables R and S are independent and that R has a beta distribution with parameters α_1 and α_2.

28. If X has a uniform distribution on the interval (α, β), show that $E(X)$ and $\mathrm{Var}(X)$ are as given in Eq. (2) of Sec. 4.10. Also, show that the characteristic function ζ of X is

$$\zeta(t) = \frac{e^{it\beta} - e^{it\alpha}}{it(\beta - \alpha)}.$$

29. Let $X_1, \ldots, X_n$ be a random sample from a uniform distribution on the interval (α, β), and let $Y = \min\{X_1, \ldots, X_n\}$ and $Z = \max\{X_1, \ldots, X_n\}$. Show that the joint p.d.f. f of the random variables Y and Z is

$$f(y, z) = \begin{cases} \dfrac{n(n - 1)(z - y)^{n-2}}{(\beta - \alpha)^n} & \text{for } \alpha < y < z < \beta, \\ 0 & \text{otherwise.} \end{cases}$$

30. If X has a Pareto distribution with parameters x_0 and α, for what positive integers k does the kth moment $E(X^k)$ exist? Show that if $\alpha > 2$, then $E(X)$ and $\mathrm{Var}(X)$ are as given in Eqs. (2) and (3) of Sec. 4.11.

31. If X has a Pareto distribution with parameters x_0 and α, show that the random variable $\log(X/x_0)$ has an exponential distribution with parameter α.

32. If X and Y are independent random variables, if X has a standard normal distribution, and if Y has a χ^2 distribution with n degrees of freedom, show that the random variable specified in expression (2) of Sec. 4.12 has a t distribution with n degrees of freedom.

33. If X has a t distribution with α degrees of freedom, for what positive integers k does the kth moment $E(X^k)$ exist? Show that if $\alpha > 2$, then $E(X)$ and $\mathrm{Var}(X)$ are as given in Eq. (4) of Sec. 4.12. *Hint:* To compute $E(X^2)$, restrict the integral to the positive half of the real line and then change the variable from x to y, where

$$y = \frac{x^2/\alpha}{1 + (x^2/\alpha)}.$$

34. If X_1 and X_2 are independent random variables and each has a standard normal distribution, show that X_1/X_2 has a Cauchy distribution.

35. A particle is emitted at the origin of the xy-plane and travels in a straight line into the half-plane for which $x > 0$. Let θ be the angle that the path of the particle makes with the positive x-axis, and let Y be the height at which the particle strikes the line $x = 1$. If it is assumed that θ is uniformly distributed on the interval $(-\pi/2, \pi/2)$, show that Y has a Cauchy distribution.

36. If X and Y are independent random variables, if X has a χ^2 distribution with α degrees of freedom, and if Y has a χ^2 distribution with β degrees of freedom, show that the random variable specified in expression (2) of Sec. 4.13 has an F distribution with α and β degrees of freedom.

37. (a) If X has an F distribution with α and β degrees of freedom, show that $Y = 1/X$ has an F distribution with β and α degrees of freedom.

(b) If X has a t distribution with α degrees of freedom, show that $Y = X^2$ has an F distribution with 1 and α degrees of freedom.

38. Suppose that X has an F distribution with α and β degrees of freedom and that

$$Y = \frac{\alpha X}{\beta + \alpha X}.$$

Show that Y has a beta distribution with parameters $\alpha/2$ and $\beta/2$.

39. If X has an F distribution with α and β degrees of freedom, for what positive integers k does the kth moment $E(X^k)$ exist? Show that if $\beta > 4$, then $E(X)$ and $\mathrm{Var}(X)$ are as given in Eqs. (4) and (5) of Sec. 4.13.

some special
multivariate
distributions

5.1 INTRODUCTION

In this chapter we shall present some of the basic properties of six special multivariate distributions that will appear in later chapters of this book. These distributions are the multinomial, the Dirichlet, the multivariate normal, the Wishart, the multivariate t, and what is called here the bilateral bivariate Pareto.   The format of this chapter is essentially the same as that of the preceding chapter on univariate distributions.   However, since it is anticipated that many readers will not be sufficiently familiar with some of the multivariate distributions, many properties of these distributions will be given here in relatively great detail.

5.2 THE MULTINOMIAL DISTRIBUTION

Consider an experiment whose outcome must belong to one of k $(k \geq 2)$ mutually exclusive and exhaustive categories, and let p_i $(0 < p_i < 1)$ be the probability that the outcome belongs to the ith category $(i = 1, \ldots, k)$.   Here $\Sigma_{i=1}^{k} p_i = 1$.   Suppose that the experiment is performed n times and that the n outcomes are independent.   Furthermore, let X_i denote the number of these outcomes that belong to category i $(i = 1, \ldots, k)$. Then the random vector $\mathbf{X} = (X_1, \ldots, X_k)'$ has a *multinomial distribu-*

tion with parameters n and $\mathbf{p} = (p_1, \ldots, p_k)'$. Let $\mathbf{x} = (x_1, \ldots, x_k)'$ be any point in R^k each of whose components x_i is a nonnegative integer $(i = 1, \ldots, k)$ such that $\Sigma_{i=1}^k x_i = n$. Then at the point $\mathbf{x}$, the value of the p.f. $f(\,\cdot\,|n, \mathbf{p})$ of $\mathbf{X}$ is specified by the equation

$$f(\mathbf{x}|n, \mathbf{p}) = \frac{n!}{x_1! \cdots x_k!} p_1^{x_1} \cdots p_k^{x_k}. \tag{1}$$

It follows from the definition of the random vector $\mathbf{X}$ that $f(\mathbf{x}|n, \mathbf{p}) = 0$ at any other point $\mathbf{x} \, \epsilon \, R^k$.

Since the probability that $\Sigma_{i=1}^k X_i = n$ is 1, one of the k random variables $X_1, \ldots, X_k$ can be eliminated, and the multinomial p.f. defined by Eq. (1) can be written as a $(k-1)$-dimensional distribution. However, this reduction introduces an asymmetry among the k categories that was not originally present. Indeed, a reduction similar to this was made when the binomial distribution was defined in Sec. 4.3. Hence, if X has a binomial distribution with parameters n and p, then the two-dimensional random vector $(X, n - X)'$ has a multinomial distribution with parameters n and $(p, 1 - p)'$.

If a random vector $\mathbf{X} = (X_1, \ldots, X_k)'$ has a multinomial distribution with parameters n and $\mathbf{p}$, then the mean vector of $\mathbf{X}$ is $E(\mathbf{X}) = n\mathbf{p}$ and the elements of the covariance matrix of $\mathbf{X}$ are as follows (see Exercise 2):

$$\begin{aligned} \mathrm{Var}(X_i) &= np_i(1 - p_i) & i &= 1, \ldots, k, \\ \mathrm{Cov}(X_i, X_j) &= -np_i p_j & i, j &= 1, \ldots, k; i \neq j. \end{aligned} \tag{2}$$

The following properties result from the description of the multinomial distribution given at the beginning of this section. If $\mathbf{X}$ has the multinomial distribution defined by Eq. (1), then the marginal distribution of any component X_i is binomial with parameters n and p_i $(i = 1, \ldots, k)$. Also, if $\mathbf{X}_1, \mathbf{X}_2, \ldots, \mathbf{X}_r$ are independent k-dimensional random vectors and if $\mathbf{X}_i$ has a multinomial distribution with parameters n_i and $\mathbf{p}$ $(i = 1, \ldots, r)$, then the sum $\mathbf{X}_1 + \cdots + \mathbf{X}_r$ has a multinomial distribution with parameters $n_1 + \cdots + n_r$ and $\mathbf{p}$ (see Exercise 3).

5.3 THE DIRICHLET DISTRIBUTION

The conventions which we shall use in introducing the p.d.f. of the Dirichlet distribution are slightly different from those which we have been using. The appropriate explanation will be given after Eq. (1) has been presented.

A random vector $\mathbf{X} = (X_1, \ldots, X_k)'$ has a *Dirichlet distribution with parametric vector* $\boldsymbol{\alpha} = (\alpha_1, \ldots, \alpha_k)'$ $(\alpha_i > 0; i = 1, \ldots, k)$ if the p.d.f. $f(\,\cdot\,|\alpha)$ of $\mathbf{X}$ satisfies the following properties: Let $\mathbf{x} = (x_1, \ldots , x_k)'$ be any point in R^k such that $x_i > 0$ for $i = 1, \ldots , k$ and $\Sigma_{i=1}^{k}x_i = 1$. Then

$$f(\mathbf{x}|\alpha) = \frac{\Gamma(\alpha_1 + \cdots + \alpha_k)}{\Gamma(\alpha_1) \cdots \Gamma(\alpha_k)}\, x_1^{\alpha_1 - 1} \cdots x_k^{\alpha_k - 1}. \tag{1}$$

Also, $f(\mathbf{x}|\alpha) = 0$ at any other point $\mathbf{x} \in R^k$.

Equation (1) requires further explanation. The function $f(\,\cdot\,|\alpha)$ is positive only on the $(k-1)$-dimensional simplex of points $\mathbf{x} = (x_1, \ldots , x_k)'$ such that $x_i > 0$ for $i = 1, \ldots , k$ and $\Sigma_{i=1}^{k}x_i = 1$. Therefore,

$$\Pr\!\left(\sum_{i=1}^{k} X_i = 1\right) = 1. \tag{2}$$

It follows from this linear relation that the k random variables $X_1, \ldots , X_k$ cannot have a joint k-dimensional p.d.f. Thus, despite its appearance, $f(\,\cdot\,|\alpha)$ is not a k-dimensional p.d.f. Rather, it gives the joint p.d.f. of any subcollection of $k - 1$ of the random variables $X_1, \ldots , X_k$ after the other one, say X_j, has been eliminated through use of the relation $X_1 + \cdots + X_k = 1$. This joint $(k - 1)$-dimensional p.d.f. is obtained by eliminating x_j from the right side of Eq. (1) through use of the relation $x_1 + \cdots + x_k = 1$. The resulting p.d.f. is positive at those points in R^{k-1} such that each of the $k - 1$ coordinates is positive and their sum is less than 1.

A comparison of the expression in Eq. (1) with the p.d.f. of the beta distribution in Eq. (1) of Sec. 4.9 reveals why the Dirichlet distribution is sometimes referred to as a multivariate beta distribution. In fact, if a random variable X has a beta distribution with parameters α and β, then the random vector $\mathbf{Y} = (X, 1 - X)'$ has a Dirichlet distribution with parametric vector $(\alpha, \beta)'$. Furthermore, if a random vector $\mathbf{X} = (X_1, \ldots , X_k)'$ has a Dirichlet distribution with parametric vector $\alpha = (\alpha_1, \ldots , \alpha_k)'$, then the marginal distribution of one component of $\mathbf{X}$, say X_j, is a beta distribution with parameters α_j and $\Sigma_{i=1}^{k}\alpha_i - \alpha_j$.

A more general result is the following: Suppose that $\mathbf{X} = (X_1, \ldots , X_k)'$ has a Dirichlet distribution as above and that the indices $1, \ldots , k$ are partitioned into m groups $J_1, \ldots , J_m$ ($2 \le m \le k$) which are nonempty, mutually exclusive, and exhaustive. Also, let $Y_i = \Sigma_{j \in J_i} X_j$ ($i = 1, \ldots , m$). Then the m-dimensional random vector $\mathbf{Y} = (Y_1, \ldots , Y_m)'$ has a Dirichlet distribution with parametric vector $\boldsymbol{\beta} = (\beta_1, \ldots , \beta_m)'$, where $\beta_i = \Sigma_{j \in J_i} \alpha_j$ ($i = 1, \ldots , m$).

If a random vector $\mathbf{X} = (X_1, \ldots , X_k)'$ has a Dirichlet distribution with parametric vector $\alpha = (\alpha_1, \ldots , \alpha_k)'$, then any moment of the form $E(X_1^{r_1} \cdots X_k^{r_k})$, where $r_1, \ldots , r_k$ are nonnegative integers, can be found as follows: Let the set $S \subset R^{k-1}$ be defined by the equation

$$S = \{(x_1, \ldots, x_{k-1})': x_i > 0 \ (i = 1, \ldots, k - 1) \text{ and } \sum_{i=1}^{k-1} x_i < 1\}. \tag{3}$$

The integral over the set S of the p.d.f. given in Eq. (1) must be unity. Therefore, if $\alpha_1, \ldots, \alpha_k$ are any positive numbers and if $x_k = 1 - \sum_{i=1}^{k-1}x_i$,

$$\int \cdots \int_S x_1^{\alpha_1-1} \cdots x_k^{\alpha_k-1} \, dx_1 \cdots dx_{k-1} = \frac{\Pi_{i=1}^k \Gamma(\alpha_i)}{\Gamma(\sum_{i=1}^k \alpha_i)}. \tag{4}$$

It follows that

$$E(X_1^{r_1} \cdots X_k^{r_k}) = \frac{\Gamma(\sum_{i=1}^k \alpha_i)}{\Pi_{i=1}^k \Gamma(\alpha_i)} \cdot \frac{\Pi_{i=1}^k \Gamma(\alpha_i + r_i)}{\Gamma[\sum_{i=1}^k (\alpha_i + r_i)]}. \tag{5}$$

Let $\alpha_0 = \sum_{i=1}^k \alpha_i$. Then it follows from Eq. (5) that for $i = 1, \ldots, k$,

$$E(X_i) = \frac{\alpha_i}{\alpha_0} \tag{6}$$

and

$$\text{Var}(X_i) = \frac{\alpha_i(\alpha_0 - \alpha_i)}{\alpha_0^2(\alpha_0 + 1)}. \tag{7}$$

These results also follow from the known facts about the beta distribution. Furthermore, when $i \neq j$,

$$\text{Cov}(X_i, X_j) = -\frac{\alpha_i \alpha_j}{\alpha_0^2(\alpha_0 + 1)}. \tag{8}$$

Further discussion of these and other interesting properties of the Dirichlet distribution is given by Wilks (1962).

5.4 THE MULTIVARIATE NORMAL DISTRIBUTION

A k-dimensional random vector $\mathbf{X} = (X_1, \ldots, X_k)'$ has a *nonsingular multivariate normal distribution with mean vector* $\mathbf{\mu}$ *and covariance matrix* $\mathbf{\Sigma}$ if $\mathbf{X}$ has an absolutely continuous distribution whose p.d.f. $f(\cdot | \mathbf{\mu}, \mathbf{\Sigma})$ is specified at any point $\mathbf{x} \in R^k$ by the equation

$$f(\mathbf{x}|\mathbf{\mu}, \mathbf{\Sigma}) = (2\pi)^{-k/2}|\mathbf{\Sigma}|^{-\frac{1}{2}} \exp\left[-\tfrac{1}{2}(\mathbf{x} - \mathbf{\mu})'\mathbf{\Sigma}^{-1}(\mathbf{x} - \mathbf{\mu})\right]. \tag{1}$$

In Eq. (1), the mean vector $\mathbf{\mu} = (\mu_1, \ldots, \mu_k)'$ is a k-dimensional vector whose components can be arbitrary real numbers; the $k \times k$ covariance matrix $\mathbf{\Sigma}$ must be symmetric and positive definite, but otherwise its components can also be arbitrary real numbers.

For reasons similar to those given in Sec. 4.7 for the univariate normal distribution, the multivariate normal distribution is very important in many phases of current work in probability and statistics. The entire body of theory known as multivariate statistical analysis deals almost exclusively with the analysis of random samples from multivariate normal distributions. An excellent book in this area is Anderson (1958).

We shall now derive the characteristic function ζ of a random vector $\mathbf{X} = (X_1, \ldots, X_k)'$ which has a multivariate normal distribution whose p.d.f. is defined by Eq. (1). For any point $\mathbf{t} = (t_1, \ldots, t_k)' \in R^k$,

$$\zeta(\mathbf{t}) = E(e^{i\mathbf{t}'\mathbf{x}})$$

$$= \int \cdots \int_{R^k} (2\pi)^{-k/2} |\mathbf{\Sigma}|^{-\frac{1}{2}} \exp\left[i\mathbf{t}'\mathbf{x}\right.$$

$$\left. - \tfrac{1}{2}(\mathbf{x} - \mathbf{\mu})'\mathbf{\Sigma}^{-1}(\mathbf{x} - \mathbf{\mu})\right] \prod_{j=1}^{k} dx_j. \quad (2)$$

It is well known that there exists a nonsingular $k \times k$ matrix $\mathbf{B}$ such that the symmetric and positive definite matrix $\mathbf{\Sigma}$ can be expressed in the form $\mathbf{\Sigma} = \mathbf{BB}'$. The variables in the integral in Eq. (2) can now be changed by means of the transformation

$$\mathbf{x} = \mathbf{\mu} + \mathbf{B}\mathbf{y}. \quad (3)$$

The Jacobian J of this transformation is $J = |\mathbf{B}| = |\mathbf{\Sigma}|^{\frac{1}{2}}$, and so Eq. (2) becomes

$$\zeta(\mathbf{t}) = \int \cdots \int_{R^k} (2\pi)^{-k/2} \exp\left[i\mathbf{t}'(\mathbf{\mu} + \mathbf{B}\mathbf{y}) - \tfrac{1}{2}\mathbf{y}'\mathbf{y}\right] \prod_{j=1}^{k} dy_j$$

$$= e^{i\mathbf{t}'\mathbf{\mu}} \int \cdots \int_{R^k} (2\pi)^{-k/2} \exp\left[i(\mathbf{B}'\mathbf{t})'\mathbf{y} - \tfrac{1}{2}\mathbf{y}'\mathbf{y}\right] \prod_{j=1}^{k} dy_j. \quad (4)$$

Next, let $\mathbf{B}'\mathbf{t} = \mathbf{s} = (s_1, \ldots, s_k)'$. Then Eq. (4) can be written in the following form:

$$\zeta(\mathbf{t}) = e^{i\mathbf{t}'\mathbf{\mu}} \prod_{j=1}^{k} \int_{-\infty}^{\infty} (2\pi)^{-\frac{1}{2}} \exp\left(is_j y_j - \tfrac{1}{2}y_j^2\right) dy_j. \quad (5)$$

Since each integral in Eq. (5) is simply the c.f. of a univariate standard normal distribution (see Exercise 15 of Chap. 4), it follows that

$$\zeta(\mathbf{t}) = e^{i\mathbf{t}'\mathbf{\mu}} \prod_{j=1}^{k} \exp\left(\frac{-s_j^2}{2}\right) = \exp\left(i\mathbf{t}'\mathbf{\mu} - \tfrac{1}{2}\mathbf{s}'\mathbf{s}\right)$$

$$= \exp\left(i\mathbf{t}'\mathbf{\mu} - \tfrac{1}{2}\mathbf{t}'\mathbf{BB}'\mathbf{t}\right), \quad (6)$$

or

$$\zeta(\mathbf{t}) = \exp\left(i\mathbf{t}'\mathbf{\mu} - \tfrac{1}{2}\mathbf{t}'\mathbf{\Sigma}\mathbf{t}\right). \quad (7)$$

An obvious analogy can be seen between the c.f. given by Eq. (7) and the c.f. of a random variable having a univariate normal distribution, as given in Exercise 15 of Chap. 4. It should be noted that Eq. (7) provides a verification of the fact that the integral over R^k of the p.d.f. given by Eq. (1) actually has the value 1, since at the point $\mathbf{t} = \mathbf{0}$ it is seen that $\zeta(\mathbf{0}) = 1$. Furthermore, it can be verified that μ is actually the mean vector and Σ is the covariance matrix of this distribution (see Exercise 7).

Equation (7) represents the c.f. of a random vector $\mathbf{X} = (X_1, \ldots, X_k)'$ even if the matrix Σ is nonnegative definite but not positive definite. The random vector $\mathbf{X}$ is said to have a *singular multivariate normal distribution* if the c.f. of $\mathbf{X}$ is given by Eq. (7), where Σ is a symmetric, nonnegative definite, singular matrix. It is still true that $E(\mathbf{X}) = \mu$ and $\text{Cov}(\mathbf{X}) = \Sigma$. However, since Σ is singular, there must be certain linear relations among the components $X_1, \ldots, X_k$, and therefore these k random variables cannot have a joint k-dimensional p.d.f. If some of the components are deleted until there are no linear relations among the ones that remain, then these remaining components will have a nonsingular multivariate normal distribution.

Suppose that $\mathbf{X} = (X_1, \ldots, X_k)'$ has a multivariate normal distribution with mean vector μ and covariance matrix Σ, where Σ might be either singular or nonsingular. Let $\mathbf{A}$ be a given $m \times k$ matrix, and let the m-dimensional random vector $\mathbf{Y}$ be defined as $\mathbf{Y} = \mathbf{AX}$. At any point $\mathbf{t} \in R^m$, the c.f. ζ_Y of $\mathbf{Y}$ has the following value:

$$\zeta_Y(\mathbf{t}) = E(e^{i\mathbf{t}'\mathbf{Y}}) = E(e^{i\mathbf{t}'\mathbf{AX}}) = E(e^{i(\mathbf{A}'\mathbf{t})'\mathbf{X}}) = \zeta(\mathbf{A}'\mathbf{t}).$$

Hence, from Eq. (7),

$$\zeta_Y(\mathbf{t}) = \exp\left(i\mathbf{t}'\mathbf{A}\mu - \tfrac{1}{2}\mathbf{t}'\mathbf{A}\Sigma\mathbf{A}'\mathbf{t}\right). \tag{8}$$

A direct comparison of Eqs. (7) and (8) reveals that $\mathbf{Y}$ itself has a multivariate normal distribution for which the mean vector is $\mathbf{A}\mu$ and the covariance matrix is $\mathbf{A}\Sigma\mathbf{A}'$.

One consequence of this derivation is the following basic result, which can be demonstrated by choosing the matrix $\mathbf{A}$ appropriately: The marginal joint distribution of any subset of the random variables $X_1, \ldots, X_k$ will again be normal, and the corresponding subvector of μ and the corresponding submatrix of Σ will be the mean vector and the covariance matrix of that distribution (see Exercise 8).

We shall now derive the conditional distribution of some of the components of $\mathbf{X}$ when the values of the other components are given. It is assumed that $\mathbf{X} = (X_1, \ldots, X_k)'$ has a multivariate normal distribution with mean vector μ and with nonsingular covariance matrix Σ.

Suppose that the k-dimensional random vector $\mathbf{X}$ is partitioned as in the form

$$\mathbf{X} = \begin{bmatrix} \mathbf{X}_1 \\ \mathbf{X}_2 \end{bmatrix}, \tag{9}$$

where $\mathbf{X}_1$ comprises the first k_1 components of $\mathbf{X}$, $\mathbf{X}_2$ comprises the last k_2 components of $\mathbf{X}$, and $k_1 + k_2 = k$. Suppose also that the mean vector $\boldsymbol{\mu}$, the covariance matrix $\boldsymbol{\Sigma}$, and its inverse $\boldsymbol{\Sigma}^{-1}$ are partitioned as follows:

$$\boldsymbol{\mu} = \begin{bmatrix} \boldsymbol{\mu}_1 \\ \boldsymbol{\mu}_2 \end{bmatrix}, \qquad \boldsymbol{\Sigma} = \begin{bmatrix} \boldsymbol{\Sigma}_{11} & \boldsymbol{\Sigma}_{12} \\ \boldsymbol{\Sigma}_{21} & \boldsymbol{\Sigma}_{22} \end{bmatrix}, \qquad \text{and} \qquad \boldsymbol{\Sigma}^{-1} = \begin{bmatrix} \mathbf{T}_{11} & \mathbf{T}_{12} \\ \mathbf{T}_{21} & \mathbf{T}_{22} \end{bmatrix}. \tag{10}$$

Here, $\boldsymbol{\mu}_1$ is a k_1-dimensional vector and $\boldsymbol{\mu}_2$ is a k_2-dimensional vector; both $\boldsymbol{\Sigma}_{11}$ and $\mathbf{T}_{11}$ are $k_1 \times k_1$ matrices; both $\boldsymbol{\Sigma}_{22}$ and $\mathbf{T}_{22}$ are $k_2 \times k_2$ matrices; both $\boldsymbol{\Sigma}_{12}$ and $\mathbf{T}_{12}$ are $k_1 \times k_2$ matrices; and $\boldsymbol{\Sigma}_{12} = \boldsymbol{\Sigma}_{21}'$ and $\mathbf{T}_{12} = \mathbf{T}_{21}'$. Note that $E(\mathbf{X}_i) = \boldsymbol{\mu}_i$ and $\text{Cov}(\mathbf{X}_i) = \boldsymbol{\Sigma}_{ii}$ for $i = 1, 2$ and that the elements of $\boldsymbol{\Sigma}_{12}$ are the covariances of the various components of $\mathbf{X}_1$ with those of $\mathbf{X}_2$. The conditional distribution of $\mathbf{X}_1$ when the value of $\mathbf{X}_2$ is given can now be developed in terms of the elements which are displayed in Eq. (10).

The joint p.d.f. of $\mathbf{X}_1$ and $\mathbf{X}_2$ is simply the p.d.f. of $\mathbf{X}$, as given by Eq. (1). Therefore, for any points $\mathbf{x}_1 \in R^{k_1}$ and $\mathbf{x}_2 \in R^{k_2}$, the value $f(\mathbf{x}_1, \mathbf{x}_2)$ of this joint p.d.f. is specified by Eq. (1) with

$$\mathbf{x} = \begin{bmatrix} \mathbf{x}_1 \\ \mathbf{x}_2 \end{bmatrix}. \tag{11}$$

Furthermore, the marginal distribution of $\mathbf{X}_2$ is a multivariate normal distribution with mean vector $\boldsymbol{\mu}_2$ and covariance matrix $\boldsymbol{\Sigma}_{22}$. Therefore, at any point $\mathbf{x}_2 \in R^{k_2}$, the value $g(\mathbf{x}_2)$ of the marginal p.d.f. of $\mathbf{X}_2$ is specified by the equation

$$g(\mathbf{x}_2) = (2\pi)^{-k_2/2} |\boldsymbol{\Sigma}_{22}|^{-\frac{1}{2}} \exp\left[-\tfrac{1}{2}(\mathbf{x}_2 - \boldsymbol{\mu}_2)' \boldsymbol{\Sigma}_{22}^{-1}(\mathbf{x}_2 - \boldsymbol{\mu}_2)\right]. \tag{12}$$

It follows from Eqs. (10) and (11) that

$$\begin{aligned}
(\mathbf{x} - \boldsymbol{\mu})' \boldsymbol{\Sigma}^{-1}(\mathbf{x} - \boldsymbol{\mu}) &= (\mathbf{x}_1' - \boldsymbol{\mu}_1', \mathbf{x}_2' - \boldsymbol{\mu}_2') \begin{bmatrix} \mathbf{T}_{11} & \mathbf{T}_{12} \\ \mathbf{T}_{21} & \mathbf{T}_{22} \end{bmatrix} \begin{bmatrix} \mathbf{x}_1 - \boldsymbol{\mu}_1 \\ \mathbf{x}_2 - \boldsymbol{\mu}_2 \end{bmatrix} \\
&= (\mathbf{x}_1 - \boldsymbol{\mu}_1)' \mathbf{T}_{11}(\mathbf{x}_1 - \boldsymbol{\mu}_1) \\
&\quad + (\mathbf{x}_1 - \boldsymbol{\mu}_1)' \mathbf{T}_{12}(\mathbf{x}_2 - \boldsymbol{\mu}_2) \\
&\quad + (\mathbf{x}_2 - \boldsymbol{\mu}_2)' \mathbf{T}_{21}(\mathbf{x}_1 - \boldsymbol{\mu}_1) \\
&\quad + (\mathbf{x}_2 - \boldsymbol{\mu}_2)' \mathbf{T}_{22}(\mathbf{x}_2 - \boldsymbol{\mu}_2). \quad (13)
\end{aligned}$$

It can be shown (see Exercise 9) that

$$\boldsymbol{\Sigma}_{22}^{-1} = \mathbf{T}_{22} - \mathbf{T}_{21}\mathbf{T}_{11}^{-1}\mathbf{T}_{12}. \tag{14}$$

This identity permits the value specified in Eq. (13) to be rewritten as

$$[(\mathbf{x}_1 - \mathbf{\mu}_1) + \mathbf{T}_{11}^{-1}\mathbf{T}_{12}(\mathbf{x}_2 - \mathbf{\mu}_2)]'\mathbf{T}_{11}[(\mathbf{x}_1 - \mathbf{\mu}_1) + \mathbf{T}_{11}^{-1}\mathbf{T}_{12}(\mathbf{x}_2 - \mathbf{\mu}_2)]$$
$$+ (\mathbf{x}_2 - \mathbf{\mu}_2)'\mathbf{\Sigma}_{22}^{-1}(\mathbf{x}_2 - \mathbf{\mu}_2). \quad (15)$$

It is also shown in Exercise 9 that

$$\mathbf{T}_{11}^{-1}\mathbf{T}_{12} = -\mathbf{\Sigma}_{12}\mathbf{\Sigma}_{22}^{-1} \quad (16)$$

and

$$\mathbf{T}_{11}^{-1} = \mathbf{\Sigma}_{11} - \mathbf{\Sigma}_{12}\mathbf{\Sigma}_{22}^{-1}\mathbf{\Sigma}_{21}. \quad (17)$$

Therefore, if we let

$$Q_1 = (\mathbf{x}_1 - \mathbf{\nu}_1)'(\mathbf{\Sigma}_{11} - \mathbf{\Sigma}_{12}\mathbf{\Sigma}_{22}^{-1}\mathbf{\Sigma}_{21})^{-1}(\mathbf{x}_1 - \mathbf{\nu}_1), \quad (18)$$

where

$$\mathbf{\nu}_1 = \mathbf{\mu}_1 + \mathbf{\Sigma}_{12}\mathbf{\Sigma}_{22}^{-1}(\mathbf{x}_2 - \mathbf{\mu}_2), \quad (19)$$

and also let

$$Q_2 = (\mathbf{x}_2 - \mathbf{\mu}_2)'\mathbf{\Sigma}_{22}^{-1}(\mathbf{x}_2 - \mathbf{\mu}_2), \quad (20)$$

then it follows that expression (15) is equal to $Q_1 + Q_2$.

Hence, for any points $\mathbf{x}_1 \in R^{k_1}$ and $\mathbf{x}_2 \in R^{k_2}$,

$$f(\mathbf{x}_1, \mathbf{x}_2) = (2\pi)^{-k/2}|\mathbf{\Sigma}|^{-\frac{1}{2}} \exp\left(-\frac{Q_1 + Q_2}{2}\right). \quad (21)$$

Furthermore, by Eqs. (12) and (20),

$$g(\mathbf{x}_2) = (2\pi)^{-k_2/2}|\mathbf{\Sigma}_{22}|^{-\frac{1}{2}} \exp\left(-\frac{Q_2}{2}\right). \quad (22)$$

It therefore follows that the conditional p.d.f. $h(\,\cdot\,|\mathbf{x}_2)$ of $\mathbf{X}_1$, when $\mathbf{X}_2 = \mathbf{x}_2$, is specified at the point $\mathbf{x}_1$ by the equation

$$h(\mathbf{x}_1|\mathbf{x}_2) = (2\pi)^{-k_1/2}\left(\frac{|\mathbf{\Sigma}_{22}|}{|\mathbf{\Sigma}|}\right)^{\frac{1}{2}} \exp\left(-\frac{Q_1}{2}\right). \quad (23)$$

By examination of Eq. (18) and a direct comparison of the conditional p.d.f. given by Eq. (23) and the multivariate normal p.d.f. given by Eq. (1), it can be seen that the conditional distribution of $\mathbf{X}_1$, when $\mathbf{X}_2 = \mathbf{x}_2$, is a k_1-dimensional multivariate normal distribution whose mean vector is $\mathbf{\nu}_1$, as given by Eq. (19), and whose covariance matrix is $\mathbf{T}_{11}^{-1}$, as given by Eq. (17).

Furthermore, by comparing the constant that appears in Eq. (23) with the constant that usually appears in the p.d.f. of a multivariate normal distribution whose covariance matrix is given by Eq. (17), we obtain the following relation:

$$|\mathbf{\Sigma}| = |\mathbf{\Sigma}_{22}| \cdot |\mathbf{\Sigma}_{11} - \mathbf{\Sigma}_{12}\mathbf{\Sigma}_{22}^{-1}\mathbf{\Sigma}_{21}|. \quad (24)$$

The *precision matrix* **T** of a nonsingular multivariate normal distribution is defined to be the inverse of the covariance matrix; that is,

$$\mathbf{T} = \mathbf{\Sigma}^{-1}. \tag{25}$$

As stated in Sec. 4.7 in regard to a univariate normal distribution, it will be more convenient in much of our later work to specify a nonsingular multivariate normal distribution by its mean vector and its precision matrix rather than by its mean vector and its covariance matrix. Thus, if a random vector $\mathbf{X} = (X_1, \ldots, X_k)'$ has a multivariate normal distribution with mean vector $\boldsymbol{\mu}$ and precision matrix **T**, its p.d.f. $g(\cdot \,|\boldsymbol{\mu}, \mathbf{T})$ is specified at any point $\mathbf{x}$ $(\mathbf{x} \in R^k)$ by the equation

$$g(\mathbf{x}|\boldsymbol{\mu}, \mathbf{T}) = (2\pi)^{-k/2}|\mathbf{T}|^{\frac{1}{2}} \exp\left[-\tfrac{1}{2}(\mathbf{x} - \boldsymbol{\mu})'\mathbf{T}(\mathbf{x} - \boldsymbol{\mu})\right]. \tag{26}$$

Note that if it is stated that a random vector has a multivariate normal distribution with a specified precision matrix, then this distribution must be nonsingular and it must have a p.d.f. of the form given in Eq. (26).

5.5 THE WISHART DISTRIBUTION

Let $\mathbf{X}_1, \ldots, \mathbf{X}_n$ be a random sample of k-dimensional random vectors from a multivariate normal distribution for which the mean vector is **0** and the $k \times k$ covariance matrix is **Σ**. Also, let **V** denote the random symmetric $k \times k$ matrix which is defined by the equation

$$\mathbf{V} = \sum_{i=1}^{n} \mathbf{X}_i\mathbf{X}_i'. \tag{1}$$

This random matrix **V** has a *Wishart distribution with n degrees of freedom and parametric matrix* **Σ**. It will be assumed that $n > k - 1$ and that the matrix **Σ** is nonsingular. Under these conditions, the Wishart distribution is called *nonsingular* and the distribution of **V** can be represented by a p.d.f. in the following way:

Since **V** is a symmetric matrix, it can be expressed in the form

$$\mathbf{V} = \begin{bmatrix} V_{11} & \cdots & V_{1k} \\ \cdots\cdots\cdots\cdots \\ V_{k1} & \cdots & V_{kk} \end{bmatrix}, \tag{2}$$

where $V_{ij} = V_{ji}$ $(i, j = 1, \ldots, k)$. Hence, the random matrix **V** is completely specified by the $k(k + 1)/2$ distinct random variables V_{ij} which lie on or above its main diagonal. When $n > k - 1$ and **Σ** is nonsingular, it can be shown that the joint distribution of the random variables V_{ij} $(i, j = 1, \ldots, k; i \leq j)$ is absolutely continuous and can therefore be represented by their joint p.d.f. in $R^{k(k+1)/2}$.

Each specification of the values v_{ij} of the random variables V_{ij} $(i, j = 1, \ldots, k; i \leq j)$ can be regarded either as characterizing a vector $(v_{11}, v_{12}, v_{22}, \ldots, v_{kk})' \in R^{k(k+1)/2}$ or, equivalently, as characterizing a symmetric $k \times k$ matrix

$$\mathbf{v} = \begin{bmatrix} v_{11} & \cdots & v_{1k} \\ \cdots & \cdots & \cdots \\ v_{1k} & \cdots & v_{kk} \end{bmatrix}. \tag{3}$$

Accordingly, the value $f(\mathbf{v}|n, \mathbf{\Sigma})$ of the joint p.d.f. of the $k(k + 1)/2$ distinct random variables V_{ij} can be specified for each symmetric matrix $\mathbf{v}$ of the form given in Eq. (3). Furthermore, it is convenient to refer to the joint p.d.f. $f(\cdot | n, \mathbf{\Sigma})$ of these random variables simply as the p.d.f. of the random symmetric matrix $\mathbf{V}$. It is important to keep in mind that although the function $f(\cdot | n, \mathbf{\Sigma})$ will be called the p.d.f. of the random matrix $\mathbf{V}$ and will be treated as a function of symmetric matrices $\mathbf{v}$, it is, in fact, the joint p.d.f. of only $k(k + 1)/2$ distinct random variables and is a function of vectors in $R^{k(k+1)/2}$.

It can be shown that when $n > k - 1$ and $\mathbf{\Sigma}$ is nonsingular, the probability is 1 that the random matrix $\mathbf{V}$ defined by Eq. (2) will be positive definite. Therefore, for any matrix $\mathbf{v}$ which is not symmetric and positive definite, $f(\mathbf{v}|n, \mathbf{\Sigma}) = 0$. Furthermore, for any $k \times k$ matrix $\mathbf{v}$ which is symmetric and positive definite,

$$f(\mathbf{v}|n, \mathbf{\Sigma}) = c|\mathbf{\Sigma}|^{-n/2}|\mathbf{v}|^{(n-k-1)/2} \exp\left[-\tfrac{1}{2} \operatorname{tr}(\mathbf{\Sigma}^{-1}\mathbf{v})\right]. \tag{4}$$

In this equation, $\operatorname{tr}(\mathbf{\Sigma}^{-1}\mathbf{v})$ denotes the trace of the matrix $\mathbf{\Sigma}^{-1}\mathbf{v}$ and the constant c has the following value:

$$c = \left[2^{nk/2}\pi^{k(k-1)/4} \prod_{j=1}^{k} \Gamma\left(\frac{n + 1 - j}{2}\right) \right]^{-1}. \tag{5}$$

Equations (4) and (5) together describe the p.d.f. of a nonsingular Wishart distribution of dimension $k \times k$, with n degrees of freedom and parametric matrix $\mathbf{\Sigma}$.

Let S denote the set of symmetric, positive definite $k \times k$ matrices. Since each matrix in S is specified by $k(k + 1)/2$ distinct elements, the set S can be regarded as a subset of $R^{k(k+1)/2}$. Since the integral of the p.d.f. $f(\cdot | n, \mathbf{\Sigma})$ over the space $R^{k(k+1)/2}$ must have the value 1, Eq. (5) is equivalent to the following equation:

$$c^{-1}|\mathbf{\Sigma}|^{n/2} = \int_S |\mathbf{v}|^{(n-k-1)/2} \exp\left[-\tfrac{1}{2} \operatorname{tr}(\mathbf{\Sigma}^{-1}\mathbf{v})\right] dv_{11} dv_{12} \cdots dv_{kk}. \tag{6}$$

For a derivation of Eqs. (4) to (6), as well as of the other properties of the Wishart distribution that are mentioned in this section, the reader is referred to Anderson (1958), Wilks (1962), or Rao (1965).

Much of the importance of the Wishart distribution stems from the following famous result. Let $X_1, \ldots, X_n$ be a random sample of k-dimensional random vectors from a multivariate normal distribution for which the mean vector is μ and the covariance matrix is Σ. Also, let the vector $\bar{X}$ and the $k \times k$ matrix S be defined as follows:

$$\bar{X} = \frac{1}{n} \sum_{i=1}^{n} X_i \quad \text{and} \quad S = \sum_{i=1}^{n} (X_i - \bar{X})(X_i - \bar{X})'.$$

Then the random vector $\bar{X}$ and the random matrix S are independent; $\bar{X}$ has a multivariate normal distribution with mean vector μ and covariance matrix Σ/n, and S has a Wishart distribution with $n - 1$ degrees of freedom and parametric matrix Σ. It is clear from this result as well as from the representation given in Eq. (1) that the Wishart distribution is essentially a multivariate generalization of the χ^2 distribution.

Suppose that a $k \times k$ random matrix V has a Wishart distribution with n degrees of freedom and parametric matrix Σ. The following three properties are easily obtained from the representation in Eq. (1) (see Exercise 15): (1) The expectation of V is given by the relation $E(V) = n\Sigma$. (2) If A is an $m \times k$ matrix, then the $m \times m$ random matrix AVA' has a Wishart distribution with n degrees of freedom and parametric matrix $A\Sigma A'$. (3) Suppose that the matrices V and Σ are partitioned as follows:

$$V = \begin{bmatrix} V_{11} & V_{12} \\ V_{21} & V_{22} \end{bmatrix} \quad \text{and} \quad \Sigma = \begin{bmatrix} \Sigma_{11} & \Sigma_{12} \\ \Sigma_{21} & \Sigma_{22} \end{bmatrix}, \tag{7}$$

where V_{11} and Σ_{11} are square matrices with the same dimension. Then the random matrix V_{11} has a Wishart distribution with n degrees of freedom and parametric matrix Σ_{11}.

Furthermore (see Exercise 16), suppose that $V_1, \ldots, V_r$ are independent $k \times k$ random matrices and that V_i has a Wishart distribution with n_i degrees of freedom and parametric matrix Σ $(i = 1, \ldots, r)$. Then the sum $V_1 + \cdots + V_r$ also has a Wishart distribution with $n_1 + \cdots + n_r$ degrees of freedom and parametric matrix Σ.

If a $k \times k$ random matrix V has a Wishart distribution whose p.d.f. is given by Eq. (4), then the characteristic function ζ of V can be given in a simple form. The correct interpretation of the function ζ involves essentially the same considerations as those involved in the interpretation of the p.d.f. given in Eq. (4). Thus, ζ is really the joint characteristic function of the $k(k + 1)/2$ distinct random variables V_{ij} which define V, and ζ is therefore a function of $k(k + 1)/2$ distinct real variables t_{ij} $(i, j = 1, \ldots, k; i \leq j)$. However, each specification of the values of the variables t_{ij} can be regarded as characterizing a symmetric matrix t

defined as follows:

$$
\mathbf{t} = \begin{bmatrix} 2t_{11} & t_{12} & \cdots & t_{1k} \\ t_{12} & 2t_{22} & \cdots & t_{2k} \\ \cdots\cdots\cdots\cdots\cdots\cdots \\ t_{1k} & t_{2k} & \cdots & 2t_{kk} \end{bmatrix}. \tag{8}
$$

Therefore, the characteristic function ζ can conveniently be regarded as a function of symmetric matrices $\mathbf{t}$ which have the form given in Eq. (8). With this convention, it can be shown that

$$
\begin{aligned}
\zeta(\mathbf{t}) &= E \left[\exp \left(i \sum_{\beta=1}^{k} \sum_{\alpha=1}^{\beta} t_{\alpha\beta} V_{\alpha\beta} \right) \right] \\
&= \left(\frac{|\mathbf{\Sigma}^{-1}|}{|\mathbf{\Sigma}^{-1} - i\mathbf{t}|} \right)^{n/2}.
\end{aligned} \tag{9}
$$

The *precision matrix* $\mathbf{T}$ of a nonsingular Wishart distribution with parametric matrix $\mathbf{\Sigma}$ is defined by the relation

$$
\mathbf{T} = \mathbf{\Sigma}^{-1}. \tag{10}
$$

As stated in connection with the normal distribution, it will be more convenient in much of our later work to specify a Wishart distribution by its degrees of freedom n and its precision matrix $\mathbf{T}$, rather than by n and $\mathbf{\Sigma}$. Thus, if a random $k \times k$ matrix $\mathbf{V}$ has a Wishart distribution with n degrees of freedom $(n > k - 1)$ and precision matrix $\mathbf{T}$, its p.d.f. $g(\cdot \mid n, \mathbf{T})$ is specified for any symmetric, positive definite $k \times k$ matrix $\mathbf{v}$ by the equation

$$
g(\mathbf{v} \mid n, \mathbf{T}) = c|\mathbf{T}|^{n/2}|\mathbf{v}|^{(n-k-1)/2} \exp \left[-\tfrac{1}{2} \operatorname{tr}(\mathbf{T}\mathbf{v}) \right]. \tag{11}
$$

The constant c is given by Eq. (5). Furthermore, for any symmetric matrix $\mathbf{v}$ which is not positive definite, $g(\mathbf{v} \mid n, \mathbf{T}) = 0$.

5.6 THE MULTIVARIATE t DISTRIBUTION

Suppose that the k-dimensional random vector $\mathbf{Y} = (Y_1, \ldots, Y_k)'$ has a multivariate normal distribution with mean vector $\mathbf{0}$ and precision matrix $\mathbf{T}$, that the random variable Z has a χ^2 distribution with n degrees of freedom, and that $\mathbf{Y}$ and Z are independent. Suppose also that $\mathbf{\mu} = (\mu_1, \ldots, \mu_k)'$ is any given vector in R^k. Let a random vector $\mathbf{X} = (X_1, \ldots, X_k)'$ be defined by the equation

$$
X_i = Y_i \left(\frac{Z}{n} \right)^{-\frac{1}{2}} + \mu_i \qquad i = 1, \ldots, k. \tag{1}
$$

Then the distribution of $\mathbf{X}$ is called a *multivariate t distribution with n degrees of freedom, location vector $\mathbf{\mu}$, and precision matrix $\mathbf{T}$*. We shall now derive the p.d.f. $f(\cdot \mid n, \mathbf{\mu}, \mathbf{T})$ of the vector $\mathbf{X}$.

At any point $y \epsilon R^k$, the value $g_Y(y)$ of the p.d.f. of the random vector Y is

$$g_Y(y) = (2\pi)^{-k/2}|T|^{\frac{1}{2}} \exp(-\tfrac{1}{2}y'Ty). \tag{2}$$

Furthermore, for any number z $(z > 0)$, the value $g_Z(z)$ of the p.d.f. of the random variable Z is

$$g_Z(z) = \left[2^{n/2}\Gamma\left(\frac{n}{2}\right)\right]^{-1} z^{(n/2)-1}e^{-z/2}. \tag{3}$$

Since Y and Z are independent, the value $g_{Y,Z}(y, z)$ of their joint p.d.f. at any point (y, z) such that $y \epsilon R^k$ and $z > 0$ is the product of the values given in Eqs. (2) and (3). Moreover, $g_{Y,Z}(y, z) = 0$ at any other point (y, z).

The joint p.d.f. $g_{X,Z}$ of the random vector X and the random variable Z can now be computed. From Eq. (1),

$$Y_i = \left(\frac{Z}{n}\right)^{\frac{1}{2}} (X_i - \mu_i) \qquad i = 1, \ldots, k. \tag{4}$$

Therefore, the Jacobian J of the transformation from the $k + 1$ random variables $\{X_1, \ldots, X_k, Z\}$ to the $k + 1$ random variables $\{Y_1, \ldots, Y_k, Z\}$ is the determinant of a triangular matrix, and its value is

$$J = \left(\frac{z}{n}\right)^{k/2}. \tag{5}$$

By substituting the values of $Y_1, \ldots, Y_k$ given by Eq. (4) in the joint p.d.f. $g_{Y,Z}$, multiplying by J, and combining terms, we obtain for any point $x \epsilon R^k$ and any number z $(z > 0)$,

$$g_{X,Z}(x, z) = c'z^{(n+k-2)/2} \exp\left\{-\tfrac{1}{2}\left[1 + \frac{1}{n}(x - \mu)'T(x - \mu)\right]z\right\}. \tag{6}$$

Here

$$c' = |T|^{\frac{1}{2}}\left[2^{(n+k)/2}(n\pi)^{k/2}\Gamma\left(\frac{n}{2}\right)\right]^{-1}. \tag{7}$$

The desired p.d.f. $f(\cdot \mid n, \mu, T)$ of X can now be obtained as a marginal p.d.f. by integrating the joint p.d.f. given in Eq. (6) over all positive values of z. From the definition of the gamma function given in Eq. (2) of Sec. 4.8, for any number Q $(Q > 0)$,

$$\int_0^\infty z^{(n+k-2)/2} \exp(-Qz) \, dz = \Gamma\left(\frac{n + k}{2}\right) Q^{-(n+k)/2}. \tag{8}$$

Hence, for any point $x \epsilon R^k$,

$$f(x \mid n, \mu, T) = c\left[1 + \frac{1}{n}(x - \mu)'T(x - \mu)\right]^{-(n+k)/2}, \tag{9}$$

where

$$c = \frac{\Gamma[(n+k)/2]\,|\mathbf{T}|^{\frac{1}{2}}}{\Gamma(n/2)(n\pi)^{k/2}}. \tag{10}$$

Equations (9) and (10) define the p.d.f. of the k-dimensional multivariate t distribution with n degrees of freedom, location vector $\mathbf{\mu}$, and precision matrix $\mathbf{T}$.

This distribution is a k-dimensional generalization of the univariate t distribution discussed in Sec. 4.12; when $k = 1$, the p.d.f. given by Eqs. (9) and (10) is simply the univariate p.d.f. given by Eq. (5) of Sec. 4.12. Of course, this result could have been anticipated from the definition of the multivariate t distribution given at the beginning of this section and the corresponding property of the univariate t distribution, as given in Exercise 32 of Chap. 4.

If $\mathbf{X}$ has a multivariate t distribution whose p.d.f. is given by Eqs. (9) and (10), it can be shown that for $n > 2$, the mean vector $E(\mathbf{X})$ and the covariance matrix $\mathrm{Cov}(\mathbf{X})$ of $\mathbf{X}$ exist and their values are (see Exercise 18)

$$E(\mathbf{X}) = \mathbf{\mu} \quad \text{and} \quad \mathrm{Cov}(\mathbf{X}) = \frac{n}{n-2}\,\mathbf{T}^{-1}. \tag{11}$$

Furthermore, the marginal distribution of any subset of the components of $\mathbf{X}$ can be obtained from the representation in Eq. (1) and the corresponding properties of the multivariate normal distribution, as discussed in Sec. 5.4. The result is as follows (see Exercise 19): Suppose that the random vector X is partitioned as in the form

$$\mathbf{X} = \begin{bmatrix} \mathbf{X}_1 \\ \mathbf{X}_2 \end{bmatrix}. \tag{12}$$

Here the dimension of $\mathbf{X}_i$ is k_i $(i = 1, 2)$ and $k_1 + k_2 = k$. Also, suppose that the location vector $\mathbf{\mu}$ and the precision matrix $\mathbf{T}$ are partitioned as in the forms

$$\mathbf{\mu} = \begin{bmatrix} \mathbf{\mu}_1 \\ \mathbf{\mu}_2 \end{bmatrix}, \qquad \mathbf{T} = \begin{bmatrix} \mathbf{T}_{11} & \mathbf{T}_{12} \\ \mathbf{T}_{21} & \mathbf{T}_{22} \end{bmatrix}. \tag{13}$$

Here the dimension of $\mathbf{\mu}_i$ is k_i $(i = 1, 2)$ and the dimension of the submatrix $\mathbf{T}_{ij}$ is $k_i \times k_j$ $(i, j = 1, 2)$. Then the marginal distribution of $\mathbf{X}_1$ is a k_1-dimensional multivariate t distribution with n degrees of freedom, its location vector is $\mathbf{\mu}_1$, and its precision matrix is $\mathbf{T}_{11} - \mathbf{T}_{12}\mathbf{T}_{22}^{-1}\mathbf{T}_{21}$.

The conditional distribution of $\mathbf{X}_1$ when $\mathbf{X}_2 = \mathbf{x}_2$ $(\mathbf{x}_2 \in R^{k_2})$ is also a k_1-dimensional multivariate t distribution, but it is rather complicated. The degrees of freedom change, and both the location vector of the conditional distribution and the precision matrix depend on the given point $\mathbf{x}_2$. Specifically (see Exercise 20), the conditional distribution has

$n + k_2$ degrees of freedom, the location vector is

$$\mu_1 - T_{11}^{-1}T_{12}(x_2 - \mu_2), \tag{14}$$

and the precision matrix is

$$\frac{n + k_2}{n + (x_2 - \mu_2)'(T_{22} - T_{21}T_{11}^{-1}T_{12})(x_2 - \mu_2)} T_{11}. \tag{15}$$

The following fact will also be used later in this book. If a random vector X has a k-dimensional multivariate t distribution whose p.d.f. is given by Eqs. (9) and (10), then the random variable

$$\frac{1}{k}(X - \mu)'T(X - \mu) \tag{16}$$

has an F distribution with k and n degrees of freedom, as defined in Sec. 4.13 (see Exercise 21).

A bibliography on multivariate normal distributions and multivariate t distributions has been compiled by Gupta (1963).

5.7 THE BILATERAL BIVARIATE PARETO DISTRIBUTION

We shall conclude this chapter with the definition of a bivariate distribution such that each of its two individual univariate marginal distributions is essentially a Pareto distribution, although one of these is on an interval of the form $(-\infty, r_1)$ while the other is on an interval of the form (r_2, ∞). Suppose that r_1, r_2, and α are three numbers such that $r_1 < r_2$ and $\alpha > 0$. We shall say that the joint distribution of the random variables X_1 and X_2 is a *bilateral bivariate Pareto distribution with parameters r_1, r_2, and α* if the distribution is absolutely continuous and the joint p.d.f. $f(\cdot | r_1, r_2, \alpha)$ is as follows:

At any point $(x_1, x_2) \in R^2$ such that $x_1 < r_1$ and $x_2 > r_2$,

$$f(x_1, x_2 | r_1, r_2, \alpha) = \frac{\alpha(\alpha + 1)(r_2 - r_1)^\alpha}{(x_2 - x_1)^{\alpha+2}}. \tag{1}$$

Furthermore, at any other point $(x_1, x_2) \in R^2$, $f(x_1, x_2 | r_1, r_2, \alpha) = 0$.

If the joint p.d.f. of X_1 and X_2 is given by Eq. (1), it can be easily verified that both the marginal distribution of $r_2 - X_1$ and the marginal distribution of $X_2 - r_1$ are univariate Pareto distributions with parameters $r_2 - r_1$ and α, as defined in Sec. 4.11 (see Exercise 23).

Thus, the following means and variances of X_1 and X_2 can be computed from Eqs. (2) and (3) of Sec. 4.11:

$$E(X_1) = \frac{\alpha r_1 - r_2}{\alpha - 1}, \qquad E(X_2) = \frac{\alpha r_2 - r_1}{\alpha - 1}, \tag{2}$$

and

$$\text{Var}(X_1) = \text{Var}(X_2) = \frac{\alpha(r_2 - r_1)^2}{(\alpha - 1)^2(\alpha - 2)}. \tag{3}$$

Furthermore, it can be shown that the correlation between X_1 and X_2 is $-1/\alpha$ (see Exercise 24).

EXERCISES

1. Suppose that $X_1, \ldots, X_k$ are independent random variables and that X_i has a Poisson distribution with mean λ_i $(i = 1, \ldots, k)$. Let n be any fixed positive integer. Show that the conditional distribution of the random vector $X = (X_1, \ldots, X_k)'$, given that $\Sigma_{i=1}^k X_i = n$, is multinomial with parameters n and $p = (p_1, \ldots, p_k)'$, where $p_i = \lambda_i/(\Sigma_{j=1}^k \lambda_j)$ for $i = 1, \ldots, k$.

2. If the random vector $X = (X_1, \ldots, X_k)'$ has a multinomial distribution with parameters n and $p = (p_1, \ldots, p_k)'$, show that its characteristic function ζ is specified at any point $t = (t_1, \ldots, t_k)' \in R^k$ by the equation

$$\zeta(t) = \Big(\sum_{j=1}^k p_j e^{it_j} \Big)^n.$$

Also, show that $E(X) = np$, that $\text{Var}(X_i) = np_i(1 - p_i)$ for $i = 1, \ldots, k$, and that $\text{Cov}(X_i, X_j) = -np_ip_j$ for $i \neq j$. Is the covariance matrix of X singular or nonsingular?

3. If $X_1, \ldots, X_r$ are independent random vectors and X_i has a multinomial distribution with parameters n_i and p $(i = 1, \ldots, r)$, show that the sum $X_1 + \cdots + X_r$ also has a multinomial distribution with parameters $n_1 + \cdots + n_r$ and the same vector p.

4. Let S be the subset of R^{k-1} which is defined by Eq. (3) of Sec. 5.3. For any point in S, define $x_k = 1 - \Sigma_{i=1}^{k-1} x_i$. Prove that the p.d.f. of the Dirichlet distribution, as defined by Eq. (1) of Sec. 5.3, is properly normed by showing that

$$\int \cdots \int_S \Big(\prod_{i=1}^k x_i^{\alpha_i - 1} \Big) dx_1 \cdots dx_{k-1} = \frac{\Gamma(\alpha_1) \cdots \Gamma(\alpha_k)}{\Gamma(\alpha_1 + \cdots + \alpha_k)}.$$

Hint: Change variables by means of the following equations:

$$y_1 = x_1,$$
$$y_2 = \frac{x_2}{1 - x_1},$$
$$y_3 = \frac{x_3}{1 - x_1 - x_2},$$
$$\cdots \cdots \cdots \cdots$$
$$y_{k-1} = \frac{x_{k-1}}{1 - x_1 - x_2 - \cdots - x_{k-2}}.$$

5. (a) Suppose that $X_1, \ldots, X_n$ are independent random variables and that X_i has a gamma distribution with parameters α_i and β $(i = 1, \ldots, n)$. Also, let $Y = (Y_1, \ldots, Y_n)'$ be a random vector defined as follows:

$$Y_i = \frac{X_i}{X_1 + \cdots + X_n} \qquad i = 1, \ldots, n.$$

Show that Y and the random variable $X_1 + \cdots + X_n$ are independent and that Y has a Dirichlet distribution with parametric vector $\alpha = (\alpha_1, \ldots, \alpha_n)'$.

(b) Suppose that the random vector $(X_1, \ldots, X_k)'$ has a Dirichlet distribution with parametric vector $\alpha = (\alpha_1, \ldots, \alpha_k)'$. For any given integer r such that $2 \leq r \leq k$, let $Y = (Y_1, \ldots, Y_r)'$ be a random vector defined as follows:

$$Y_i = \frac{X_i}{X_1 + \cdots + X_r} \qquad i = 1, \ldots, r.$$

Show that Y and the random variable $X_1 + \cdots + X_r$ are independent and that Y has a Dirichlet distribution with parametric vector $\alpha_r = (\alpha_1, \ldots, \alpha_r)'$.

6. Suppose that $X_1, \ldots, X_n$ is a random sample from the uniform distribution on the interval $(0, 1)$, and let $Y_1 \leq \cdots \leq Y_n$ be the order statistics of this sample. Also, let $Z = (Z_1, \ldots, Z_{n+1})'$ be the random vector defined as follows:

$$Z_1 = Y_1,$$
$$Z_2 = Y_2 - Y_1,$$
$$\cdots \cdots \cdots \cdots$$
$$Z_n = Y_n - Y_{n-1},$$
$$Z_{n+1} = 1 - Y_n.$$

Show that Z has a Dirichlet distribution.

7. Suppose that $X = (X_1, \ldots, X_k)'$ is a random vector whose p.d.f. is given by Eq. (1) of Sec. 5.4. Prove that $E(X) = \mu$ and that $\text{Cov}(X) = \Sigma$.

8. Suppose that the random vector $X = (X_1, \ldots, X_k)'$ has a multivariate normal distribution with mean vector μ and covariance matrix Σ. Let r be any integer such that $1 \leq r \leq k$. Also, let μ_r be the subvector of μ comprising the first r components of μ, and let Σ_r be the submatrix of Σ comprising the elements in the first r rows and first r columns of Σ. Show that the random vector $X_r = (X_1, \ldots, X_r)'$ also has a normal distribution with mean vector μ_r and covariance matrix Σ_r.

9. Suppose that a nonsingular matrix Σ and its inverse Σ^{-1} have been partitioned as in Eq. (10) of Sec. 5.4. Show that

$$T_{11}^{-1}T_{12} = -\Sigma_{12}\Sigma_{22}^{-1}$$

and that

$$T_{11}^{-1} = \Sigma_{11} - \Sigma_{12}\Sigma_{22}^{-1}\Sigma_{21}.$$

10. Assuming that the joint distribution of the random variables $X_1, \ldots, X_k$ is multivariate normal and that the correlation between any two of them is 0, show that $X_1, \ldots, X_k$ are independent.

11. Suppose that $X = (X_1, \ldots, X_k)'$, where $X_1, \ldots, X_k$ are independent random variables each of which has a normal distribution with variance σ^2. Let A be an orthogonal $k \times k$ matrix (i.e., a matrix A such that $A' = A^{-1}$), and let the random vector $Y = (Y_1, \ldots, Y_k)'$ be defined by the transformation $Y = AX$. Show that $Y_1, \ldots, Y_k$ are also independent random variables each of which has a normal distribution with the same variance σ^2.

12. Suppose that a k-dimensional random vector X has a multivariate normal distribution with mean vector μ and nonsingular covariance matrix Σ. Show that the random variable $(X - \mu)'\Sigma^{-1}(X - \mu)$ has a χ^2 distribution with k degrees of freedom.

13. Suppose that in a certain population of married couples, the heights of husbands and wives have a bivariate normal distribution. Express, in terms of the univariate standard normal distribution, the probability that if a couple is chosen at random, the husband's height will be greater than his wife's height. Assuming that

the wife's height is y, find a similar expression for the conditional probability of this event. For what value of y are the two probabilities equal?

14. Suppose that a random matrix V has a Wishart distribution whose p.d.f. is given by Eqs. (4) and (5) of Sec. 5.5. Show that for any positive integer r,

$$E(|V|^r) = 2^{rk} \left\{ \prod_{j=1}^{k} \frac{\Gamma[(n+1-j+2r)/2]}{\Gamma[(n+1-j)/2]} \right\} |\Sigma|^r.$$

15. Suppose that a $k \times k$ random matrix V has a Wishart distribution with n degrees of freedom and parametric matrix Σ. Prove the following three properties:

(a) $E(V) = n\Sigma$.

(b) If A is any $m \times k$ matrix of constants, then the $m \times m$ random matrix AVA' has a Wishart distribution with n degrees of freedom and parametric matrix $A\Sigma A'$.

(c) If the matrices V and Σ are partitioned as in Eq. (7) of Sec. 5.5, then the random matrix V_{11} has a Wishart distribution with n degrees of freedom and parametric matrix Σ_{11}.

16. Suppose that the $k \times k$ random matrices $V_1, \ldots, V_r$ are independent and that V_i has a Wishart distribution with n_i degrees of freedom and parametric matrix Σ ($i = 1, \ldots, r$). Show that the sum $V_1 + \cdots + V_r$ also has a Wishart distribution with $n_1 + \cdots + n_r$ degrees of freedom and parametric matrix Σ.

17. Let $g(\cdot \mid n, T)$ denote the p.d.f. of the Wishart distribution with n degrees of freedom and precision matrix T, as given by Eq. (11) of Sec. 5.5. Let $\Sigma = T^{-1}$, and assume that Eq. (6) of Sec. 5.5 holds even when the matrix T is replaced by any matrix of complex numbers whose real parts form a positive definite matrix. Under this assumption, derive the characteristic function ζ of the Wishart distribution, as given by Eq. (9) of Sec. 5.5.

18. Suppose that a random vector X has a multivariate t distribution with n degrees of freedom, location vector μ, and precision matrix T. Prove that for $n > 2$, the mean vector $E(X)$ and the covariance matrix $Cov(X)$ exist and that their values are given by Eq. (11) of Sec. 5.6. Hint: To compute $E[(X - \mu)(X - \mu)']$, use the representation in Eq. (1) of Sec. 5.6 and the independence of Y and Z.

19. Suppose that a k-dimensional random vector X has a multivariate t distribution with n degrees of freedom, location vector μ, and precision matrix T. If X, μ, and T are partitioned as in Eqs. (12) and (13) of Sec. 5.6, show that the marginal distribution of X_1 is a multivariate t distribution with n degrees of freedom, location vector μ_1, and precision matrix $T_{11} - T_{12}T_{22}^{-1}T_{21}$. Hint: From the discussion at the beginning of Sec. 5.6 and the discussion in Sec. 5.4, it follows that the conditional distribution of X_1 when Z is given is a multivariate normal distribution having the appropriate mean vector and the appropriate precision matrix.

20. Suppose that the k-dimensional random vector X has a multivariate t distribution as specified in Exercise 19, and suppose that X, μ, and T are partitioned as also specified in that exercise. Show that the conditional distribution of X_1, when $X_2 = x_2$ is given ($x_2 \in R^{k_2}$), is a multivariate t distribution with $n + k_2$ degrees of freedom and that the location vector and the precision matrix are given by the expressions (14) and (15), respectively, of Sec. 5.6. Hint: Rewrite the quadratic form that appears in the joint p.d.f. of X_1 and X_2 as was done in Eq. (15) of Sec. 5.4.

21. If a random vector X has a multivariate t distribution whose p.d.f. is given by Eqs. (9) and (10) of Sec. 5.6, show that the random variable which is displayed in the expression (16) of Sec. 5.6 has an F distribution with k and n degrees of freedom. Hint: Use the representation in Eq. (1) of Sec. 5.6.

22. Suppose that the random vector $\mathbf{X} = (X_1, \ldots, X_k)'$ has a k-dimensional multivariate t distribution with n degrees of freedom, location vector $\mathbf{\mu}$, and precision matrix $\mathbf{T}$. Also, let $\mathbf{A}$ be an $m \times k$ matrix such that $\mathbf{AT}^{-1}\mathbf{A}'$ is nonsingular, and let the random vector $\mathbf{U} = (U_1, \ldots, U_m)'$ be defined as $\mathbf{U} = \mathbf{AX}$. Show that $\mathbf{U}$ has an m-dimensional multivariate t distribution with n degrees of freedom, location vector $\mathbf{A\mu}$, and precision matrix $(\mathbf{AT}^{-1}\mathbf{A}')^{-1}$.

23. Suppose that the joint distribution of two random variables X_1 and X_2 is a bilateral bivariate Pareto distribution whose p.d.f. is given by Eq. (1) of Sec. 5.7. Show that both the marginal distribution of $r_2 - X_1$ and the marginal distribution of $X_2 - r_1$ are Pareto distributions with parameters $r_2 - r_1$ and α.

24. (a) Suppose that the random variables X_1 and X_2 have a bilateral bivariate Pareto distribution with parameters r_1, r_2, and α. Show that if $\alpha > 2$,

$$E[(X_2 - X_1)^2] = \frac{\alpha(\alpha + 1)(r_2 - r_1)^2}{(\alpha - 1)(\alpha - 2)}.$$

(b) Show that if $\alpha > 2$, the correlation between X_1 and X_2 is $-1/\alpha$.

25. Suppose that X_1 and X_2 have a bilateral bivariate Pareto distribution with parameters r_1, r_2, and α. For fixed values of r_1 and r_2, describe the limiting behavior of the joint distribution of X_1 and X_2 as $\alpha \to \infty$.

subjective probability and utility

subjective probability

6.1 INTRODUCTION

Much work in probability and statistics deals with the derivation of the probabilities of certain complicated events from the specified probabilities of simpler events, with the study of the manner in which certain specified probabilities change in the light of new information, and with procedures for making effective decisions in certain situations that can be character-ized in terms of specified probability distributions. In the remaining chapters of this book, other than the present one, we shall consider the solutions of problems in these areas. It should be noted, however, that each problem involves the manipulation of the specified probabilities of certain fundamental events. In this chapter, we shall consider the following question: How does the statistician initially assign to these events the probabilities on which all of his subsequent calculations are based?

In many problem areas, the assignment of a probability distribution over the relevant σ-field of events has become routine and standardized among workers in those areas. This assignment is usually based on a combination of tradition and experience. In some other problems, there are natural probability models that represent the situations. Some typical examples were mentioned briefly in Chaps. 4 and 5. Many other examples can be found in the references given at the end of Chap. 1. Hence, suitable probabilities can often be assigned objectively and quickly because of wide agreement on the appropriateness of a specific distribution for a certain type of problem.

On the other hand, there are some situations for which it would be very difficult to find even two people who would agree on the appropriateness of any specific probability distribution. In such a situation, the statistician's assignment of probabilities must be highly subjective and must reflect his own information and beliefs. We shall now discuss in detail the conditions under which the statistician can represent his information and beliefs in terms of probability distributions.

6.2 RELATIVE LIKELIHOOD

Consider a sample space S together with a σ-field α of events, and suppose that it is desired to assign a probability to each event in α. A fundamental concept is that of one event being at least as likely to occur as another event, and this concept will be accepted as a primitive one in the development to be presented here. Thus it will be assumed that a person, when considering any two events in α, can decide whether he regards one of them as being more likely to occur than the other or whether he regards the two events as being equally likely to occur.

When any two events A and B are compared, we write $A < B$ to indicate that B is more likely to occur than A and we write $A \sim B$ to indicate that A and B are equally likely to occur. Furthermore, by analogy with the inequality relation between real numbers, we write $A \precsim B$ to indicate that B is at least as likely to occur as A or, equivalently, that A is not more likely to occur than B. Hence, if $A \precsim B$, then either $A < B$ or $A \sim B$. Finally, we define $A > B$ to mean the same as $B < A$, and we define $A \succsim B$ to mean the same as $B \precsim A$.

Since the probability of an event is supposed to be a numerical measure of the likelihood that the event will occur, any probability distribution P which is assigned to the events in the σ-field α should have the property that $P(A) \leq P(B)$ if, and only if, $A \precsim B$. A probability distribution P which has this property is said to *agree* with the relation $\precsim$. We shall now investigate the conditions which must be satisfied by the relation $\precsim$ in order for there to exist a unique probability distribution that agrees with it. The assumptions that will be made in regard to the relation $\precsim$ are suggested by our intuitive use of the concept of the relative likelihood of occurrence of two events and by the mathematical properties of probability distributions, as given in Chap. 2.

The basic assumption is the following.

Assumption SP_1 *For any two events A and B, exactly one of the following three relations must hold: $A < B$, $A > B$, $A \sim B$.*

The next assumption has a simple interpretation which makes it intuitively plausible.

Assumption SP_2 *If A_1, A_2, B_1, and B_2 are four events such that $A_1 A_2 = B_1 B_2 = \emptyset$ and $A_i \lesssim B_i$ for $i = 1, 2$, then $A_1 \cup A_2 \lesssim B_1 \cup B_2$. If, in addition, either $A_1 < B_1$ or $A_2 < B_2$, then $A_1 \cup A_2 < B_1 \cup B_2$.*

This assumption can be interpreted as follows: Suppose that each of the events A and B can occur in either of two mutually exclusive ways. If each way that leads to A is not more likely to occur than the corresponding way that leads to B, then $A \lesssim B$. Furthermore, if at least one of the ways that leads to A is actually less likely than the corresponding way that leads to B, then $A < B$.

Several results can be derived from these two elementary assumptions. One interesting and important result is the transitivity of the relation $\lesssim$, as proved in Theorem 1 below. Before proceeding to that theorem, we shall present a simple lemma.

Lemma 1 *Suppose that A, B, and D are events such that $AD = BD = \emptyset$. Then $A \lesssim B$ if, and only if, $A \cup D \lesssim B \cup D$.*

Proof Suppose $A \lesssim B$. Then the desired result follows from Assumption SP_2. Conversely, suppose that $A > B$. Then, again by Assumption SP_2, $A \cup D > B \cup D$.∎

Theorem 1 *If A, B, and D are events such that $A \lesssim B$ and $B \lesssim D$, then $A \lesssim D$.*

Proof Consider the seven disjoint events shown in Fig. 6.1 whose union is $A \cup B \cup D$. Since $A \lesssim B$, it follows from Lemma 1 that

$$AB^c D^c \cup AB^c D \lesssim A^c BD^c \cup A^c BD. \tag{1}$$

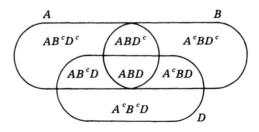

Fig. 6.1 The partitioning of $A \cup B \cup D$.

Similarly, since $B \precsim D$, it follows from Lemma 1 that

$$ABD^c \cup A^cBD^c \precsim AB^cD \cup A^cB^cD. \tag{2}$$

Since the left sides of the relations (1) and (2) are disjoint and the right sides are also disjoint, it follows from Assumption SP_2 that

$$AB^cD^c \cup AB^cD \cup ABD^c \cup A^cBD^c$$
$$\precsim A^cBD^c \cup A^cBD \cup AB^cD \cup A^cB^cD. \tag{3}$$

If the common event $AB^cD \cup A^cBD^c$ is eliminated from both sides of this relation, it follows from Lemma 1 that

$$AB^cD^c \cup ABD^c \precsim A^cBD \cup A^cB^cD. \tag{4}$$

It can now be seen from Fig. 6.1 and Lemma 1 that $A \precsim D$.∎

It follows from Assumption SP_1 and the transitivity property exhibited in Theorem 1 that the relation $\precsim$ yields a complete ordering of the events in $\mathcal{Q}$. Of course, in this ordering it is possible that two distinct events A and B will be equivalent, that is, $A \sim B$.

The next result is an extension of Assumption SP_2 to unions of any finite number of disjoint events.

Theorem 2 *If $A_1, \ldots, A_n$ are n disjoint events and $B_1, \ldots, B_n$ are also n disjoint events such that $A_i \precsim B_i$ for $i = 1, \ldots, n$, then $\bigcup_{i=1}^n A_i \precsim \bigcup_{i=1}^n B_i$. If, in addition, $A_i < B_i$ for at least one value of i ($i = 1, \ldots, n$), then $\bigcup_{i=1}^n A_i < \bigcup_{i=1}^n B_i$.*

This theorem is easily verified by an induction argument, and its proof is omitted.

The next theorem reflects a basic property of relative likelihood which has previously been noted in Chap. 2 for probabilities.

Theorem 3 *For any events A and B, $A \precsim B$ if, and only if, $A^c \succsim B^c$.*

The proof of this theorem serves as Exercise 1b at the end of this chapter.

We now make the natural assumption that no event is less likely than the empty set $\emptyset$, and we avoid a completely trivial situation by assuming that the entire sample space S is actually more likely than $\emptyset$.

Assumption SP_3 *If A is any event, then $\emptyset \precsim A$. Furthermore, $\emptyset < S$.*

This assumption leads to the following fundamental property of relative likelihoods, which has also been previously noted for probabilities (see Exercise 1c).

Theorem 4 *If A and B are events such that $A \subset B$, then $A \precsim B$. In particular, if A is any event, then $\emptyset \precsim A \precsim S$.*

We now introduce an assumption whose distinguishing feature is that it involves an infinite sequence of events.

Assumption SP_4 *If $A_1 \supset A_2 \supset \cdots$ is a decreasing sequence of events and B is some fixed event such that $A_i \succsim B$ for $i = 1, 2, \ldots$, then $\bigcap_{i=1}^{\infty} A_i \succsim B$.*

To help clarify the meaning of Assumption SP_4, consider an example in which the events under consideration are subsets of the real line. Suppose that each infinite interval of the form (n, ∞), for $n = 1, 2, \ldots$, is regarded as more likely than some fixed but small subset B of the line. Since the intersection of all the infinite intervals is the empty set $\emptyset$, it then follows from Assumption SP_4 that the small subset B must itself be equivalent to $\emptyset$. In other words, if B is any set such that $B \succ \emptyset$, then regardless of how small the set B is, it is impossible for every infinite interval (n, ∞) to be at least as likely as B. A property of this type distinguishes a probability distribution that is countably additive (i.e., one that satisfies property 2 in the definition of a probability distribution given in Sec. 2.3) from one that is only finitely additive.

The following theorem is the dual of Assumption SP_4. The statement of this theorem could equally well have been taken as Assumption SP_4 and the statement in that assumption then derived as a theorem.

Theorem 5 *If $A_1 \subset A_2 \subset \cdots$ is an increasing sequence of events and B is some fixed event such that $A_i \precsim B$ for $i = 1, 2, \ldots$, then $\bigcup_{i=1}^{\infty} A_i \precsim B$.*

Proof It follows from the hypotheses of the theorem and from Theorem 3 that $A_1^c \supset A_2^c \supset \cdots$ is a decreasing sequence of events such that $A_i^c \succsim B^c$ for $i = 1, 2, \ldots$. Hence, by Assumption SP_4, $(\bigcup_{i=1}^{\infty} A_i)^c = \bigcap_{i=1}^{\infty} A_i^c \succsim B^c$. The desired result now follows from another application of Theorem 3.■

The next result extends Theorem 2 to unions of infinite sequences of disjoint events.

Theorem 6 *If $A_1, A_2, \ldots$ is an infinite sequence of disjoint events and $B_1, B_2, \ldots$ is another infinite sequence of disjoint events such that $A_i \precsim B_i$ for $i = 1, 2, \ldots$, then $\bigcup_{i=1}^{\infty} A_i \precsim \bigcup_{i=1}^{\infty} B_i$. If, in addition, $A_i \prec B_i$ for at least one value of i $(i = 1, 2, \ldots)$, then $\bigcup_{i=1}^{\infty} A_i \prec \bigcup_{i=1}^{\infty} B_i$.*

Proof It follows from Theorem 2 that $\bigcup_{i=1}^{n} A_i \precsim \bigcup_{i=1}^{n} B_i$ for any value of n $(n = 1, 2, \ldots)$. Hence, by Theorem 4,

$$\bigcup_{i=1}^{n} A_i \precsim \bigcup_{i=1}^{\infty} B_i \qquad n = 1, 2, \ldots . \tag{5}$$

Since the left side of the relation (5) yields an increasing sequence of events for $n = 1, 2, \ldots ,$ it follows from Theorem 5 that $\bigcup_{i=1}^{\infty} A_i \precsim \bigcup_{i=1}^{\infty} B_i$.

If, in addition, $A_j < B_j$ for at least one value of j $(j = 1, 2, \ldots)$, then it follows from Theorem 3 that for $n \geq j$,

$$\bigcup_{i=1}^{n} A_i < \bigcup_{i=1}^{n} B_i. \tag{6}$$

Furthermore, it follows from the first part of Theorem 6, which has just been proved, that

$$\bigcup_{i=n+1}^{\infty} A_i \precsim \bigcup_{i=n+1}^{\infty} B_i. \tag{7}$$

Hence, from the relations (6) and (7) and Assumption SP_2, we can obtain the following result:

$$\bigcup_{i=1}^{\infty} A_i = (\bigcup_{i=1}^{n} A_i) \cup (\bigcup_{i=n+1}^{\infty} A_i) < (\bigcup_{i=1}^{n} B_i) \cup (\bigcup_{i=n+1}^{\infty} B_i) = \bigcup_{i=1}^{\infty} B_i. \blacksquare \tag{8}$$

It should be clear that the relation $\precsim$ must satisfy Assumptions SP_1 to SP_4 if there is to exist any probability distribution which agrees with it. However, these four assumptions are not sufficient to guarantee the existence of such a probability distribution. In the next section we shall introduce a fifth assumption on the basis of which it is possible to construct a unique probability distribution that agrees with the relation $\precsim$.

Further Remarks and References

Sometimes the statistician's decision as to which of two events A and B is more likely to occur can be facilitated by auxiliary considerations. For instance, he may ask himself whether he would prefer to participate in a contest in which he would receive a valuable prize if the event A occurred and receive nothing if A did not occur or a contest in which he would receive the same valuable prize if the event B occurred and receive nothing if B did not occur. Although the statistician would typically prefer to participate in the contest which he considered had the greater likelihood of yielding the prize, there are limitations and dangers in this reasoning. An ordinary person would consider it more likely that he will be exterminated in a nuclear war within the next ten years than that he will become President of the United States within that period. But if he were given a choice, he would certainly prefer the promise of a prize at his inauguration rather than the promise of the same prize at his extermination. This example is clearly extreme, but it illustrates the difficulties which can arise when one must compare events that are not, according to Ramsey (1926), ethically neutral. These difficulties reveal the advantage of making, whenever possible, a direct comparison of the relative likelihoods of two events without considering consequences which depend on their occurrence or nonoccurrence.

The question of whether there always exists at least one probability distribution that agrees with the relation $\lesssim$, when this relation satisfies Assumptions SP_1 to SP_4, was raised by de Finetti and Savage [see Savage (1954), p. 40]. It was answered in the negative by Kraft, Pratt, and Seidenberg (1959), who constructed an example using a sample space S with just a finite number of points.

6.3 THE AUXILIARY EXPERIMENT

In this section we shall introduce one final assumption which will make it possible to assign a probability to each event in an unambiguous manner. Any assignment of a probability distribution indicates not only which of two events is more likely to occur but also how much more likely its occurrence is. It is not always possible to give a meaningful numerical evaluation of these relative likelihoods solely on the basis of the assumptions which have been made in Sec. 6.2. For example, consider an experiment or, equivalently, a sample space which has just two possible outcomes, A and A^c. The statistician may feel that $A > A^c$, but it will clearly be impossible for him to assign meaningful numerical probabilities to these two events without additional considerations. In particular, he must be able to compare the relative likelihoods of A and A^c, not only with each other but also with many other events whose various probabilities have already been established.

The effect of these remarks, in informal terms, is that it must be assumed that there exists a class $\mathcal{B}$ of events having the following two properties: (1) Each event in the class $\mathcal{B}$ has a known probability, and (2) for any number p ($0 \leq p \leq 1$), there exists an event $B \epsilon \mathcal{B}$ whose probability is p. Hence, when assigning a probability to some event A in which he is interested, the statistician simply finds an event $B \epsilon \mathcal{B}$ such that $A \sim B$ and assigns to A the same probability as that of B.

The assumption which is needed, however, cannot properly be formulated in this way since this description provides no indication of the manner in which the probabilities of the events in the class $\mathcal{B}$ were chosen by the statistician or became known to him. The assumption can be stated more precisely in terms of the existence of a particular type of random variable.

It will be recalled from Sec. 3.1 that a random variable X is an α-measurable function whose value is specified at each point $s \in S$. Hence, for any random variable X and any intervals I_1 and I_2 of the real line, the events $\{X \in I_1\}$ and $\{X \in I_2\}$ belong to the σ-field α, and by Assumption SP_1, either $\{X \in I_1\} \precsim \{X \in I_2\}$ or $\{X \in I_1\} \succsim \{X \in I_2\}$.

For any interval I with finite end points a and b ($a \leq b$), let $\lambda(I) = b - a$ denote the length of I. Note that $\lambda(I)$ has the same value regardless of whether or not either of the end points a or b is included in I.

The interval with end points a and b will be denoted by (a, b) if neither a nor b is included in the interval, by $[a, b)$ if a but not b is included in the interval, by $(a, b]$ if b but not a is included in the interval, and by $[a, b]$ if both a and b are included in the interval.

We can now define what is meant by a uniformly distributed random variable in the present context. Let X be a random variable such that $0 \leq X(s) \leq 1$ for every $s \in S$. Then the random variable X is said to have a *uniform distribution on the interval* $[0, 1]$ if the following property holds: For any two subintervals I_1 and I_2 of the interval $[0, 1]$, $\{X \in I_1\} \precsim \{X \in I_2\}$ if, and only if, $\lambda(I_1) \leq \lambda(I_2)$.

Note that this definition of a uniform distribution does not mention probabilities. The final assumption can now be formulated very simply as follows:

Assumption SP_5 *There exists a random variable which has a uniform distribution on the interval* $[0, 1]$.

Since Assumption SP_5, or one similar to it, underlies all work on the construction of explicit subjective probabilities, some further remarks in regard to this assumption may be helpful. As mentioned above, many interesting experiments have only a finite number of outcomes, and all of these are not necessarily equally likely to occur. Clearly, it is not possible to define on the sample space of such an experiment a random variable which has a uniform distribution on the interval $[0, 1]$. The statistician must enlarge the sample space by considering, along with the original experiment, an auxiliary experiment in which the value of a random variable having the appropriate uniform distribution is observed. Hence, each point in the enlarged sample space of this composite experiment comprises an outcome of the original experiment together with an outcome of the auxiliary experiment. It is assumed that Assumptions SP_1 to SP_4 concerning the relation $\precsim$ still hold in the composite experiment.

It is not necessary that an auxiliary experiment actually be performed, or even that it be possible for such an experiment to be performed. The statistician need only be able to imagine an ideal auxiliary experiment in which a uniformly distributed random variable X can be generated, and be able to compare the relative likelihood of any event A which was originally of interest to him with that of any event of the form $\{X \in I\}$.

Further Remarks and References

The statistician may think of the value of the random variable X as being determined by some sort of randomization device, such as a spinning pointer which selects "at random" a point on a circle having unit circumference. Alternatively, he may think of a "fair" coin tossed repeatedly. Suppose that for any finite number n of tosses, any possible sequence of n heads and tails has the same likelihood of being obtained as any other sequence. A random variable X having a uniform distribution on the interval $[0, 1]$ can then be constructed as follows:

For $n = 1, 2, \ldots$, let T_n be the random variable such that $T_n = 1$ if the result of the nth toss is heads and $T_n = 0$ if the result of that toss is tails. Let X be the random variable defined by the equation $X = \sum_{n=1}^{\infty} 2^{-n} T_n$. Then it can be shown [see, e.g., Feller (1966), p. 34] that X has a uniform distribution on the interval $[0, 1]$.

6.4 CONSTRUCTION OF THE PROBABILITY DISTRIBUTION

Suppose now that the sample space S, the σ-field $\mathcal{C}$ of events, and the relation $\precsim$ satisfy Assumptions SP_1 to SP_6. It will be shown in this section and the next one that under these assumptions there exists a unique probability distribution P that agrees with the relation $\precsim$.

By Assumption SP_6, there exists a random variable X which has a uniform distribution on the interval $[0, 1]$. For any subinterval (a, b) of the interval $[0, 1]$, let $G(a, b)$ denote the event that the random variable X lies in the interval (a, b). Then for any two intervals (a_1, b_1) and (a_2, b_2), with $0 \leq a_i \leq b_i \leq 1$ for $i = 1, 2$, it follows that $G(a_1, b_1) \precsim G(a_2, b_2)$ if, and only if, $b_1 - a_1 \leq b_2 - a_2$. Moreover, $G(a_1, b_1) \sim G(a_1, b_1] \sim G[a_1, b_1) \sim G[a_1, b_1]$.

Theorem 1 *If A is any event, there exists a unique number a^* ($0 \leq a^* \leq 1$) such that $A \sim G[0, a^*]$.*

Proof For any event A, let $U(A)$ be the subset of the interval $[0, 1]$ defined by the following equation:

$$U(A) = \{a: G[0, a] \succsim A\}. \tag{1}$$

Since $G[0, 1] = S \succsim A$, the number 1 belongs to the set $U(A)$, and hence,

$U(A)$ is not empty. Let $a^* = \inf \{a: a \epsilon U(A)\}$. If $a_1 \geq a_2 \geq \cdots$ is any decreasing sequence of values from $U(A)$ that converges to the value a^*, then $G[0, a^*] = \cap_{i=1}^{\infty} G[0, a_i]$. Hence it follows from Assumption SP_4 that

$$G[0, a^*] \gtrsim A. \tag{2}$$

If $a^* = 0$, then since $G[0, 0] \sim \emptyset \lesssim A$, it follows from this inequality and the inequality (2) that $G[0, 0] \sim A$.

Suppose then that $a^* > 0$. It follows from the definition of a^* that $G[0, a] < A$ for any number a such that $0 \leq a < a^*$. Furthermore, if $a_1 < a_2 < \cdots$ is any strictly increasing sequence of values that converges to the value a^*, then $G[0, a^*) = \cup_{i=1}^{\infty} G[0, a_i]$. Hence, by Theorem 5 of Sec. 6.2, it follows that $G[0, a^*] \sim G[0, a^*) \lesssim A$. This inequality together with the inequality (2) again yields the result that $G[0, a^*] \sim A$.

The value of a^* must be unique since if a_1 and a_2 are any other numbers such that $a_1 < a^* < a_2$, it follows that $G[0, a_1] < G[0, a^*] < G[0, a_2]$. Hence, only one of these three events can be equivalent to the event A.■

The desired probability distribution P can now be defined directly from Theorem 1. If A is any event, then $P(A)$ is defined to be the number a^* specified in Theorem 1. Hence, $P(A)$ can be determined by the relation

$$A \sim G[0, P(A)]. \tag{3}$$

In the next theorem, it is shown that with this definition, P agrees with the relation $\lesssim$.

Theorem 2 *Let A and B be any two events. Then $A \lesssim B$ if, and only if, $P(A) \leq P(B)$.*

Proof By Eq. (3), $A \lesssim B$ if, and only if,

$$G[0, P(A)] \lesssim G[0, P(B)]. \tag{4}$$

Furthermore, it follows from the definition of the uniform distribution that the relation (4) holds if, and only if, $P(A) \leq P(B)$.■

6.5 VERIFICATION OF THE PROPERTIES OF A PROBABILITY DISTRIBUTION

It must now be verified that the function P, as defined above, actually meets all the requirements of a probability distribution as given in Sec. 2.3. It is clear from the definition of P that $P(A) \geq 0$ for any event A.

Furthermore, since $S = G[0, 1]$, it follows that $P(S) = 1$. To complete the verification, it is only necessary to show that for any sequence of disjoint events $A_1, A_2, \ldots ,$

$$P(\bigcup_{i=1}^{\infty} A_i) = \sum_{i=1}^{\infty} P(A_i). \tag{1}$$

The next theorem shows that Eq. (1) is correct at least for the union of two disjoint events.

Theorem 1 *If A and B are any two events such that $AB = \emptyset$, then $P(A \cup B) = P(A) + P(B)$.*

Proof By Eq. (3) of Sec. 6.4, $A \sim G[0, P(A)]$ and $A \cup B \sim G[0, P(A \cup B)]$. Furthermore, since $A \lesssim A \cup B$, then $P(A) \le P(A \cup B)$.

It will now be shown that

$$B \sim G(P(A), P(A \cup B)]. \tag{2}$$

Suppose first that $B < G(P(A), P(A \cup B)]$. Then it follows from Assumption SP_2 that

$$A \cup B < G[0, P(A)] \cup G(P(A), P(A \cup B)] = G[0, P(A \cup B)], \tag{3}$$

which is a contradiction. A similar contradiction is obtained when it is supposed that $B > G(P(A), P(A \cup B)]$. Hence, Eq. (2) must be correct. Since

$$G(P(A), P(A \cup B)] \sim G[0, P(A \cup B) - P(A)] \tag{4}$$

and since by definition $B \sim G[0, P(B)]$, it now follows from Eqs. (2) and (4) that

$$P(A \cup B) - P(A) = P(B). \blacksquare \tag{5}$$

By an elementary induction argument, which will not be given here, the result of Theorem 1 can now be extended to any finite number of disjoint events.

Corollary 1 *If $A_1, \ldots , A_n$ are any disjoint events, then*

$$P(\bigcup_{i=1}^{n} A_i) = \sum_{i=1}^{n} P(A_i).$$

Theorem 2 *Let $A_1 \supset A_2 \supset \cdots$ be any decreasing sequence of events such that $\bigcap_{i=1}^{\infty} A_i = \emptyset$. Then*

$$\lim_{n \to \infty} P(A_n) = 0. \tag{6}$$

Proof Since $A_1 \supset A_2 \supset \cdots$, then $P(A_1) \geq P(A_2) \geq \cdots$. Hence, the numbers $P(A_n)$ must converge to some nonnegative limit b as $n \to \infty$. Since $P(A_i) \geq b$ for $i = 1, 2, \ldots$, it follows that $A_i \succsim G[0, b]$ for $i = 1, 2, \ldots$. By Assumption SP_4, this implies that

$$\emptyset = \bigcap_{i=1}^{\infty} A_i \succsim G[0, b]. \tag{7}$$

If b were any positive number, then it would be true that

$$G[0, b] \succ G\left[0, \frac{b}{2}\right] \succsim \emptyset. \tag{8}$$

Since this relation contradicts relation (7), it must be true that $b = 0$.∎

The combination of Corollary 1 and Theorem 2 can now be used to prove that Eq. (1) is correct.

Theorem 3 *The function P is a probability distribution.*

Proof Suppose that $A_1, A_2, \ldots$ is any sequence of disjoint events. It follows from Corollary 1 that

$$P(\bigcup_{i=1}^{\infty} A_i) = \sum_{i=1}^{n} P(A_i) + P(\bigcup_{i=n+1}^{\infty} A_i) \qquad n = 1, 2, \ldots . \tag{9}$$

Since the events A_i $(i = 1, 2, \ldots)$ are disjoint, the sequence of events $B_n = \bigcup_{i=n+1}^{\infty} A_i$ $(n = 1, 2, \ldots)$ is a decreasing sequence such that $\bigcap_{n=1}^{\infty} B_n = \emptyset$. Hence, it follows from Theorem 2 that $\lim_{n \to \infty} P(B_n) = 0$. Therefore, by taking the limit of the right side of Eq. (9) as $n \to \infty$, we obtain Eq. (1).∎

By showing that P is the only probability distribution which satisfies Theorem 2 of Sec. 6.4, we can now establish the following theorem, which summarizes the results of this section.

Theorem 4 *If the relation $\precsim$ satisfies Assumptions SP_1 to SP_5, then the function P as defined by Eq. (3) of Sec. 6.4 is the unique probability distribution which agrees with the relation $\precsim$.*

Proof All parts of this theorem except that relating to the uniqueness of P have already been established.

Consider any probability distribution P' which agrees with the relation $\precsim$. It can be shown from the properties of the uniform distribution that for any event of the form $G[0, a]$, it must be true that

$$P'\{G[0, a]\} = a.$$

If A is any event, it was shown in Theorem 1 of Sec. 6.4 that $A \sim$ $G[0, a^*]$ for some number a^*. Since P' agrees with the relation $\precsim$, then

$$P'(A) = P'\{G[0, a^*]\} = a^* = P(A). \tag{10}$$

Hence, $P'(A) = P(A)$ for any event A, and the uniqueness of P is established.■

Further Remarks and References

The presentation given here is related to that of Villegas (1964), and some of the exercises at the end of this chapter are based on his work. An excellent bibliography on subjective probability is given by Kyburg and Smokler (1964), who also reprint important papers by Ramsey (1926), de Finetti (1937), Koopman (1940), and others. Savage (1954) gives a thorough development of subjective probability and a highly informative bibliography. Other interesting contributions are by Anscombe and Aumann (1963), Fishburn (1967b), and Scott (1964). Fishburn (1964) also discusses subjective probability. Several of the famous books on this topic were mentioned in Chap. 1. Other references are given at the end of Sec. 6.6.

6.6 CONDITIONAL LIKELIHOODS

Suppose now that the relation $\precsim$ satisfies Assumptions SP_1 to SP_5 given in preceding sections and there exists a unique probability distribution P which has the properties described in Theorem 4 of Sec. 6.5. In this section we shall extend the concept of the relation $\precsim$ so that we can consider not only the relation $A \precsim B$ but also the relation $(A|D) \precsim (B|D)$ for any three events A, B, and D. This latter relation, which compares conditional likelihoods, has the following meaning: The event B is at least as likely to occur as the event A when it is known that the event D has occurred.

Hence, we must now ascertain what further conditions should be imposed on the relation $\precsim$, in this extended sense, in order that all the conditional probability distributions which are constructed from P will also agree with this relation. In other words, we now seek conditions under which, for any three events A, B, and D, the relation $(A|D) \precsim$ $(B|D)$ will hold if, and only if, $P(A|D) \leq P(B|D)$. Clearly, this equivalence can be required only for events D such that $P(D) > 0$, since conditional probability is defined only for such events.

For any event D such that $P(D) > 0$, the inequality $P(A|D) \leq$ $P(B|D)$ is equivalent to the inequality $P(AD) \leq P(BD)$. Furthermore, this inequality is equivalent to the relation $AD \precsim BD$. These remarks naturally lead to the next assumption.

Assumption CP *For any three events* A, B, *and* D, $(A|D) \precsim (B|D)$ *if, and only if,* $AD \precsim BD$.

As we remarked above, when Assumptions SP_1 to SP_5 are made and there exists a probability distribution P which can be uniquely specified, it is sufficient for most purposes to apply Assumption CP only to events D such that $P(D) > 0$. However, in order to keep the assumptions independent of each other, we have imposed the slightly stronger requirement that Assumption CP applies to all events D. Furthermore, it now follows from Assumptions SP_1 and CP that for any three events A, B, and D, either $(A|D) \precsim (B|D)$ or $(B|D) \precsim (A|D)$. It is possible, of course, that both relations will be correct. In fact, both relations will always be correct when D is an event such that $P(D) = 0$ (see Exercise 6a).

By combining Assumption CP with Theorem 4 of Sec. 6.5, we obtain the following general theorem.

Theorem 1   *If the relation* $\precsim$ *satisfies Assumptions* SP_1 *to* SP_5 *and Assumption* CP, *then the function* P *defined by Eq. (3) of Sec. 6.4 is the unique probability distribution which has the following property: For any three events* A, B, *and* D *such that* $P(D) > 0$, $(A|D) \precsim (B|D)$ *if, and only if,* $P(A|D) \leq P(B|D)$.

Further Remarks and References

Luce and Suppes (1965) give a good review of subjective probability, including the relatively small amount of work that has been done on the experimental measurement of subjective probabilities. An interesting and controversial group of papers dealing with the existence and measurement of subjective probabilities are those by Ellsberg (1961), Fellner (1961, 1963), Raiffa (1961), Brewer (1963), and Roberts (1963). The book by Fellner (1965) is also of interest here.

Exercises 7 to 9 at the end of this chapter require the actual evaluation of the reader's subjective probabilities of some specific events.

EXERCISES

1. Prove that the following results follow from Assumptions SP_1 to SP_3:

    (a) If A, B, and D are any three events such that $A \precsim B$ and $B \prec D$, then $A \prec D$.

    (b) If A and B are any two events, then $A \precsim B$ if, and only if, $A^c \succsim B^c$.

(c) If A and B are events such that $A \subset B$, then $A \lesssim B$.

(d) If $A_1, \ldots, A_n$ and $B_1, \ldots, B_n$ are events such that $B_i B_j = \emptyset$ for $i \neq j$ and $A_i \lesssim B_i$ for $i = 1, \ldots, n$, then $\cup_{i=1}^n A_i \lesssim \cup_{i=1}^n B_i$. If, in addition, $A_i \prec B_i$ for at least one value of i ($i = 1, \ldots, n$), then $\cup_{i=1}^n A_i \prec \cup_{i=1}^n B_i$. (This is an extension of Theorem 2 of Sec. 6.2 since it is not assumed here that $A_1, \ldots, A_n$ are disjoint.) Furthermore, the statements remain true if the condition that $B_i B_j = \emptyset$ for $i \neq j$ is replaced by the weaker condition that $B_i B_j \sim \emptyset$ for $i \neq j$.

2. Suppose that Assumptions SP_1 to SP_3 are made. Let $A_1, \ldots, A_m$ be disjoint events, and let $B_1, \ldots, B_n$ also be disjoint events such that $\cup_{i=1}^m A_i = \cup_{i=1}^n B_i = S$, $A_1 \lesssim \cdots \lesssim A_m$ and $B_1 \lesssim \cdots \lesssim B_n$, and $m \leq n$. Prove that $B_1 \lesssim A_m$. Prove also that if $m < n$, then $B_1 \prec A_m$.

3. Prove that the following results follow from Assumptions SP_1 to SP_4:

(a) If $A_1 \supset A_2 \supset \cdots$ is a decreasing sequence of events and if B is an event such that $\cap_{i=1}^\infty A_i \prec B$, then $A_i \gtrsim B$ for only a finite number of values of i ($i = 1, 2, \ldots$). State and prove the analogous result for an increasing sequence of events $A_1 \subset A_2 \subset \cdots$.

(b) If $A_1 \subset A_2 \subset \cdots$ is an increasing sequence of events and $B_1 \supset B_2 \supset \cdots$ is a decreasing sequence of events such that $A_i \lesssim B_i$ for $i = 1, 2, \ldots$, then $\cup_{i=1}^\infty A_i \lesssim \cap_{i=1}^\infty B_i$.

(c) If $A_1 \supset A_2 \supset \cdots$ is a decreasing sequence of events and $B_1 \supset B_2 \supset \cdots$ is also a decreasing sequence of events such that $A_i \lesssim B_i$ for $i = 1, 2, \ldots$, then $\cap_{i=1}^\infty A_i \lesssim \cap_{i=1}^\infty B_i$. Furthermore, if $A_i \sim B_i$ for $i = 1, 2, \ldots$, then $\cap_{i=1}^\infty A_i \sim \cap_{i=1}^\infty B_i$. State and prove the analogous results for increasing sequences of events $A_1 \subset A_2 \subset \cdots$ and $B_1 \subset B_2 \subset \cdots$.

(d) If $A_1, A_2, \ldots$ and $B_1, B_2, \ldots$ are sequences of events such that $B_i B_j \sim \emptyset$ for $i \neq j$ and $A_i \lesssim B_i$ for $i = 1, 2, \ldots$, then $\cup_{i=1}^\infty A_i \lesssim \cup_{i=1}^\infty B_i$. If, in addition, $A_i \prec B_i$ for at least one value of i ($i = 1, 2, \ldots$), then $\cup_{i=1}^\infty A_i \prec \cup_{i=1}^\infty B_i$. (Note that this is an extension of Theorem 6 of Sec. 6.2, since it is not assumed here that the events $A_1, A_2, \ldots$ are disjoint and the condition that $B_1, B_2, \ldots$ are disjoint is replaced by the weaker condition that $B_i B_j \sim \emptyset$ for $i \neq j$.)

(e) If $A_1, A_2, \ldots$ is a sequence of disjoint events and B is an event such that $B \succ \emptyset$, then $A_i \gtrsim B$ for only a finite number of values of i ($i = 1, 2, \ldots$).

4. Let P^* be a given probability distribution on the set S of positive integers $\{1, 2, \ldots\}$ such that each integer in S is assigned a positive probability. Let $S_1 = \{1, 3, 5, \ldots\}$, and let $S_2 = \{2, 4, 6, \ldots\}$. Hence, every subset A of S can be expressed in the form $A = (AS_1) \cup (AS_2)$. Suppose that a relation $\lesssim$ is defined between subsets of S as follows: If A and B are any two subsets of S, then $A \lesssim B$ if either $P^*(AS_1) < P^*(BS_1)$ or $P^*(AS_1) = P^*(BS_1)$ and $P^*(AS_2) \leq P^*(BS_2)$. Show that the relation $\lesssim$ satisfies Assumptions SP_1 to SP_3 but not Assumption SP_4.

5. Consider the problem described in Exercise 4, but suppose now that the relation $\lesssim$ between subsets of S is defined as follows: If A and B are any two subsets of S, then $A \lesssim B$ if either $P^*(A) < P^*(B)$ or $P^*(A) = P^*(B)$ and $P^*(AS_1) \leq P^*(BS_1)$. Show that the relation $\lesssim$ satisfies Assumptions SP_1 to SP_3 but not Assumption SP_4.

6. Prove that the following results can be derived from Assumptions SP_1 to SP_4 and Assumption CP:

(a) If D is any event, then $D \sim \emptyset$ if, and only if, $(A|D) \sim (B|D)$ for all events A and B.

(b) If A and D are any events, then $(\emptyset|D) \lesssim (A|D) \lesssim (D|D)$.

(c) Let A, B, D, and E be events such that $ADE \sim BDE \sim \emptyset$. Then $(A|E) \lesssim (B|E)$ if, and only if, $[(A \cup D)|E] \lesssim [(B \cup D)|E]$.

(d) If A, B, D, and E are events such that $(A|E) \precsim (B|E)$ and $(B|E) \precsim (D|E)$, then $(A|E) \precsim (D|E)$.

(e) If A and B are events such that $A \subset B$, then $(A|D) \precsim (B|D)$ for any event D.

(f) If A, B, and D are any events, then $(A|D) \precsim (B|D)$ if, and only if, $(A^c|D) \succsim (B^c|D)$.

(g) If $A_1 \supset A_2 \supset \cdots$ is a decreasing sequence of events and if B and D are any given events such that $(A_i|D) \succsim (B|D)$ for $i = 1, 2, \ldots$, then $(\cap_{i=1}^{\infty} A_i|D) \succsim (B|D)$.

(h) If A_1, A_2, ... and B_1, B_2, ... and D are events such that $B_i B_j D \sim \emptyset$ for $i \neq j$ and $(A_i|D) \precsim (B_i|D)$ for $i = 1, 2, \ldots$, then $(\cup_{i=1}^{\infty} A_i|D) \precsim (\cup_{i=1}^{\infty} B_i|D)$. If, in addition, $(A_i|D) \prec (B_i|D)$ for at least one value of i ($i = 1, 2, \ldots$), then $(\cup_{i=1}^{\infty} A_i|D) \prec (\cup_{i=1}^{\infty} B_i|D)$.

(i) Let D_1, D_2, ... be events such that $D_i D_j \sim \emptyset$ for $i \neq j$ and $\cup_{i=1}^{\infty} D_i \sim S$. If A and B are events such that $(A|D_i) \precsim (B|D_i)$ for $i = 1, 2, \ldots$, then $A \precsim B$. If, in addition, $(A|D_i) \prec (B|D_i)$ for at least one value of i ($i = 1, 2, \ldots$), then $A \prec B$.

Special Note: The purpose of Exercises 7 to 9 is to provide specific examples involving the evaluation of subjective probabilities and showing how these probabilities change when additional information is gained. In order that Exercises 7 and 8 may be effective, each part of an exercise must be worked out without assuming any knowledge obtained from a later part. Accordingly, you should place a sheet of paper so as to expose each exercise to your view one line at a time. You should be able to see only the line you are currently reading and earlier lines, while the remainder of the exercise is still covered.

7. (a) Consider four events A_1, A_2, A_3, and A_4, which are described as follows: A_1 is the event that the year of birth of John Tyler, former President of the United States, was no later than 1750; A_2 is the event that John Tyler was born during one of the years 1751 to 1775, inclusive; A_3 is the event that he was born during one of the years 1776 to 1800, inclusive; A_4 is the event that he was born during the year 1801 or later. Allow yourself sufficient time to evaluate the relative likelihoods of these four events, but do not use any outside sources of information. Which of these four events do you consider the most likely? Which do you consider the least likely? What probability do you assign to the most likely event?

(b) You are now given the information that John Tyler was the tenth President of the United States. Use this information to reevaluate the relative likelihoods of the events A_1, A_2, A_3, and A_4. Before going on to part c, answer again the questions that were asked in part a.

(c) You are now given the information that George Washington, the first President of the United States, was born in 1732. Before going on to part d, answer again the questions that were asked in part a.

(d) You are now given the information that John Tyler was inaugurated as President in 1841. Answer again the questions that were asked in part a.

8. (a) Consider four events A_1, A_2, A_3, and A_4, which are described as follows: A_1 is the event that the area of the state of Pennsylvania is less than 5,000 sq miles; A_2 is the event that the area of Pennsylvania is between 5,000 and 50,000 sq miles; A_3 is the event that the area is between 50,000 and 100,000 sq miles; A_4 is the event that the area is greater than 100,000 sq miles. Which of these four events do you consider the most likely? Which do you consider the least likely? What probability do you assign to the most likely event?

(b) You are now given the following information: The area of Alaska, the largest of the 50 states, is 586,400 sq miles, and the area of Rhode Island, the smallest of the 50 states, is 1,214 sq miles. Before going on to part c, answer again the questions that were asked in part a.

(c) You are now given the information that when area is considered, Pennsylvania is the thirty-third largest of the 50 states. Before going on to part d, answer again the questions that were asked in part a.

(d) You are now given the information that the area of New York, the thirtieth largest of the 50 states, is 49,576 sq miles. Answer again the questions that were asked in part a.

9. Think of a fixed site outside the building in which you are at this moment. Let X be the temperature at that site at noon tomorrow. Choose a number x_1 such that:

(a) $P(X < x_1) = P(X > x_1) = \frac{1}{2}$.

Next, choose a number x_2 such that:

(b) $P(X < x_2) = P(x_2 < X < x_1) = \frac{1}{4}$.

Finally, choose numbers x_3 and x_4 ($x_3 < x_1 < x_4$) such that:

(c) $P(X < x_3) + P(X > x_4) = P(x_3 < X < x_1) = P(x_1 < X < x_4) = \frac{1}{3}$.

Using the values of x_1 and x_2 that you have chosen and tables of the standard normal distribution, find the unique normal distribution for X that satisfies the relations in parts a and b. Assuming that X has this normal distribution, find from the tables the values which x_3 and x_4 must have in order to satisfy the relation in part c and compare them with the values that you have chosen. Decide whether or not your distribution for X can be represented approximately by a normal distribution.

utility

7.1 PREFERENCES AMONG REWARDS

A statistician's subjective probabilities, as defined and developed in the preceding chapter, are numerical representations of his beliefs and information. His *utilities*, as they will be defined and developed in this chapter, are numerical representations of his tastes and preferences.

We shall consider a situation in which the statistician is to receive a reward r from a set R of possible rewards. The elements of R, which for convenience of language we have called rewards, may in general be quite complicated quantities. They are not necessarily monetary rewards, nor are they necessarily desirable rewards. In some developments of utility theory, these elements have also been referred to as consequences, a term which has a more neutral flavor. The essential requirement is that R must be a well-defined set of elements. Some examples of sets which might be of interest are a set of tickets to different musical events; a set of commodity bundles or market baskets; a set of possible economic states of an individual person at some fixed time in the future, as defined in terms of how much money he gains or loses between now and then; and a set of possible economic states of a nation at some fixed time in the future, as defined in terms of gross national product, per capita income, unemployment rate, or other variables.

The statistician will have preferences among the rewards in any set R. In some situations, these preferences will be evident. Thus, if the rewards are monetary, it is generally true that the greater the reward, the more preferable it is. Again, the preference may be clear when the rewards are tickets to different musical events which all take place on the same evening at equally convenient concert halls with equally good seats, since the choice would depend almost entirely on a preference for certain music or for certain musicians. If, however, it is necessary to consider also the relative convenience of different evenings and the relative quality of different seats, then the preference becomes less clear. A decision is much more difficult when the rewards are different states of the economy and the choice must be made by a very complex governmental organization rather than by a single individual.

Even an individual's preferences among certain commodity bundles, market baskets, or states of the economy are seldom clear, for the reason that such rewards are essentially multidimensional elements. When it is necessary to compare two vectorial rewards of this type and each component of the first vector is considered more desirable than the corresponding component of the second vector, then it is generally true that the first vector will be chosen in preference to the second vector. If, however, the first vector is preferable when only some of the components are considered and the second vector is preferable when only the remaining components are considered, then the overall preference between the two vectors is not obvious. Presumably, the final choice would be made by assigning relative weights to the various components.

No matter how complex, subtle, or sensitive may be the processes that lead to the preference for one reward in a specified set, it is a common observation that individuals and organizations have such preferences. These preferences are revealed through the decisions and choices which must regularly be made. It may therefore be assumed that the statistician will have preferences among the rewards in any given set R.

When any two rewards $r_1 \in R$ and $r_2 \in R$ are compared, we write $r_1 <^* r_2$ to indicate that r_2 is preferred to r_1 and we write $r_1 \sim^* r_2$ to indicate that r_1 and r_2 are equivalent, i.e., equally preferred. Furthermore, we write $r_1 \lesssim^* r_2$ to indicate that r_1 is not preferred to r_2. Hence, if $r_1 \lesssim^* r_2$, then either $r_1 <^* r_2$ or $r_1 \sim^* r_2$. Finally, we define $r_1 >^* r_2$ to mean the same as $r_2 <^* r_1$, and we define $r_1 \gtrsim^* r_2$ to mean the same as $r_2 \lesssim^* r_1$.

It is assumed that on the basis of his preferences among the rewards, the statistician can specify a complete ordering of R. In other words, the relation $\lesssim^*$ is assumed to satisfy the following two properties:

1. If r_1 and r_2 are any rewards in R, then exactly one of the following three relations must hold: $r_1 <^* r_2$, $r_1 >^* r_2$, $r_1 \sim^* r_2$.

2.   If r_1, r_2, and r_3 are any rewards in R such that $r_1 \lesssim^* r_2$ and $r_2 \lesssim^* r_3$, then $r_1 \lesssim^* r_3$.

Finally, it will be assumed that not all the rewards in R are equivalent to each other.   That is, we shall eliminate the possibility of a completely trivial situation by assuming that $s_0 \prec^* t_0$ for some pair of rewards $s_0 \epsilon R$ and $t_0 \epsilon R$.

7.2   PREFERENCES AMONG PROBABILITY DISTRIBUTIONS

In most problems that are of interest, the statistician does not have complete freedom in choosing the reward which he receives.   Typically, he can only specify, within a certain limited class of possible distributions, a probability distribution on R according to which his reward will be selected.   For example, a government cannot choose a particular state of the economy, but by applying certain economic instruments and policies that are at its disposal, it does specify a probability distribution for various possible states.   As another example, a marksman cannot choose the point at which his bullet will strike the target area, but by aiming at a certain point, he does specify a probability distribution over the area. (Note that the probability distributions in these two examples are highly subjective.)   In another type of problem, a particular industrial process is to be selected from among two or more available processes.   Although the rewards can be expressed in terms of productivity and costs, the production characteristics of the different processes can be known only in terms of probability distributions.

The basic problems of statistics provide many examples.   A statistician attempts to acquire information about the value of some parameter. His reward is the amount of information, appropriately defined, that he obtains about this value through experimentation.   He must select an experiment from some available class of experiments, but the information that he will obtain from any particular experiment is random.   In any problem of this kind, the statistician's choice is not directly among rewards in the set R but is among probability distributions over R.   Choices of probability distributions in situations of the type illustrated in these examples constitute decision problems of precisely the type that will be of principal interest to us throughout this book.

Since we shall be considering probability distributions on the set R, we must, in order to be precise, specify an appropriate σ-field $\mathcal{Q}$ of subsets of R and consider only probabilities of sets of rewards that belong to $\mathcal{Q}$. The development of utility that will be given in this chapter is completely general, and results will be expressed in terms of distributions and integrals on an abstract set R with a specified σ-field $\mathcal{Q}$ of subsets.   However, throughout this book and in almost all practical applications, it will be possible to choose the set R in a natural way so that it will be a subset of m-dimensional Euclidean space R^m for some suitable value of m. Therefore, the reader can assume that the set R is of this special form whenever making such an assumption would help him understand any part of this chapter.

In particular, for any probability distribution P over the set R, any subset $A \in \alpha$, and any integrable function g on R, the abstract integral $\int_A g(r) \, dP(r)$ can safely be interpreted as

$$\int_A g(r) \, dF(r). \tag{1}$$

Here F is the m-dimensional distribution function on the set $R \subset R^m$ which corresponds to the probability distribution P. The integral (1) was defined in Sec. 3.3.

In many problems, the set R is countable or even finite. In such problems, every probability distribution must necessarily be discrete and every integral can be reduced to a simple sum.

Consider now a fixed set R of rewards, together with a specific σ-field α of subsets of R, and let $\mathcal{P}$ be the class of all probability distributions P on the set R. The statistician's preferences among the rewards in R will lead him to have preferences among the probability distributions in $\mathcal{P}$. Thus, if he is given a choice between one lottery that yields a reward from R according to the probability distribution P_1 and another lottery that yields a reward from R according to the probability distribution P_2, he will typically prefer one of the two lotteries. The allusion to lotteries is very important here because it properly conveys the following notion: When two probability distributions on R are to be compared, only the probabilities of receiving the various rewards are relevant, and the statistician need not take into consideration the particular events of the lottery that generated these probabilities.

In other words, each probability distribution on R should be regarded as having been generated by an auxiliary experiment all of whose possible outcomes are ethically neutral. Any system which might be used in a lottery, such as spinning a wheel of fortune, can serve as an appropriate auxiliary experiment of this type.

The same notation will be used to represent preferences among the probability distributions in $\mathcal{P}$ as was introduced in Sec. 7.1 to represent preferences among the rewards in R. Hence, if $P_1 \in \mathcal{P}$ and $P_2 \in \mathcal{P}$ are any two probability distributions, we write $P_1 \prec^* P_2$ to indicate that P_2 is preferred to P_1, $P_1 \precsim^* P_2$ to indicate that P_1 is not preferred to P_2, and $P_1 \sim^* P_2$ to indicate that P_1 and P_2 are equivalent.

We shall not distinguish in our notation between a particular reward

$r_0 \in R$ and the degenerate probability distribution $P_0 \in \mathcal{P}$ which yields the reward r_0 with probability 1. Hence, for any two rewards r_1 and r_2, the relation $r_1 \precsim^* r_2$ defined in Sec. 7.1 can now be interpreted to mean that this relation holds between the corresponding degenerate probability distributions.

If r_1 and r_2 are any two rewards such that $r_1 \precsim^* r_2$, the *interval* $[r_1, r_2]$ is the subset of R defined by the equation

$$[r_1, r_2] = \{r: r_1 \precsim^* r \precsim^* r_2\}. \tag{2}$$

It is assumed that every interval $[r_1, r_2]$ belongs to the σ-field $\mathcal{A}$. Hence, for any distribution $P \in \mathcal{P}$, the probability $P([r_1, r_2])$ is a well-defined number.

A distribution $P \in \mathcal{P}$ is defined to be *bounded* if there exist rewards r_1 and r_2 such that $r_1 \precsim^* r_2$ and $P([r_1, r_2]) = 1$. In other words, a distribution is bounded if all of the probability is concentrated in some interval $[r_1, r_2]$.

The assumptions made in Sec. 7.1 in regard to the statistician's preferences among the rewards in R will now be strengthened as follows:

Let $\mathcal{P}_B$ denote the class of all bounded distributions in $\mathcal{P}$. It is assumed that on the basis of his preferences among these distributions, the statistician can specify a complete ordering of $\mathcal{P}_B$. In other words, the relation $\precsim^*$ is assumed to satisfy the following two properties:

1. If P_1 and P_2 are any two distributions in $\mathcal{P}_B$, then exactly one of the following three relations must hold: $P_1 \prec^* P_2$, $P_1 \succ^* P_2$, $P_1 \sim^* P_2$.
2. If P_1, P_2, and P_3 are any distributions in $\mathcal{P}_B$ such that $P_1 \precsim^* P_2$ and $P_2 \precsim^* P_3$, then $P_1 \precsim^* P_3$.

Note that no assumptions have as yet been made in regard to preferences involving any unbounded distributions in $\mathcal{P}$.

7.3 THE DEFINITION OF A UTILITY FUNCTION

For any distribution $P \in \mathcal{P}$ and any real-valued function g on the set R, we shall let $E(g|P)$ denote the expectation of g (if this expectation exists) when P is the distribution on R. That is,

$$E(g|P) = \int_R g(r) \, dP(r). \tag{1}$$

A real-valued function U defined on the set R is said to be a *utility function* if it has the following property: Let $P_1 \in \mathcal{P}$ and $P_2 \in \mathcal{P}$ be any two distributions such that both $E(U|P_1)$ and $E(U|P_2)$ exist. Then $P_1 \precsim^* P_2$

if, and only if, $E(U|P_1) \leq E(U|P_2)$. For each reward $r \in R$, the number $U(r)$ is called the *utility of r*. Hence, one probability distribution will be preferred to another if, and only if, the expected utility of the reward to be received is larger under the first distribution than it is under the other distribution.

For any distribution $P \in \mathcal{P}$, the number $E(U|P)$, when it exists, is often called simply the *utility of P*. Hence, the utility of a probability distribution is equal to the expected utility of the reward that will be received under that distribution. For this reason, the hypothesis that there exists a utility function is often called the *expected-utility hypothesis*.

Two consequences of the existence of a utility function will be mentioned here.

First, if $r_1 \in R$ and $r_2 \in R$ are any two rewards, then $r_1 \preceq^* r_2$ if, and only if, $U(r_1) \leq U(r_2)$. This result follows from the definition of a utility function if each reward r is identified with the degenerate probability distribution which yields that reward with probability 1.

Second, the existence of a utility function guarantees that a wide class of distributions in $\mathcal{P}$, including all the bounded distributions, will be comparable. If $\mathcal{P}_E$ denotes the class of all distributions P such that $E(U|P)$ exists, then any two distributions in $\mathcal{P}_E$ can be compared. The class $\mathcal{P}_E$ must include the class $\mathcal{P}_B$ of all bounded distributions, as the following argument shows:

Let $P \in \mathcal{P}$ be any bounded distribution. Then there exist rewards $r_1 \in R$ and $r_2 \in R$ such that $r_1 \preceq^* r_2$ and $P([r_1, r_2]) = 1$. By the first property mentioned above, $U(r_1) \leq U(r) \leq U(r_2)$ for any reward r in the interval $[r_1, r_2]$. It follows that $U(r_1) \leq E(U|P) \leq U(r_2)$. Hence, $P \in \mathcal{P}_E$.

Because of this fact, the assumption that the class $\mathcal{P}_B$ is completely ordered, which was made in Sec. 7.2, is well articulated with the existence of a utility function.

The main purpose of this chapter is to specify conditions on the relation $\preceq^*$ which ensure the existence of a utility function. The next lemma shows that if a utility function does exist, then certain linear transformations of this function will also be utility functions.

Lemma 1 *Let U be a utility function on R. Then any function V of the form $V = aU + b$, where a and b are constants $(a > 0)$, is also a utility function.*

Proof For any distribution $P \in \mathcal{P}$, the expectation $E(U|P)$ exists if, and only if, $E(V|P)$ exists. Let $P_1 \in \mathcal{P}$ and $P_2 \in \mathcal{P}$ be any two distributions for which these expectations exist. Then, since U is a utility function, $P_1 \preceq^* P_2$ if, and only if, $E(U|P_1) \leq E(U|P_2)$. But

$$E(V|P_i) = E(aU + b|P_i) = aE(U|P_i) + b \qquad i = 1, 2.$$

Since $a > 0$, it follows that $E(U|P_1) \leq E(U|P_2)$ if, and only if, $E(V|P_1) \leq E(V|P_2)$. Hence, V is a utility function.∎

It will be shown a little later that under fairly simple conditions a utility function does exist and that, except for increasing linear transformations of the type exhibited in Lemma 1, such a function is unique. First we shall discuss some of the properties of utility functions.

7.4 SOME PROPERTIES OF UTILITY FUNCTIONS

Our discussion of various properties of utility functions in this section and the next two sections will be largely based on specific examples. Two such examples follow.

EXAMPLE 1 Consider a given set R containing just three distinct rewards r_1, r_2, and r_3. These rewards might be, for example, different books or theater tickets. Each probability distribution in the class $\mathcal{P}$ may be represented by a triple of probabilities (p_1, p_2, p_3) such that $p_i \geq 0$ for $i = 1, 2, 3$ and $p_1 + p_2 + p_3 = 1$. It is to be understood here that for any triple $(p_1, p_2, p_3) \in \mathcal{P}$, p_i is the probability of receiving the reward r_i $(i = 1, 2, 3)$. It can be assumed, without any loss of generality, that $r_1 \succ^* r_2 \succ^* r_3$. We shall assume that there exists a utility function U on the set R, and we shall study the consequences of this assumption in regard to preferences among the distributions in $\mathcal{P}$.

As indicated in Lemma 1 of Sec. 7.3, the utility function U can be subjected to an arbitrary increasing linear transformation. Therefore, without loss of generality, we can assume that $U(r_1) = 1$ and $U(r_3) = 0$. If these assumptions are made, all preferences among the distributions in $\mathcal{P}$ will depend on the single number $U(r_2)$, which we shall call u. The number u must be in the interval $0 < u < 1$ and must be the unique number such that the probability distribution $(u, 0, 1 - u)$ will be equivalent to the distribution $(0, 1, 0)$. In other words, u is the unique number having the following property: A lottery ticket that yields the reward r_1 with probability u and yields the reward r_3 with probability $1 - u$ is equivalent to a ticket that yields the reward r_2 as a sure thing.

Furthermore, suppose that $P = (p_1, p_2, p_3)$ and $Q = (q_1, q_2, q_3)$ are any two distributions. Then $P \precsim^* Q$ if, and only if,

$$E(U|P) - E(U|Q) = (p_1 - q_1) + (p_2 - q_2)u \leq 0. \tag{1}$$

It follows that the preference relation between P and Q depends only on the differences $(p_1 - q_1)$ and $(p_2 - q_2)$ between their corresponding components. Hence, the mere existence of a utility function implies, for instance, that

$$(0.2, 0.5, 0.3) \lesssim^* (0.4, 0.2, 0.4) \tag{2}$$

if, and only if,

$$(0.3, 0.3, 0.4) \lesssim^* (0.5, 0, 0.5). \tag{3}$$

The explicit value of u would determine the common direction of the preferences in the relations (2) and (3).

EXAMPLE 2 (a) Suppose that you had your choice of either of the following two gambles:

> In gamble 1 the reward would be $500,000 with certainty.
> In gamble 2 the possible rewards would be either
> $2,500,000 with probability 0.10,
> $500,000 with probability 0.89,
> or $0 with probability 0.01.

Which gamble would you prefer? You cannot lose under any circumstances, and it should be assumed that all rewards are tax-free.

(b) Now suppose that instead of the above choice, you had your choice of either of the following two gambles:

> In gamble 3 the possible rewards would be either
> $500,000 with probability 0.11
> or $0 with probability 0.89.
> In gamble 4 the possible rewards would be either
> $2,500,000 with probability 0.10
> or $0 with probability 0.90.

Which of these gambles would you prefer?

This example was given first by Allais (1953) and was reproduced and discussed by Savage (1954). It has been found that many people would prefer gamble 1 to gamble 2 and gamble 4 to gamble 3. When a person prefers gamble 1 to gamble 2, he apparently reasons as follows: Since $500,000 is such a large amount of money, he should not pass up the opportunity of certainly receiving this amount in order to gamble on the possibility of receiving a much larger amount, because there is a small chance of receiving nothing. When a person prefers gamble 4 to gamble 3, he apparently reasons as follows: The likelihood of receiving nothing is only slightly greater for gamble 4 than for gamble 3, whereas the possible gain under gamble 4 is very much larger.

The interesting feature of this example is the following. Only three distinct amounts of money appear in the four gambles; hence, the discussion of Example 1 is relevant. Suppose that a person's utility function U on the set of possible winnings satisfies the inequalities

$$U(\$2{,}500{,}000) > U(\$500{,}000) > U(\$0).$$

Then, from the discussion given at the end of Example 1, it follows that he will prefer gamble 1 to gamble 2 if, and only if, he prefers gamble 3 to gamble 4.

Since the result just obtained conflicts with the observed preferences of many people, as stated earlier in this example, the contradiction would seem to indicate that a utility function does not exist for many people making a decision under the assumed conditions. However, it probably would be more appropriate to conclude that a person's utility function in this example depends on more than just his monetary winnings. Briefly, if a person chooses gamble 2 instead of gamble 1 and he unfortunately wins $0, then his reward is not simply $0 and a continuation of his life as it was before he was offered the choice. Rather his reward is the continuation of his life with his present resources but with the certain knowledge that if he had chosen otherwise, he would have been $500,000 wealthier. Hence, when considering whether or not to choose gamble 2, a person not only must recognize the possibility that he will win $0 but also must consider the utility of such factors as the remorse that he will feel because he passed up the opportunity for certain wealth and the ridicule from others to which he will be subjected. Considerations like these, which may be largely absent in making the choice between gamble 3 and gamble 4, would tend to be additional reasons for preferring gamble 1 to gamble 2.

The purpose of the preceding discussion is to stress the need for caution in regard to the assumption that an individual's utility function, when he is making a choice, depends only on some simple set of rewards, such as the amounts of possible monetary gains. The individual making the choice may consider his possible rewards to be much more complex, and his utility function will be determined by several factors. This is probably true when a person makes choices in an artificial experiment of the type considered in Example 2, in the paper by Ellsberg (1961) mentioned in Sec. 6.6, and in much of the work dealing with the experimental measurement of utility. As another illustration of these remarks, when a person decides whether or not to spend an evening gambling, he considers not only the utility of what he might win or lose but also the utility of spending the evening in this way.

7.5 THE UTILITY OF MONETARY REWARDS

In the preceding section, an attempt was made to point out some of the difficulties associated with the concept of the utility of a monetary gain, as well as with its measurement. However, if the possible monetary gains and losses in a given situation are not extremely large and if consideration of other factors can be minimized, then it is not unreasonable to discuss utility functions defined on sets of monetary rewards. Such utility functions have been widely used, and they will be discussed in this section and the next.

The St. Petersburg Paradox

The example which we shall now give was one of the earliest illustrations of the fact that a person's utility function, considered as a function of possible monetary rewards, will not be a linear function. Suppose that a person is to be given the opportunity of playing the following game: A fair coin such that heads and tails are equally probable is to be tossed repeatedly until a head is obtained for the first time. If the first head is obtained on the nth toss, then the person's reward will be 2^n dollars ($n = 1, 2, \ldots$). How much should the person be willing to pay for the opportunity of playing this game once?

Since the probability that the first head will be obtained on the nth toss is $(\frac{1}{2})^n$ for $n = 1, 2, \ldots$, the expected monetary gain can be represented by the infinite series $\sum_{n=1}^{\infty} 2^n (\frac{1}{2})^n$. Hence, the expected gain is infinite. If a person's utility function is linear, then he should be willing to pay any arbitrarily large amount for the opportunity of playing the game. In fact, however, each person is willing to pay only a specific finite amount which depends on his own utility function. This fact was called the St. Petersburg paradox, although the concept of utility shows that there is no paradox.

This game was discussed by Daniel Bernoulli, who, in the early eighteenth century, was among the first men to consider the expected-utility hypothesis.

The Experimental Measurement of Utility

We shall now consider the problem of specifying a person's utility function. A simple method will be presented which is helpful in determining this function over a specified interval of possible monetary rewards, which will be taken here to be the interval between $0 and $1,000. For the purposes of this discussion, let $\langle a, b \rangle$ denote, for any real numbers a and b, a lottery ticket that yields a reward of a dollars with probability $\frac{1}{2}$ and a reward of b dollars with probability $\frac{1}{2}$. Consider the following sequence of four steps:

> *Step 1* is to find a number x_1 such that receiving x_1 dollars with certainty is regarded by the person as equivalent to receiving a reward from the lottery ticket $\langle 0, 1{,}000 \rangle$.
>
> *Step 2* is to find a number x_2 such that receiving x_2 dollars with certainty is regarded as equivalent to receiving a reward from the lottery ticket $\langle 0, x_1 \rangle$.
>
> *Step 3* is to find a number x_3 such that receiving x_3 dollars with certainty is regarded as equivalent to receiving a reward from the lottery ticket $\langle x_1, 1{,}000 \rangle$.
>
> *Step 4* is to find a number x_4 such that receiving x_4 dollars with certainty is regarded as equivalent to receiving a reward from the lottery ticket $\langle x_2, x_3 \rangle$.

One experimental technique for determining these four numbers is indicated in Exercise 2 at the end of this chapter.

Suppose that the person's utility function U is a strictly increasing function of monetary reward. Then, without loss of generality, we can set $U(0) = 0$ and $U(1{,}000) = 1$. It follows that $U(x_1) = \frac{1}{2}$, $U(x_2) = \frac{1}{4}$, $U(x_3) = \frac{3}{4}$, and $U(x_4) = \frac{1}{2}$. Hence, $x_1 = x_4$. When strictly interpreted, this equality means that if the numbers found in steps 1 and 4 are not equal, then the results do not meet the requirements of a utility function. In a practical case, these two numbers will seldom be equal in the first trial. However, by studying the discrepancy between x_1 and x_4 and by repeating the four steps with different numbers as often as necessary, it is possible eventually to decide on a final value x^* for which $U(x^*) = \frac{1}{2}$. This procedure can also be used to study decisions based on lottery tickets which involve various rewards, and probabilities other than $\frac{1}{2}$, in order to determine the gain x for which $U(x) = u$ for any fixed value of u ($0 < u < 1$). Hence, the procedure provides a technique for approximately determining a person's utility function.

Further Remarks and References

The procedure described in this section is based on the paper by Becker, DeGroot, and Marschak (1964). Some other work on the experimental measurement of utility is by Mosteller and Nogee (1951) and by Davidson, Suppes, and Siegel (1957).

7.6 CONVEX AND CONCAVE UTILITY FUNCTIONS

Convex Functions

It is said that a real-valued function g defined on an interval (a, b) of the real line is *convex* on (a, b) if, for any two points x and y in (a, b) and any number α such that $0 < \alpha < 1$,

$$g[\alpha x + (1 - \alpha)y] \leq \alpha g(x) + (1 - \alpha)g(y). \tag{1}$$

Geometrically, the relation (1) states that the line segment joining any two points on the curve $y = g(x)$ does not fall below the curve anywhere between those two points. It is said that the function g is *strictly convex* on (a, b) if strict inequality holds in the relation (1) for every pair of distinct points x and y in (a, b). Geometrically, this means that the function g is convex and that the curve $y = g(x)$ cannot contain any linear segments. Examples of convex functions are sketched in Fig. 7.1.

We shall now present a basic theorem, known as *Jensen's inequality*, which relates to the expectations of convex functions. This theorem applies whether the interval (a, b) on which the convex function is defined is finite or infinite. Also, if either end point is finite, the interval may be closed or open at that end.

Theorem (Jensen's inequality) *Let g be a convex function on the interval (a, b), and let X be a random variable such that $\Pr\{X \in (a, b)\} = 1$ and the expectations $E(X)$ and $E[g(X)]$ exist. Then*

$$E[g(X)] \geq g[E(X)]. \tag{2}$$

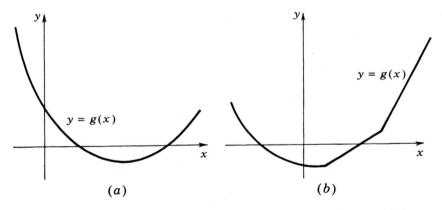

Fig. 7.1 (a) A strictly convex function g. (b) A convex function g which is not strictly convex.

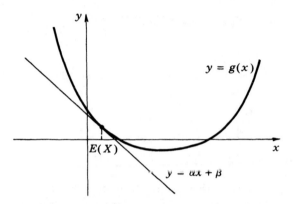

Fig. 7.2 The line of support.

Furthermore, if g is strictly convex and $\Pr\{X = E(X)\} \neq 1$, *then there is strict inequality in the relation* (2).

Proof If the distribution of X is degenerate and concentrated on a single value, then the two expectations in the relation (2) are equal. Suppose that the distribution is not degenerate. Then it must be true that $a < E(X) < b$. It is a property of convex functions that there exists a line of support $y = \alpha x + \beta$ which touches the curve $y = g(x)$ at the point where $x = E(X)$ and which is not higher than the curve at any point in the interval (a, b). This line of support is sketched in Fig. 7.2. In more precise terms,

$$\alpha E(X) + \beta = g[E(X)] \tag{3}$$

and

$$\alpha x + \beta \leq g(x) \qquad \text{for } x \in (a, b). \tag{4}$$

If x is replaced by the random variable X in the relation (4) and if the expectation is taken on each side of the inequality, we obtain

$$\alpha E(X) + \beta \leq E[g(X)]. \tag{5}$$

The relation (2) now follows directly from the relations (5) and (3).

Furthermore, if the function g is strictly convex, then there is strict inequality in the relation (4) at every point $x \in (a, b)$ except $x = E(X)$. Since, by hypothesis, $\Pr\{X \neq E(X)\} > 0$, it follows that this strict inequality will be preserved when expectations are taken. Hence, there will be strict inequality in the relation (5). This result, together with Eq. (3), will yield strict inequality in the relation (2). ∎

Now consider a situation in which the set R of rewards is an interval (a, b) of monetary gains, and suppose that the utility function U of some person is strictly convex over (a, b). It follows from Jensen's inequality that this person will choose a gamble which yields a random monetary gain X in the interval (a, b) in preference to a definite gain $E(X)$ which is certain. In particular, if the interval (a, b) includes the value 0, then a fair gamble from which the expected gain is 0 would be preferred to not gambling at all. For these reasons, a person with a strictly convex utility function is said to be a risk taker or, even more strongly, a risk lover. Clearly, two risk takers will enjoy making even-money bets with each other on whether a coin will fall heads or tails, since both will prefer making the bet to not betting.

Concave Functions

It is said that a real-valued function g defined on an interval (a, b) of the real line is *concave* on (a, b) if the negative of g is convex on (a, b). Hence, g is concave on (a, b) if, for any two points x and y in (a, b) and any number α such that $0 < \alpha < 1$,

$$g[\alpha x + (1 - \alpha)y] \geq \alpha g(x) + (1 - \alpha)g(y). \tag{6}$$

It is said that the function g is *strictly concave* on (a, b) if the negative of g is strictly convex.

Now suppose that the utility function U of some person is strictly concave over an interval (a, b) of monetary gains, as illustrated in Fig. 7.3. It again follows from Jensen's inequality that such a person will choose a fixed gain x in the interval (a, b) in preference to a gamble which yields a random gain X in (a, b) such that $E(X) = x$. A person with a strictly concave utility function is called a risk averter.

The utility functions of most people tend to be concave, at least for large gains or large losses. People purchase insurance because they prefer to pay a fixed premium $x > 0$ rather than risk a large loss h with very small probability ϵ. In other words, when a person purchases

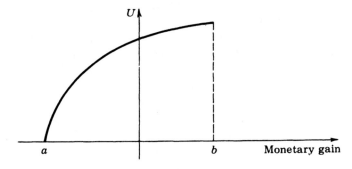

Fig. 7.3 A strictly concave utility function.

insurance, he chooses a fixed gain $-x$ in preference to a gamble that yields $-h$ with probability ϵ and no gain or loss with probability $1 - \epsilon$. This preference is expressed in the relation

$$U(-x) > \epsilon U(-h) + (1 - \epsilon)U(0). \tag{7}$$

The function U need not be concave over the entire interval in order for the relation (7) to be true for a particular value of x.

An insurance company makes many gambles (since it sells many policies) and has very large resources. Hence, in regard to any particular policy, the utility function U_1 of an insurance company is approximately a linear function of its monetary gain. Therefore, U_1 can be chosen so that $U_1(y) = y$ over a suitable interval. When the company sells an insurance policy, it does so because its expected gain from the sale is greater than 0, which would be its gain from no sale. Since the expected gain from the sale is the premium x which is received minus the expected amount ϵh which must be paid under the policy, the insurance company fixes the premium x high enough to make

$$x - \epsilon h > 0. \tag{8}$$

When both relations (7) and (8) are satisfied, an insurance policy is sold.

Further Remarks and References

Friedman and Savage (1948, 1952) and Pratt (1964) discuss some problems involving convex and concave utility functions. Because of the variability in an individual's preferences that is often observed in actual situations, various stochastic models of choice behavior and of utility have been proposed [e.g., Luce (1959), Suppes and Walsh (1959), and Becker, DeGroot, and Marschak (1963)]. Luce and Suppes (1965) give an excellent survey of utility, covering the work that has been done on the experimental measurement of utility, on stochastic models, and on the axiomatic development of utility, which will be discussed beginning in the next section.

Before we proceed to the axiomatic development of utility, one final comment should be made. Under the present theory, when a person has a choice of one of several gambles, and especially if the gamble is to be taken only once, he should prefer the one with the highest expected utility. His choice will not depend on any "long-run" arguments of the type sometimes invoked when expectations are considered, but it will be simply a consequence of the definition of utility. If gambles are to be taken repeatedly, then the total gain from the entire sequence of gambles should be considered as the reward from one large composite gamble, and the preferred sequence of gambles should be the one that leads to the highest expected utility of the total gain. Usually, such a sequence of gambles will be different from the sequence that would be obtained by simply taking, at each stage, the gamble with the highest expected utility of the gain at that stage. Problems of this type are sequential decision problems, which will be systematically discussed in Chaps. 12 to 14.

7.7 THE AXIOMATIC DEVELOPMENT OF UTILITY

It was assumed in Sec. 7.2 that the relation $\precsim^*$ can be used to establish a complete ordering of the class $\mathcal{P}_B$ of all bounded probability distributions over the set of rewards R. In this section we shall begin investigating the further conditions which must be imposed on the relation $\precsim^*$ in order for there to exist a function U which is defined on the set R and which has the following property Π:

Property Π *Let P_1 and P_2 be any two distributions in the class $\mathcal{P}_B$. Then $P_1 \precsim^* P_2$ if, and only if, $E(U|P_1) \leq E(U|P_2)$.* In other words, the function U is a utility function for the distributions in the class $\mathcal{P}_B$.

The following notation will be used in the discussion. Let P_1 and P_2 be any two distributions in $\mathcal{P}$, and let α be any number such that $0 < \alpha < 1$. Then the distribution which assigns the probability $\alpha P_1(A) + (1 - \alpha)P_2(A)$ to each set of rewards A ($A \in \mathcal{A}$) will be denoted by the expression $\alpha P_1 + (1 - \alpha)P_2$. In particular, if r_1 and r_2 are any rewards in R and α is any number such that $0 < \alpha < 1$, then $\alpha r_1 + (1 - \alpha)r_2$ denotes the distribution in $\mathcal{P}$ such that the reward r_1 is received with probability α and the reward r_2 is received with probability $1 - \alpha$. Furthermore, it should be noted that if both P_1 and P_2 are distributions which belong to the class $\mathcal{P}_B$, then the distribution $\alpha P_1 + (1 - \alpha)P_2$ also belongs to $\mathcal{P}_B$.

We can now state the first assumption.

Assumption U_1 *Suppose that P_1, P_2, and P are any three distributions in the class $\mathcal{P}_B$, and suppose that α is any number such that $0 < \alpha < 1$. Then $P_1 \prec^* P_2$ if, and only if, $\alpha P_1 + (1 - \alpha)P \prec^* \alpha P_2 + (1 - \alpha)P$.*

This assumption formalizes the following property: Consider two lotteries in which the reward will be determined according to the distributions $\alpha P_1 + (1 - \alpha)P$ and $\alpha P_2 + (1 - \alpha)P$. In both lotteries, there is probability $1 - \alpha$ that the reward will be determined according to the distribution P. Hence, this common aspect of the two lotteries is irrelevant to the preference relation between them.

Three consequences of this assumption can be established.

Lemma 1   *Let P_1, P_2, Q_1, and Q_2 be any four distributions in the class $\mathcal{P}_B$ such that $P_1 \lesssim^* Q_1$ and $P_2 \lesssim^* Q_2$, and let α be any number such that $0 < \alpha < 1$. Then $\alpha P_1 + (1 - \alpha)P_2 \lesssim^* \alpha Q_1 + (1 - \alpha)Q_2$. Furthermore, if either $P_1 \prec^* Q_1$ or $P_2 \prec^* Q_2$, then $\alpha P_1 + (1 - \alpha)P_2 \prec^* \alpha Q_1 + (1 - \alpha)Q_2$.*

Proof   By two applications of Assumption U_1,

$$\alpha P_1 + (1 - \alpha)P_2 \lesssim^* \alpha Q_1 + (1 - \alpha)P_2 \lesssim^* \alpha Q_1 + (1 - \alpha)Q_2. \quad (1)$$

Furthermore, if $P_1 \prec^* Q_1$ or $P_2 \prec^* Q_2$, then at least one of the preferences in relation (1) must be a strict preference.∎

Lemma 2   *Let r_1 and r_2 be any two rewards in R such that $r_1 \prec^* r_2$, and let α be any number such that $0 < \alpha < 1$. Then $r_1 \prec^* \alpha r_2 + (1 - \alpha)r_1 \prec^* r_2$.*

Proof   Each reward r_i can be written in the form $r_i = \beta r_i + (1 - \beta)r_i$ for any number β. It follows that

$$r_1 = \alpha r_1 + (1 - \alpha)r_1 \prec^* \alpha r_2 + (1 - \alpha)r_1$$
$$\prec^* \alpha r_2 + (1 - \alpha)r_2 = r_2.∎ \quad (2)$$

Lemma 3   *Let r_1 and r_2 be any two rewards in R such that $r_1 \prec^* r_2$, and let α and β be any numbers such that $0 \leq \alpha \leq 1$ and $0 \leq \beta \leq 1$. Then*

$$\alpha r_2 + (1 - \alpha)r_1 \prec^* \beta r_2 + (1 - \beta)r_1 \quad (3)$$

if, and only if, $\alpha < \beta$.

Proof   It is sufficient to prove that if $\alpha < \beta$, then the relation (3) is true. Suppose then that $\alpha < \beta \leq 1$, and let $\gamma = (1 - \beta)/(1 - \alpha)$. Hence, $0 \leq \gamma < 1$, and it can be verified that

$$\beta r_2 + (1 - \beta)r_1 = \gamma[\alpha r_2 + (1 - \alpha)r_1] + (1 - \gamma)r_2. \quad (4)$$

It follows that

$$\alpha r_2 + (1 - \alpha)r_1 = \gamma[\alpha r_2 + (1 - \alpha)r_1] + (1 - \gamma)[\alpha r_2 + (1 - \alpha)r_1]$$
$$\prec^* \gamma[\alpha r_2 + (1 - \alpha)r_1] + (1 - \gamma)r_2$$
$$= \beta r_2 + (1 - \beta)r_1.∎ \quad (5)$$

The next assumption states that a sufficiently slight change in probabilities will not reverse a strict preference.

Assumption U_2   *Let P_1, P_2, and P be any three distributions in the class $\mathcal{P}_B$ such that $P_1 \prec^* P \prec^* P_2$. Then there exist numbers α and β $(0 < \alpha < 1, 0 < \beta < 1)$ such that*

$$P \prec^* \alpha P_2 + (1 - \alpha)P_1 \quad \text{and} \quad P \succ^* \beta P_2 + (1 - \beta)P_1. \quad (6)$$

One of the consequences of Assumption U_2 is the following: Suppose that r and r_1 are rewards such that $r >^* r_1$. Then no reward $r_2 \epsilon R$ can be so desirable that assigning to it any arbitrary positive probability β, no matter how small, will make $\beta r_2 + (1 - \beta)r_1 >^* r$. Similarly, suppose that r and r_2 are rewards such that $r <^* r_2$. Then no reward $r_1 \epsilon R$ can be so undesirable that assigning to it any arbitrary positive probability $1 - \alpha$, no matter how small, will make $\alpha r_2 + (1 - \alpha)r_1 <^* r$. In other words, in an ordinary problem there is neither "heaven" nor "hell" in R. If such extremes are possible, then Assumption U_2 states that the statistician is willing to reduce the chance of gaining heaven and to risk a small chance of gaining hell in order to have a large chance of gaining desirable mundane rewards. With respect to this property, Assumption U_2 seems to have strong empirical support.

Theorem 1 *Let r, r_1, and r_2 be any rewards in R such that $r_1 <^* r_2$ and $r_1 \lesssim^* r \lesssim^* r_2$. Then there exists a unique number v $(0 \leq v \leq 1)$ such that $r \sim^* v r_2 + (1 - v)r_1$.*

Proof If $r \sim^* r_1$, then $v = 0$. Also, if $r \sim^* r_2$, then $v = 1$. Suppose then that $r_1 <^* r <^* r_2$, and let S_1 and S_2 be subsets of the unit interval defined as follows:

$$S_1 = \{\alpha: r <^* \alpha r_2 + (1 - \alpha)r_1\},$$
$$S_2 = \{\alpha: r >^* \alpha r_2 + (1 - \alpha)r_1\}.$$

By Lemma 3, if $\alpha_1 \epsilon S_1$ and $\alpha_2 > \alpha_1$, then $\alpha_2 \epsilon S_1$; and if $\alpha_1 \epsilon S_2$ and $\alpha_2 < \alpha_1$, then $\alpha_2 \epsilon S_2$. Also, neither S_1 nor S_2 is empty since $1 \epsilon S_1$ and $0 \epsilon S_2$. Therefore, S_1 must be an interval of the form $(\beta,1]$ and S_2 must be an interval of the form $[0,\alpha)$. Furthermore, it follows from Assumption U_2 that if $\beta_1 \epsilon S_1$, then there must exist a slightly smaller number β_2 which also belongs to S_1. Hence, the end point β cannot belong to S_1. It follows in a similar manner that the end point α cannot belong to S_2.

Since S_1 and S_2 are disjoint, in accordance with their definitions, then $\alpha \leq \beta$. Hence, there exists a number v such that $\alpha \leq v \leq \beta$. Since any such number v belongs neither to S_1 nor to S_2, it must be true that $r \sim^* v r_2 + (1 - v)r_1$. Therefore, there will always be a number v having the desired property, and by Lemma 3 that number will be unique. ∎

7.8 CONSTRUCTION OF THE UTILITY FUNCTION

Let s_0 and t_0 be any two specific rewards in R such that $s_0 <^* t_0$. We can now construct the function U that will serve as our utility function. If r is any reward in R such that $s_0 \lesssim^* r \lesssim^* t_0$, then $U(r)$ is defined to be the unique number in the interval $0 \leq U(r) \leq 1$ which satisfies the relation

$$r \sim^* U(r)t_0 + [1 - U(r)]s_0. \tag{1}$$

The existence and uniqueness of the number $U(r)$ are guaranteed by Theorem 1 of Sec. 7.7. In particular, $U(s_0) = 0$ and $U(t_0) = 1$.

For any reward $r \, \epsilon \, R$ such that $r <^* s_0$, there must exist a unique number α $(0 < \alpha < 1)$ which satisfies the relation

$$s_0 \sim^* \alpha t_0 + (1 - \alpha)r. \tag{2}$$

When the relation (2) is satisfied, we would like the utility function U to have the following property:

$$U(s_0) = \alpha U(t_0) + (1 - \alpha)U(r). \tag{3}$$

Hence, in order to satisfy Eq. (3) when $U(s_0) = 0$ and $U(t_0) = 1$, as specified above, we define $U(r)$ by the equation

$$U(r) = \frac{-\alpha}{1 - \alpha}. \tag{4}$$

Finally, for any reward $r \, \epsilon \, R$ such that $r >^* t_0$, there must exist a unique number α $(0 < \alpha < 1)$ which satisfies the relation

$$t_0 \sim^* \alpha r + (1 - \alpha)s_0. \tag{5}$$

When the relation (5) is satisfied, we would like the utility function U to have the following property:

$$U(t_0) = \alpha U(r) + (1 - \alpha)U(s_0). \tag{6}$$

Hence, in order to satisfy Eq. (6), we define $U(r)$ by the equation

$$U(r) = \frac{1}{\alpha}. \tag{7}$$

Together, Eqs. (1), (2), (4), (5), and (7) define U on the whole set R.

The next theorem proves that the linearity of the function U indicated in Eqs. (3) and (6) actually holds throughout the set R.

Theorem 1 *Let r_1, r_2, and r_3 be any three rewards in R such that for some value of α $(0 \leq \alpha \leq 1)$,*

$$r_2 \sim^* \alpha r_3 + (1 - \alpha)r_1. \tag{8}$$

Then

$$U(r_2) = \alpha U(r_3) + (1 - \alpha)U(r_1). \tag{9}$$

Proof Among the five rewards r_1, r_2, r_3, s_0, and t_0, let s_1 be the least preferred one and let t_1 be the most preferred one. Then $s_1 <^* t_1$ For any reward r in R such that $s_1 \lesssim^* r \lesssim^* t_1$, let $U_1(r)$ be defined as the unique number in the interval $0 \le U_1(r) \le 1$ which satisfies the relation

$$r \sim^* U_1(r)t_1 + [1 - U_1(r)]s_1. \tag{10}$$

Since the relation (8) must be true by hypothesis and since the relation (10) must be true for $r = r_3$ and for $r = r_1$, it follows from Lemma 1 of Sec. 7.7 that

$$r_2 \sim^* [\alpha U_1(r_3) + (1 - \alpha)U_1(r_1)]t_1$$
$$+ [1 - \alpha U_1(r_3) - (1 - \alpha)U_1(r_1)]s_1. \tag{11}$$

By Lemma 3 of Sec. 7.7, the coefficient of t_1 in the relation (11) must be equal to the coefficient of t_1 in the relation (10) when $r = r_2$. Therefore,

$$U_1(r_2) = \alpha U_1(r_3) + (1 - \alpha)U_1(r_1). \tag{12}$$

Although Eq. (12) is the desired relation, it contains the function U_1 rather than U. To complete the proof, we must show that $U_1(r_i)$ is simply a linear function of $U(r_i)$ for $i = 1, 2, 3$, in which case Eq. (12) will also be correct for the function U.

Let r_i be any one of the three rewards r_1, r_2, r_3. Suppose first that $s_0 \lesssim^* r_i \lesssim^* t_0$. Then, by Eq. (1),

$$r_i \sim^* U(r_i)t_0 + [1 - U(r_i)]s_0.$$

Accordingly, in Eqs. (8) and (12), let r_2 be replaced by r_i, r_3 by t_0, r_1 by s_0, and α by $U(r_i)$. From these replacements, we obtain the equation

$$U_1(r_i) = U(r_i)U_1(t_0) + [1 - U(r_i)]U_1(s_0). \tag{13}$$

Next, suppose that $r_i <^* s_0$. Then, by Eqs. (2) and (4),

$$s_0 \sim^* - \frac{U(r_i)}{1 - U(r_i)} t_0 + \frac{1}{1 - U(r_i)} r_i.$$

When appropriate replacements are made in Eq. (12), we obtain the equation

$$U_1(s_0) = - \frac{U(r_i)}{1 - U(r_i)} U_1(t_0) + \frac{1}{1 - U(r_i)} U_1(r_i). \tag{14}$$

Although the terms are arranged differently, Eq. (14) is precisely the same as Eq. (13).

Finally, suppose that $r_i >^* t_0$. Then, by Eqs. (5) and (7),

$$t_0 \sim^* \frac{1}{U(r_i)} r_i + \left[1 - \frac{1}{U(r_i)}\right] s_0.$$

When appropriate replacements are made in Eq. (12), we obtain the equation

$$U_1(t_0) = \frac{1}{U(r_i)} U_1(r_i) + \left[1 - \frac{1}{U(r_i)}\right] U_1(s_0). \tag{15}$$

By rearranging the terms, Eq. (15) can also be made precisely the same as Eq. (13).

Hence, $U_1(r_i)$ for $i = 1$, 2, 3 can be replaced in Eq. (12) by the expression given in Eq. (13). The result is then precisely the desired Eq. (9).∎

7.9   VERIFICATION OF THE PROPERTIES OF A UTILITY FUNCTION

In this section two further assumptions will be made which guarantee that the function U will satisfy the required property II stated at the beginning of Sec. 7.7. The property II involves the expectation $E(U|P)$ for each distribution $P \in \mathcal{P}_B$, and the first assumption which will be made here ensures that the function U will be measurable with respect to the σ-field $\mathcal{Q}$ of subsets of R. This measurability property is needed in order to be certain that the expectation $E(U|P)$ is meaningful for every $P \in \mathcal{P}_B$. The second assumption which will be made ensures that certain simple equivalence relations involving the value $E(U|P)$ will be satisfied.

Measurability

It was assumed in Sec. 7.2 that if r_1 and r_2 are any rewards in R such that $r_1 \lesssim^* r_2$, then the interval $[r_1, r_2] \in \mathcal{Q}$. A slightly stronger assumption will now be made.

Assumption U_3   *For any three rewards r_1, r_2, and r_3 in R and any numbers α and β $(0 \leq \alpha \leq 1, 0 \leq \beta \leq 1)$, it must be true that*

$$\{r: \alpha r + (1 - \alpha)r_1 \lesssim^* \beta r_2 + (1 - \beta)r_3\} \in \mathcal{Q}. \tag{1}$$

Lemma 1   *The function U is measurable with respect to $\mathcal{Q}$.*

Proof   It must be shown that for each real number x,

$$\{r: U(r) \leq x\} \in \mathcal{Q}. \tag{2}$$

Suppose first that $x < 0$. If $U(r) \leq x$ for some reward r, then $r \prec^* s_0$. Hence, it follows from Eqs. (2) and (4) of Sec. 7.8 and from Lemma 3 of Sec. 7.7 that $U(r) \leq x$ if, and only if,

$$s_0 \gtrsim^* -\frac{x}{1 - x} t_0 + \frac{1}{1 - x} r. \tag{3}$$

But, by Assumption U_3, the set of rewards r which satisfy the relation (3) belongs to $\mathcal{Q}$. Hence, the relation (2) is correct.

Next, suppose that $0 \le x \le 1$.   Then $U(r) \le x$ if, and only if,

$$r \precsim^* xt_0 + (1 - x)s_0. \tag{4}$$

Hence, by Assumption U_3, the relation (2) is correct.

Finally, suppose that $x > 1$.   Then it follows from Eqs. (5) and (7) of Sec. 7.8 that $U(r) \le x$ if, and only if,

$$t_0 \succsim^* \frac{1}{x} r + \left(1 - \frac{1}{x}\right) s_0. \tag{5}$$

Again by Assumption U_3, the relation (2) is correct. ∎

Equivalence Relations for Bounded Distributions

Let r_1 and r_2 be any two rewards in R such that $r_1 \prec^* r_2$, and consider the interval $[r_1, r_2]$ as defined by Eq. (2) of Sec. 7.2.   By Theorem 1 of Sec. 7.7, for any reward $r \epsilon [r_1, r_2]$, there exists a number $\alpha(r)$ such that

$$r \sim^* [\alpha(r)]r_2 + [1 - \alpha(r)]r_1. \tag{6}$$

In fact, the value of $\alpha(r)$ can be given explicitly.   By Theorem 1 of Sec. 7.8,

$$U(r) = \alpha(r)U(r_2) + [1 - \alpha(r)]U(r_1). \tag{7}$$

Therefore,

$$\alpha(r) = \frac{U(r) - U(r_1)}{U(r_2) - U(r_1)}. \tag{8}$$

The assumption which will be made here can be explained as follows: Consider any distribution P such that $P\{[r_1, r_2]\} = 1$.   Since each reward $r \epsilon [r_1, r_2]$ is equivalent to a lottery which yields the reward r_2 with probability $\alpha(r)$ and the reward r_1 with probability $1 - \alpha(r)$, then P itself will be equivalent to a lottery which yields either r_2 or r_1 with appropriate probabilities.   In fact, consider a two-stage lottery in which one of the lotteries of the form $[\alpha(r)]r_2 + [1 - \alpha(r)]r_1$ is selected first according to the distribution P over the interval $[r_1, r_2]$ and then one of the rewards r_2 or r_1 is selected according to this lottery.   The overall probability β that the reward r_2 will be selected in this two-stage lottery is given by the equation

$$\beta = \int_{[r_1, r_2]} \alpha(r) \, dP(r). \tag{9}$$

The overall probability that the reward r_1 will be selected is $1 - \beta$.   Note that the integral in Eq. (9) exists since, by Eq. (8), the function α is simply a linear function of the measurable function U and its values on the interval $[r_1, r_2]$ must be bounded.

Assumption U_4 *Let $P \epsilon \mathcal{P}$ be any distribution such that $P\{[r_1, r_2]\} = 1$ for some rewards $r_1 \epsilon R$ and $r_2 \epsilon R$.   For each reward $r \epsilon [r_1, r_2]$, let the number $\alpha(r)$ be defined by Eq. (6), and let the number β be defined by Eq. (9).   Then $P \sim^* \beta r_2 + (1 - \beta)r_1$.*

It follows from Eqs. (8) and (9) that the equivalence stated in Assumption U_4 can be written in the form

$$P \sim^* \left[\frac{E(U|P) - U(r_1)}{U(r_2) - U(r_1)}\right] r_2 + \left[\frac{U(r_2) - E(U|P)}{U(r_2) - U(r_1)}\right] r_1. \tag{10}$$

The basic results can now be established.   The first theorem shows that the function U satisfies the desired property II.   The second theorem shows that the only other functions which do so are increasing linear transformations of U.

Theorem 1   *Let P_1 and P_2 be any two distributions in the class $\mathcal{P}_B$.   Then $P_1 \lesssim^* P_2$ if, and only if, $E(U|P_1) \leq E(U|P_2)$.*

Proof   Since P_1 and P_2 are both bounded distributions, there exist rewards $r_1 \epsilon R$ and $r_2 \epsilon R$ such that $r_1 <^* r_2$ and

$$P_1\{[r_1, r_2]\} = P_2\{[r_1, r_2]\} = 1.$$

Hence, both P_1 and P_2 are equivalent to distributions of the form given in Eq. (10).   It now follows from Lemma 3 of Sec. 7.7 that $P_1 <^* P_2$ if, and only if, $E(U|P_1) \leq E(U|P_2)$.∎

Theorem 2   *Suppose that U and U^* are real-valued functions on R each of which has the property stated in Theorem 1.   Then there exist constants $a > 0$ and b such that $U(r) = aU^*(r) + b$ for every $r \epsilon R$.*

Proof   Since $s_0 <^* t_0$, it follows from the assumed property of U and U^* that $U(s_0) < U(t_0)$ and $U^*(s_0) < U^*(t_0)$.   Therefore, it is possible to find a linear transformation of the desired type which makes the two functions U and U^* equal at both s_0 and t_0.   Therefore, it is sufficient to prove that if $U(s_0) = U^*(s_0)$ and $U(t_0) = U^*(t_0)$, then $U(r) = U^*(r)$ for every $r \epsilon R$.

Consider first any reward $r \epsilon R$ such that $s_0 \lesssim^* r \lesssim^* t_0$.   Then there must exist a number γ $(0 \leq \gamma \leq 1)$ such that $r \sim^* \gamma t_0 + (1 - \gamma)s_0$. Hence, from the assumed properties of U and U^*, we can obtain the following equation:

$$\begin{aligned}
U(r) &= E[U|\gamma t_0 + (1 - \gamma)s_0] = \gamma U(t_0) + (1 - \gamma)U(s_0) \\
&= \gamma U^*(t_0) + (1 - \gamma)U^*(s_0) = E[U^*|\gamma t_0 + (1 - \gamma)s_0] \\
&= U^*(r).
\end{aligned}$$

Next, consider any reward $r \epsilon R$ such that $r \succ^* t_0$. Then there must exist a number γ $(0 \leq \gamma \leq 1)$ such that $t_0 \sim^* \gamma r + (1 - \gamma)s_0$. A similar computation again yields $U(r) = U^*(r)$.

Finally, if $r \prec^* s_0$, the result $U(r) = U^*(r)$ can be obtained in a similar way.■

Bounded Utility Functions

In many axiomatic developments of the theory of utility (see the references at the end of Sec. 7.10), assumptions are made which are stronger in certain respects than those which have been made here. These strengthened assumptions make it possible to conclude that the utility function U must be a bounded function on the set R.

In particular, in such a development it is assumed that the relation $\precsim^*$ specifies a complete ordering of *all* the distributions in $\mathcal{P}$. Since the utility function U is bounded, the expectation $E(U|P)$ will be finite for *every* distribution $P \epsilon \mathcal{P}$. Hence, for *any* two distributions $P_1 \epsilon \mathcal{P}$ and $P_2 \epsilon \mathcal{P}$, it will be true that $P_1 \precsim^* P_2$ if, and only if, $E(U|P_1) \leq E(U|P_2)$.

A development which necessarily leads to a bounded utility function has not been presented in this chapter because in many statistical problems it is common practice, and it is extremely convenient, to work with a utility function which is not bounded. For example, in a large class of problems, several of which will be treated later in this book, the set R can be chosen to be the real line and the utility function is chosen to be either a linear function or a quadratic function.

Thus far in this chapter, the properties of the utility function U have been derived only for the distributions in the class $\mathcal{P}_B$. However, it is often necessary to compute the expected utility $E(U|P)$ for a distribution P which is not bounded. Hence, in the next section, the fundamental property Π of the function U will be extended to the class $\mathcal{P}_E$ of all distributions P such that $E(U|P)$ is finite.

It is very important to note that if the set R contains both a most preferred reward and a least preferred reward, then every distribution in $\mathcal{P}$ is bounded and the development which will be given in the next section is unnecessary. Even if the actual set R does not contain a most preferred or a least preferred reward, it is sometimes possible to adjoin to R hypothetical rewards which will serve the purpose. The difficulty lies in being able to adjoin such rewards without invalidating any of the assumptions which have been made. In order to be certain that none of these assumptions will be violated, we must be able to assign a finite utility to both the most preferred reward and the least preferred reward. In other words, to satisfy this requirement, the utility function U must be a bounded function on the set R. However, as we have seen, the function

U may be unbounded.   In such a problem, the extension to unbounded distributions which will be developed in Sec. 7.10 is needed.

7.10 EXTENSION OF THE PROPERTIES OF A UTILITY FUNCTION TO THE CLASS $\mathcal{P}_E$

In this section we shall extend the property Π of the function U described in Theorem 1 of Sec. 7.9 to the class $\mathcal{P}_E$ of all distributions P such that $E(U|P)$ is finite.   In order to do this, the assumptions which have been made about the relation $\precsim$ * for distributions in the class $\mathcal{P}_B$ must now be assumed to be true for distributions in the larger class $\mathcal{P}_E$.   In particular, it will now be assumed that on the basis of his preferences among the distributions in $\mathcal{P}_E$, the statistician can specify a complete ordering of the class $\mathcal{P}_E$.   In other words, it is assumed that for all distributions in the class $\mathcal{P}_E$, the relation $\precsim$ * satisfies properties 1 and 2 given at the end of Sec. 7.2.   Furthermore, it will now be assumed that Assumptions U_1 and U_2 of Sec. 7.7 are satisfied for all distributions in the class $\mathcal{P}_E$.

We shall also need two more assumptions.   By analogy with real numbers, we can define, for any reward $r_0 \epsilon R$, the following subsets of R:

$$(-\infty, r_0] = \{r: r \precsim {}^* r_0\} \quad \text{and} \quad [r_0, \infty) = \{r: r \succsim {}^* r_0\}. \quad (1)$$

It follows from Assumption U_3 that for any reward $r \epsilon R$, both the set $(-\infty, r]$ and the set $[r, \infty)$ will belong to the σ-field $\mathcal{C}$.   Since no reward in the set $(-\infty, r_0]$ can be preferred to any reward in the set $[r_0, \infty)$, the following assumption appears plausible.

Assumption U_5 *Suppose that P and Q are any distributions in the class $\mathcal{P}_E$ such that for some reward $r \epsilon R$, $P\{(-\infty, r]\} = Q\{[r, \infty)\} = 1$.   Then $P \precsim {}^* Q$.*

The statement of Assumption U_5 can actually be proved for any bounded distributions P and Q by using Assumption U_4 and Lemma 3 of Sec. 7.7.   Here we are assuming that it is also true for any unbounded distributions in $\mathcal{P}_E$.

For any distribution $P \epsilon \mathcal{P}_E$ and any set $B \epsilon \mathcal{C}$ such that $P(B) > 0$, we shall let P_B denote the conditional distribution obtained by restricting P to the set B.   In other words, for any set $A \epsilon \mathcal{C}$,

$$P_B(A) = \frac{P(AB)}{P(B)}. \quad (2)$$

Every such distribution P_B will also belong to the class $\mathcal{P}_E$.

Throughout the remaining part of this section, we shall let $s_1 \succsim {}^* s_2 \succsim {}^* \cdots$ and $t_1 \precsim {}^* t_2 \precsim {}^* \cdots$ be two sequences of rewards with the following property: For any reward $r \in R$, there is a reward s_m in the first sequence such that $s_m \lesssim^* r$ and there is a reward t_n in the second sequence such that $r \lesssim^* t_n$. Sequences having this property are called *cofinal with R*, and we assume here that such sequences exist. It should be noted, however, that there are some completely ordered sets which do not have any cofinal sequences [see, e.g., Wilder (1965), p. 134]. The standard examples are certain sets of ordinal numbers.

For any rewards s_m and t_n in the cofinal sequences defined above, consider the sets $[s_m, \infty)$ and $(-\infty, t_n]$. If P is any distribution in $\mathcal{P}_E$ such that $P\{[s_m, \infty)\} > 0$ and $P\{(-\infty, t_n]\} > 0$, let $P_{[s_m, \infty)}$ and $P_{(-\infty, t_n]}$ be the conditional distributions defined by Eq. (2). These distributions will be used repeatedly in the development to be given here. For convenience, we shall let P_m denote the distribution $P_{[s_m, \infty)}$ and P^n denote the distribution $P_{(-\infty, t_n]}$.

Lemma 1 *For any distribution $P \in \mathcal{P}_E$,*

$$E(U|P) = \lim_{m \to \infty} E(U|P_m) = \lim_{n \to \infty} E(U|P^n). \tag{3}$$

Proof Let $A_m = [s_m, \infty)$ for $m = 1, 2, \ldots$. Then it follows from the properties of the cofinal sequences that $A_1 \subset A_2 \subset \cdots$ and that $\bigcup_{m=1}^{\infty} A_m = R$. Hence, $\lim_{m \to \infty} P(A_m) = 1$, and it follows that

$$\lim_{m \to \infty} E(U|P_m) = \lim_{m \to \infty} \left[\int_{A_m} \frac{U(r) \, dP(r)}{P(A_m)} \right]$$

$$= \lim_{m \to \infty} \int_{A_m} U(r) \, dP(r) = \int_R U(r) \, dP(r) = E(U|P).$$

The proof for the distributions P^n is similar.■

Lemma 2 *Let P be any distribution in $\mathcal{P}_E$, and let m and n be any integers such that $P\{[s_m, \infty)\} > 0$ and $P\{(-\infty, t_n]\} > 0$. Also, let m_1 and n_1 be any integers such that $m < m_1$ and $n < n_1$. Then*

$$P^n \lesssim^* P^{n_1} \lesssim^* P \lesssim^* P_{m_1} \lesssim^* P_m. \tag{4}$$

Proof We shall prove first that $P \lesssim^* P_m$. Let $A = [s_m, \infty)$, and let $B = A^c$. If $P(B) = 0$, then $P_m = P$. Suppose therefore that $P(B) > 0$. Then the distribution P can be represented as the following mixture:

$$P = [P(A)]P_m + [P(B)]P_B. \tag{5}$$

Since $P_B\{(-\infty, s_m]\} = P_m\{[s_m, \infty)\} = 1$, it follows from Assumption U_5 that $P_B \lesssim^* P_m$. Therefore, $P \lesssim^* P_m$, by Lemma 1 of Sec. 7.7.

When m is replaced by m_1, the proof just given shows that $P \lesssim^* P_{m_1}$. When P is replaced by P_{m_1}, a similar proof shows that $P_{m_1} \lesssim^* P_m$.

The proofs of those preferences in the relation (4) which involve n and n_1 are also similar.■

We shall now present the final assumption, which guarantees continuity in the preference relationship when unbounded distributions are approximated by bounded ones.

Assumption U_6   *Let P and Q be any two distributions in $\mathcal{P}_E$. If there exists an integer m_0 such that $P_m \gtrsim^* Q$ for every value of m ($m \geq m_0$), then $P \gtrsim^* Q$. Also, if there exists an integer n_0 such that $P^n \lesssim^* Q$ for every value of n ($n \geq n_0$), then $P \lesssim^* Q$.*

In the next lemma, it is established that the function U has the desired property II for at least those distributions in $\mathcal{P}_E$ which are bounded from below but may be unbounded from above.

Lemma 3   *Let P and Q be any two distributions in $\mathcal{P}_E$ for which there exist rewards $r_1 \, \epsilon \, R$ and $r_2 \, \epsilon \, R$ such that*

$$P\{[r_1, \infty)\} = Q\{[r_2, \infty)\} = 1. \tag{6}$$

Then $P \lesssim^* Q$ if, and only if, $E(U|P) \leq E(U|Q)$.

Proof   The distributions P^n and Q^n will be well defined for sufficiently large values of n. Hence, by omitting a finite number of elements of the sequence $\{t_n\}$ and reindexing the remaining elements of the sequence, we can assume without loss of generality that these distributions exist for all values of n ($n = 1, 2, \ldots$).

For $n = 1, 2, \ldots$, let $a_n = E(U|P^n)$ and let $b_n = E(U|Q^n)$. It follows from Eq. (6) that all the distributions P^n and Q^n are bounded. Hence, from Lemma 2 and from the known properties of U for bounded distributions, it follows that $a_1 \leq a_2 \leq \cdots$ and $b_1 \leq b_2 \leq \cdots$.

Now suppose that $P <^* Q$. By Assumption U_6, there must exist a positive integer n_0 such that $P <^* Q^{n_0}$. Also, by Lemma 2, $P^{n_0} \lesssim^* P$. Therefore, by Assumption U_2, there must exist a number α ($0 < \alpha < 1$) such that

$$P <^* \alpha Q^{n_0} + (1 - \alpha)P^{n_0}.$$

It now follows from Lemma 2 that, for $n = 1, 2, \ldots$,

$$P^n <^* \alpha Q^{n_0} + (1 - \alpha)P^{n_0}. \tag{7}$$

Since the distributions in the relation (7) are bounded, it follows from the known properties of the function U that

$$a_n < \alpha b_{n_0} + (1 - \alpha)a_{n_0} \qquad n = 1, 2, \ldots. \tag{8}$$

If we now take the limit in the relation (8) as $n \to \infty$, it follows from Lemma 1 that

$$E(U|P) \leq \alpha b_{n_0} + (1 - \alpha)a_{n_0}. \tag{9}$$

Moreover, since $P^{n_0} \prec^* Q^{n_0}$ and both of these distributions are bounded, then $a_{n_0} < b_{n_0}$. Hence,

$$\alpha b_{n_0} + (1 - \alpha)a_{n_0} < b_{n_0}. \tag{10}$$

Since the sequence $\{b_n\}$ is nondecreasing, it follows from Lemma 1 that

$$\alpha b_{n_0} + (1 - \alpha)a_{n_0} < E(U|Q). \tag{11}$$

From the relations (9) and (11), we may now conclude that $E(U|P) < E(U|Q)$.

To complete the proof of this lemma, it must now be shown that if $E(U|P) < E(U|Q)$, then $P \prec^* Q$. Accordingly, suppose that $E(U|P) < E(U|Q)$. Then there is a positive integer n_0 such that $E(U|P) < b_{n_0}$ and, hence, such that $a_{n_0} \leq E(U|P) < b_{n_0}$. It follows that there exists a number α ($0 < \alpha < 1$) such that for $n = 1, 2, \ldots$,

$$a_n \leq E(U|P) < \alpha b_{n_0} + (1 - \alpha)a_{n_0} < b_{n_0}. \tag{12}$$

Because all the distributions P^n and Q^n are bounded, it follows from relation (12) and from the known properties of the function U for bounded distributions that for $n = 1, 2, \ldots$,

$$P^n \prec^* \alpha Q^{n_0} + (1 - \alpha)P^{n_0} \prec^* Q^{n_0}. \tag{13}$$

By Assumption U_6 and Lemma 2, we can now obtain the following relation:

$$P \precsim^* \alpha Q^{n_0} + (1 - \alpha)P^{n_0} \prec^* Q^{n_0} \precsim^* Q. \tag{14}$$

Hence, $P \prec^* Q$. ∎

The final step in the demonstration that U is a utility function is to extend the property II to all distributions in the class $\mathcal{P}_E$. This will be done in the next theorem.

Theorem 1 *Let P and Q be any two distributions in the class $\mathcal{P}_E$. Then $P \precsim^* Q$ if, and only if, $E(U|P) \leq E(U|Q)$.*

Proof The distributions P_n and Q_n will be well defined for sufficiently large values of n. Hence, as in Lemma 3, we can assume without loss of generality that these distributions exist for all values of n ($n = 1, 2, \ldots$).

For $n = 1, 2, \ldots$, let $c_n = E(U|P_n)$ and $d_n = E(U|Q_n)$. Then $c_1 \geq c_2 \geq \cdots$ and $d_1 \geq d_2 \geq \cdots$. Furthermore, since all the distributions P_n and Q_n are bounded from below, they are subject to the conclusions of Lemma 3.

Now suppose that $P \prec^* Q$. By Assumption U_6, there must exist a positive integer n_0 such that $P_{n_0} \prec^* Q$. Also, by Lemma 2, $Q \precsim^* Q_{n_0}$. Therefore, by Assumption U_2 and Lemma 2, there exists a number α $(0 < \alpha < 1)$ such that for $n = 1, 2, \ldots,$

$$P_{n_0} \prec^* \alpha Q_{n_0} + (1 - \alpha) P_{n_0} \precsim^* Q \precsim^* Q_n. \tag{15}$$

Hence, by Lemma 3, for $n = 1, 2, \ldots,$

$$c_{n_0} < \alpha d_{n_0} + (1 - \alpha) c_{n_0} \leq d_n. \tag{16}$$

It now follows from Lemmas 1 and 2 that

$$E(U|P) \leq c_{n_0} < \alpha d_{n_0} + (1 - \alpha) c_{n_0} \leq E(U|Q). \tag{17}$$

The relation (17) establishes that $E(U|P) < E(U|Q)$.

Conversely, suppose that $E(U|P) < E(U|Q)$. Then there is a positive integer n_0 such that $c_{n_0} < E(U|Q) \leq d_{n_0}$. Hence, there must exist a number α $(0 < \alpha < 1)$ such that for $n = 1, 2, \ldots,$

$$c_{n_0} < \alpha d_{n_0} + (1 - \alpha) c_{n_0} < E(U|Q) \leq d_n. \tag{18}$$

It follows from Lemma 3 that for $n = 1, 2, \ldots,$

$$P_{n_0} \prec^* \alpha Q_{n_0} + (1 - \alpha) P_{n_0} \precsim^* Q_n. \tag{19}$$

By Assumption U_6 and Lemma 2, we can now obtain the following relation:

$$P \precsim^* P_{n_0} \prec^* \alpha Q_{n_0} + (1 - \alpha) P_{n_0} \precsim^* Q. \tag{20}$$

Hence, $P \prec^* Q.\blacksquare$

It was established in Theorem 2 of Sec 7.9 that the function U which has been constructed in this chapter and increasing linear transformations of U are the only functions which have the property described in Theorem 1.

Further Remarks and References

It should be remarked that no conditions have been imposed on the distributions in the class $\mathcal{P}$ which do not belong to $\mathcal{P}_E$. For some distributions $P \in \mathcal{P}$, the value of $E(U|P)$ can be regarded as $+\infty$. For such a distribution P, it is natural to assume that $P \succsim^* Q$ for any distribution $Q \in \mathcal{P}_E$, but it is not obvious how two such distributions should be compared with each other. Similar remarks apply to those distributions $P \in \mathcal{P}$ such that the value of $E(U|P)$ can be regarded as $-\infty$. However, there may still exist distributions $P \epsilon \mathcal{P}$ such that $E(U|P)$ does not exist, either as a finite or as an infinite number. Although it could be assumed that the relation $\lesssim^*$ specifies a complete ordering of all distributions in $\mathcal{P}$, and further assumptions could be introduced which would assign a position in this ordering to each distribution outside of $\mathcal{P}_E$, it does not appear worthwhile to do so, and this subject will not be explored further. It should be emphasized that when the function U is bounded, then the classes $\mathcal{P}$, $\mathcal{P}_E$, and $\mathcal{P}_B$ are identical and the above questions do not arise.

The first axiomatic development of utility was given by Von Neumann and Morgenstern (1947). Other developments are due to Marschak (1950), Herstein and Milnor (1953), and Debreu (1960). Derivations are also given in the texts of Blackwell and Girshick (1954) and Chernoff and Moses (1959). Other interesting discussions and developments are given by Fishburn (1964, 1967a), Luce and Raiffa (1957), Luce and Suppes (1965), Pratt, Raiffa, and Schlaifer (1964), and Savage (1954). An indication of research in the theory of utility which emphasizes various aspects that have not been considered here is given in several of the papers in the collection edited by Thrall, Coombs, and Davis (1954) and in the papers of Aumann (1964), Koopmans, Diamond, and Williamson (1964), and Radner (1964).

Fishburn (1968) has published a bibliography on utility theory containing 315 items.

EXERCISES

1. Consider the possibility of your winning or losing any amount of money up to \$1,000. By using a technique of the type described in Sec. 7.5, determine several different points on your utility function and sketch that function as a curve over the interval from $-1,000$ to $+1,000$.

2. Consider an experiment in which a person first specifies a number of dollars x and then observes the value of a random variable Y. Suppose that if $Y \geq x$, he receives Y dollars as his reward. If $Y < x$, he receives a random reward X having some given probability distribution. It is assumed that the random variables X and Y are independent. Show that in order to maximize the expected utility of his reward, the person should specify a number x such that $U(x) = E[U(X)]$.

3. Suppose that two people, J and K, desire to make a bet. Person J will pay \$1 to person K if a specific event A occurs, and person K will pay x dollars to person J if the event A does not occur. Suppose that both J and K agree that $P(A) = p(0 < p < 1)$ and that the utility functions U_J and U_K of both persons are strictly increasing functions of monetary gain. Show that there is an amount x which is mutually agreeable to both J and K if, and only if,

$$U_J^{-1} \left[\frac{U_J(0) - pU_J(-1)}{1 - p} \right] < -U_K^{-1} \left[\frac{U_K(0) - pU_K(1)}{1 - p} \right].$$

Show also that when this relation is satisfied, any value of x which is between these two numbers will result in a mutually satisfactory bet.

4. Let $R = \{r_1, r_2, \ldots \}$ be a countable set of rewards, and let U be a utility function on R. Let $P_1, P_2, \ldots$ be a sequence of probability distributions on

R. For each distribution P_i in this sequence and each reward $r_j \in R$, let $p_{ij} = P_i\{$receiving $r_j\}$. Also, let $\alpha_1, \alpha_2, \ldots$ be a sequence of numbers such that $\alpha_i \geq 0$ for $i = 1, 2, \ldots$ and $\Sigma_{i=1}^{\infty}\alpha_i = 1$, and let

$$p_j = \sum_{i=1}^{\infty} \alpha_i p_{ij} \qquad j = 1, 2, \ldots$$

Finally, let $\tilde{P}$ denote the probability distribution on R such that $p_j = \tilde{P}\{$receiving $r_j\}$ for $j = 1, 2, \ldots$.

Verify that $\tilde{P}$ is a probability distribution. Also, show that if U is a bounded function, then the distribution $\tilde{P}$ cannot be preferred to each of the distributions P_i ($i = 1, 2, \ldots$). Give a simple example to show that if U is not bounded, then it might be true that $E(U|P_i)$ is finite for $i = 1, 2, \ldots$ but $E(U|\tilde{P})$ is infinite.

5. For any real numbers r_1 and r_2 and for any probability p ($0 \leq p \leq 1$), let $\langle r_1, p; r_2, 1 - p \rangle$ denote a lottery which yields a reward of r_1 dollars with probability p and a reward of r_2 dollars with probability $1 - p$. Without referring to your utility function as sketched in Exercise 1, decide which lottery you prefer in each of the following pairs:

   (a) $\langle 250, \frac{1}{4}; 0, \frac{3}{4} \rangle$ or $\langle 40, \frac{1}{2}; 70, \frac{1}{2} \rangle$;
   (b) $\langle 400, \frac{1}{2}; -100, \frac{1}{2} \rangle$ or $\langle 150, \frac{2}{3}; 0, \frac{1}{3} \rangle$;
   (c) $\langle 1,000, \frac{1}{2}; -1,000, \frac{1}{2} \rangle$ or $\langle 50, \frac{1}{2}; -50, \frac{1}{2} \rangle$.

Now find which lottery would be preferred in parts a to c, as determined by the sketch of your utility function made in Exercise 1. If these preferences do not agree with the decisions you have just made, revise your sketch.

6. Suppose that g is a real-valued function over some interval of the real line and that g can be differentiated twice throughout the interval. Prove that g is concave if, and only if, $g''(x) \leq 0$ at every point x in the interval.

7. Consider two boxes each of which contains both red balls and green balls. It is known that one-half the balls in box 1 are red and the other half are green. In box 2, the proportion X of red balls is not known with certainty, but this proportion has a probability distribution over the interval $0 \leq X \leq 1$.

   (a) Suppose that a person is to select one ball at random from either box 1 or box 2. If that ball is red, he wins \$1; if it is green, he wins nothing. Show that under any utility function which is an increasing function of monetary gain, the person should prefer selecting the ball from box 1 if, and only if, $E(X) < \frac{1}{2}$.

   (b) Suppose that a person can select n balls ($n \geq 2$) at random from either of the boxes but that all n balls must be selected from the same box; suppose that each selected ball will be put back in the box before the next ball is selected; and suppose that he will receive \$1 for each red ball selected and nothing for each green ball. Also, suppose that his utility function U of monetary gain is strictly concave over the interval $[0, n]$, and suppose that $E(X) = \frac{1}{2}$. Show that the person should prefer to select the balls from box 1.

   Hint: Show that if the balls are selected from box 2, then for any given value $X = x$, $E(U|x)$ is a concave function of x on the interval $0 \leq x \leq 1$. This can be done by showing that

$$\frac{d^2}{dx^2} E(U|x)$$

$$= n(n - 1) \sum_{i=0}^{n-2} [U(i) - 2U(i + 1) + U(i + 2)] \binom{n - 2}{i} x^i (1 - x)^{n-2-i} < 0.$$

Then apply Jensen's inequality to $E(U|X)$.

(c) Consider again the problem in part (b), with the same utility function and the same distribution of X. Suppose now, however, that instead of receiving \$1 for each red ball and nothing for each green ball, the person will receive \$1 for each green ball and nothing for each red ball. Show that he should still prefer to select the balls from box 1.

8. Let p and m be fixed numbers such that $0 < p < 1$ and $m > 0$, and consider the following situation: A person is given a stake of m dollars which he can allocate between an event A of probability p and its complement A^c. He keeps as his reward the amount which he has allocated to either A or A^c, whichever actually occurs. In other words, for all values of x in the interval $0 \leq x \leq m$, he can choose among all lotteries of the form $\langle x, p; m - x, 1 - p \rangle$, as defined in Exercise 5. Being careful in each case to consider every possible pair of values of p and m, find the preferred allocation of the m dollars when the person's utility function U is defined on the interval $[0, m]$ of monetary gains as follows:

(a) $U(r) = r^\alpha$, where $\alpha > 1$.

(b) $U(r) = r$.

(c) $U(r) = r^\alpha$, where $0 < \alpha < 1$.

(d) $U(r) = \log r$.

(e) $U(r) = 2(r/m_0) - (r/m_0)^2$, where $m_0 \geq m$.

9. Consider a roulette wheel which is partitioned into k disjoint events A_1, ..., A_k such that $P(A_i) = p_i$, for $i = 1, \ldots, k$, and $\Sigma_{i=1}^k p_i = 1$. Suppose that a person is given a stake of m dollars which he can allocate among the k events A_1, ..., A_k and that he receives as his reward the amount x_i which he has allocated to the event A_i that actually occurs. In other words, he can choose among all lottery tickets which yield a reward x_i with probability p_i ($i = 1, \ldots, k$), subject to the restriction that $x_i \geq 0$ for $i = 1, \ldots, k$ and $\Sigma_{i=1}^k x_i = m$. Suppose also that the person's utility function U, as a function of positive monetary rewards r, is defined as $U(r) = \log r$. Show that his preferred allocation is given by $x_i = m p_i$ for $i = 1, \ldots, k$. (Note that this exercise is a generalization of Exercise 8d.)

10. Suppose that a person's current fortune is x_0 dollars and that he has an opportunity to purchase a lottery ticket which yields either a reward of r dollars ($r > 0$) or a reward of 0 dollars with equal probabilities. Suppose also that for any value of x ($x > 0$), his utility $U(x)$ of having a fortune of x dollars is $U(x) = \log x$. Show that he should be willing to purchase the lottery ticket for any amount b such that

$$b < x_0 + \tfrac{1}{2}[r - (r^2 + 4x_0^2)^{\frac{1}{2}}].$$

11. Suppose that Mr. A and Mr. B are equally wealthy, each having a current fortune of x_0 dollars. Suppose also that both men have the same utility function U and that for any number x, their utility of having a fortune of x dollars is $U(x) = (x - x_0)^{\frac{1}{2}}$.

Suppose now that one of the two men receives, as a gift, a lottery ticket which yields either a reward of r dollars ($r > 0$) or a reward of 0 dollars with equal probabilities. Show that there exists a number $b > 0$ having the following property: Regardless of which man receives the lottery ticket, he can sell it to the other man for b dollars and the sale will be advantageous to both men.

12. Let R be an abstract set of rewards on which a utility function U is defined which has the property stated in Theorem 1 of Sec. 7.10. For any reward $r \in R$, let the set $[r, \infty)$ be defined by Eq. (1) of Sec. 7.10. Suppose that P and Q

are distributions in the class $\mathcal{P}_R$ such that for every reward $r \in R$,

$$P\{[r, \infty)\} \leq Q\{[r, \infty)\}.$$

Show that $P \precsim^* Q$.

This exercise can be rephrased as follows: Let X and Y be random variables with d.f.'s F and G, respectively. Suppose that both $E(X)$ and $E(Y)$ exist, and suppose that $F(x) \leq G(x)$ for every number x. Show that $E(X) \geq E(Y)$. [See Lehmann (1959), p. 73.]

13. Suppose that a person is going to sell Fizzy Cola at a football game and must decide in advance how much to order. Suppose that he makes a gain of m cents on each quart that he sells at the game but suffers a loss of c cents on each quart that he has ordered but does not sell. If it is assumed that the demand for Fizzy Cola at the game, as measured in quarts, is an absolutely continuous random variable X with p.d.f. f and d.f. F, show that his expected profit will be maximized if he orders an amount α such that $F(\alpha) = m/(m + c)$.

statistical decision
problems

decision problems

8.1 ELEMENTS OF A DECISION PROBLEM

Problems of the type considered in the preceding chapter are examples of decision problems in which the statistician must choose a most preferred distribution from a given class of probability distributions on a set R of rewards. In this chapter we shall specify in more detail the structure of a broad class of decision problems of this type.

Consider a particular experiment whose possible outcomes w belong to a space Ω. Suppose that the statistician, without knowing the outcome of the experiment, must make a decision the consequences of which will depend on the outcome of the experiment. Let D be the space of all possible decisions d which might be made by the statistician, and let R be the space of all possible rewards r which might be received as a result of the statistician's decision d and the outcome w of the experiment. Specifically, we shall denote by $\sigma(w, d)$ the reward in R that would be received if the statistician makes the decision d and the outcome of the experiment is w.

We shall suppose that all the assumptions in Chap. 6 are satisfied. Hence, we shall assume that there exists a probability distribution P on the space Ω of outcomes whose value $P(A)$ is specified for each event A

belonging to an appropriate σ-field $\mathcal{Q}$ of subsets of Ω. We shall let W denote the unknown outcome. Then, for any event $A \epsilon \mathcal{Q}$, it may be said that $\Pr(W \epsilon A) = P(A)$.

Furthermore, we shall suppose that the statistician's preferences among the rewards in R satisfy all the assumptions in Chap. 7. Hence, we shall assume that there exists a utility function U on the set R. It is assumed that U is a measurable function with respect to an appropriate σ-field $\mathcal{B}$ of subsets of R.

For each fixed decision $d \epsilon D$, the function σ induces a probability distribution P_d on the set R of rewards. For any subset $B \epsilon \mathcal{B}$, the value $P_d(B)$ is specified as follows:

$$P_d(B) = \Pr[\sigma(W, d) \epsilon B] = P\{w: \sigma(w, d) \epsilon B\}. \tag{1}$$

In order that the distribution P_d defined by Eq. (1) will be meaningful, the following requirement must be met. For each subset $B \epsilon \mathcal{B}$, the set $\{w: \sigma(w, d) \epsilon B\}$ must belong to the σ-field $\mathcal{Q}$. We shall assume that this requirement is met for every decision $d \epsilon D$. Then, for each probability distribution P_d for which the utility function U can be integrated, the expected utility $E(U|P_d)$ can be computed as follows:

$$E(U|P_d) = \int_R U(\dot{r}) \, dP_d(r) = \int_\Omega U[\sigma(w, d)] \, dP(w). \tag{2}$$

The statistician should choose, if possible, a decision d which maximizes $E(U|P_d)$.

In a context like the present one, where a decision must be made without knowledge of the outcome W of some experiment, W is called a *parameter* and the set Ω of all possible values of W is called the *parameter space*. Furthermore, in these decision problems, it has become standard to specify for each reward $r \epsilon R$ the negative of its utility, rather than its utility, and to call this number the *loss*. Hence, for each outcome $w \epsilon \Omega$ and each decision $d \epsilon D$, the loss $L(w, d)$ is defined by the equation

$$L(w, d) = -U[\sigma(w, d)]. \tag{3}$$

It follows from this discussion that the elements of a decision problem may be regarded as a parameter space Ω, a decision space D, and a real-valued loss function L which is defined on the product space $\Omega \times D$. For any point $(w, d) \epsilon \Omega \times D$, the number $L(w, d)$ represents the loss when the value of the parameter W is w and the statistician chooses decision d. It is assumed that for each $d \epsilon D$, the loss $L(\cdot, d)$ is an $\mathcal{Q}$-measurable function on the space Ω.

Let P be any given probability distribution of the parameter W. For any decision $d \epsilon D$, the expected loss, or *risk*, $\rho(P, d)$ is specified by the equation

$$\rho(P, d) = \int_\Omega L(w, d) \, dP(w). \tag{4}$$

It will be assumed that the integral in Eq. (4) is finite for every $d \, \epsilon \, D$. Any decision d for which this assumption is not true can usually be eliminated from the set D. It follows from Eqs. (2) and (3) that the statistician should choose, if possible, a decision d which minimizes the risk $\rho(P, d)$.

8.2 BAYES RISK AND BAYES DECISIONS

Consider now a decision problem defined by a parameter space Ω, a decision space D, and a loss function L. For any distribution P of the parameter W, the *Bayes risk* $\rho^*(P)$ is defined to be the greatest lower bound for the risks $\rho(P, d)$ for all the decisions $d \, \epsilon \, D$. Hence, $\rho^*(P)$ is specified by the equation

$$\rho^*(P) = \inf_{d \, \epsilon \, D} \rho(P, d). \tag{1}$$

Any decision d^* whose risk is equal to the Bayes risk is called a *Bayes decision against the distribution P*. Hence, a decision d^* is a Bayes decision against the distribution P if, and only if, $\rho(P, d^*) = \rho^*(P)$.

If the distribution of the parameter W is P, any Bayes decision against P will be an optimal decision for the statistician because the risk cannot be smaller for any other decision. It is possible, however, that no decision in the space D is a Bayes decision. This situation occurs when the greatest lower bound in Eq. (1) is not actually attained for any decision $d \, \epsilon \, D$. In such a situation, the statistician should select any decision $d \, \epsilon \, D$ for which the risk $\rho(P, d)$ is sufficiently close to the Bayes risk. Since difficulties of this kind are not central to the theory or practice of decision making, it will typically be assumed in the following discussion that for any distribution P which might be considered, the Bayes risk $\rho^*(P)$ is attained for some decision $d \, \epsilon \, D$.

We shall now discuss three examples.

EXAMPLE 1 Suppose that the parameter space Ω contains just the two numbers 0 and 1 and that the decision space D contains all the numbers d in the interval $0 \leq d \leq 1$. Suppose also that the loss function L is defined, for any values $w \, \epsilon \, \Omega$ and $d \, \epsilon \, D$, by the equation

$$L(w, d) = |w - d|^\alpha. \tag{2}$$

Here, α is a given positive integer. Finally, it is assumed that the probability distribution P of W is such that $\Pr(W = 0) = \frac{3}{4}$ and $\Pr(W = 1) = \frac{1}{4}$.

Consider now the problem in which $\alpha = 1$ in the loss function defined by Eq. (2). Then, for any decision $d \in D$, the risk $\rho(P, d)$ is given by the equation

$$\rho(P, d) = L(0, d) \Pr(W = 0) + L(1, d) \Pr(W = 1)$$
$$= \tfrac{3}{4}d + \tfrac{1}{4}(1 - d) = \tfrac{1}{2}d + \tfrac{1}{4}. \tag{3}$$

It follows that $\rho(P, d)$ is minimized when $d = 0$. Hence, the value $d = 0$ is the unique Bayes decision, and the Bayes risk $\rho^*(P)$ has the value $\frac{1}{4}$.

The following point should be noted. If the decision space D is defined to be the interval $0 < d \leq 1$, where the end point 0 has now been deleted from D, then the Bayes risk $\rho^*(P)$ will still have the value $\frac{1}{4}$ but no decision in D will be a Bayes decision.

EXAMPLE 2 Consider again the problem in Example 1, but now suppose that $\alpha > 1$ in the loss function defined by Eq. (2). Then, for any decision $d \in D$,

$$\rho(P, d) = \tfrac{3}{4}d^\alpha + \tfrac{1}{4}(1 - d)^\alpha. \tag{4}$$

The value of d which minimizes the risk in Eq. (4) may be obtained by elementary differentiation. In this way, the unique Bayes decision d^* is found to have the value

$$d^* = (1 + 3^{1/(\alpha-1)})^{-1}. \tag{5}$$

EXAMPLE 3 Consider a problem in which the parameter is a vector $\mathbf{W} = (W_1, W_2)'$ which takes values in the plane R^2. Suppose that the decision space D contains just two decisions d_1 and d_2, and suppose that for any point $(w_1, w_2)' \in R^2$, the loss function L is defined by the equations

$$L(w_1, w_2, d_1) = w_1^2, \qquad L(w_1, w_2, d_2) = w_2^2. \tag{6}$$

Then, for any joint distribution P of W_1 and W_2, the Bayes decision against P is d_1 if $E(W_1^2) < E(W_2^2)$ and is d_2 if $E(W_1^2) > E(W_2^2)$. If $E(W_1^2) = E(W_2^2)$, then both d_1 and d_2 are Bayes decisions against P.

8.3 NONNEGATIVE LOSS FUNCTIONS

Suppose that the distribution of the parameter W in some decision problem is P. Let a be a given constant ($a > 0$), and let λ be a real-valued function on the parameter space Ω such that the integral $\int_\Omega \lambda(w) \, dP(w)$ is finite. Consider a new loss function L_0, which is defined in terms of the original loss function L by the relation

$$L_0(w, d) = aL(w, d) + \lambda(w) \qquad w \, \epsilon \, \Omega, \, d \, \epsilon \, D. \tag{1}$$

For any decision $d \, \epsilon \, D$, let $\rho(P, d)$ denote the risk which results from the original loss function L, as defined by Eq. (4) of Sec. 8.1, and let $\rho_0(P, d)$ denote the risk which results from the new loss function L_0. Then, for any two decisions $d_1 \, \epsilon \, D$ and $d_2 \, \epsilon \, D$, it follows that $\rho_0(P, d_1) \leq \rho_0(P, d_2)$ if, and only if, $\rho(P, d_1) \leq \rho(P, d_2)$. In particular, a decision d^* is a Bayes decision against P in the original problem with loss function L if, and only if, d^* is a Bayes decision against P in the new problem with loss function L_0.

Now consider the function λ_0 which is defined at each point $w \, \epsilon \, \Omega$ by the equation

$$\lambda_0(w) = \inf_{d \, \epsilon \, D} L(w, d). \tag{2}$$

If the integral of λ_0 satisfies the condition given at the beginning of this section, then we can replace L by a new loss function L_0, which is defined for any values $w \, \epsilon \, \Omega$ and $d \, \epsilon \, D$ by the equation

$$L_0(w, d) = L(w, d) - \lambda_0(w). \tag{3}$$

The loss function L_0 has the following properties for any values $w \, \epsilon \, \Omega$ and $d \, \epsilon \, D$:

$$L_0(w, d) \geq 0 \qquad \text{and} \qquad \inf_{d \, \epsilon \, D} L_0(w, d) = 0. \tag{4}$$

It has been found convenient in many problems to work with nonnegative loss functions of this type, although the use of such functions makes it appear that the statistician must continually choose decisions from which he can never realize a positive gain.

It should be noted that the loss functions in Exercises 1 to 3 at the end of this chapter all satisfy the conditions given in relation (4).

8.4 CONCAVITY OF THE BAYES RISK

We shall now show that in any decision problem, the Bayes risk $\rho^*(P)$ must be a concave function of the distribution P of the parameter W. Here, as in Chap. 7, for any two distributions P_1 and P_2 of W and for any number α such that $0 \leq \alpha \leq 1$, we shall let $\alpha P_1 + (1 - \alpha)P_2$ denote the distribution which assigns the probability $\alpha P_1(A) + (1 - \alpha)P_2(A)$ to each event $A \, \epsilon \, \mathbb{Q}$. Hence, we must prove the following theorem.

Theorem 1 *For any distributions P_1 and P_2 of W and for any number α*

such that $0 \leq \alpha \leq 1$,

$$\rho^*[\alpha P_1 + (1 - \alpha)P_2] \geq \alpha\rho^*(P_1) + (1 - \alpha)\rho^*(P_2). \tag{1}$$

Proof From the definition of the risk ρ given by Eq. (4) of Sec. 8.1, it follows that for any decision $d \, \epsilon \, D$,

$$\rho[\alpha P_1 + (1 - \alpha)P_2, d] = \alpha\rho(P_1, d) + (1 - \alpha)\rho(P_2, d). \tag{2}$$

Therefore,

$$\rho^*[\alpha P_1 + (1 - \alpha)P_2] = \inf_{d \, \epsilon \, D} \rho[\alpha P_1 + (1 - \alpha)P_2, d]$$

$$= \inf_{d \, \epsilon \, D} [\alpha\rho(P_1, d) + (1 - \alpha)\rho(P_2, d)]. \tag{3}$$

Since the greatest lower bound of the sum of two functions can never be smaller than the sum of their individual greatest lower bounds, it follows from Eq. (3) that

$$\rho^*[\alpha P_1 + (1 - \alpha)P_2] \geq \alpha \inf_{d \, \epsilon \, D} \rho(P_1, d) + (1 - \alpha) \inf_{d \, \epsilon \, D} \rho(P_2, d)$$

$$= \alpha\rho^*(P_1) + (1 - \alpha)\rho^*(P_2). \blacksquare \tag{4}$$

It can be seen from Eqs. (2) and (3) that the concave function $\rho^*(P)$ is the greatest lower bound of the family of linear functions $\rho(P, d)$ generated by the different decisions in the space D. These concepts can be illustrated for a problem in which the parameter W can take just two values: $\Omega = \{w_1, w_2\}$. In this problem, any distribution of W is characterized by the number $p = \Pr(W = w_1)$.

In Fig. 8.1, the Bayes risk $\rho^*(p)$ is sketched for a problem in which

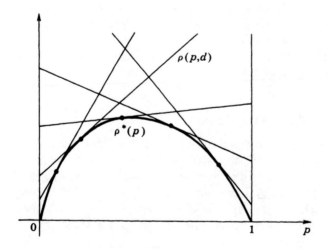

Fig. 8.1 A strictly concave Bayes risk.

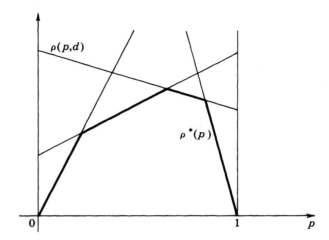

Fig. 8.2 The Bayes risk when D is finite.

the decision space D contains an infinite number of decisions and no decision in D is a Bayes decision against two distinct values of p. In this problem, the Bayes risk is a strictly concave function on the interval $0 \leq p \leq 1$.

In Fig. 8.2, the Bayes risk $\rho^*(p)$ is sketched for a problem in which the decision space D contains only a finite number of decisions. In this problem, the Bayes risk is again a concave function on the interval $0 \leq p \leq 1$ but is not strictly concave.

The Consequences of Using an Inappropriate Distribution of the Parameter

The above discussion and Fig. 8.1 reveal another interesting feature of Bayes decisions and the Bayes risk. Suppose that the statistician chooses a decision d_0 which is a Bayes decision against a certain value p_0 of p, but suppose that the actual value of p is some other number p_1. Let h denote the difference between the risk $\rho(p_1, d_0)$ which he actually incurs and the risk $\rho^*(p_1)$ which he could have attained by choosing a Bayes decision against p_1.

This increment h in the risk is illustrated in Fig. 8.3. The straight line $y = \rho(p, d_0)$ is tangent to the concave curve $y = \rho^*(p)$ at the point $p = p_0$. More precisely, the straight line is a line of support to the curve at that point. Therefore, if the curve $y = \rho^*(p)$ is sufficiently smooth in a small interval around the value $p = p_0$ which includes the value $p = p_1$, then h will be small. This property indicates that the effectiveness of a Bayes decision typically is relatively insensitive to small changes in the probability distribution of the parameter.

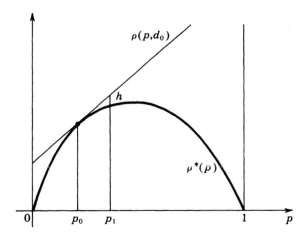

Fig. 8.3   The increment of the Bayes risk.

If the function ρ^* is piecewise linear, as illustrated in Fig. 8.2, then the value of h will actually be 0 whenever both p_0 and p_1 are contained in an interval on which ρ^* is linear.   However, if the function ρ^* has a sharp bend between p_0 and p_1, then the value of h will be relatively large.

These properties can be extended to decision problems which involve a more general parameter space Ω and a greater variety of distributions P of the parameter, but we shall not consider the matter further here.

8.5   RANDOMIZATION AND MIXED DECISIONS

It is often useful (and certainly realistic) to recognize explicitly that in any decision problem, the statistician may wish to choose a decision from D by means of an auxiliary randomization procedure of some sort, such as by tossing a coin.   In other words, the statistician may wish to make a *mixed*, or *randomized*, decision d by assigning probabilities $p_1, p_2, \ldots$ to the elements of a sequence $d_1, d_2, \ldots$ of decisions from D and then selecting one of the decisions d_i on the basis of these probabilities.   If the loss $L(w, d)$ associated with the mixed decision d exists for a given point $w \in \Omega$, then, by the expected-utility hypothesis, its value is specified by the equation

$$L(w, d) = \sum_{i=1}^{\infty} p_i L(w, d_i). \tag{1}$$

For any distribution P of the parameter W and any mixed decision d, the risk $\rho(P, d)$, whenever it exists, can again be defined by Eq. (4) of Sec. 8.1.

In problems where the decision space D has more than a countable number of elements, a mixed decision can be defined more generally by an arbitrary probability distribution on some appropriate σ-field of subsets of D.

Let M denote the set of all mixed decisions in a given problem. In this context, the decisions in D are called *pure* decisions. Each pure decision d can also be regarded as a mixed decision by identifying d with a trivial randomization in which the pure decision d must be selected with probability 1. Hence, with this identification, $D \subset M$. By recognizing that the statistician may make a mixed decision, we have replaced his decision space D by the larger space M. However, as we shall now indicate, this enlargement of the decision space provides no additional reduction of the risk to the statistician, and he need not consider any decision which was not in the original decision space D.

For any distribution P of the parameter W, the statistician should choose, if possible, a decision $d \in M$ which minimizes the risk $\rho(P, d)$. By Eq. (1), the loss function for any mixed decision d is a weighted average of the loss functions for pure decisions d_i ($i = 1, 2, \ldots$). Therefore, whenever the risk $\rho(P, d)$ for the mixed decision exists, its value must be a weighted average of the risks $\rho(P, d_i)$ for pure decisions d_i. It follows that

$$\inf_{d \in M} \rho(P, d) = \inf_{d \in D} \rho(P, d) = \rho^*(P). \tag{2}$$

Equation (2) states that none of the mixed decisions in M can reduce the risk below the minimum value $\rho^*(P)$ which can be attained from the pure decisions in D. Furthermore, Eq. (2) is correct regardless of whether or not the value of the Bayes risk $\rho^*(P)$ is finite and regardless of whether or not this risk is actually attained at some decision $d \in M$. If the Bayes risk $\rho^*(P)$ is finite and is attained for a mixed decision in M, then it follows from the above comments that this risk must also be attained for some pure decision in D.

This discussion supports the intuitive notion that the statistician should not base an important decision on the outcome of the toss of a coin. When two or more pure decisions each yield the Bayes risk, an auxiliary randomization can be used to select one of these Bayes decisions. However, the randomization is irrelevant in this situation since any method of selecting one of the Bayes decisions is acceptable. In any other situation, when the statistician makes use of a randomization procedure, there is a chance that the final decision may not be a Bayes decision.

Nevertheless, randomization has an important use in statistical work. The concepts of selecting a random sample and of assigning different treatments to experimental units at random are basic ones for the performance of effective experimentation. These comments do not really conflict with those in the preceding paragraph which indicate that the statistician need never use randomized decisions, as the following argument indicates:

The statement that the statistician need never use randomized decisions was made within the framework of a specific decision problem having a fixed reward space and loss function. The statement that random sampling is a useful statistical technique recognizes the possibility that the consequences of performing an experiment with a random sample may differ greatly from the consequences of performing the experiment with a sample selected in some nonrandom way. No matter which sampling method is used, the statistician must build a probability model representing the experiment in order to provide a formal basis for any subsequent statistical analysis.

When a random sample is taken, this model will be relatively simple. Consequently, there will seldom be serious disagreement among statisticians in regard to the appropriateness of the model for the experiment and the use of the model in the subsequent statistical analysis. As a result, any aspects of the model and the analysis that are highly subjective can be separated from those aspects on which there is general agreement. For this reason, when the data gathered in a large random sample are analyzed in accordance with a model on which there is general agreement, there will be a tendency to greatly narrow the differences among diverse subjective views.

On the other hand, when a nonrandom sample is taken, the probability model which the statistician must build will normally be highly subjective. Consequently, the subsequent analysis will also be highly subjective, and it may be difficult for the statistician to convince others of the validity of his results. Furthermore, if the statistician has any erroneous opinions, it is possible that they will become strengthened through the use of a subjective model that reflects his prejudices. Neyman (1967) states that "without randomization there is no guarantee that the experimental data will be free from a bias that no test of significance can detect."

We shall now resume the discussion of some specific decision problems, and we have seen that the statistician can always restrict his choice in such problems to the space D of pure decisions. However, it will be shown in Sec. 8.7 that consideration of the space M of all mixed decisions provides further insight into the nature of the optimal pure decisions.

8.6 CONVEX SETS

In the next section, we shall discuss decision problems in which both the parameter space Ω and the decision space D contain a finite number of points. This assumption will permit us to give a simple geometric interpretation of Bayes decisions based on the theory of convex sets in a finite dimensional space R^k ($k \geq 2$).   The relevant features of this theory will be reviewed in this section.   For further details, the reader should refer to any text on convexity, such as Eggleston (1958).

A set G in R^k is *convex* if, whenever two points $\mathbf{x}_1$ and $\mathbf{x}_2$ belong to G, all points of the form $\alpha\mathbf{x}_1 + (1 - \alpha)\mathbf{x}_2$ ($0 < \alpha < 1$) also belong to G.   In other words, the entire line segment joining $\mathbf{x}_1$ and $\mathbf{x}_2$ also belongs to G. The *convex hull* of any given set of points $\mathbf{x}_1, \ldots, \mathbf{x}_m$ in R^k is the smallest convex set in R^k that contains all m points.   The convex hull will consist of all points that can be expressed as linear combinations of the form $\sum_{i=1}^{m} \alpha_i\mathbf{x}_i$, where $\alpha_i \geq 0$ ($i = 1, \ldots, m$) and $\sum_{i=1}^{m}\alpha_i = 1$.

A linear combination $\sum_{i=1}^{m}\alpha_i\mathbf{x}_i$ whose coefficients satisfy these requirements is called a *convex combination* of the points $\mathbf{x}_1, \ldots, \mathbf{x}_m$.

A point $\mathbf{x}$ in a convex set G is an *extreme point* of G if $\mathbf{x}$ does not lie in the interior of any line segment joining two points of G.   Every extreme point of G must be a boundary point of G; but if the boundary of G contains linear segments, not every boundary point will be an extreme point.

If $\mathbf{a} \,\epsilon\, R^k$ is any vector at least one component of which is not 0 and if c is any constant, the set of points $\mathbf{x} \,\epsilon\, R^k$ satisfying the linear equation $\mathbf{a}'\mathbf{x} = c$ is called a *hyperplane*.   In R^2 a hyperplane is a line, and in R^3 it is a plane.   Now suppose that $\mathbf{x}^*$ is a boundary point of the convex set G. Then the hyperplane $\mathbf{a}'\mathbf{x} = c$ is a *supporting hyperplane* to the set G at the point $\mathbf{x}^*$ if $\mathbf{a}'\mathbf{x}^* = c$ and $\mathbf{a}'\mathbf{x} \geq c$ for all points $\mathbf{x} \,\epsilon\, G$; that is, $\mathbf{x}^*$ lies on the hyperplane and all the points in G lie on the same side of the hyperplane.   At every boundary point of a convex set, there exists a supporting hyperplane to the set.

We shall be interested in certain parts of the boundary of a convex set G.   A point $\mathbf{x} = (x_1, \ldots, x_k)'$ in G belongs to the *admissible boundary* of G if there is no point $\mathbf{y} = (y_1, \ldots, y_k)'$ in G such that $y_i \leq x_i$ for $i = 1, \ldots, k$ and $y_i < x_i$ for at least one value of i.

A point $\mathbf{x} = (x_1, \ldots, x_k)'$ in G belongs to the *Bayes boundary* of G if there is no point $\mathbf{y} = (y_1, \ldots, y_k)'$ in G such that $y_i < x_i$ for $i = 1, \ldots, k$.   The Bayes boundary of G must contain the admissible boundary.   It can be shown that if $\mathbf{x}^*$ is a point on the Bayes boundary of G, then there must be a supporting hyperplane $\mathbf{a}'\mathbf{x} = c$ to G at $\mathbf{x}^*$ such that each component of the coefficient vector $\mathbf{a}$ is nonnegative.

These concepts are illustrated in Fig. 8.4 for a convex set in R^2. In that figure, the convex set G is the convex hull of the nine points $\mathbf{x}_1, \ldots, \mathbf{x}_9$.   The seven points $\mathbf{x}_1, \ldots, \mathbf{x}_7$ are the extreme points of the set G.   The admissible boundary of G consists of the line segments from $\mathbf{x}_2$ to $\mathbf{x}_3$ and from $\mathbf{x}_3$ to $\mathbf{x}_4$, their end points being included.   The Bayes boundary also includes the vertical segment from $\mathbf{x}_1$ to $\mathbf{x}_2$ and the horizontal segment from $\mathbf{x}_4$ to $\mathbf{x}_5$, the points $\mathbf{x}_1$ and $\mathbf{x}_5$ being included.   It

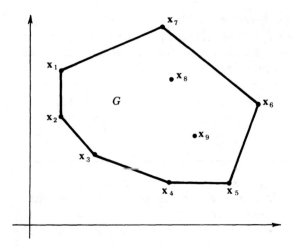

Fig. 8.4 A convex set G.

can be seen that a line of support can be drawn through any point on the Bayes boundary. This line will be vertical at any point between x_1 and x_2; it will be horizontal at any point between x_4 and x_5; and it will have a negative slope at any point between x_2 and x_3 or at any point between x_3 and x_4. In all cases, the line of support has an equation $a_1x_1 + a_2x_2 = c$, where $a_1 \geq 0$ and $a_2 \geq 0$.

8.7 DECISION PROBLEMS IN WHICH Ω AND D ARE FINITE

Consider now a specific decision problem in which the parameter space Ω contains k points $(k \geq 2)$ and the decision space D contains m points $(m \geq 2)$. Hence, $\Omega = \{w_1, \ldots, w_k\}$ and $D = \{d_1, \ldots, d_m\}$. Any mixed decision d is specified by the probabilities $p_1, \ldots, p_m$ of choosing the different decisions in D. These probabilities must satisfy the requirements that $p_j \geq 0$ for $j = 1, \ldots, m$ and $\sum_{j=1}^{m} p_j = 1$. The losses for any such mixed decision are defined by the equation

$$L(w_i, d) = \sum_{j=1}^{m} p_j L(w_i, d_j) \qquad i = 1, \ldots, k. \tag{1}$$

As before, let M denote the space of all mixed decisions. For any decision $d \,\epsilon\, M$, let $\mathbf{y}(d)$ be the k-dimensional vector of losses defined by the equation

$$\mathbf{y}(d) = [L(w_1, d), \ldots, L(w_k, d)]'. \tag{2}$$

Also, let G be the set in R^k containing all such vectors $\mathbf{y}(d)$. It follows from Eq. (1) that the set G will be the convex hull of the m points $\mathbf{y}(d_i)$ which are computed from the pure decisions $d_1, \ldots, d_m$ in D.

A decision $d \in M$ is defined to be *admissible* if there is no other decision $d' \in M$ such that $L(w_i, d') \leq L(w_i, d)$ for $i = 1, \ldots, k$ and $L(w_i, d') < L(w_i, d)$ for at least one value of i. In other words, a decision d is admissible if, and only if, the vector $\mathbf{y}(d)$ belongs to the admissible boundary of the set G.

Now suppose that $\boldsymbol{\xi} = (\xi_1, \ldots, \xi_k)'$ is any given probability distribution of the parameter W, where $\xi_i = \Pr(W = w_i)$ for $i = 1, \ldots, k$. For any decision $d \in M$, the risk $\rho(\boldsymbol{\xi}, d)$ is given by the equation

$$\rho(\boldsymbol{\xi}, d) = \boldsymbol{\xi}'\mathbf{y}(d).$$

Thus, a decision $d^* \in M$ is a Bayes decision against $\boldsymbol{\xi}$ if, and only if, it minimizes the value $\boldsymbol{\xi}'\mathbf{y}(d)$ among all decisions $d \in M$. The next theorem shows that for any Bayes decision d^*, the vector $\mathbf{y}(d^*)$ must belong to the Bayes boundary of the convex set G.

Theorem 1 *Let d^* be any decision in M. Then the vector $\mathbf{y}(d^*)$ belongs to the Bayes boundary of the set G if, and only if, there exists some distribution $\boldsymbol{\xi}$ of the parameter W such that d^* is a Bayes decision against $\boldsymbol{\xi}$.*

Proof Suppose first that d^* is a Bayes decision against some distribution $\boldsymbol{\xi} = (\xi_1, \ldots, \xi_k)'$. Let $\mathbf{y}(d^*) = (y_1^*, \ldots, y_k^*)'$, and for any other decision $d \in M$, let $\mathbf{y}(d) = \mathbf{y} = (y_1, \ldots, y_k)'$. Then $\boldsymbol{\xi}'\mathbf{y}(d^*) \leq \boldsymbol{\xi}'\mathbf{y}$. Since $\xi_i \geq 0$ for $i = 1, \ldots, k$ and $\xi_i > 0$ for at least one value of i, it is not possible that $y_i < y_i^*$ for $i = 1, \ldots, k$. Hence, from the definition, $\mathbf{y}(d^*)$ lies on the Bayes boundary of G.

Conversely, suppose that $\mathbf{y}(d^*)$ lies on the Bayes boundary of the set G. Then there exists a supporting hyperplane $\mathbf{a}'\mathbf{x} = c$ to G at $\mathbf{y}(d^*)$ such that all the components of the coefficient vector $\mathbf{a} = (a_1, \ldots, a_k)'$ are nonnegative. Since $\mathbf{a} \neq \mathbf{0}$, we can divide all these components and the constant c by $\sum_{i=1}^{k} a_i$ without changing the hyperplane. It can therefore be assumed that the equation of the supporting hyperplane to G at the point $\mathbf{y}(d^*)$ is of the form $\mathbf{a}'\mathbf{x} = c$, where the components of $\mathbf{a}$ are nonnegative and their sum is 1. In other words, the vector $\mathbf{a}$ can be regarded as a probability distribution of the parameter W.

It follows from the definition of a supporting hyperplane that $\mathbf{a}'\mathbf{y}(d^*) = c$ and $\mathbf{a}'\mathbf{y} \geq c$ for any vector $\mathbf{y} \in G$. These relations indicate that $\mathbf{y}(d^*)$ minimizes the value $\mathbf{a}'\mathbf{y}$ among all vectors $\mathbf{y} \in G$. Hence, d^* is a Bayes decision against the distribution $\mathbf{a}$. ∎

Since the admissible boundary of G is a subset of the Bayes boundary of G, it follows that every admissible decision in M must be a Bayes decision against some distribution of the parameter W. [These properties are not necessarily true if Ω has an infinite number of points (see Exercise 9).] On the other hand, it is possible that a decision will be Bayes against some distribution $\xi = (\xi_1, \ldots, \xi_k)'$ but will not be admissible. In such a case, at least one of the components of ξ must vanish. If $\xi_i > 0$ for $i = 1, \ldots, k$, then any Bayes decision against ξ will be admissible.

The above results can be given the following geometric interpretation. Suppose that it is desired to find a Bayes decision against a given distribution ξ. For any fixed value of c, the points $\mathbf{x} \, \epsilon \, R^k$ which satisfy the equation $\xi'\mathbf{x} = c$ lie on a hyperplane, and as the value of c is varied, a family of parallel hyperplanes is generated. Let c^* be the smallest value of c such that the hyperplane is in contact with the convex set G. Any point of contact $\mathbf{y}(d^*)$ is the loss vector, as defined by Eq. (2), of a Bayes decision d^* against ξ. Furthermore, the value of the Bayes risk $\rho^*(\xi)$ is c^*.

It should be noted that the contact between the hyperplane $\xi'\mathbf{x} = c^*$ and the set G will be made at a single point unless one of the flat surfaces of the Bayes boundary of G has the same slope as the hyperplane. In every case, at least one point of contact must be an extreme point of G. Hence, we have a geometric verification of the fact that there must always be a pure decision that is Bayes against ξ. When there is more than one point of contact, then there must be at least two pure decisions which are Bayes against ξ.

AN EXAMPLE The properties described above can be illustrated by the following simple example, in which the parameter W can take just two different values. Consider a decision problem in which $\Omega = \{w_1, w_2\}$, $D = \{d_1, \ldots, d_6\}$, and the loss function L is specified by Table 8.1.

Table 8.1

	d_1	d_2	d_3	d_4	d_5	d_6
w_1	10	8	4	2	0	0
w_2	0	1	2	5	6	10

For $i = 1, \ldots, 6$, the points $\mathbf{y}_i = \mathbf{y}(d_i)$ are plotted in Fig. 8.5. The convex hull G of these six points is also sketched in Fig. 8.5. The admissible boundary of G consists of the line segment joining the points $\mathbf{y}_1$ and $\mathbf{y}_3$ and the line segment joining the points $\mathbf{y}_3$ and $\mathbf{y}_5$. Therefore, the only decisions in the set D which are admissible are d_1, d_3, and d_5. The Bayes boundary of G consists of the admissible boundary and the line segment

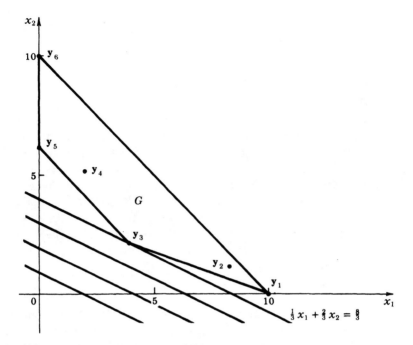

Fig. 8.5 The location of the Bayes decision.

joining the points $\mathbf{y}_5$ and $\mathbf{y}_6$. Hence, although the decision d_6 is not admissible, it is a Bayes decision against the special distribution such that $\Pr(W = w_1) = 1$.

Now suppose that it is desired to find a Bayes decision against the distribution such that $\Pr(W = w_1) = \frac{1}{3}$ and $\Pr(W = w_2) = \frac{2}{3}$. A straight line of the form $\frac{1}{3}x_1 + \frac{2}{3}x_2 = c$, that is, a line whose slope is $-\frac{1}{2}$, must be found which is a line of support at some point on the Bayes boundary of the set G. This line is sketched in Fig. 8.5. Since $\mathbf{y}_3$ is the unique point of contact of the line with the set G, it follows that the decision d_3 is the unique Bayes decision against the specified distribution of W. Furthermore, since $c = \frac{8}{3}$ in the line of support, the Bayes risk for this distribution must also be $\frac{8}{3}$.

Now consider any distribution of W such that $\Pr(W = w_1) = p_1$ and $\Pr(W = w_2) = p_2$. It follows from the above remarks that there are exactly three distributions against which the Bayes decision will not be unique. For each of these distributions, the slope $-p_1/p_2$ of the line of support coincides with the slope of one of the three line segments on the Bayes boundary of G. For example, if $p_1 = p_2 = \frac{1}{2}$, then the line of support makes contact with the set G along the entire line segment between $\mathbf{y}_3$ and $\mathbf{y}_5$. Hence, both d_3 and d_5 are Bayes decisions against this distribution, and any mixed decision which must select either d_3 or d_5 will also be a Bayes decision.

Other Methods of Selecting a Decision

We shall conclude this section by considering briefly other methods of selecting a decision from the set M. Most of the methods that have been proposed are based on the following idea. Selecting a decision d from M is equivalent to selecting a loss vector $\mathbf{y}(d)$ from the convex set G. If the undesirability of each loss vector can be represented by some single number, then the statistician should select a vector such that this measure of undesirability is minimized. For example, a Bayes decision against some distribution ξ can be found by minimizing the value $\xi' \mathbf{y}(d)$.

The best known of the methods which do not require the specification of a distribution ξ is to select a *minimax* decision, i.e., a decision $d \in M$ which minimizes the value

$$\max \{L(w_1, d), \ . \ . \ . \ , L(w_k, d)\}.$$

In many problems, however, the only minimax decision is a mixed decision rather than a pure one. For example, in the problem illustrated by Fig. 8.5, the two-dimensional loss vector $\mathbf{y}(d)$ of the unique minimax decision has components $(3, 3)$. This vector lies on the line segment joining the points $\mathbf{y}_3$ and $\mathbf{y}_5$, and can be represented by the convex combination $\frac{3}{4}\mathbf{y}_3 + \frac{1}{4}\mathbf{y}_5$. Hence, it follows that the unique minimax decision is the mixed decision which selects the decision d_3 with probability $\frac{3}{4}$ and selects the decision d_5 with probability $\frac{1}{4}$.

The theory of minimax decisions was introduced and developed extensively by Von Neumann and Morgenstern (1947). This theory will not be treated any further in this book.

8.8 DECISION PROBLEMS WITH OBSERVATIONS

In the remainder of this chapter we shall consider decision problems in which the statistician, before choosing a decision from the set D, has the opportunity of observing the value of a random variable or random vector X that is related to the parameter W. The observation of X provides the statistician with some information about the value of W which may be helpful to him in choosing a good decision. We shall assume that the conditional distribution of X when $W = w$ can be specified for each possible value $w \in \Omega$. A problem of this type is called a *statistical decision problem*. The essential components of a statistical decision problem are a parameter space Ω, a decision space D, a loss function L, and a family of conditional g.p.d.f.'s $\{f(\ \cdot \ |w), w \in \Omega\}$ of an observation X whose value will be available to the statistician when he chooses a decision.

Let S be the sample space of all possible values of the observation X. Since the decision chosen by the statistician will depend on the observed value of X, the statistician in effect must choose a *decision function* δ which specifies, for each possible value $x \in S$, a decision $\delta(x) \in D$. In many problems, the statistician need not really choose an entire decision function δ. If he waits until an observed value $x \in S$ is obtained, he will only have to consider the problem of choosing an appropriate decision $\delta(x)$. In other problems, the statistician must compare the information which he will gain from the observation of X with the cost of making that observation. In these problems, the statistician must specify his entire decision function δ since the properties of δ will determine whether or not he should pay the price of the observation. In any problem, there is no loss in applicability, and there is some gain in theoretical simplicity, if we assume that the statistician always chooses a decision function δ.

In a given problem, the class of all decision functions δ will be denoted by Δ. As the need arises, we shall have to impose certain measurability conditions on the decision functions which can be considered. We might also have considered those decision functions δ which specify a randomized decision $\delta(x)$ when certain values $x \in S$ are observed. However, for reasons similar to those given in Sec. 8.5, the statistician has no need to make use of randomized decisions in statistical decision problems of the type being studied here, and we shall not consider the use of randomized decisions any further in this book.

For any g.p.d.f. ξ of the parameter W and any decision function $\delta \in \Delta$, the *risk* $\rho(\xi, \delta)$ is defined by the equation

$$\rho(\xi, \delta) = E\{L[W, \delta(X)]\}$$
$$= \int_\Omega \int_S L[w, \delta(x)]f(x|w)\xi(w)\, d\mu(x)\, d\nu(w). \tag{1}$$

It is assumed in Eq. (1) that for each value $w \in \Omega$, the function $L[w, \delta(\cdot)]$ is measurable and integrable over the set S. The differentials $d\mu(x)$ and $d\nu(w)$ indicate that each of the integrals in Eq. (1) may actually be either the integral of a p.d.f. or the sum of the values of a discrete p.f.

The term "risk" is used here, as in Sec. 8.1, to denote expected loss. For any particular decision $d \in D$, $\rho(\xi, d)$ will again denote the risk of the decision d against the g.p.d.f. ξ, as defined by the equation

$$\rho(\xi, d) = \int_\Omega L(w, d)\xi(w)\, d\nu(w). \tag{2}$$

For any particular value $w \in \Omega$, $\rho(w, \delta)$ will denote the risk of the decision function δ when $W = w$, as defined by the equation

$$\rho(w, \delta) = \int_S L[w, \delta(x)]f(x|w)\, d\mu(x). \tag{3}$$

For any decision function $\delta \epsilon \Delta$, the function $\rho(\cdot , \delta)$ whose values are specified by Eq. (3) is called the *risk function* of δ. It follows from Eqs. (1) and (3) that

$$\rho(\xi, \delta) = \int_\Omega \rho(w, \delta) \xi(w) \, d\nu(w). \tag{4}$$

Let $\delta^* \epsilon \Delta$ be a decision function such that

$$\rho(\xi, \delta^*) = \inf_{\delta \epsilon \Delta} \rho(\xi, \delta) = \rho^*(\xi). \tag{5}$$

Then it is said that δ^* is a *Bayes decision function against* ξ, and $\rho^*(\xi)$ is again called the *Bayes risk*. For any given g.p.d.f. ξ of the parameter W, the statistician should choose a decision function δ which is a Bayes decision function against ξ.

Further Remarks and References

The theory of statistical decision functions was initiated and developed by Wald (1950).   This theory followed the development of the theory of games by Von Neumann and Morgenstern (1947).   Some standard texts on statistical decision theory at both introductory and advanced levels are Blackwell and Girshick (1954), Chernoff and Moses (1959), Ferguson (1967), Hadley (1967), Pratt, Raiffa, and Schlaifer (1965), Raiffa and Schlaifer (1961), Schlaifer (1961), and Weiss (1961).

Statistical procedures based on the notion that a distribution can be assigned to any parameter in a statistical decision problem, as we are assuming here, are called *Bayesian statistical methods*.   The scope of these methods has been the subject of much discussion in the statistical literature; see, e.g., the monograph by Savage and others (1962).

The biographical sketches of Bayes by Barnard (1958) and Holland (1962) are of historical interest.

8.9 CONSTRUCTION OF BAYES DECISION FUNCTIONS

Suppose now that for a given g.p.d.f. ξ of the parameter W, it is desired to construct a decision function δ which minimizes the risk $\rho(\xi, \delta)$ as defined by Eq. (1) of Sec. 8.8. We shall assume that the order of integration in the integral in that equation can be reversed.   In particular, if the loss function L is either a nonnegative function or a bounded function, this reversal will be permissible for every g.p.d.f. ξ and every decision function δ.

When the order of integration is reversed, the risk $\rho(\xi, \delta)$ becomes

$$\rho(\xi, \delta) = \int_S \left\{ \int_\Omega L[w, \delta(x)] f(x|w) \xi(w) \, d\nu(w) \right\} d\mu(x). \tag{1}$$

Therefore, a decision function δ that minimizes this risk can be found by minimizing, for each fixed value $x \in S$, the inner integral in Eq. (1). In other words, a Bayes decision function δ^* against ξ can be constructed as follows: For each value $x \in S$, let $\delta^*(x) = d^*$, where d^* is any decision in D which minimizes the integral

$$\int_\Omega L(w, d) f(x|w) \xi(w) \, d\nu(w). \tag{2}$$

This result has an interesting interpretation. For any value $x \in S$, let $f_1(x)$ be defined by the equation

$$f_1(x) = \int_\Omega f(x|w) \xi(w) \, d\nu(w). \tag{3}$$

Since f_1 is the marginal g.p.d.f. of X, the value of $f_1(x)$ can be 0 only on a set of points x which has probability 0. Instead of finding a decision d^* that minimizes the integral (2), the statistician can find equivalently a decision d^* which minimizes the integral

$$\int_\Omega L(w, d) \left[\frac{f(x|w) \xi(w)}{f_1(x)} \right] d\nu(w). \tag{4}$$

Since the quantity inside the brackets in the integral (4) is the conditional g.p.d.f. of W when $X = x$, it follows that the value of the integral is the conditional expectation $E[L(W, d)|x]$. Hence, any minimizing decision d^* is simply a decision which yields the smallest expected loss under the conditional distribution of W when the observed value of X is x. In other words, d^* is a Bayes decision against the conditional distribution of W when $X = x$.

In a statistical decision problem, the marginal distribution of W is called the *prior distribution*, or *a priori distribution*, of W because it is the distribution of W before X has been observed. Also, the conditional distribution of W when the value of X is known is called the *posterior distribution*, or *a posteriori distribution*, of W because it is the distribution of W after X has been observed.

It is helpful to think of a Bayes decision function in the following way. If a decision were to be chosen without any observation, a Bayes decision against the prior distribution of W would be optimal. Now suppose that X is to be observed before a decision is chosen. Then, once X has been observed, the decision problem for the statistician is basically the same as it was in the first case. There is a difference only because the distribution of W has changed from the prior to the posterior distribution. Hence, a Bayes decision against the posterior distribution of W is now optimal.

It is clear from this discussion that the decision $\delta^*(x_0)$ which is specified by a Bayes decision function δ^* when a certain value x_0 is observed can be found without computing the decision $\delta^*(x)$ which should be made when any other value x is observed. Furthermore, a Bayes decision function δ^* against the g.p.d.f. ξ can be found without computing the value of the Bayes risk $\rho^*(\xi)$.

EXAMPLE 1   As an illustration of the results developed thus far in connection with statistical decision problems, we shall consider a problem in which $\Omega = \{w_1, w_2\}$, $D = \{d_1, d_2\}$, and the loss function L is as given in Table 8.2. Suppose that the statistician can observe a random variable

Table 8.2

	d_1	d_2
w_1	0	5
w_2	10	0

X with the following conditional distributions:

$$\Pr(X = 1|W = w_1) = \tfrac{3}{4}, \qquad \Pr(X = 0|W = w_1) = \tfrac{1}{4};$$
$$\Pr(X = 1|W = w_2) = \tfrac{1}{3}, \qquad \Pr(X = 0|W = w_2) = \tfrac{2}{3}. \tag{5}$$

It is required to construct a Bayes decision function if the prior distribution of the parameter W is specified as

$$\Pr(W = w_1) = \xi, \qquad \Pr(W = w_2) = 1 - \xi. \tag{6}$$

Here ξ is a given number $(0 \le \xi \le 1)$.

For $x = 0, 1$, let $\xi(x)$ denote the posterior probability that $W = w_1$ if the value x of X has been observed; that is,

$$\xi(x) = \Pr(W = w_1|X = x). \tag{7}$$

It follows from Eqs. (5) and (6) and Bayes' theorem that

$$\xi(1) = \frac{\tfrac{3}{4}\xi}{\tfrac{3}{4}\xi + \tfrac{1}{3}(1 - \xi)},$$
$$\xi(0) = \frac{\tfrac{1}{4}\xi}{\tfrac{1}{4}\xi + \tfrac{2}{3}(1 - \xi)}. \tag{8}$$

After the value x of X has been observed, a choice must be made between the two decisions d_1 and d_2. Table 8.2 reveals that the risk from d_1 is $10[1 - \xi(x)]$ and the risk from d_2 is $5\xi(x)$. Hence, d_2 is the Bayes decision if $\xi(x) < \tfrac{2}{3}$, d_1 is the Bayes decision if $\xi(x) > \tfrac{2}{3}$, and both d_1 and d_2 are Bayes decisions if $\xi(x) = \tfrac{2}{3}$. By using these numbers with the posterior probabilities in Eq. (8), we obtain the following results:

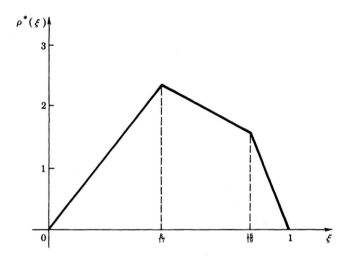

Fig. 8.6 The Bayes risk ρ^* in Example 1.

If the value $X = 1$ is observed, then the Bayes decision function δ^* specifies that $\delta^*(1) = d_2$ whenever $\xi(1) < \frac{2}{3}$ or, equivalently, whenever $\xi < \frac{8}{17}$. It follows that $\delta^*(1) = d_1$ if $\xi > \frac{8}{17}$. Furthermore, if $\xi = \frac{8}{17}$ and $X = 1$, then both decisions d_1 and d_2 are Bayes.

Similarly, if the value $X = 0$ is observed, then $\delta^*(0) = d_2$ whenever $\xi(0) < \frac{2}{3}$ or, equivalently, whenever $\xi < \frac{16}{19}$. Also, $\delta^*(0) = d_1$ if $\xi > \frac{16}{19}$. Finally, if $\xi = \frac{16}{19}$ and $X = 0$, then both decisions d_1 and d_2 are Bayes.

We shall now compute the value of the Bayes risk $\rho^*(\xi)$ for any given prior probability ξ. If $0 \leq \xi \leq \frac{8}{17}$, then it follows from the above remarks that decision d_2 will be a Bayes decision regardless of the observed value of X. Hence, by Table 8.2, $\rho^*(\xi) = 5\xi$.

If $\frac{8}{17} < \xi < \frac{16}{19}$, then $\delta^*(0) = d_2$ and $\delta^*(1) = d_1$. Therefore, by Eq. (5) and Table 8.2,

$$
\begin{aligned}
\rho^*(\xi) &= \xi\rho(w_1, \delta^*) + (1 - \xi)\rho(w_2, \delta^*) \\
&= \xi[(0)(\tfrac{3}{4}) + (5)(\tfrac{1}{4})] + (1 - \xi)[(10)(\tfrac{1}{3}) + (0)(\tfrac{2}{3})] \\
&= \tfrac{5}{4}\xi + \tfrac{10}{3}(1 - \xi).
\end{aligned}
\tag{9}
$$

If $\frac{16}{19} \leq \xi \leq 1$, then decision d_1 will be a Bayes decision regardless of the observed value of X. Hence, by Table 8.2, $\rho^*(\xi) = 10(1 - \xi)$.

The Bayes risk ρ^* is sketched in Fig. 8.6.

Further Remarks and References

The method which has been described here for the construction of a Bayes decision function is called by Raiffa and Schlaifer (1961, chap. 1) the *extensive* form of analysis. The method of selecting a Bayes decision function against a prior distribution ξ by computing the risk $\rho(\xi, \delta)$ for every decision function δ and then finding a decision function for which this risk is minimized is called by Raiffa and Schlaifer the *normal* form of analysis. In statistical decision problems, the extensive form of analysis is the more useful of these two methods.

8.10 THE COST OF OBSERVATION

In many statistical decision problems, a sampling cost must be associated with the observation of a random variable X, and this cost must be considered when the statistician evaluates the risk of any decision function which makes use of the observed value of X. This cost is particularly important when the statistician has to decide which one of several random variables should be observed, or when he has to decide either to make an observation or to choose a decision without making any observation at all. Let $c(w, x)$ denote the cost of observing the value x of X when $W = w$. Then, if ξ is the g.p.d.f. of W, the expected cost of observation is

$$E[c(W, X)] = \int_\Omega \int_S c(w, x) f(x|w) \xi(w) \, d\mu(x) \, d\nu(w). \tag{1}$$

We assume here that the expected-utility hypothesis is applicable to the cost $c(W, X)$. In other words, we assume that this cost is expressed in the appropriate units of negative utility so that, as with other random costs, the expectation is the only relevant feature of the probability distribution of $c(W, X)$.

The *total risk* of observing X and using a decision function δ is defined as the sum of the risk $\rho(\xi, \delta)$ and the expected cost of observation $E[c(W, X)]$. The statistician must select an observation X from some available class of observable random variables and a corresponding Bayes decision function δ which minimize this total risk.

When we express the total risk as the sum of the risk of the decision function δ and the expected cost of observation, we are making an assumption, in effect, about the additivity of the statistician's utilities. Essentially, all work in statistical decision theory is based on this additive form for the total risk, and we shall use this form throughout the book. We shall not probe into this matter any further here. However, Raiffa and Schlaifer (1961, chap. 4) discuss the propriety of the assumption in some detail.

In many problems in which the statistician can select the size of a random sample which is to be taken, the cost of observation depends only on the size of the sample. In other words, this cost is not affected by the value of W or by the magnitudes of the observed values.

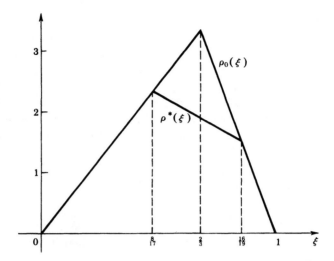

Fig. 8.7 The risks ρ^* and ρ_0 in Example 1.

EXAMPLE 1 Consider again Example 1 of Sec. 8.9 and suppose now that the cost of observing the random variable X is $c > 0$. The statistician either can choose a decision without observing X or can pay the cost c and observe X before choosing a decision. For any given prior probability ξ, how large an amount c should he be willing to pay?

To solve this problem, it is necessary to compare the minimum risk $\rho^*(\xi)$, excluding the cost c, that can be attained by using the observation X with the minimum risk $\rho_0(\xi)$ that can be attained from a Bayes decision without making any observation. The function ρ^* has already been computed and has been sketched in Fig. 8.6. The function ρ_0 is found from Table 8.2 to be as follows:

$$\rho_0(\xi) = \begin{cases} 5\xi & \text{for } 0 \le \xi \le \tfrac{2}{3}, \\ 10(1 - \xi) & \text{for } \tfrac{2}{3} < \xi \le 1. \end{cases} \tag{2}$$

The functions ρ^* and ρ_0 are sketched together in Fig. 8.7.

It can be seen from Fig. 8.7 that $\rho^*(\xi) = \rho_0(\xi)$ either if $\xi \le \tfrac{8}{17}$ or if $\xi \ge \tfrac{16}{19}$. Hence, for any prior probability ξ in either of these intervals, the statistician can do just as well without any observation as he can by observing X. If $\tfrac{8}{17} < \xi < \tfrac{16}{19}$, it can be seen that $\rho^*(\xi) < \rho_0(\xi)$. The statistician should be willing to pay any amount c such that $c < [\rho_0(\xi) - \rho^*(\xi)]$ for the opportunity of observing X before choosing his decision. The reduction in risk $[\rho_0(\xi) - \rho^*(\xi)]$ is greatest when $\xi = \tfrac{2}{3}$, at which point it is $\tfrac{25}{18}$.

EXAMPLE 2   Now suppose that in the decision problem of Example 1, the statistician can actually select the size of a random sample of observations which will be taken.   That is, the statistician is permitted to observe the values of n random variables $X_1, \ldots, X_n$.   For either given value $W = w_i$ $(i = 1, 2)$, the variables $X_1, \ldots, X_n$ are independent and identically distributed, and each has the same conditional p.f. $f(\cdot|w_i)$ as the single observation X in Example 1.   The conditional p.f.'s $f(\cdot|w_i)$ are defined by Eq. (5) of Sec. 8.9 and can be expressed in the following form, for $x = 0,1$:

$$f(x|w_1) = \frac{3^x}{4}, \qquad f(x|w_2) = \frac{2^1 \cdot x}{3}. \tag{3}$$

For any sequence of values $x_1, \ldots, x_n$ of the observations $X_1, \ldots, X_n$, let $y = \sum_{i=1}^{n} x_i$.   Then the value $g(x_1, \ldots, x_n|w_i)$ of the conditional joint p.f. of $X_1, \ldots, X_n$ when $W = w_i$ is

$$g(x_1, \ldots, x_n|w_1) = \frac{3^y}{4^n}, \qquad g(x_1, \ldots, x_n|w_2) = \frac{2^{n-y}}{3^n}. \tag{4}$$

If the cost of each observation is $c > 0$, how large a sample size n should the statistician select?

For any prior probability $\xi = \Pr(W = w_1)$, let $\xi(x_1, \ldots, x_n)$ denote the posterior probability that $W = w_1$ when a sample of size n is selected and $X_1 = x_1, \ldots, X_n = x_n$.   It follows from Bayes' theorem that

$$\xi(x_1, \ldots, x_n) = \frac{\xi(3^y/4^n)}{\xi(3^y/4^n) + (1 - \xi)(2^{n-y}/3^n)}$$

$$= \left[1 + \left(\frac{1 - \xi}{\xi}\right)\left(\frac{8}{3}\right)^n\left(\frac{1}{6}\right)^y\right]^{-1} \tag{5}$$

As was shown in Example 1 of Sec. 8.9, decision d_2 is a Bayes decision if $\xi(x_1, \ldots, x_n) < \frac{2}{3}$ and d_1 is a Bayes decision otherwise.   Therefore, from Eq. (5), there is a Bayes decision function δ_n which specifies that decision d_2 should be chosen if

$$y < \frac{\log\left[2(1 - \xi)/\xi\right] + n \log \frac{8}{3}}{\log 6}. \tag{6}$$

Otherwise, the decision function δ_n specifies that decision d_1 should be chosen.

Let k_n denote the value of the right side of the relation (6).   From Table 8.2 it follows that the risk $\rho(\xi, \delta_n)$ of the Bayes decision function δ_n is given by the equation

$$\rho(\xi, \delta_n) = 5\xi \Pr\left(\sum_{i=1}^{n} X_i < k_n | W = w_1\right)$$

$$+ 10(1 - \xi)\Pr\left(\sum_{i=1}^{n} X_i > k_n | W = w_2\right). \tag{7}$$

The conditional distribution of the sum $\Sigma_{i=1}^n X_i$ when $W = w_1$ is binomial with parameters n and $\frac{3}{4}$ (see Sec. 4.3). Also, its conditional distribution when $W = w_2$ is binomial with parameters n and $\frac{1}{3}$. Hence, the risk $\rho(\xi, \delta_n)$ can be evaluated for any prior probability ξ and any moderate sample size n from tables of the binomial distribution. If n is large, $\rho(\xi, \delta_n)$ can be evaluated from the normal approximation to the binomial distribution [see, e.g., Feller (1957, chap. 7)] and from tables of the normal distribution.

To obtain the total risk of the Bayes decision function δ_n for a sample of size n, the sampling cost nc must be added to the risk $\rho(\xi, \delta_n)$. The optimal sample size against the prior probability ξ will be the value of n which minimizes the total risk $\rho_t(\xi, \delta_n)$ as given by the equation

$$\rho_t(\xi, \delta_n) = \rho(\xi, \delta_n) + nc. \tag{8}$$

Table 8.3 gives some values of the risk $\rho(\xi, \delta_n)$ and the total risk $\rho_t(\xi, \delta_n)$ when the prior probability ξ is $\frac{2}{3}$ and the cost per observation c is 0.01.

Table 8.3 Values of the risk when $\xi = \frac{2}{3}$ and $c = 0.01$

n	$\rho(\xi, \delta_n)$	$\rho_t(\xi, \delta_n)$
21	0.1394	0.3494
22	0.1370	0.3570
23	0.1116	0.3416
24	0.1056	0.3456
25	0.0903	0.3403
26	0.0822	0.3422
27	0.0738	0.3438
28	0.0646	0.3446
29	0.0609	0.3509
30	0.0513	0.3513
31	0.0506	0.3606
32	0.0411	0.3611
33	0.0402	0.3702
34	0.0334	0.3734
35	0.0313	0.3813

It should be noted that the risk $\rho(\xi, \delta_n)$ must be a decreasing function of n. However, because the observations have a discrete distribution, the total risk $\rho_t(\xi, \delta_n)$ can have many small upward and downward movements as a function of n. It is seen from Table 8.3 that the optimal sample size for this problem is $n = 25$ and that the minimum total risk is 0.3403.

8.11 STATISTICAL DECISION PROBLEMS IN WHICH BOTH Ω AND D CONTAIN TWO POINTS

Consider a statistical decision problem in which $\Omega = \{w_1, w_2\}$, $D = \{d_1, d_2\}$, and the loss function L has the form given in Table 8.4, where $a_i > 0$

Table 8.4

	d_1	d_2
w_1	0	a_1
w_2	a_2	0

$(i = 1, 2)$. We shall assume that a random variable, or random vector, X can be observed before the statistician must choose a decision. The examples which were discussed in Secs. 8.9 and 8.10 satisfied these conditions. In this section, we shall describe the Bayes decision functions for a general problem of this type.

For any decision function δ, let $\alpha(\delta)$ denote the conditional probability that decision d_2 will be chosen when $W = w_1$. Also, let $\beta(\delta)$ denote the conditional probability that decision d_1 will be chosen when $W = w_2$. In other words, $\alpha(\delta)$ and $\beta(\delta)$ are the probabilities that δ will lead to the wrong decision when $W = w_1$ and $W = w_2$, respectively. Suppose that the prior distribution of W is specified by the equation $\Pr(W = w_1) = \xi$, where $0 < \xi < 1$. Then it follows from Table 8.4 that the risk $\rho(\xi, \delta)$ of the decision function δ is given by the equation

$$\rho(\xi, \delta) = a_1 \xi \alpha(\delta) + a_2(1 - \xi)\beta(\delta). \tag{1}$$

It is seen from Eq. (1) that in any particular problem, a Bayes decision function must be found which minimizes a linear combination of the form $a\alpha(\delta) + b\beta(\delta)$, where $a = a_1\xi$ and $b = a_2(1 - \xi)$ are given positive numbers. The next theorem describes a decision function which yields this minimum value. One version of this result is known in statistics as the *Neyman-Pearson lemma*.

For $i = 1, 2$, let f_i denote the conditional g.p.d.f. of the observation X when $W = w_i$.

Theorem 1 *For any constants $a > 0$ and $b > 0$, let δ^* be a decision function such that*

$$\delta^*(x) = d_1 \qquad if \ af_1(x) > bf_2(x) \tag{2}$$

and

$$\delta^*(x) = d_2 \qquad if \ af_1(x) < bf_2(x). \tag{3}$$

The value of $\delta^*(x)$ may be either d_1 or d_2 if $af_1(x) = bf_2(x)$. Then, for any other decision function δ,

$$a\alpha(\delta^*) + b\beta(\delta^*) \leq a\alpha(\delta) + b\beta(\delta). \tag{4}$$

Proof For any decision function δ, let S_1 and S_2 be the subsets of the sample space S which are defined as follows:

$$S_1 = \{x: \delta(x) = d_1\}, \qquad S_2 = S_1{}^c = \{x: \delta(x) = d_2\}. \tag{5}$$

Then we have

$$a\alpha(\delta) + b\beta(\delta) = a \int_{S_2} f_1(x)\, d\mu(x) + b \int_{S_1} f_2(x)\, d\mu(x)$$

$$= a + \int_{S_1} [bf_2(x) - af_1(x)]\, d\mu(x). \tag{6}$$

Since any decision function δ can be characterized by the set S_1 on which it specifies that decision d_1 should be chosen, it follows that finding a decision function δ which minimizes the linear combination $a\alpha(\delta) + b\beta(\delta)$ is equivalent to finding a set S_1 for which the final integral in Eq. (6) is minimized. This integral will be minimized if the set S_1 includes every point $x \in S$ for which the integrand is negative and excludes every point $x \in S$ for which the integrand is positive. It is irrelevant whether the set S_1 includes or excludes any point $x \in S$ for which the integrand vanishes. The decision function δ^* which is determined by this choice of the set S_1 satisfies the relations (2) and (3).∎

It can be seen from Theorem 1 that a Bayes decision function in a problem of this type depends in a very simple way on the ratio $f_2(x)/f_1(x)$.

8.12 COMPUTATION OF THE POSTERIOR DISTRIBUTION WHEN THE OBSERVATIONS ARE MADE IN MORE THAN ONE STAGE

Consider now a problem in which both the random variables, or random vectors, X and Y are to be observed. Suppose that X and Y have a joint conditional g.p.d.f. $f(x, y|w)$ for each value $w \in \Omega$. The posterior g.p.d.f. $\xi(\cdot\,|x, y)$ of W when $X = x$ and $Y = y$ is then specified at any value w by the equation

$$\xi(w|x, y) = \frac{f(x, y|w)\xi(w)}{\int_\Omega f(x, y|w')\xi(w')\, d\nu(w')}. \tag{1}$$

Suppose, however, that X and Y are not observed at the same time and that X is observed before Y. Let $g(\cdot\,|w)$ denote the conditional g.p.d.f. of X when $W = w$. After observing the value $X = x$, we can compute the posterior g.p.d.f. $\xi(\cdot \,|x)$ of W, before Y is observed, as follows:

$$\xi(w|x) = \frac{g(x|w)\xi(w)}{\int_\Omega g(x|w')\xi(w')\,d\nu(w')}. \tag{2}$$

Furthermore, the conditional g.p.d.f. $h(\cdot \,|w,\, x)$ of Y when $W = w$ and $X = x$ is

$$h(y|w,\, x) = \frac{f(x,\, y|w)}{g(x|w)}. \tag{3}$$

Hence, for the second stage of the experiment, which involves the observation of Y, the g.p.d.f. given by Eq. (2) can be regarded as the prior g.p.d.f. of W, and the conditional g.p.d.f.'s given by Eq. (3) for each value $w \in \Omega$ form the appropriate family of distributions of Y. The posterior g.p.d.f. $\xi(\cdot \,|x,\, y)$ of W when $Y = y$ can now be computed as follows:

$$\xi(w|x,\, y) = \frac{h(y|w,\, x)\xi(w|x)}{\int_\Omega h(y|w',\, x)\xi(w'|x)\,d\nu(w')}. \tag{4}$$

If the g.p.d.f.'s given by Eqs. (2) and (3) are substituted in Eq. (4), the result is Eq. (1). This means that if the observations are made in more than one stage, then the posterior distribution can be computed in different stages by letting the posterior distribution after each stage serve as the prior distribution for the next stage. It also follows from this derivation that if the posterior distribution of W when $X = x$ and $Y = y$ is computed in two stages, the final result is the same regardless of whether X or Y is observed first.

The decision-making process can now be described in the following highly simplified, but helpful, way. At any given time, the statistician has a probability distribution for the parameter W. As time progresses, the statistician gains information about W from various sources and uses this information to revise his distribution for W. From time to time, when he must choose a decision whose consequences are related to W, he will select a decision that is optimal against his current distribution for W.

Some aspects of this description of decision making are realistic. As we live our lives, we revise our beliefs about various parameters as we learn about them, and when a decision must be chosen, we obviously make the choice on the basis of our current beliefs. However, in some situations, the choice of a decision at a certain time may affect the information that will become available and, hence, may affect the decisions that the statistician can choose at future times. Problems in which the statistician must consider the future and plan ahead appropriately are called *sequential decision problems*. Such problems will be treated in Chaps. 12 to 14.

EXERCISES

1. Consider a decision problem in which the parameter space $\Omega = \{w_1, w_2, w_3, w_4\}$, the decision space $D = \{d_1, d_2, d_3\}$, and the loss function L is specified by the accompanying table. Suppose that the p.f. ξ of the parameter W is such that $\xi(w_1) = \frac{1}{8}$, $\xi(w_2) = \frac{3}{8}$, $\xi(w_3) = \frac{1}{4}$, and $\xi(w_4) = \frac{1}{4}$. Show that decision d_3 is the Bayes decision against ξ.

Table for Exercise 1

	d_1	d_2	d_3
w_1	0	2	3
w_2	1	0	2
w_3	3	4	0
w_4	1	2	0

2. Consider a decision problem in which the parameter space $\Omega = \{w_1, w_2\}$, the decision space $D = \{d_1, d_2, d_3\}$, and the loss function L is specified by the accompanying table. Prove that decision d_3 is a Bayes decision against a given distribution of the parameter W if, and only if, $\frac{1}{3} \leq \Pr(W = w_1) \leq \frac{5}{9}$.

Table for Exercise 2

	d_1	d_2	d_3
w_1	0	10	4
w_2	8	0	3

3. Consider a decision problem in which the parameter space Ω contains all the numbers w in the interval $0 \leq w \leq 1$, the decision space D is the real line R^1, and for any numbers $w \in \Omega$ and $d \in D$, the loss $L(w, d)$ is defined by the equation

$$L(w, d) = 100(w - d)^2.$$

Suppose that the p.d.f. ξ of the parameter W is specified by the equation

$$\xi(w) = 2w \qquad 0 \leq w \leq 1.$$

Show that the value $d = \frac{2}{3}$ is the Bayes decision against ξ and that the Bayes risk is $\frac{50}{9}$.

4. Consider the decision problem described in Exercise 1, but suppose that the loss function L specified in that exercise is replaced by the new loss function L_0 specified by the table on page 150. Prove that against any distribution of the parameter W, the Bayes decisions when the loss function is L_0 are the same as the Bayes decisions when the loss function is L.

Table for Exercise 4

	d_1	d_2	d_3
w_1	4	6	7
w_2	0	-1	1
w_3	0	1	-3
w_4	1	2	0

5. Suppose that in an arbitrary decision problem, a certain decision d^* is a Bayes decision against two distinct distributions P_1 and P_2 of the parameter W. For any number α such that $0 < \alpha < 1$, show that d^* must also be a Bayes decision against the distribution $\alpha P_1 + (1 - \alpha)P_2$.

6. Consider the decision problem presented in Exercise 3. Suppose that statistician A believes that the p.d.f. of W is ξ, as specified in Exercise 3, but that statistician B believes that the p.d.f. of W is ξ_B, as specified by the equation

$$\xi_B(w) = 3w^2 \qquad 0 \le w \le 1.$$

How much additional risk does B think that A will incur because of A's incorrect belief about the distribution of W?

7. Consider a decision problem in which $\Omega = \{w_1, w_2\}$, $D = \{d_1, \ldots, d_5\}$, and the loss function L is given by the accompanying table. Against what distributions of the parameter W is the Bayes decision not unique?

Table for Exercise 7

	d_1	d_2	d_3	d_4	d_5
w_1	0	4	2	1	5
w_2	4	5	0	1	0

8. Consider a decision problem in which $\Omega = \{w_1, w_2\}$, $D = \{d_1, \ldots, d_7\}$, and the loss function L is given by the accompanying table. For each distribution ξ of the parameter W, find all Bayes decisions against ξ.

Table for Exercise 8

	d_1	d_2	d_3	d_4	d_5	d_6	d_7
w_1	1	6	0	2	7	3	4
w_2	10	1	13	8	0	5	4

9. Consider a decision problem in which both Ω and D have an infinite number of elements. Suppose that $\Omega = \{w_1, w_2, \ldots\}$, $D = \{d^*, d_1, d_2, \ldots\}$, and the loss function L is given by the accompanying table. Prove that d^* is the only admissi-

ble decision in the set D but that d^* is not a Bayes decision against any distribution of the parameter W.

Table for Exercise 9

	d^*	d_1	d_2	d_3	d_4	$\cdots$
w_1	$\frac{1}{2}$	0	0	0	0	$\cdots$
w_2	$\frac{1}{2}$	1	0	0	0	$\cdots$
w_3	$\frac{1}{2}$	1	1	0	0	$\cdots$
w_4	$\frac{1}{2}$	1	1	1	0	$\cdots$
w_5	$\frac{1}{2}$	1	1	1	1	$\cdots$
$\cdots$		.	.	.	.	$\cdots$

10. Suppose that there is probability $\frac{1}{10}$ that a signal is present in a certain system at any given time and probability $\frac{9}{10}$ that no signal is present in the system. Suppose that a measurement made on the system when a signal is present is normally distributed with mean 50 and precision 1 and a measurement made on the system when no signal is present is normally distributed with mean 52 and precision 1. Suppose that a measurement made on the system at a certain time has the value x. Show that the posterior probability that a signal is present is greater than the posterior probability that no signal is present if, and only if, $x < 51 - \frac{1}{2} \log 9$.

11. Consider a decision problem in which $\Omega = \{w_1, w_2\}$, $D = \{d_1, d_2\}$, and the loss function L is given by Table 8.2. Suppose that the statistician can observe a random variable Z whose conditional distribution when $W = w_1$ is normal with mean 0 and variance 1 and whose conditional distribution when $W = w_2$ is normal with mean 1 and variance 1. For any given prior probability $\xi = \Pr(W = w_1)$, let δ be the decision function such that $\delta(z) = d_1$ if

$$z \le \frac{1}{2} + \log \frac{\xi}{2(1 - \xi)}$$

and $\delta(z) = d_2$ otherwise. (a) Show that δ is a Bayes decision function against ξ. (b) Sketch the Bayes risk $\rho^*(\xi)$ as a function of ξ over the interval $0 \le \xi \le 1$.

12. Suppose that a statistician must decide whether a certain random variable has a uniform distribution on the interval $(0, 1)$ or a uniform distribution on the interval $(0, \frac{1}{2})$. Each of these two distributions has probability $\frac{1}{2}$ of being the correct distribution. Suppose that the loss is 0 if a correct decision is made and is $a > 0$ if an incorrect decision is made. Suppose that the statistician can select the number of observations which will be made on the random variable but that the cost of each observation is $c > 0$. Show that the statistician should make n^* observations, where n^* is the nonnegative integer which minimizes the value $a2^{-(n+1)} + nc$.

13. Consider a decision problem in which $\Omega = \{w_1, w_2\}$, $D = \{d_1, d_2, d_3\}$, and the loss function L is given by the table on page 152. Suppose that an observation X is available with the following conditional distributions:

$$\Pr(X = 1 | W = w_1) = \tfrac{3}{4}, \qquad \Pr(X = 0 | W = w_1) = \tfrac{1}{4};$$
$$\Pr(X = 1 | W = w_2) = \tfrac{1}{4}, \qquad \Pr(X = 0 | W = w_2) = \tfrac{3}{4}.$$

Suppose that $\xi = \Pr(W = w_1)$. Find a Bayes decision function against each value of ξ ($0 \le \xi \le 1$), and sketch the Bayes risk $\rho^*(\xi)$ as a function of ξ.

Table for Exercise 13

	d_1	d_2	d_3
w_1	0	10	3
w_2	10	0	3

14. Suppose that for the conditions in Exercise 13, before the statistician chooses a decision, he can observe the values of the random variables $X_1, \ldots, X_n$ which, for any given value $W = w_i$ ($i = 1, 2$), are a random sample from the conditional distribution of X. For any given sample size n, find a Bayes decision function against each value of ξ.

15. Suppose that for the conditions in Exercise 14, $\xi = \frac{1}{2}$ and the cost c of each observation is $\frac{1}{10}$. Show that the optimal sample size n is 8 and that the minimum total risk which can be attained is 1.33.

16. Suppose that for the conditions in Exercises 13 and 14, the cost of each observation is not constant but that any observation whose value is 1 costs 0.15 and any observation whose value is 0 costs 0.05. Show that when $\xi = \frac{1}{2}$, the optimal sample size and minimum total risk will be precisely the same as those in Exercise 15.

17. Consider a decision problem in which $\Omega = \{w_1, w_2\}$, $D = \{d_1, d_2\}$, and the loss function L is given by the accompanying table. Suppose that $\Pr(W = w_1) = \Pr(W = w_2) = \frac{1}{2}$. Suppose also that the conditional distribution of an observation X when $W = w_1$ is normal with mean -1 and variance 9 and that the conditional distribution of X when $W = w_2$ is normal with mean 1 and variance 9. Suppose, further, that before the statistician chooses a decision, he can observe values of the random variables $X_1, \ldots, X_n$ which, for any given value of W, are a random sample from the conditional distribution of X just given. If each observation in the sample costs 1 unit, show that the optimal number of observations n is 42 and that the minimum total risk is 57.4.

Table for Exercise 17

	d_1	d_2
w_1	0	1,000
w_2	1,000	0

18. Consider k random variables $X_1, \ldots, X_k$. Suppose that the p.d.f. of exactly one of these random variables is g and the p.d.f. of each of the other $k - 1$ random variables is h but that it is not known which one of the random variables has p.d.f. g. For $i = 1, \ldots, k$, let ξ_i be the prior probability that X_i is the random variable whose p.d.f. is g. Here $\xi_i > 0$ for $i = 1, \ldots, k$ and $\sum_{i=1}^{k} \xi_i = 1$. (a) Suppose that the random variable X_1 is observed and found to have the value x. Find the posterior probability that the p.d.f. of X_1 is g. (b) Suppose that the random variable X_2 is observed and found to have the value x. Find the posterior probability that the p.d.f. of X_1 is g.

19. Consider two boxes A and B each of which contains both red balls and

green balls. It is known that in one of the boxes, $\frac{1}{2}$ of the balls are red and $\frac{1}{2}$ of the balls are green and that in the other box, $\frac{1}{4}$ of the balls are red and $\frac{3}{4}$ of the balls are green. Let the box in which $\frac{1}{2}$ of the balls are red be denoted as box W, and suppose that it is not known with certainty whether $W = A$ or $W = B$. Assume that $\Pr(W = A) = \xi$ and $\Pr(W = B) = 1 - \xi$, where ξ is a given number such that $0 < \xi < 1$.

Suppose that the statistician may select one ball at random from either box A or box B and that after observing its color, he must decide whether $W = A$ or $W = B$. Prove that if $\frac{1}{2} < \xi < \frac{2}{3}$, then in order to maximize the probability of making a correct decision, he should select the ball from box B. Prove also that if $\frac{2}{3} \leq \xi \leq 1$, then it does not matter from which box the ball is selected.

20. Consider a decision problem in which $\Omega = \{w_1, w_2\}$, $D = \{d_1, d_2\}$, and the loss function L is given by Table 8.4. Suppose that the statistician can observe either a random variable X or a random variable Y whose conditional distributions are as follows:

$$\Pr(X = 1 | W = w_1) = \tfrac{2}{3}, \qquad \Pr(X = 0 | W = w_1) = \tfrac{1}{3},$$
$$\Pr(X = 1 | W = w_2) = \tfrac{1}{2}, \qquad \Pr(X = 0 | W = w_2) = \tfrac{1}{2},$$

and

$$\Pr(Y = 1 | W = w_1) = \tfrac{3}{4}, \qquad \Pr(Y = 0 | W = w_1) = \tfrac{1}{4},$$
$$\Pr(Y = 1 | W = w_2) = \tfrac{1}{2}, \qquad \Pr(Y = 0 | W = w_2) = \tfrac{1}{2}.$$

Suppose also that the cost of observing X is the same as the cost of observing Y. Show that for any prior distribution of W and any values of the losses a_1 and a_2, the statistician should observe Y rather than X.

21. Let W be a parameter which takes the values w_1 and w_2 with prior probabilities specified by the equations $\Pr(W = w_1) = \xi$ and $\Pr(W = w_2) = 1 - \xi$. Suppose that an observation X is to be taken with conditional g.p.d.f.'s $f(\cdot | w_i)$ for $i = 1, 2$. Let $\xi(x)$ denote the posterior probability that $W = w_1$ when $X = x$. Prove that $E[\xi(X)] = \xi$, where the expectation is computed by assuming that W has the specified prior distribution.

22. For the conditions specified in Exercise 21, if it is assumed that $W = w_1$, prove that $E[\xi(X)] \geq \xi$. *Note:* This exercise can be interpreted as stating that, on the average, the posterior distribution will assign greater probability to the correct value of W than the prior distribution assigned to that value.

23. For the conditions specified in Exercise 21, suppose that the prior distribution of W is such that $\xi = 1 - \xi = \frac{1}{2}$. If it is assumed that $W = w_1$, prove that for any number ϵ $(0 < \epsilon < 1)$,

$$\Pr[\xi(X) \leq \epsilon] \leq \frac{\epsilon}{1 - \epsilon}.$$

Note: This exercise can be interpreted as stating that there is only a small probability that the posterior distribution will assign a small probability to the correct value of W.

24. Consider a set $\Omega = \{w_1, w_2, w_3\}$ containing three points, and let $\mathcal{P}$ be the set of all probability distributions (p_1, p_2, p_3) such that $p_i \geq 0$ $(i = 1, 2, 3)$ and $p_1 + p_2 + p_3 = 1$. Let $\mathcal{C}$ be the set of points either inside or on the boundary of an equilateral triangle which has a unit height. Also, let $v_1, v_2,$ and v_3 denote the vertices of that triangle, and for $i = 1, 2, 3$, let S_i denote the side of the triangle that is opposite the vertex v_i. Show that the sum of the distances from any point in the triangle to the three sides of the triangle is 1. Then show that there is a one-to-one correspondence between the sets $\mathcal{P}$ and $\mathcal{C}$, in which any point $(p_1, p_2, p_3) \in \mathcal{P}$ corresponds to the point $\mathbf{x} \in \mathcal{C}$ whose distance from side S_i is p_i $(i = 1, 2, 3)$.

25. Suppose that in Exercise 24, corresponding to any fixed point (p_1, p_2, p_3) ϵ $\mathcal{P}$, a mass p_i is placed at the vertex v_i $(i = 1, 2, 3)$. Find the center of gravity $\mathbf{x}$ of this system of three mass points, and show that the position of $\mathbf{x}$ defines a one-to-one correspondence between $\mathcal{P}$ and $\mathcal{C}$. [For any point $\mathbf{x} \, \epsilon \, \mathcal{C}$, the corresponding values (p_1, p_2, p_3) are called the *barycentric coordinates* of $\mathbf{x}$.]

26. Show that the correspondence between $\mathcal{P}$ and $\mathcal{C}$ defined in Exercise 25 is the same as the correspondence defined in Exercise 24.

27. Suppose that each of k statisticians has his own prior distribution for a certain parameter W, and let ξ_i be the g.p.d.f. which statistician i assigns to W $(i = 1, \ldots, k)$. Suppose also that an executive forms his opinion about W from the opinions of the k statisticians and that he assigns to W the g.p.d.f. ξ^* defined, at each point $w \, \epsilon \, \Omega$, as follows:

$$\xi^*(w) = \alpha_1 \xi_1(w) + \cdots + \alpha_k \xi_k(w).$$

Here $\alpha_1, \ldots, \alpha_k$ are weights such that $\alpha_i \geq 0$ $(i = 1, \ldots, k)$ and $\alpha_1 + \cdots + \alpha_k = 1$. The value of α_i reflects the relative weight that the executive gives to the opinion of statistician i. Suppose further that the k statisticians and the executive observe together the value of a random variable X whose conditional g.p.d.f. when $W = w$ is $f(\cdot \, | w)$. Show that the posterior g.p.d.f. of the executive will again be a linear combination of the posterior g.p.d.f.'s of the k statisticians, with new weights $\beta_1, \ldots, \beta_k$ which will depend on the observed value of X. Also, discuss the conditions under which the weight β_1 in the posterior g.p.d.f. will be greater than the weight α_1 in the prior g.p.d.f.

conjugate prior distributions

9.1 SUFFICIENT STATISTICS

Consider a statistical problem in which a large amount of experimental data has been collected. The treatment of the data is often simplified if the statistician computes a few numerical values, or statistics, and considers these values as summaries of the relevant information in the data. In some problems, a statistical analysis that is based on these few summary values can be just as effective as any analysis that could be based on all the observed values. In this chapter we shall consider problems for which fully informative summaries of this type are available. Such summaries are known as *sufficient statistics*.

Suppose that W is a parameter which takes values in the space Ω. Also, suppose that X is a random variable, or random vector, which takes values in the sample space S. We shall let $f(\cdot \mid w)$ denote the conditional g.p.d.f. of X when $W = w$ $(w \in \Omega)$. It is assumed that the observed value of X will be available for making inferences and decisions relating to the parameter W. In this context, any function T of the observation X, whether or not T is a real-valued function, is called a *statistic*.

Loosely speaking, a statistic T is called a sufficient statistic if, for any prior distribution of W, its posterior distribution depends on the observed value of X only through $T(X)$. More formally, for any prior g.p.d.f. ξ of W and any observed value $x \in S$, let $\xi(\cdot \, | x)$ denote the posterior g.p.d.f. of W. For simplicity, it will be assumed in this section that for every value of $x \in S$ and every prior g.p.d.f. ξ, the posterior g.p.d.f. $\xi(\cdot \, | x)$ exists and is specified by Bayes' theorem. Then it is said that a statistic T is a *sufficient statistic* for the family of g.p.d.f.'s $\{ f(\cdot \, | w), \, w \in \Omega \}$ if $\xi(\cdot \, | x_1) = \xi(\cdot \, | x_2)$ for any prior g.p.d.f. ξ and any two points $x_1 \in S$ and $x_2 \in S$ such that $T(x_1) = T(x_2)$. There is a good reason for saying that a statistic which has this property is sufficient. In order to be able to compute the posterior distribution of W from any prior distribution, the statistician needs only the value of $T(X)$. He does not need the value of X itself, which may be a vector of high dimension. It should be emphasized here that the values of any g.p.d.f. can be arbitrarily changed on any set of points having probability 0. Hence, when we say that the g.p.d.f.'s $\xi(\cdot \, | x)$ in the preceding definition or the g.p.d.f.'s $f(\cdot \, | w)$ in the following theorem have certain properties, we mean that there are versions of these g.p.d.f.'s which have those properties.

The following theorem, which is known as the *factorization criterion*, provides an easy way of recognizing sufficient statistics.

Theorem 1   *A statistic T is sufficient for a family of g.p.d.f.'s $\{ f(\cdot \, | w), w \in \Omega \}$ if, and only if, $f(x | w)$ can be factored as follows for all values $x \in S$ and $w \in \Omega$:*

$$f(x | w) = u(x) v[T(x), w]. \tag{1}$$

Here, the function u is positive and does not depend on w and the function v is nonnegative and depends on x only through $T(x)$.

Proof   Suppose first that the factorization indicated in Eq. (1) is correct. Then, for any prior g.p.d.f. ξ of W and any points $w \in \Omega$ and $x \in S$, the posterior g.p.d.f. of W is

$$\xi(w | x) = \frac{v[T(x), w] \xi(w)}{\int_{\Omega} v[T(x), w'] \xi(w') \, d\nu(w')}. \tag{2}$$

Since the right side of Eq. (2) depends on the observed value x only through the value $T(x)$, it follows that T is a sufficient statistic.

Conversely, suppose that T is a sufficient statistic. Let ξ be any prior g.p.d.f. of W such that $\xi(w) > 0$ at every point $w \in \Omega$. The posterior g.p.d.f. $\xi(w | x)$ is specified at any points $w \in \Omega$ and $x \in S$ as follows:

$$\xi(w | x) = \frac{f(x | w) \xi(w)}{\int_{\Omega} f(x | w') \xi(w') \, d\nu(w')}. \tag{3}$$

Since T is a sufficient statistic, then $\xi(w|x) = r[T(x), w]$, where the function r involves only $T(x)$ and w. Hence, it follows from Eq. (3) that

$$f(x|w) = \left[\int_\Omega f(x|w') \xi(w') \, d\nu(w') \right] \frac{r[T(x), w]}{\xi(w)}. \tag{4}$$

Equation (4) exhibits a factorization of the form indicated in Eq. (1). ■

Throughout the remainder of this chapter and in many other parts of this book, we shall need to consider random variables $X_1, \ldots, X_n$ whose joint distribution depends on the value w of some parameter W in the following way: For any given value w of W ($w \in \Omega$), the variables $X_1, \ldots, X_n$ form a random sample from a specified distribution whose g.p.d.f. is $f(\cdot | w)$. Hence, the conditional joint g.p.d.f. $f_n(\cdot | w)$ of $X_1, \ldots, X_n$, when $W = w$, is specified by the equation

$$f_n(x_1, \ldots, x_n|w) = f(x_1|w) \cdots f(x_n|w). \tag{5}$$

When the distribution of $X_1, \ldots, X_n$ satisfies these conditions, we shall say that $X_1, \ldots, X_n$ is a *random sample from the specified distribution with an unknown value of the parameter* W. We shall now present three examples of sufficient statistics which will also illustrate this terminology.

EXAMPLE 1 Suppose that $X_1, \ldots, X_n$ is a random sample from a Bernoulli distribution with an unknown value of the parameter W. In other words, the only possible values of each random variable X_i are 0 and 1, and for any given value w of W such that $0 < w < 1$, the joint p.f. $f_n(\cdot | w)$ of $X_1, \ldots, X_n$ is specified by the equation

$$f_n(x_1, \ldots, x_n|w) = w^y(1 - w)^{n-y}. \tag{6}$$

Here $y = \sum_{i=1}^n x_i$. Hence, the joint p.f. given in Eq. (6) depends on the values of the random variables $X_1, \ldots, X_n$ only through their sum.

Accordingly, let T be the statistic defined by the equation

$$T(X_1, \ldots, X_n) = \sum_{i=1}^n X_i. \tag{7}$$

It follows that T is a sufficient statistic for the family of p.f.'s specified by Eq. (6) for $0 < w < 1$.

EXAMPLE 2 Suppose that $X_1, \ldots, X_n$ is a random sample from a normal distribution with an unknown value of the mean and an unknown value of the variance. Then, for any given values μ and σ^2 of the mean and variance such that $-\infty < \mu < \infty$ and $\sigma^2 > 0$, the conditional joint p.d.f. $f_n(\cdot \mid \mu, \sigma^2)$ of $X_1, \ldots, X_n$ is specified by the equation

$$f_n(x_1, \ldots, x_n \mid \mu, \sigma^2) = (2\pi\sigma^2)^{-n/2} \exp\left[-\frac{1}{2\sigma^2} \sum_{i=1}^{n} (x_i - \mu)^2 \right]. \tag{8}$$

If we let $\bar{x} = (1/n)\sum_{i=1}^{n} x_i$, then

$$\sum_{i=1}^{n} (x_i - \mu)^2 = \sum_{i=1}^{n} (x_i - \bar{x})^2 + n(\bar{x} - \mu)^2. \tag{9}$$

Therefore, the p.d.f. defined by Eq. (8) depends on the values of the observations $X_1, \ldots, X_n$ only through the two values $\bar{x}$ and $\sum_{i=1}^{n}(x_i - \bar{x})^2$.

Accordingly, let $\mathbf{T}$ be the two-dimensional vector statistic defined by the equation

$$\mathbf{T}(X_1, \ldots, X_n) = \{\bar{X}, \sum_{i=1}^{n} (X_i - \bar{X})^2\}. \tag{10}$$

It follows that $\mathbf{T}$ is a sufficient statistic for the family of p.d.f.'s specified by Eq. (8). It is sometimes said that the two components of the vector $\mathbf{T}(X_1, \ldots, X_n)$ are *jointly sufficient statistics*.

EXAMPLE 3 Suppose that $X_1, \ldots, X_n$ is a random sample from the uniform distribution on the interval $(0, W)$, where the value of the parameter W is unknown. For any given value w of W such that $w > 0$, the conditional p.d.f. $f(\cdot \mid w)$ of any single observation X_i is specified by the equation

$$f(x \mid w) = \begin{cases} \dfrac{1}{w} & \text{for } x < w, \\ 0 & \text{otherwise.} \end{cases} \tag{11}$$

The sample space S of each observation is assumed to be the set of positive numbers. It follows that the conditional joint p.d.f. $f_n(\cdot \mid w)$ of $X_1, \ldots, X_n$, when $W = w$ ($w > 0$), is as follows:

$$f_n(x_1, \ldots, x_n \mid w) = \begin{cases} \dfrac{1}{w^n} & \text{for } x_i < w \ (i = 1, \ldots, n), \\ 0 & \text{otherwise.} \end{cases} \tag{12}$$

This p.d.f. can be rewritten in the form

$$f_n(x_1, \ldots, x_n \mid w) = \begin{cases} \dfrac{1}{w^n} & \text{for max } \{x_1, \ldots, x_n\} < w, \\ 0 & \text{otherwise.} \end{cases} \tag{13}$$

This function depends on the values of the observations $X_1, \ldots, X_n$ only through the value of max $\{x_1, \ldots, x_n\}$.

Accordingly, let T be the statistic defined by the equation

$$T(X_1, \ldots, X_n) = \max \{X_1, \ldots, X_n\}. \tag{14}$$

It follows that T is a sufficient statistic for the family of p.d.f.'s specified by Eq. (12).

Other examples of sufficient statistics are given in Exercises 1 to 9 at the end of this chapter.

Further Remarks and References

The concept of a sufficient statistic was introduced by Fisher (1922). A rigorous measure-theoretic presentation of this concept is given by Halmos and Savage (1949). Sufficient statistics have also been studied by Lehmann and Scheffé (1950), Bahadur (1954), and Dynkin (1961) and are discussed in the books by Savage (1954), Lehmann (1959), and Raiffa and Schlaifer (1961).

9.2 CONJUGATE FAMILIES OF DISTRIBUTIONS

In each of the three examples presented in Sec. 9.1 and in each of the examples presented in Exercises 1 to 7, there is a sufficient statistic which can be represented by one or two real-valued functions of the observations $X_1, \ldots, X_n$ in a random sample, and the dimension of this sufficient statistic remains fixed regardless of the size n of the sample. The planning and analysis of an experiment become much easier for the statistician if it can be assumed that the observations are to be drawn from a family of distributions for which there is a sufficient statistic of fixed dimension. One important simplification results from the fact that under these conditions there must exist a standard family of distributions of the parameter W which has the following property: If the prior distribution of W belongs to this family, then for any sample size n and any values of the observations in the sample, the posterior distribution of W must also belong to the same family. A family of distributions with this property is said to be *closed under sampling*. The family is also called a *conjugate family of distributions* because of the special relationship which must exist between this family of distributions of the parameter and the family of distributions of the observations.

Therefore, when there exists a sufficient statistic of fixed dimension, it is possible for the statistician to handle only prior and posterior distributions which belong to a relatively small conjugate family. In order for this simplification to be of much use, however, the conjugate family of distributions of W must still be rich enough to permit the statistician to find within the family, in a wide variety of situations, a distribution which will adequately represent his prior distribution of W.

As an example, suppose that $X_1, \ldots, X_n$ is a random sample from the Bernoulli distribution with an unknown value of the parameter W. Suppose also that the prior distribution of W is a beta distribution with specified values of the parameters α and β such that $\alpha > 0$ and $\beta > 0$. Then the prior p.d.f. ξ of W is

$$\xi(w) \propto w^{\alpha-1}(1 - w)^{\beta-1} \qquad \text{for } 0 < w < 1. \tag{1}$$

Here we are utilizing the proportionality symbol $\propto$ to indicate that the function ξ is basically as given on the right side of the relation (1) but that there may be a factor which does not involve w.

In general, if the prior g.p.d.f. of W is ξ and the conditional g.p.d.f. of X when $W = w$ is $f(\cdot \,|w)$, then the posterior g.p.d.f. $\xi(\cdot \,|x)$ of W when $X = x$ is

$$\xi(w|x) \propto \xi(w)f(x|w) \qquad \text{for } w \in \Omega. \tag{2}$$

The use of the proportionality symbol in the relation (2) is justified since the conditional g.p.d.f. of W is equal to the right side of (2) divided by the factor $\int_\Omega f(x|w')\xi(w')\,d\nu(w')$, which does not involve w.

We shall now return to our example. The conditional joint p.f. $f_n(\cdot \,|w)$ of $X_1, \ldots, X_n$ when $W = w$ is given by Eq. (6) of Sec. 9.1. Therefore, from the relations (1) and (2), the posterior p.d.f. $\xi(\cdot \,|x_1, \ldots, x_n)$ of W when $X_i = x_i$ $(i = 1, \ldots, n)$ is

$$\xi(w|x_1, \ldots, x_n) \propto w^{\alpha+y-1}(1 - w)^{\beta+n-y-1}. \tag{3}$$

Here $y = \sum_{i=1}^{n} x_i$. It can be seen from the relation (3) that the posterior distribution of W is a beta distribution with parameters $\alpha + y$ and $\beta + n - y$. We have proved the following theorem.

Theorem 1 *Suppose that $X_1, \ldots, X_n$ is a random sample from a Bernoulli distribution with an unknown value of the parameter W. Suppose also that the prior distribution of W is a beta distribution with parameters α and β such that $\alpha > 0$ and $\beta > 0$. Then the posterior distribution of W when $X_i = x_i$ $(i = 1, \ldots, n)$ is a beta distribution with parameters $\alpha + y$ and $\beta + n - y$, where $y = \sum_{i=1}^{n} x_i$.*

In other words, the family of beta distributions is a conjugate family for samples from a Bernoulli distribution. It should be emphasized here that the term "parameter" is now doing double duty. The random variable W, whose value is not known by the statistician, is a parameter whose possible values w index the family of g.p.d.f.'s $f(\cdot \,|w)$ of each observation X which is available. The parameters α and β, whose values are specified, index the conjugate family of p.d.f.'s of W.

The construction of conjugate families of distributions will be studied more systematically in the next section.

Further Remarks and References

A family of g.p.d.f.'s $\{f(\cdot \mid w), w \in \Omega\}$, each of which is defined on a given sample space S, is said to be an *exponential family* if the g.p.d.f.'s are of the following form for any points $x \in S$ and $w \in \Omega$:

$$f(x|w) = a(w)b(x) \exp \left[\sum_{i=1}^{k} g_i(w)h_i(x) \right]. \tag{4}$$

Consider an exponential family of this type and suppose that $X_1, \ldots, X_n$ are random variables, or random vectors, all of whose values lie in the sample space S and whose joint g.p.d.f. $f_n(\cdot \mid w)$, for any point $w \in \Omega$, is specified by the equation

$$f_n(x_1, \ldots, x_n|w) = \prod_{j=1}^{n} f(x_i|w). \tag{5}$$

Let $\mathbf{T}$ be the k-dimensional vector defined by the equation

$$\mathbf{T}(X_1, \ldots, X_n) = \left\{ \sum_{j=1}^{n} h_1(X_j), \ldots, \sum_{j=1}^{n} h_k(X_j) \right\}. \tag{6}$$

Then the statistic $\mathbf{T}$ is a sufficient statistic of fixed dimension k for each sample size n.

Darmois (1935), Koopman (1936), and Pitman (1936) have shown that among families of distributions which satisfy certain regularity conditions, a sufficient statistic of fixed dimension will exist only for exponential families [see also Fraser (1963)]. Almost all the examples which we have considered in this chapter involve exponential families (see Exercise 11). A family of uniform distributions is not an exponential family, but there is a sufficient statistic of fixed dimension for such a family. However, one of the regularity conditions which is not satisfied by a family of uniform distributions is the condition that the set of points x such that $f(x|w) > 0$ must be the same set for every value $w \in \Omega$.

9.3 CONSTRUCTION OF THE CONJUGATE FAMILY

Now consider again the example summarized in Theorem 1 of Sec. 9.2. For any positive constants α and β, let $g(\cdot \mid \alpha, \beta)$ denote the p.d.f. of a beta distribution with parameters α and β. In that example, the family of beta distributions was found to be the appropriate conjugate family because of the following two properties:

First, consider any observed values $x_1, \ldots, x_n$ of the variables $X_1, \ldots, X_n$. The conditional joint p.f. $f_n(x_1, \ldots, x_n|w)$ of $X_1, \ldots, X_n$ is specified by Eq. (6) of Sec. 9.1. If this function is regarded as a function of w, then it follows from the definition of the p.d.f. of a beta distribution that

$$f_n(x_1, \ldots, x_n|w) \propto g(w|y+1, n-y+1). \tag{1}$$

Therefore, for any observed values $x_1, \ldots, x_n$, the function $f_n(x_1, \ldots, x_n|w)$ is proportional to the p.d.f. of a beta distribution.

Second, if $g(\cdot|\alpha_1, \beta_1)$ and $g(\cdot|\alpha_2, \beta_2)$ are the p.d.f.'s of any two beta distributions, then there is another p.d.f. $g(\cdot|\alpha_3, \beta_3)$ such that for $0 < w < 1$,

$$g(w|\alpha_3, \beta_3) \propto g(w|\alpha_1, \beta_1)g(w|\alpha_2, \beta_2). \tag{2}$$

In fact, it follows from the definition of the p.d.f. of a beta distribution that

$$g(w|\alpha_3, \beta_3) \propto w^{\alpha_1+\alpha_2-2}(1-w)^{\beta_1+\beta_2-2}. \tag{3}$$

Since the right side of the relation (3) is proportional to the p.d.f. of a beta distribution with parameters $\alpha_1 + \alpha_2 - 1$ and $\beta_1 + \beta_2 - 1$, we obtain the equations

$$\alpha_3 = \alpha_1 + \alpha_2 - 1 \quad \text{and} \quad \beta_3 = \beta_1 + \beta_2 - 1. \tag{4}$$

A family of p.d.f.'s which satisfies the relation (2) is said to be *closed under multiplication*.

If the prior distribution of W is taken to be a beta distribution with p.d.f. $g(\cdot|\alpha, \beta)$, then the posterior p.d.f. $\xi(\cdot|x_1, \ldots, x_n)$ of W will satisfy the relation

$$\xi(w|x_1, \ldots, x_n) \propto f_n(x_1, \ldots, x_n|w)g(w|\alpha, \beta). \tag{5}$$

It now follows from the relations (1) to (4) that the posterior p.d.f. of W must be that of a beta distribution with parameters $\alpha + y$ and $\beta + n - y$. This result is the one that was presented in Theorem 1 of Sec. 9.2.

This development suggests a method for determining a conjugate family of distributions in any problem for which there exists a sufficient statistic of fixed dimension. The statistician need only determine a family of p.d.f.'s of the parameter W such that (1) for any sample size n and any observed values $x_1, \ldots, x_n$, the conditional joint g.p.d.f. $f_n(x_1, \ldots, x_n|w)$, regarded as a function of w, is proportional to one of the p.d.f.'s in the family, and (2) the family is closed under multiplication.

We shall now show that whenever the family of g.p.d.f.'s $\{f_n(\,\cdot\,|w),$ $w \in \Omega\}$ has a sufficient statistic $T_n(X_1, \ldots, X_n)$ of fixed dimension k $(k \geq 1)$ for every sample size n, there must exist a simple conjugate family of this type. It follows from Theorem 1 of Sec. 9.1 that for each value of n, there is a function v_n such that

$$f_n(x_1, \ldots, x_n|w) \propto v_n[T_n(x_1, \ldots, x_n), w]. \tag{6}$$

If we let $T_n(x_1, \ldots, x_n) = t$ and assume that $\int_\Omega v_n(t, w)\, d\nu(w) < \infty$, then there exists a p.d.f. $g(\,\cdot\,|t, n)$ on the space Ω such that

$$g(w|t, n) \propto v_n(t, w). \tag{7}$$

Consider the family of p.d.f.'s $g(\,\cdot\,|t, n)$ for all possible sample sizes n and all possible values t of the statistic $T_n(X_1, \ldots, X_n)$. It follows from the relations (6) and (7) that $f_n(x_1, \ldots, x_n|w)$ must be proportional to one of the p.d.f.'s in this family. Furthermore, as will now be demonstrated, this family of p.d.f.'s is closed under multiplication.

Consider any two p.d.f.'s $g(\,\cdot\,|s, m)$ and $g(\,\cdot\,|t, n)$ which belong to the family. Then there must exist observed values $x_1, \ldots, x_m$ and $y_1, \ldots, y_n$ of samples of size m and size n such that $T_m(x_1, \ldots, x_m) = s$ and $T_n(y_1, \ldots, y_n) = t$. If these observations are combined, they form a sample of size $m + n$ and their g.p.d.f.'s satisfy the equation

$$f_{m+n}(x_1, \ldots, x_m, y_1, \ldots, y_n|w)$$
$$= f_m(x_1, \ldots, x_m|w)f_n(y_1, \ldots, y_n|w). \tag{8}$$

If we let $u = T_{m+n}(x_1, \ldots, x_m, y_1, \ldots, y_n)$, then it follows from the relations (6) to (8) that

$$g(w|u, m + n) \propto g(w|s, m)g(w|t, n). \tag{9}$$

Therefore, the family is closed under multiplication and it satisfies the properties of a conjugate family.

It is often convenient to choose a slightly larger family of p.d.f.'s than the one we have just constructed for the conjugate family. For example, it can be seen from the relation (1) that for samples from a Bernoulli distribution, the function $f_n(x_1, \ldots, x_n|w)$ must always be proportional to the p.d.f. of a beta distribution for which both parameters α and β are positive integers. It follows that this subfamily of the family of beta distributions will be closed under multiplication, and it could have been chosen as the conjugate family. However, we found it convenient to choose the whole family of beta distributions as the conjugate family and to verify that it also was closed under multiplication.

In the subsequent sections of this chapter we shall present several theorems which identify conjugate families for samples from various distributions. Each of these theorems was discovered by writing down the function $f_n(x_1, \ldots, x_n|w)$ and then recognizing it as being proportional to a p.d.f. which belongs to one of the standard families of distributions described in Chaps. 4 and 5. When this family has been identified and the theorem has been formulated, its proof consists only of the verification that the family is closed under sampling.

When the function $f_n(x_1, \ldots, x_n|w)$ is regarded as a function of w, for given values $x_1, \ldots, x_n$ of the observations, it is called the *likelihood function*. Thus, the likelihood function will be of fundamental importance in our development of conjugate families of distributions.

Further Remarks and References

The concept of a conjugate family of distributions was formalized by Raiffa and Schlaifer (1961), who also studied in detail many of the families which will be presented here. Other theories of statistical inference in which the likelihood function is of fundamental importance are discussed by Fisher (1956), Barnard (1949, 1962, 1967), Birnbaum (1962), Barnard, Jenkins, and Winsten (1962), and Stein (1962b). These theories are related to, but distinct from, the Bayesian approach which is being presented in this book.

9.4 CONJUGATE FAMILIES FOR SAMPLES FROM VARIOUS STANDARD DISTRIBUTIONS

In this section we shall present conjugate families of distributions for samples from Poisson, negative binomial, and exponential distributions.

Theorem 1 *Suppose that $X_1, \ldots, X_n$ is a random sample from a Poisson distribution with an unknown value of the mean W. Suppose also that the prior distribution of W is a gamma distribution with parameters α and β such that $\alpha > 0$ and $\beta > 0$. Then the posterior distribution of W when $X_i = x_i$ $(i = 1, \ldots, n)$ is a gamma distribution with parameters $\alpha + \Sigma_{i=1}^n x_i$ and $\beta + n$.*

Proof Let $f_n(x_1, \ldots, x_n|w)$ denote the value of the likelihood function when $W = w$ and $X_i = x_i$ $(i = 1, \ldots, n)$, and let ξ denote the prior p.d.f. of W. If $y = \Sigma_{i=1}^n x_i$, then it follows from the hypotheses of the theorem that for $w > 0$,

$$f_n(x_1, \ldots, x_n|w) \propto w^y e^{-nw} \qquad (1)$$

and

$$\xi(w) \propto w^{\alpha-1} e^{-\beta w}. \qquad (2)$$

If $\xi(\cdot \,|x_1, \ldots, x_n)$ denotes the posterior p.d.f. of W when $X_i = x_i$
$(i = 1, \ldots, n)$, then

$$\xi(w|x_1, \ldots, x_n) \propto f_n(x_1, \ldots, x_n|w)\xi(w). \tag{3}$$

It now follows from the relations (1) to (3) that

$$\xi(w|x_1, \ldots, x_n) \propto w^{\alpha+y-1}e^{-(\beta+n)w}. \tag{4}$$

It can be seen from the relation (4) that the posterior p.d.f. of W is that
of a gamma distribution with parameters $\alpha + y$ and $\beta + n$.∎

The coefficient of variation, as defined by the ratio (5) of Sec. 3.6, is
commonly used as a measure of the dispersion of the distribution of a
positive random variable. It follows from Eq. (4) of Sec. 4.8 and from
Theorem 1 that the coefficient of variation of the posterior distribution
of W is $(\alpha + \Sigma_{i=1}^n x_i)^{-\frac{1}{2}}$.

Let ϵ be a fixed positive number, and suppose that observations are
to be drawn from the Poisson distribution until the coefficient of variation
of the posterior distribution of W is not greater than ϵ. Then sampling
must be continued until the inequality $\alpha + \Sigma_{i=1}^n x_i \geq 1/\epsilon^2$ has been
established.

The next theorem describes a conjugate family of distributions for
a sample from a negative binomial distribution. The proof of this theorem serves as Exercise 17 at the end of this chapter.

Theorem 2 *Suppose that $X_1, \ldots, X_n$ is a random sample from a negative binomial distribution with parameters r and W, where r has a specified
value $(r > 0)$ and the value of W is unknown. Suppose also that the prior
distribution of W is a beta distribution with parameters α and β such that
$\alpha > 0$ and $\beta > 0$. Then the posterior distribution of W when $X_i = x_i$
$(i = 1, \ldots, n)$ is a beta distribution with parameters $\alpha + rn$ and
$\beta + \Sigma_{i=1}^n x_i$.*

Theorem 2, together with Theorem 1 of Sec. 9.2, provides an opportunity to illustrate, for samples from a Bernoulli distribution, an important property which was discussed briefly in Sec. 8.12. Suppose that
each item in a large population of manufactured items can be classified as
either defective or nondefective. Suppose also that the proportion W of
defective items in the population is unknown but that W has a specified
prior distribution. Suppose further that a sample of items is to be
selected from the population and inspected. Consider the following
four sampling methods, any one of which might be used in obtaining the

sample:

1. A random sample of n items is selected from the population, where n is a fixed positive integer.
2. Items are selected at random from the population one at a time until exactly y defective items have been obtained, where y is a fixed positive integer.
3. Items are selected at random from the population one at a time until the inspector is called away to another problem.
4. Items are selected at random from the population until the inspector feels that he has accumulated sufficient information about W.

For any one of these four methods, let x denote the vector of all observed values which were found during the sampling process, and let $g(x|w)$ denote the value of the likelihood function when $W = w$. Furthermore, let n denote the total number of items which were inspected, and let y denote the number of these items which were defective. Then, regardless of which method of sampling was used, the following basic relation must hold:

$$g(x|w) \propto w^y (1 - w)^{n-y}. \tag{5}$$

It follows that the posterior distribution of W will be the same regardless of which method of sampling was used. In other words, the posterior distribution of W will depend only on the total number n of items which were inspected and the number y which were defective, and not on the sampling method which led to these results.

The next theorem describes a conjugate family of distributions for a sample from an exponential distribution. The proof of this theorem serves as Exercise 20.

Theorem 3 *Suppose that $X_1, \ldots, X_n$ is a random sample from an exponential distribution with an unknown value of the parameter W. Suppose also that the prior distribution of W is a gamma distribution with parameters α and β such that $\alpha > 0$ and $\beta > 0$. Then the posterior distribution of W when $X_i = x_i$ ($i = 1, \ldots, n$) is a gamma distribution with parameters $\alpha + n$ and $\beta + \sum_{i=1}^{n} x_i$.*

9.5 CONJUGATE FAMILIES FOR SAMPLES FROM A NORMAL DISTRIBUTION

We shall begin by considering a normal distribution for which the value of the precision or, equivalently, the value of the variance is specified.

Theorem 1 *Suppose that $X_1, \ldots, X_n$ is a random sample from a normal distribution with an unknown value of the mean W and a specified value of the precision r $(r > 0)$. Suppose also that the prior distribution of W is a normal distribution with mean μ and precision τ such that $-\infty < \mu < \infty$ and $\tau > 0$. Then the posterior distribution of W when $X_i = x_i$ $(i = 1, \ldots, n)$ is a normal distribution with mean μ' and precision $\tau + nr$, where*

$$\mu' = \frac{\tau\mu + nr\bar{x}}{\tau + nr}. \tag{1}$$

Proof For $-\infty < w < \infty$, the likelihood function $f_n(x_1, \ldots, x_n|w)$ satisfies the following relation:

$$f_n(x_1, \ldots, x_n|w) \propto \exp\left[-\frac{r}{2}\sum_{i=1}^{n}(x_i - w)^2\right]. \tag{2}$$

However,

$$\sum_{i=1}^{n}(x_i - w)^2 = n(w - \bar{x})^2 + \sum_{i=1}^{n}(x_i - \bar{x})^2. \tag{3}$$

Since the final term in Eq. (3) does not involve w, we may rewrite relation (2) as follows:

$$f_n(x_1, \ldots, x_n|w) \propto \exp\left[-\frac{nr}{2}(w - \bar{x})^2\right]. \tag{4}$$

The prior p.d.f. ξ of W satisfies the relation

$$\xi(w) \propto \exp\left[-\frac{\tau}{2}(w - \mu)^2\right], \tag{5}$$

and the posterior p.d.f. $\xi(\cdot|x_1, \ldots, x_n)$ of W will be proportional to the product of the functions specified by the relations (4) and (5). However, it can be shown that

$$\tau(w - \mu)^2 + nr(w - \bar{x})^2 = (\tau + nr)(w - \mu')^2 + \frac{\tau nr(\bar{x} - \mu)^2}{\tau + nr}. \tag{6}$$

Since the final term in Eq. (6) does not involve w, it can be included in the proportionality factor, and we obtain the relation

$$\xi(w|x_1, \ldots, x_n) \propto \exp\left[-\frac{\tau + nr}{2}(w - \mu')^2\right]. \tag{7}$$

Here μ' is specified by Eq. (1). It follows from the relation (7) that the posterior distribution of W is a normal distribution with mean μ' and precision $\tau + nr$.∎

Theorem 1 reveals the advantages of expressing our results in terms of the precision rather than the variance. The mean μ' of the posterior distribution of W can be written in the following form:

$$\mu' = \frac{nr}{\tau + nr}\, \bar{x} + \frac{\tau}{\tau + nr}\, \mu. \tag{8}$$

It is seen that μ' is a weighted average of $\bar{x}$ and μ, where $\bar{x}$ is the value of the sample mean and μ is the mean of the prior distribution of W. Therefore, we may conveniently regard the mean of the posterior distribution as a weighted average of an estimate of W formed from the sample and an estimate of W formed from the prior distribution. The weights of $\bar{x}$ and μ in this average are proportional to nr and τ, where nr is the precision of the conditional distribution of the sample mean for any given value of W and τ is the precision of the prior distribution of W. The larger the sample size n and the higher the precision r of each observation, the greater will be the weight that is given to $\bar{x}$.

The form of the precision of the posterior distribution of W is particularly simple. The precision increases by the amount r with each observation that is taken, regardless of the observed values. Therefore, as the number of observations increases, the distribution of W becomes more concentrated around its mean. Moreover, the concentration must increase in a fixed, predetermined way, while the values of the mean will depend on the observed values.

In the next theorem, we shall consider a normal distribution for which the value of the mean is specified but the value of the precision is unknown. The proof of this theorem serves as Exercise 24.

Theorem 2 *Suppose that $X_1, \ldots, X_n$ is a random sample from a normal distribution with a specified value of the mean m $(-\infty < m < \infty)$ and an unknown value of the precision W. Suppose also that the prior distribution of W is a gamma distribution with parameters α and β such that $\alpha > 0$ and $\beta > 0$. Then the posterior distribution of W when $X_i = x_i$ $(i = 1, \ldots, n)$ is a gamma distribution with parameters $\alpha + (n/2)$ and β', where*

$$\beta' = \beta + \tfrac{1}{2} \sum_{i=1}^{n} (x_i - m)^2. \tag{9}$$

For a gamma distribution with parameters α and β, the coefficient of variation is $\alpha^{-\frac{1}{2}}$. Therefore, it follows from Theorem 2 that the coefficient of variation of the posterior distribution of W must decrease in a fixed, predetermined way as the sample size n increases.

9.6 SAMPLING FROM A NORMAL DISTRIBUTION WITH UNKNOWN MEAN AND UNKNOWN PRECISION

We shall now consider the important problem of sampling from a normal distribution for which both the mean and the precision are unknown.

A conjugate family for this problem must be a family of bivariate distributions.

Theorem 1 *Suppose that $X_1, \ldots, X_n$ is a random sample from a normal distribution with an unknown value of the mean M and an unknown value of the precision R. Suppose also that the prior joint distribution of M and R is as follows: The conditional distribution of M when $R = r$ $(r > 0)$ is a normal distribution with mean μ and precision τr such that $-\infty < \mu < \infty$ and $\tau > 0$, and the marginal distribution of R is a gamma distribution with parameters α and β such that $\alpha > 0$ and $\beta > 0$. Then the posterior joint distribution of M and R when $X_i = x_i$ $(i = 1, \ldots, n)$ is as follows: The conditional distribution of M when $R = r$ is a normal distribution with mean μ' and precision $(\tau + n)r$, where*

$$\mu' = \frac{\tau\mu + n\bar{x}}{\tau + n}, \tag{1}$$

and the marginal distribution of R is a gamma distribution with parameters $\alpha + (n/2)$ and β', where

$$\beta' = \beta + \frac{1}{2} \sum_{i=1}^{n} (x_i - \bar{x})^2 + \frac{\tau n(\bar{x} - \mu)^2}{2(\tau + n)}. \tag{2}$$

Proof For $-\infty < m < \infty$ and $r > 0$, let $f_n(x_1, \ldots, x_n|m, r)$ denote the value of the likelihood function when $M = m$, $R = r$, and $X_i = x_i$ $(i = 1, \ldots, n)$, and let ξ denote the prior p.d.f. of M and R. Then

$$f_n(x_1, \ldots, x_n|m, r) \propto r^{n/2} \exp\left[-\frac{r}{2} \sum_{i=1}^{n} (x_i - m)^2 \right] \tag{3}$$

and

$$\xi(m, r) \propto r^{\frac{1}{2}} e^{-(\tau r/2)(m-\mu)^2} r^{\alpha-1} e^{-\beta r}. \tag{4}$$

The posterior p.d.f. $\xi(\cdot|x_1, \ldots, x_n)$ of M and R will be proportional to the product of the right sides of the relations (3) and (4). It follows from Eqs. (3) and (6) of Sec. 9.5 that this p.d.f. can be specified by the relation

$$\xi(m, r|x_1, \ldots, x_n)$$
$$\propto \left\{ r^{\frac{1}{2}} \exp\left[-\frac{(\tau + n)r}{2} (m - \mu')^2 \right] \right\} (r^{\alpha+n/2-1} e^{-\beta' r}). \tag{5}$$

Here, μ' is defined by Eq. (1) and β' by Eq. (2).

The function inside the braces in relation (5), when regarded as a function of m, must be proportional to the conditional p.d.f. of M when R is known since the variable m does not appear inside the set of parentheses on the right. However, for each fixed value of r, the function inside the braces is proportional to the p.d.f. of a normal distribution for which the mean and the precision are as given in the statement of the theorem. It now follows that the function inside the parentheses on the right must be proportional to the marginal p.d.f. of R. Therefore, the marginal distribution of R is a gamma distribution for which the parameters are as given in the statement of the theorem.∎

When the joint p.d.f. ξ of M and R is a *normal-gamma* p.d.f., as specified by the relation (4), the conditional distribution of M for any given value $R = r$ will be normal but the marginal distribution of M will not be normal. The marginal p.d.f. ξ_M of M is defined by the equation

$$\xi_M(m) = \int_0^\infty \xi(m, r) \, dr \qquad \text{for } -\infty < m < \infty. \tag{6}$$

Therefore, if we make use of the proportionality symbol and drop all factors which do not involve m, it follows from the relation (4) that ξ_M has the form

$$\xi_M(m) \propto \left[\beta + \frac{\tau}{2} (m - \mu)^2 \right]^{-\alpha-\frac{1}{2}} \tag{7}$$

or

$$\xi_M(m) \propto \left[1 + \frac{1}{2\alpha} \frac{\alpha\tau(m - \mu)^2}{\beta} \right]^{-(2\alpha+1)/2}. \tag{8}$$

From a comparison of the function given in the relation (8) with the p.d.f. of the t distribution specified by Eq. (5) of Sec. 4.12, it is seen that the marginal distribution of M is a t distribution with 2α degrees of freedom, location parameter μ, and precision $\alpha\tau/\beta$. The posterior marginal distribution of M is obtained by replacing μ, τ, α, and β by their posterior values as given in Theorem 1. Hence, although the number of degrees of freedom $2\alpha + n$ of the posterior t distribution will not depend on the observed values $x_1, \ldots, x_n$, both the location parameter and the precision of the posterior distribution will depend on these values.

An interesting feature of the conjugate family of joint distributions of M and R specified by Theorem 1 is the following: For any normal-gamma distribution in this family, the variables M and R are dependent. There is no joint distribution in the family such that M has a normal distribution, R has a gamma distribution, and M and R are independent. This is not an important deficiency of the family. In fact, even if the prior distribution of M and R specified that these variables were independent, their posterior distribution after the value of a single observation had been noted would specify that they were dependent.

Another interesting feature is that for each distribution in this conjugate family, the precision of the conditional distribution of M when $R = r$ must be proportional to r. In other words, the prior distribution of M and R must have the following property: If the statistician is suddenly told that the value of R is r, where r is a large number, then because of this knowledge about R, he can now also specify the value of M with high precision. On the other hand, if the statistician is suddenly told that the value of R is r', where r' is a small number, then the statistician's distribution for M will still have a very large variance. Again, this is not an important deficiency of the conjugate family, for suppose that the statistician learns the value of even a single observation drawn at random from the normal distribution. If the precision R of the distribution is large, then the observation provides very precise information about the value of the mean M; whereas if the precision R is small, then the observation provides little information about the value of M.

A Numerical Example

Consider a normal distribution in which the values of the mean M and the precision R are unknown, and suppose that a statistician wishes to select a normal-gamma distribution from the conjugate family to represent the prior distribution of M and R. If he specifies that $E(M) = 2$, $\text{Var}(M) = 5$, $E(R) = 3$, and $\text{Var}(R) = 3$, what values should be selected for the parameters μ, τ, α, and β of the prior distribution? Since R has a gamma distribution with parameters α and β, the values $\alpha = 3$ and $\beta = 1$ can be found from Eq. (4) of Sec. 4.8. Furthermore, $\mu = 2$ since $E(M) = \mu$. Finally, since M has a t distribution with 2α degrees of freedom and precision $\alpha\tau/\beta$, it follows from Eq. (4) of Sec. 4.12 that

$$\text{Var}(M) = \frac{\beta}{\tau(\alpha - 1)}. \tag{9}$$

Hence, $\tau = 0.1$. This value completes the specification of the prior distribution.

Now suppose that a random sample of 10 observations is taken from the given normal distribution and it is found that for these 10 values, $\bar{x} = 4.20$ and $\Sigma_{i=1}^{10}(x_i - \bar{x})^2 = 5.40$. Then, by Theorem 1, the parameters μ', τ', α', and β' of the posterior distribution of M and R are $\mu' = 4.18$, $\tau' = 10.1$, $\alpha' = 8$, and $\beta' = 3.94$. It follows from these values that the means and variances of M and R are now $E(M) = 4.18$, $\text{Var}(M) = 0.056$, $E(R) = 2.03$, and $\text{Var}(R) = 0.515$.

Next, suppose that 10 more observations are now taken from the given normal distribution and it is found that for these 10 values, $\bar{x} = 4.48$ and $\Sigma_{i=1}^{10}(x_i - \bar{x})^2 = 5.82$. By considering the posterior distribution which we just found as the prior distribution for these new observations, the parameters μ'', τ'', α'', and β'' of the new posterior distribution of M and R become $\mu'' = 4.33$, $\tau'' = 20.1$, $\alpha'' = 13$, and $\beta'' = 7.08$. Therefore, the means and variances of M and R now have the values $E(M) = 4.33$, $\text{Var}(M) = 0.029$, $E(R) = 1.84$, and $\text{Var}(R) = 0.260$.

Since M now has a t distribution with 26 degrees of freedom, location parameter 4.33, and precision 36.9, it follows from tables of the t distribution that

$$\Pr[-2.056 \leq (36.9)^{\frac{1}{2}}(M - 4.33) \leq 2.056] = 0.95. \tag{10}$$

Upon simplification, Eq. (10) reduces to the equation

$$\Pr(3.99 \leq M \leq 4.67) = 0.95. \tag{11}$$

9.7 SAMPLING FROM A UNIFORM DISTRIBUTION

In this section we shall describe conjugate families of distributions for samples from a uniform distribution such that either the value of one end point or the values of both end points are unknown.

Theorem 1 *Suppose that $X_1, \ldots, X_n$ is a random sample from a uniform distribution on the interval $(0, W)$, where the value of W is unknown. Suppose also that the prior distribution of W is a Pareto distribution with parameters w_0 and α such that $w_0 > 0$ and $\alpha > 0$. Then the posterior distribution of W when $X_i = x_i$ $(i = 1, \ldots, n)$ is a Pareto distribution with parameters w_0' and $\alpha + n$, where*

$$w_0' = \max \{w_0, x_1, \ldots, x_n\}. \tag{1}$$

Proof For $w > w_0$, the prior p.d.f. ξ of W is of the following form:

$$\xi(w) \propto \frac{1}{w^{\alpha+1}}. \tag{2}$$

Furthermore, $\xi(w) = 0$ for $w \leq w_0$. The likelihood function $f_n(x_1, \ldots, x_n|w)$ is specified by Eq. (13) of Sec. 9.1.

It follows from these relations that the posterior p.d.f. $\xi(w|x_1, \ldots, x_n)$ of W will be positive only for values of w such that $w > w_0$ and $w > \max \{x_1, \ldots, x_n\}$. Therefore, $\xi(w|x_1, \ldots, x_n) > 0$ only if $w > w_0'$, where w_0' is defined by Eq. (1). Furthermore, for $w > w_0'$,

$$\xi(w|x_1, \ldots, x_n) \propto \frac{1}{w^{\alpha+n+1}}. \tag{3}$$

It is seen from the relation (3) that the posterior distribution of W must be a Pareto distribution whose parameters are as specified in the statement of the theorem. ∎

In the next theorem, we shall consider a uniform distribution in which both end points are unknown. The proof of this theorem serves as Exercise 32.

Theorem 2 *Suppose that $X_1, \ldots, X_n$ is a random sample from a uniform distribution on the interval (W_1, W_2), where the values of W_1 and W_2 are unknown. Suppose also that the prior joint distribution of W_1 and W_2 is a bilateral bivariate Pareto distribution with parameters $r_1, r_2,$ and α such that $r_1 < r_2$ and $\alpha > 0$. Then the posterior joint distribution of W_1 and W_2 when $X_i = x_i$ $(i = 1, \ldots, n)$ is a bilateral bivariate Pareto distribution with parameters $r_1', r_2',$ and $\alpha + n$, where*

$$r_1' = \min \{r_1, x_1, \ldots, x_n\}$$
$$and \qquad r_2' = \max \{r_2, x_1, \ldots, x_n\}. \qquad (4)$$

Let us now consider a numerical example which illustrates the use of Theorem 2. Suppose that if a certain type of atomic particle is passed through a container of water, the horizontal deflection of the particle, as measured in appropriate units, has a uniform distribution on the interval (W_1, W_2), where the values of W_1 and W_2 are unknown. Suppose also that it is known that $W_1 < -0.4$ and $W_2 > 0.1$. If bounds of this type cannot be specified in advance by the statistician, then they can certainly be specified after the horizontal deflections y_1 and y_2 of two particles have been observed. In fact, since y_1 and y_2 must lie in the interval (W_1, W_2), then, if we assume that $y_1 < y_2$, it follows that $W_1 < y_1$ and $W_2 > y_2$.

Suppose further that the statistician wishes to select a bilateral bivariate Pareto distribution to represent the prior distribution of W_1 and W_2 and that he expects the length $W_2 - W_1$ of the interval (W_1, W_2) to be about 2.5 units. What values of the parameters $r_1, r_2,$ and α of the prior distribution should be selected? It follows from the above bounds on W_1 and W_2 that $r_1 = -0.4$ and $r_2 = 0.1$. Furthermore, by Eq. (2) of Sec. 5.7, if $\alpha > 1$, then

$$E(W_2 - W_1) = \frac{(\alpha + 1)(r_2 - r_1)}{\alpha - 1}. \qquad (5)$$

If it is assumed that $E(W_2 - W_1) = 2.5$, then $\alpha = 1.5$. This value completes the specification of the prior distribution.

Suppose now that the deflections of five particles are observed and found to have the values $-0.27, -0.45, -0.36, -0.12,$ and 0.47. Since the minimum of these five values is -0.45 and the maximum is 0.47, it follows from Theorem 2 that the values of the parameters $r_1', r_2',$ and α' of the posterior distribution of W are $r_1' = -0.45$, $r_2' = 0.47$, and $\alpha' = 6.5$. Therefore, it is now known that $W_1 < -0.45$ and $W_2 > 0.47$. Furthermore, by Eq. (5), the expected length $E(W_2 - W_1)$ is now 1.25.

Next, suppose that the deflections of five more particles are found to have the values $-0.39, -0.07, 0.43, 0.01,$ and -0.14. The parameters $r_1'', r_2'',$ and α'' of the new posterior distribution of W_1 and W_2 are $r_1'' = r_1'$,

$r_2'' = r_2'$, and $\alpha'' = 11.5$. Therefore, by Eq. (5), the expected length $E(W_2 - W_1)$ is now 1.10.

Since the interval $(-0.45, 0.47)$, whose length is 0.92, must be wholly contained in the interval (W_1, W_2) and since the expected length of the interval (W_1, W_2) is itself 1.10, the statistician now has relatively precise information about the values of W_1 and W_2. In fact, by Eq. (3) of Sec. 5.7, it follows that the variances of W_1 and W_2 now have the common value 0.0093.

9.8 A CONJUGATE FAMILY FOR MULTINOMIAL OBSERVATIONS

In the next theorem, it is shown that the family of Dirichlet distributions is a conjugate family for observations which have a multinomial distribution.

Theorem 1 *Suppose that the random vector* $\mathbf{X} = (X_1, \ldots, X_k)'$ *has a multinomial distribution with parameters* n *and* $\mathbf{W} = (W_1, \ldots, W_k)'$, *where* n *is a specified positive integer and the values of the components of the vector* $\mathbf{W}$ *are unknown. Suppose also that the prior distribution of* $\mathbf{W}$ *is a Dirichlet distribution with parametric vector* $\boldsymbol{\alpha} = (\alpha_1, \ldots, \alpha_k)'$ *such that* $\alpha_i > 0$ $(i = 1, \ldots, k)$. *Then the posterior distribution of* $\mathbf{W}$ *when* $X_i = x_i$ $(i = 1, \ldots, k)$ *is a Dirichlet distribution with parametric vector* $\boldsymbol{\alpha}^* = (\alpha_1 + x_1, \ldots, \alpha_k + x_k)'$.

Proof Let Ω denote the set of points $\mathbf{w} = (w_1, \ldots, w_k)'$ such that $w_i > 0$ $(i = 1, \ldots, k)$ and $w_1 + \cdots + w_k = 1$. Then for any given value $\mathbf{w}$ of $\mathbf{W}$ such that $\mathbf{w} \in \Omega$, the likelihood function $f(x_1, \ldots, x_k | \mathbf{w})$ satisfies the following relation:

$$f(x_1, \ldots, x_k | \mathbf{w}) \propto \prod_{i=1}^{k} w_i^{x_i}. \tag{1}$$

Furthermore, for $\mathbf{w} \in \Omega$, the prior p.d.f. ξ of $\mathbf{W}$ satisfies the relation

$$\xi(\mathbf{w}) \propto \prod_{i=1}^{k} w_i^{\alpha_i - 1}. \tag{2}$$

Therefore, for $\mathbf{w} \in \Omega$, the posterior p.d.f. $\xi(\cdot | x_1, \ldots, x_k)$ of $\mathbf{W}$ must satisfy the relation

$$\xi(\mathbf{w} | x_1, \ldots, x_k) \propto \prod_{i=1}^{k} w_i^{\alpha_i + x_i - 1}. \tag{3}$$

The function in the relation (3) is proportional to the p.d.f. of a Dirichlet distribution whose parametric vector is as specified in the statement of the theorem. ∎

As an example, suppose that in a large shipment of manufactured items, there are items of k different types. For $i = 1, \ldots, k$, let W_i denote the proportion of items which are of type i, and assume that the prior distribution of the vector $\mathbf{W} = (W_1, \ldots, W_k)'$ is a Dirichlet distribution with parametric vector $\boldsymbol{\alpha} = (\alpha_1, \ldots, \alpha_k)'$. If items are selected at random from the shipment, one at a time, then it follows from Theorem 1 that the posterior distribution of $\mathbf{W}$ at each stage will be a Dirichlet distribution, and for $i = 1, \ldots, k$, the ith component of the parametric vector $\boldsymbol{\alpha}$ will be increased by 1 unit each time an item of type i is selected.

9.9 CONJUGATE FAMILIES FOR SAMPLES FROM A MULTIVARIATE NORMAL DISTRIBUTION

In the remainder of this chapter we shall consider problems in which samples are taken from a nonsingular, k-dimensional multivariate normal distribution ($k \geq 1$). In each problem, the mean vector of the distribution must be a k-dimensional vector, i.e., a point in R^k, and the precision matrix of the distribution must be a symmetric $k \times k$ positive definite matrix. Any observation $\mathbf{X}$ from this distribution will be a k-dimensional random vector whose value $\mathbf{x}$ will lie in the space R^k. The results to be derived will be generalizations of those obtained in Secs. 9.5 and 9.6, in which we dealt with samples from a univariate normal distribution. We shall begin by discussing the problem of sampling from a distribution for which the precision matrix $\mathbf{r}$ is specified. If $\mathbf{x}_1, \ldots, \mathbf{x}_n$ are the values of a sample of observations $\mathbf{X}_1, \ldots, \mathbf{X}_n$, we shall, as usual, let $\bar{\mathbf{x}}$ denote the sample mean vector as defined by the equation

$$\bar{\mathbf{x}} = \frac{1}{n} \sum_{i=1}^{n} \mathbf{x}_i. \tag{1}$$

Theorem 1 *Suppose that $\mathbf{X}_1, \ldots, \mathbf{X}_n$ is a random sample from a multivariate normal distribution with an unknown value of the mean vector $\mathbf{M}$ and a specified precision matrix $\mathbf{r}$. Suppose also that the prior distribution of $\mathbf{M}$ is a multivariate normal distribution with mean vector $\boldsymbol{\mu}$ and precision matrix $\boldsymbol{\tau}$ such that $\boldsymbol{\mu} \in R^k$ and $\boldsymbol{\tau}$ is a symmetric positive definite matrix. Then the posterior distribution of $\mathbf{M}$ when $\mathbf{X}_i = \mathbf{x}_i$ ($i = 1, \ldots, n$) is a multivariate normal distribution with mean vector $\boldsymbol{\mu}^*$ and precision matrix $\boldsymbol{\tau} + n\mathbf{r}$, where*

$$\boldsymbol{\mu}^* = (\boldsymbol{\tau} + n\mathbf{r})^{-1}(\boldsymbol{\tau}\boldsymbol{\mu} + n\mathbf{r}\bar{\mathbf{x}}). \tag{2}$$

Proof For $\mathbf{M} = \mathbf{m}$ and $\mathbf{X}_i = \mathbf{x}_i$ ($i = 1, \ldots, n$), the likelihood function

$f_n(\mathbf{x}_1, \ldots, \mathbf{x}_n | \mathbf{m})$ satisfies the following relation:

$$f_n(\mathbf{x}_1, \ldots, \mathbf{x}_n | \mathbf{m}) \propto \exp\left[-\tfrac{1}{2} \sum_{i=1}^{n} (\mathbf{x}_i - \mathbf{m})' \mathbf{r}(\mathbf{x}_i - \mathbf{m}) \right]. \qquad (3)$$

However,

$$\sum_{i=1}^{n} (\mathbf{x}_i - \mathbf{m})' \mathbf{r}(\mathbf{x}_i - \mathbf{m})$$
$$= \sum_{i=1}^{n} (\mathbf{x}_i - \bar{\mathbf{x}})' \mathbf{r}(\mathbf{x}_i - \bar{\mathbf{x}}) + n(\mathbf{m} - \bar{\mathbf{x}})' \mathbf{r}(\mathbf{m} - \bar{\mathbf{x}}). \qquad (4)$$

Therefore, the relation (3) can be rewritten in the following form:

$$f_n(\mathbf{x}_1, \ldots, \mathbf{x}_n | \mathbf{m}) \propto \exp\left[-\tfrac{1}{2}(\mathbf{m} - \bar{\mathbf{x}})'(n\mathbf{r})(\mathbf{m} - \bar{\mathbf{x}}) \right]. \qquad (5)$$

The prior p.d.f. ξ of $\mathbf{M}$ satisfies the relation

$$\xi(\mathbf{m}) \propto \exp\left[-\tfrac{1}{2}(\mathbf{m} - \boldsymbol{\mu})' \boldsymbol{\tau}(\mathbf{m} - \boldsymbol{\mu}) \right]. \qquad (6)$$

The posterior p.d.f. $\xi(\,\cdot\,|\mathbf{x}_1, \ldots, \mathbf{x}_n)$ of $\mathbf{M}$ is proportional to the product of the functions specified in the relations (5) and (6). However, it can be verified that

$$(\mathbf{m} - \boldsymbol{\mu})' \boldsymbol{\tau}(\mathbf{m} - \boldsymbol{\mu}) + (\mathbf{m} - \bar{\mathbf{x}})'(n\mathbf{r})(\mathbf{m} - \bar{\mathbf{x}})$$
$$= (\mathbf{m} - \boldsymbol{\mu}^*)'(\boldsymbol{\tau} + n\mathbf{r})(\mathbf{m} - \boldsymbol{\mu}^*)$$
$$+ \text{(terms which do not involve } \mathbf{m}). \qquad (7)$$

Since the terms in Eq. (7) which do not involve $\mathbf{m}$ can be absorbed in the proportionality factor, we obtain the following relation:

$$\xi(\mathbf{m} | \mathbf{x}_1, \ldots, \mathbf{x}_n) \propto \exp\left[-\tfrac{1}{2}(\mathbf{m} - \boldsymbol{\mu}^*)'(\boldsymbol{\tau} + n\mathbf{r})(\mathbf{m} - \boldsymbol{\mu}^*) \right]. \qquad (8)$$

The p.d.f. specified by the relation (8) is that of a multivariate normal distribution for which the mean vector and the precision matrix are as specified in the statement of the theorem.∎

The analogy between the multivariate results of this theorem and the univariate results of Theorem 1 of Sec. 9.5 is evident, and the discussion which was given following that theorem is relevant here.

Now suppose that we are sampling from a multivariate normal distribution with a specified mean vector but an unknown precision matrix.

Theorem 2 *Suppose that* $\mathbf{X}_1, \ldots, \mathbf{X}_n$ *is a random sample from a multivariate normal distribution with a specified mean vector* $\mathbf{m}$ *and an unknown value of the precision matrix* $\mathbf{R}$. *Suppose also that the prior distribution of* $\mathbf{R}$ *is a Wishart distribution with* α *degrees of freedom and precision matrix* $\boldsymbol{\tau}$ *such that* $\alpha > k - 1$ *and* $\boldsymbol{\tau}$ *is a symmetric positive definite matrix. Then the posterior distribution of* $\mathbf{R}$ *when* $\mathbf{X}_i = \mathbf{x}_i$ $(i = 1, \ldots, n)$ *is a Wishart* *distribution with* $\alpha + n$ *degrees of freedom and precision matrix* τ^*, *where*

$$\tau^* = \tau + \sum_{i=1}^{n} (\mathbf{x}_i - \mathbf{m})(\mathbf{x}_i - \mathbf{m})'. \tag{9}$$

Proof The likelihood function $f_n(\mathbf{x}_1, \ldots, \mathbf{x}_n | \mathbf{r})$ satisfies the following relation:

$$f_n(\mathbf{x}_1, \ldots, \mathbf{x}_n | \mathbf{r}) \propto |\mathbf{r}|^{n/2} \exp\left[-\tfrac{1}{2} \sum_{i=1}^{n} (\mathbf{x}_i - \mathbf{m})'\mathbf{r}(\mathbf{x}_i - \mathbf{m})\right]. \tag{10}$$

The exponent of e in relation (10) is a real number which may be regarded as a 1×1 matrix. Therefore, by Eqs. (1) and (2) of Sec. 3.5, we obtain the following relation:

$$\sum_{i=1}^{n} (\mathbf{x}_i - \mathbf{m})'\mathbf{r}(\mathbf{x}_i - \mathbf{m}) = \operatorname{tr}\left[\sum_{i=1}^{n} (\mathbf{x}_i - \mathbf{m})'\mathbf{r}(\mathbf{x}_i - \mathbf{m})\right]$$

$$= \operatorname{tr}\left\{\left[\sum_{i=1}^{n} (\mathbf{x}_i - \mathbf{m})(\mathbf{x}_i - \mathbf{m})'\right]\mathbf{r}\right\}. \tag{11}$$

Furthermore, the p.d.f. ξ of $\mathbf{R}$ satisfies the relation

$$\xi(\mathbf{r}) \propto |\mathbf{r}|^{(\alpha-k-1)/2} \exp\left[-\tfrac{1}{2} \operatorname{tr}(\tau\mathbf{r})\right]. \tag{12}$$

Since the posterior p.d.f. of $\mathbf{R}$ is proportional to the product of the functions given in the relations (10) and (12), it follows from Eq. (11) that the posterior distribution of $\mathbf{R}$ must be a Wishart distribution as specified in the statement of the theorem.∎

This theorem is a straightforward multivariate generalization of Theorem 2 of Sec. 9.5. The analogy between the two theorems is slightly obscured, however, by the fact that a one-dimensional Wishart distribution with α degrees of freedom and precision matrix β, where β is simply a positive number, is the same as a gamma distribution with parameters $\alpha/2$ and $\beta/2$. Although a slightly different definition of the parameters of one of these distributions would have removed this discrepancy, the definitions which we are using here are the traditional ones.

It should be noted that the Wishart p.d.f. is well defined by the relation (12) even when the number of degrees of freedom α is not an integer, provided that $\alpha > k - 1$. However, if α is chosen to be an integer in the prior distribution of $\mathbf{R}$, then the number of degrees of freedom of the posterior distribution will also be an integer.

9.10 MULTIVARIATE NORMAL DISTRIBUTIONS WITH UNKNOWN MEAN VECTOR AND UNKNOWN PRECISION MATRIX

We shall now extend Theorem 1 of Sec. 9.6 by considering the problem of sampling from a multivariate normal distribution for which both the mean vector and the precision matrix are unknown. If $\mathbf{x}_1, \ldots, \mathbf{x}_n$ are the values of a sample of observations $\mathbf{X}_1, \ldots, \mathbf{X}_n$, we shall let $\mathbf{s}$ denote the symmetric $k \times k$ nonnegative definite matrix which is defined by the equation

$$\mathbf{s} = \sum_{i=1}^{n} (\mathbf{x}_i - \bar{\mathbf{x}})(\mathbf{x}_i - \bar{\mathbf{x}})'. \tag{1}$$

Theorem 1 *Suppose that $\mathbf{X}_1, \ldots, \mathbf{X}_n$ is a random sample from a multivariate normal distribution with an unknown value of the mean vector $\mathbf{M}$ and an unknown value of the precision matrix $\mathbf{R}$. Suppose also that the prior joint distribution of $\mathbf{M}$ and $\mathbf{R}$ is as follows: The conditional distribution of $\mathbf{M}$ when $\mathbf{R} = \mathbf{r}$ is a multivariate normal distribution with mean vector $\mathbf{\mu}$ and precision matrix $\nu\mathbf{r}$ such that $\mathbf{\mu} \in R^k$ and $\nu > 0$, and the marginal distribution of $\mathbf{R}$ is a Wishart distribution with α degrees of freedom and precision matrix τ such that $\alpha > k - 1$ and τ is a symmetric positive definite matrix. Then the posterior joint distribution of $\mathbf{M}$ and $\mathbf{R}$ when $\mathbf{X}_i = \mathbf{x}_i \ (i = 1, \ldots, n)$ is as follows: The conditional distribution of $\mathbf{M}$ when $\mathbf{R} = \mathbf{r}$ is a multivariate normal distribution with mean vector $\mathbf{\mu}^*$ and precision matrix $(\nu + n)\mathbf{r}$, where*

$$\mathbf{\mu}^* = \frac{\nu\mathbf{\mu} + n\bar{\mathbf{x}}}{\nu + n}, \tag{2}$$

and the marginal distribution of $\mathbf{R}$ is a Wishart distribution with $\alpha + n$ degrees of freedom and precision matrix τ^*, where

$$\tau^* = \tau + \mathbf{s} + \frac{\nu n}{\nu + n}(\mathbf{\mu} - \bar{\mathbf{x}})(\mathbf{\mu} - \bar{\mathbf{x}})'. \tag{3}$$

Proof When $\mathbf{M} = \mathbf{m}, \mathbf{R} = \mathbf{r}$, and $\mathbf{X}_i = \mathbf{x}_i \ (i = 1, \ldots, n)$, the likelihood function $f_n(\mathbf{x}_1, \ldots, \mathbf{x}_n | \mathbf{m}, \mathbf{r})$ is proportional to the function specified by the right side of relation (10) of Sec. 9.9. Furthermore, it follows from Eqs. (4) and (11) of Sec. 9.9 and from Eq. (1) that the sum which appears in the exponent of this function can be written in the form

$$\sum_{i=1}^{n} (\mathbf{x}_i - \mathbf{m})'\mathbf{r}(\mathbf{x}_i - \mathbf{m}) = n(\mathbf{m} - \bar{\mathbf{x}})'\mathbf{r}(\mathbf{m} - \bar{\mathbf{x}}) + \mathrm{tr}\ (\mathbf{sr}). \tag{4}$$

The prior joint p.d.f. ξ of $\mathbf{M}$ and $\mathbf{R}$ satisfies the following relation:

$$\xi(\mathbf{m}, \mathbf{r}) \propto |\mathbf{r}|^{\frac{1}{2}} \exp \left[-\frac{\nu}{2}(\mathbf{m} - \mathbf{\mu})'\mathbf{r}(\mathbf{m} - \mathbf{\mu}) \right]$$
$$\times |\mathbf{r}|^{(\alpha - k - 1)/2} \exp \left[-\tfrac{1}{2}\,\mathrm{tr}\ (\tau\mathbf{r}) \right]. \tag{5}$$

It can be verified that

$$\nu(\mathbf{m} - \mathbf{\mu})'\mathbf{r}(\mathbf{m} - \mathbf{\mu}) + n(\mathbf{m} - \bar{\mathbf{x}})'\mathbf{r}(\mathbf{m} - \bar{\mathbf{x}})$$
$$= (\nu + n)(\mathbf{m} - \mathbf{\mu}^*)'\mathbf{r}(\mathbf{m} - \mathbf{\mu}^*) + \frac{\nu n}{\nu + n}(\mathbf{\mu} - \bar{\mathbf{x}})'\mathbf{r}(\mathbf{\mu} - \bar{\mathbf{x}}). \tag{6}$$

Furthermore, the final term in Eq. (6) can be rewritten as follows:

$$\frac{vn}{v + n} (\mathbf{\mu} - \bar{\mathbf{x}})' \mathbf{r} (\mathbf{\mu} - \bar{\mathbf{x}}) = \text{tr} \left[\frac{vn}{v + n} (\mathbf{\mu} - \bar{\mathbf{x}})(\mathbf{\mu} - \bar{\mathbf{x}})' \mathbf{r} \right]. \qquad (7)$$

From relation (10) of Sec. 9.9 and from (4) to (7), we can obtain the following relation for the posterior joint p.d.f. $\xi(\mathbf{m}, \mathbf{r}|\mathbf{x}_1, \ldots, \mathbf{x}_n)$ of $\mathbf{M}$ and $\mathbf{R}$:

$$\xi(\mathbf{m}, \mathbf{r}|\mathbf{x}_1, \ldots, \mathbf{x}_n) \propto \left\{ |\mathbf{r}|^{\frac{1}{2}} \exp \left[- \frac{v + n}{2} (\mathbf{m} - \mathbf{\mu}^*)' \mathbf{r} (\mathbf{m} - \mathbf{\mu}^*) \right] \right\}$$
$$\times \{ |\mathbf{r}|^{(\alpha+n-k-1)/2} \exp \left[-\tfrac{1}{2} \text{tr} \, (\mathbf{\tau}^* \mathbf{r}) \right] \}. \qquad (8)$$

The function inside the first set of braces in the relation (8), when regarded as a function of $\mathbf{m}$, must be proportional to the conditional p.d.f. of $\mathbf{M}$ when $\mathbf{R} = \mathbf{r}$, since the variable $\mathbf{m}$ does not appear inside the second set of braces. This function is proportional to the p.d.f. of a multivariate normal distribution for which the mean vector and the precision matrix are as specified in the statement of the theorem. It now follows that the function inside the second set of braces in the relation (8) is proportional to the marginal p.d.f. of $\mathbf{R}$, and it is proportional to the p.d.f. of the Wishart distribution specified in the statement of the theorem.∎

9.11 THE MARGINAL DISTRIBUTION OF THE MEAN VECTOR

We shall now find the marginal distribution of $\mathbf{M}$ when the joint distribution of $\mathbf{M}$ and $\mathbf{R}$ is a *multivariate normal-Wishart* distribution of the form specified in Theorem 1 of Sec. 9.10. If the analogy with the univariate results given in Theorem 1 of Sec. 9.6 can be extended far enough, then the discussion following that theorem would lead us to conclude that the marginal distribution of $\mathbf{M}$ here will be a multivariate t distribution. Such a conclusion is correct, as we shall now show.

Suppose, as in Theorem 1 of Sec. 9.10, that the conditional distribution of $\mathbf{M}$ when $\mathbf{R} = \mathbf{r}$ is a multivariate normal distribution with mean vector $\mathbf{\mu}$ and precision matrix $v\mathbf{r}$ and that the marginal distribution of $\mathbf{R}$ is a Wishart distribution with α degrees of freedom and precision matrix $\mathbf{\tau}$. Then the joint p.d.f. ξ of $\mathbf{M}$ and $\mathbf{R}$, as specified by the relation (5) of Sec. 9.10, can be rewritten as follows:

$$\xi(\mathbf{m}, \mathbf{r}) \propto |\mathbf{r}|^{(\alpha-k)/2} \exp \left(-\tfrac{1}{2} \text{tr} \{ [\mathbf{\tau} + v(\mathbf{m} - \mathbf{\mu})(\mathbf{m} - \mathbf{\mu})'] \mathbf{r} \} \right). \qquad (1)$$

The marginal p.d.f. ξ_M of $\mathbf{M}$ is obtained by integrating the $k(k + 1)/2$ distinct variables in the symmetric matrix $\mathbf{r}$ over the set of all values such that $\mathbf{r}$ is positive definite. For any positive integer n $(n \geq k)$ and any positive definite matrix $\mathbf{T}$, the p.d.f. of the Wishart distribution specified by Eq. (11) of Sec. 5.5 must integrate to unity over this set. Hence, by integrating the function on the right side of relation (1) over this set, we obtain the relation

$$\xi_M(\mathbf{m}) \propto |\boldsymbol{\tau} + \nu(\mathbf{m} - \boldsymbol{\mu})(\mathbf{m} - \boldsymbol{\mu})'|^{-(\alpha+1)/2}. \tag{2}$$

A standard result in the theory of determinants (see Exercise 40) may be stated as follows: If $\mathbf{A}$ is any $k \times k$ nonsingular matrix and $\mathbf{v}$ is any k-dimensional column vector, then

$$|\mathbf{A} + \mathbf{vv}'| = |\mathbf{A}|(1 + \mathbf{v}'\mathbf{A}^{-1}\mathbf{v}). \tag{3}$$

From the relations (2) and (3), we can obtain the following result:

$$\xi_M(\mathbf{m}) \propto [1 + \nu(\mathbf{m} - \boldsymbol{\mu})'\boldsymbol{\tau}^{-1}(\mathbf{m} - \boldsymbol{\mu})]^{-(\alpha+1)/2}. \tag{4}$$

When the function on the right side of (4) is rewritten so as to bring it into the form of the p.d.f. of a multivariate t distribution specified by Eq. (9) of Sec. 5.6, it can be seen that $\mathbf{M}$ has a multivariate t distribution with $\alpha - k + 1$ degrees of freedom, location vector $\boldsymbol{\mu}$, and precision matrix $\nu(\alpha - k + 1)\boldsymbol{\tau}^{-1}$. This t distribution is the marginal distribution of $\mathbf{M}$ under the prior joint distribution of $\mathbf{M}$ and $\mathbf{R}$ specified in Theorem 1 of Sec. 9.10. The posterior marginal distribution of $\mathbf{M}$ is obtained by replacing $\boldsymbol{\mu}$, ν, α, and $\boldsymbol{\tau}$ in the t distribution by their posterior values as given in that same theorem. Hence, the number of degrees of freedom $\alpha + n - k + 1$ of the posterior distribution does not depend on the observed values $\mathbf{x}_1, \ldots, \mathbf{x}_n$ in the sample. However, the location vector of the posterior distribution depends on the value of the sample mean vector $\bar{\mathbf{x}}$, and the precision matrix of the posterior distribution depends on both the vector $\bar{\mathbf{x}}$ and the matrix $\mathbf{s}$ formed from the sample.

9.12   THE DISTRIBUTION OF A CORRELATION

Suppose that X_1 and X_2 are two random variables whose joint distribution is a bivariate normal distribution with precision matrix $\mathbf{R}$. Let the elements of the 2×2 matrix $\mathbf{R}$ be defined by the equation

$$\mathbf{R} = \begin{pmatrix} R_{11} & R_{12} \\ R_{12} & R_{22} \end{pmatrix}. \tag{1}$$

Since the precision matrix $\mathbf{R}$ is the inverse of the covariance matrix of X_1 and X_2, it follows that the correlation P of X_1 and X_2 is specified by the equation

$$P = \frac{-R_{12}}{(R_{11}R_{22})^{\frac{1}{2}}}. \tag{2}$$

Suppose now that the value of the precision matrix $\mathbf{R}$ is unknown and that in accordance with the results developed in the preceding sections, it is assumed that the distribution of $\mathbf{R}$ is a Wishart distribution with α degrees of freedom and precision matrix τ such that $\alpha > 1$ and τ is a symmetric, positive definite matrix whose elements are specified by the equation

$$\tau = \begin{bmatrix} \tau_{11} & \tau_{12} \\ \tau_{12} & \tau_{22} \end{bmatrix}. \tag{3}$$

In this section, we shall derive the distribution of the correlation P. It will be shown that, in general, this distribution is quite complicated, that it is not one of the standard distributions which have been discussed in this book, and that the p.d.f. of P involves a certain integral which cannot be evaluated directly. However, if $\tau_{12} = 0$ in the distribution of $\mathbf{R}$, then these complications are not present and the p.d.f. of P has a simple form.

Let c be the constant defined by Eq. (5) of Sec. 5.5 with $n = \alpha$ and $k = 2$. The joint p.d.f. $f(r_{11}, r_{22}, r_{12})$ of the three random variables R_{11}, R_{22}, and R_{12} is positive for any values r_{11}, r_{22}, and r_{12} such that $r_{11} > 0$, $r_{22} > 0$, and $r_{11}r_{22} - r_{12}^2 > 0$. For such values,

$$f(r_{11}, r_{22}, r_{12}) = c|\tau|^{\alpha/2}(r_{11}r_{22} - r_{12}^2)^{(\alpha-3)/2}$$
$$\times \exp\left[-\tfrac{1}{2}(\tau_{11}r_{11} + \tau_{22}r_{22} + 2\tau_{12}r_{12})\right]. \tag{4}$$

Let Y be a random variable defined as follows:

$$Y = \left(\frac{\tau_{22}R_{22}}{\tau_{11}R_{11}}\right)^{\frac{1}{2}}. \tag{5}$$

We shall now compute the joint p.d.f. of the three random variables R_{11}, Y, and P. The joint p.d.f. $g(r_{11}, y, \rho)$ of these variables will be positive for any values r_{11}, y, and ρ such that $r_{11} > 0$, $y > 0$, and $|\rho| < 1$. The original variables r_{22} and r_{12} can be expressed in terms of the variables r_{11}, y, and ρ as follows:

$$r_{22} = \frac{\tau_{11}}{\tau_{22}} r_{11}y^2 \quad \text{and} \quad r_{12} = -\left(\frac{\tau_{11}}{\tau_{22}}\right)^{\frac{1}{2}} r_{11}y\rho. \tag{6}$$

Therefore, when the transformation from the three variables r_{11}, y, and ρ to the three variables r_{11}, r_{22}, and r_{12} is considered, the Jacobian J of this transformation will be the determinant of a 3×3 matrix each of whose elements on one side of the main diagonal is 0. Therefore, the value of J is specified by the equation

$$J = \frac{\partial r_{11}}{\partial r_{11}} \cdot \frac{\partial r_{22}}{\partial y} \cdot \frac{\partial r_{12}}{\partial \rho} = -2\left(\frac{\tau_{11}}{\tau_{22}}\right)^{\frac{1}{2}} r_{11}^2 y^2. \tag{7}$$

By changing the variables in Eq. (4) to r_{11}, y, and ρ and multiplying the result by $|J|$ as specified by Eq. (7), we obtain the following equation:

$$g(r_{11}, y, \rho) = 2c \left(\frac{\tau_{11}}{\tau_{22}} |\tau|\right)^{\alpha/2} (r_{11}y)^{\alpha-1}(1 - \rho^2)^{(\alpha-3)/2}$$

$$\times \exp\left\{-\frac{r_{11}}{2}\left[\tau_{11}(1 + y^2) - 2\tau_{12}\left(\frac{\tau_{11}}{\tau_{22}}\right)^{\frac{1}{2}} y\rho\right]\right\}. \quad (8)$$

The marginal joint p.d.f. $h(y, \rho)$ of Y and P can be obtained by integrating the p.d.f. in Eq. (8) over all positive values of r_{11}. Let

$$c' = 2^{\alpha+1}c\Gamma(\alpha)\left(\frac{|\tau|}{\tau_{11}\tau_{22}}\right)^{\alpha/2}. \quad (9)$$

Then the result of this integration is

$$h(y, \rho) = \frac{c'(1 - \rho^2)^{(\alpha-3)/2}}{y[y + y^{-1} - 2\tau_{12}(\tau_{11}\tau_{22})^{-\frac{1}{2}}\rho]^\alpha}. \quad (10)$$

Finally, for any value ρ such that $|\rho| < 1$, the marginal p.d.f. $h_P(\rho)$ of P is defined by the equation

$$h_P(\rho) = \int_0^\infty h(y, \rho)\, dy. \quad (11)$$

However, in general, this integration cannot be carried out in closed form.

Suppose, however, that $\tau_{12} = 0$ in the precision matrix of the original Wishart distribution of R. Then it can be seen from Eq. (10) that the joint p.d.f. $h(y, \rho)$ can be factored into the product of a function of y and a function of ρ. Therefore, the random variables Y and P are independent, and the p.d.f. $h_P(\rho)$ of P must have the following simple form:

$$h_P(\rho) = c''(1 - \rho^2)^{(\alpha-3)/2}. \quad (12)$$

Here, c'' is a constant which makes the integral of the p.d.f. (12) equal to unity.

The distribution of the correlation has also been studied by Lindley (1965), sec. 8.2.

9.13 PRECISION MATRICES HAVING AN UNKNOWN FACTOR

In this section we shall consider the problem of sampling from a multivariate normal distribution for which the value of the mean vector M is unknown and the precision matrix is the product of a specified matrix and an unknown positive number W. In other words, the precision matrix is of the form Wr, where r is a specified symmetric $k \times k$ positive definite matrix. A commonly occurring situation of this type involves sampling from a multivariate normal distribution for which it is known that the components are independent and have the same variance but where the value of this variance is unknown. In such a case, W^{-1} is the unknown common variance and the precision matrix is of the form $W\mathbf{I}$, where $\mathbf{I}$ is the identity matrix. The proof of the next theorem is to be developed as Exercise 44.

Theorem 1 *Suppose that* $\mathbf{X}_1, \ldots, \mathbf{X}_n$ *is a random sample from a multivariate normal distribution for which the mean vector* $\mathbf{M}$ *has an unknown value and the precision matrix is of the form* $W\mathbf{r}$, *where* $\mathbf{r}$ *is a specified positive definite matrix and the value of* W *is unknown. Suppose also that the prior joint distribution of* $\mathbf{M}$ *and* W *is as follows: The conditional distribution of* $\mathbf{M}$ *when* $W = w$ *is a multivariate normal distribution with mean vector* $\mathbf{\mu}$ *and precision matrix* $w\mathbf{r}$ *such that* $\mathbf{\mu} \in R^k$ *and* $\mathbf{r}$ *is a symmetric* $k \times k$ *positive definite matrix, and the marginal distribution of* W *is a gamma distribution with parameters* α *and* β *such that* $\alpha > 0$ *and* $\beta > 0$. *Then the posterior joint distribution of* $\mathbf{M}$ *and* W *when* $\mathbf{X}_i = \mathbf{x}_i$ ($i = 1, \ldots, n$) *is as follows: The conditional distribution of* $\mathbf{M}$ *when* $W = w$ *is a multivariate normal distribution with mean vector* $\mathbf{\mu}^*$ *and precision matrix* $w(\mathbf{r} + n\mathbf{r})$, *where*

$$\mathbf{\mu}^* = (\mathbf{r} + n\mathbf{r})^{-1}(\mathbf{r}\mathbf{\mu} + n\mathbf{r}\bar{\mathbf{x}}), \tag{1}$$

and the marginal distribution of W *is a gamma distribution with parameters* $\alpha + (nk/2)$ *and* β^*, *where*

$$\beta^* = \beta + \tfrac{1}{2} \sum_{i=1}^{n} (\mathbf{x}_i - \bar{\mathbf{x}})'\mathbf{r}(\mathbf{x}_i - \bar{\mathbf{x}}) + \tfrac{1}{2}(\mathbf{\mu}^* - \mathbf{\mu})'\mathbf{r}(\bar{\mathbf{x}} - \mathbf{\mu}). \tag{2}$$

It can be shown (see Exercise 45) that when the joint distribution of $\mathbf{M}$ and W is a *multivariate normal-gamma* distribution as specified in Theorem 1, the marginal distribution of the mean vector $\mathbf{M}$ is a multivariate t distribution with 2α degrees of freedom, location vector $\mathbf{\mu}$, and precision matrix $(\alpha/\beta)\mathbf{r}$.

Further Remarks and References

Posterior distributions for the parameters of multivariate normal distributions have been studied by Ando and Kaufman (1965), Geisser (1964; 1965a, b; 1966), and Geisser and Cornfield (1963). Probability models related to the one presented in Theorem 1 will be used in Chap. 11 for studying some problems of regression and analysis of variance.

EXERCISES

1. Suppose that $X_1, \ldots, X_n$ is a random sample from a Poisson distribution with an unknown value of the mean. For any given value λ of the mean such that

$\lambda > 0$, let $f_n(\cdot |\lambda)$ denote the conditional joint p.f. of $X_1, \ldots, X_n$. Let T be the statistic defined by the equation

$$T(X_1, \ldots, X_n) = \sum_{i=1}^{n} X_i.$$

Show that T is a sufficient statistic for the family of p.f.'s $f_n(\cdot |\lambda)$.

2. Suppose that $X_1, \ldots, X_n$ is a random sample from a normal distribution with an unknown value of the mean and a specified value $\sigma^2 > 0$ of the variance. For any given value μ of the mean such that $-\infty < \mu < \infty$, let $f_n(\cdot |\mu)$ denote the conditional joint p.d.f. of $X_1, \ldots, X_n$. If the statistic T is defined as in Exercise 1, show that T is a sufficient statistic for the family of p.d.f.'s $f_n(\cdot |\mu)$.

3. As in Example 2 of Sec. 9.1, suppose that $X_1, \ldots, X_n$ is a random sample from a normal distribution with an unknown value of the mean and an unknown value of the variance. Let V be the two-dimensional vector defined by the equation

$$V(X_1, \ldots, X_n) = \left\{ \sum_{i=1}^{n} X_i, \sum_{i=1}^{n} X_i^2 \right\}.$$

Show that V is a sufficient statistic for the family of p.d.f.'s $f_n(\cdot |\mu, \sigma^2)$ specified by Eq. (8) of Sec. 9.1.

4. Suppose that $X_1, \ldots, X_n$ is a random sample from a gamma distribution with parameters α and W, where the value of α is specified ($\alpha > 0$) and the value of the parameter W is unknown. For any given value β of the parameter W such that $\beta > 0$, let $f_n(\cdot |\beta)$ denote the conditional joint p.d.f. of $X_1, \ldots, X_n$. If the statistic T is as defined in Exercise 1, show that T is a sufficient statistic for the family of p.d.f.'s $f_n(\cdot |\beta)$.

5. Suppose that $X_1, \ldots, X_n$ is a random sample from a gamma distribution with parameters W_1 and W_2, both of whose values are unknown. For any given values α and β of the parameters W_1 and W_2 such that $\alpha > 0$ and $\beta > 0$, let $f_n(\cdot |\alpha, \beta)$ denote the conditional joint p.d.f. of $X_1, \ldots, X_n$. Let T be the two-dimensional vector defined by the equation

$$T(X_1, \ldots, X_n) = \left\{ \sum_{i=1}^{n} X_i, \sum_{i=1}^{n} \log X_i \right\}.$$

Show that T is a sufficient statistic for the family of p.d.f.'s $f_n(\cdot |\alpha, \beta)$.

6. Suppose that $X_1, \ldots, X_n$ is a random sample from a uniform distribution on the interval (W_1, W_2), where the values of the parameters W_1 and W_2 are unknown. For any given values w_1 and w_2 of W_1 and W_2 such that $-\infty < w_1 < w_2 < \infty$, let $f_n(\cdot |w_1, w_2)$ denote the conditional joint p.d.f. of $X_1, \ldots, X_n$. Let T be the two-dimensional vector defined by the equation

$$T(X_1, \ldots, X_n) = \{\min (X_1, \ldots, X_n), \max (X_1, \ldots, X_n)\}.$$

Show that T is a sufficient statistic for the family of p.d.f.'s $f_n(\cdot |w_1, w_2)$.

7. Suppose that $X_1, \ldots, X_n$ is a random sample from a uniform distribution on the interval $(W - 1, W + 1)$, where the value of the parameter W is unknown. For any given value w of W such that $-\infty < w < \infty$, let $f_n(\cdot |w)$ denote the conditional joint p.d.f. of $X_1, \ldots, X_n$. If the two-dimensional statistic T is as defined in Exercise 6, show that T is a sufficient statistic for the family of p.d.f.'s $f_n(\cdot |w)$.

8. Suppose that $X_1, \ldots, X_n$ is a random sample of vectors from a k-dimensional multivariate normal distribution with an unknown value of the mean vector

W and a specified nonsingular covariance matrix Σ. For any given value w of W such that $w \epsilon R^k$, let $f_n(\cdot \ |w)$ denote the conditional joint p.d.f. of the random vectors $X_1, \ldots, X_n$. Let

$$\bar{X} = \frac{1}{n} \sum_{i=1}^{n} X_i.$$

Show that $\bar{X}$ is a sufficient statistic for the family of p.d.f.'s $f_n(\cdot \ |w)$.

9. Suppose that $X_1, \ldots, X_n$ is a random sample from a k-dimensional multivariate normal distribution with an unknown value of the mean vector and an unknown value of the covariance matrix. For any given value μ of the mean vector and any given value Σ of the covariance matrix such that $\mu \epsilon R^k$ and Σ is a symmetric $k \times k$ positive definite matrix, let $f_n(\cdot \ |\mu, \Sigma)$ denote the conditional joint p.d.f. of the random vectors $X_1, \ldots, X_n$. Let the statistic V be specified by the equation

$$V(X_1, \ldots, X_n) = \left\{ \bar{X}, \sum_{i=1}^{n} (X_i - \bar{X})(X_i - \bar{X})' \right\}.$$

Show that V is a sufficient statistic for the family of p.d.f.'s $f_n(\cdot \ |\mu, \Sigma)$.

10. Let $\{f(\cdot \ |w), w \epsilon \Omega\}$ be a family of g.p.d.f.'s each of which is defined on the sample space S. Let T_1 and T_2 be two statistics which have the following property: For any two points $x_1 \epsilon S$ and $x_2 \epsilon S$, $T_1(x_1) = T_1(x_2)$ if, and only if, $T_2(x_1) = T_2(x_2)$. Show that T_1 is a sufficient statistic for the specified family of g.p.d.f.'s if, and only if, T_2 is a sufficient statistic for that family.

11. Show that each of the following families of distributions is an exponential family:

(a) The family of Bernoulli distributions with an unknown value of the parameter

(b) The family of Poisson distributions with an unknown value of the mean

(c) The family of normal distributions with an unknown value of the mean and a specified value of the variance

(d) The family of normal distributions with an unknown value of the mean and an unknown value of the variance

(e) The family of gamma distributions with unknown values of the parameters α and β

(f) The family of k-dimensional multivariate normal distributions with an unknown value of the mean vector and an unknown value of the covariance matrix

12. Show that if the mean of a beta distribution is μ and the variance is σ^2, then $\sigma^2 < \mu(1 - \mu)$.

13. Let μ and σ^2 be numbers such that $0 < \mu < 1$, $\sigma^2 > 0$, and $\sigma^2 < \mu(1 - \mu)$. Also, let α and β be the parameters of the unique beta distribution which has mean μ and variance σ^2. Show that

$$\alpha = \mu \left[\frac{\mu(1 - \mu)}{\sigma^2} - 1 \right] \quad \text{and} \quad \beta = (1 - \mu) \left[\frac{\mu(1 - \mu)}{\sigma^2} - 1 \right].$$

14. Let W denote the unknown probability that a certain machine will produce a defective item, and suppose that the prior distribution of W is a uniform distribution on the interval $(0, 1)$. Suppose that after the items in a random sample produced by the machine have been inspected, the posterior distribution of W is a beta distribution with parameters $\alpha = 7$ and $\beta = 95$. Show that 100 items were inspected and that six of them were defective.

15. Show that for any given positive numbers μ and σ^2, there is a unique gamma distribution which has mean μ and variance σ^2.

16. Suppose that when magnetic recording tape is manufactured by a certain process, the mean number W of defects on a 1,200-ft roll of tape is unknown, and suppose that the prior distribution of W is a gamma distribution whose mean is 2 and whose variance is 1. Suppose also that the number of defects on any roll of tape when $W = w$ i. s a Poisson distribution with mean w. Suppose further that after a random sample of rolls of tape has been taken and the number of defects on each roll has been counted, the mean of the posterior distribution of W is 1.6 and the variance is 0.16. Show that eight rolls of tape were included in the random sample and that the average number of defects per roll in the sample was 1.5.

17. Prove Theorem 2 of Sec. 9.4.

18. An unknown proportion W of the items produced by a certain machine are defective. Suppose that the prior distribution of W is a beta distribution with parameters $\alpha = 1$ and $\beta = 99$. Suppose also that items produced by the machine are selected at random and observed one at a time until exactly five defective items have been found. If, when sampling terminates, the mean of the posterior distribution of W is 0.02, show that 195 nondefective items were observed during the sampling process.

19. Suppose that in a large population of voters, the proportion W who belong to the Liberal Party is unknown, and suppose that the prior distribution of W is a beta distribution with parameters $\alpha = 1$ and $\beta = 10$.

(a) If, in a random sample of 1,000 voters, it is found that 123 belong to the Liberal Party, what is the posterior distribution of W?

(b) Suppose that instead of taking a random sample as in part a, voters are selected one at a time until exactly 123 have been found who belong to the Liberal Party. Suppose that a total of 1,000 voters had to be selected in order to accomplish this. What is the posterior distribution of W?

20. Prove Theorem 3 of Sec. 9.4.

21. The length of life of a lamp manufactured by a certain process has an exponential distribution with an unknown value of the parameter W. Suppose that the prior distribution of W is a gamma distribution for which the coefficient of variation is 0.5. A random sample of lamps is to be tested, and the length of life of each of the lamps is to be noted. If the coefficient of variation of the posterior distribution of W must be reduced to the value 0.1, show that 96 lamps should be tested.

22. Extend Theorem 3 of Sec. 9.4 so that it covers the case of a random sample from a gamma distribution with parameters α and W, where the value of α is specified $(\alpha > 0)$ and the value of W is unknown.

23. Consider a normal distribution for which the value of the mean W is unknown and the variance is 4, and suppose that the prior distribution of W is a normal distribution whose variance is 9. How large a random sample must be taken from the given normal distribution in order to be able to specify an interval having a length of 1 unit such that the probability that W lies in this interval is at least 0.95? (*Answer: n* = 62.)

24. Prove Theorem 2 of Sec. 9.5.

25. Suppose that the value of the precision W of a normal distribution is unknown, and suppose that the distribution of W is a gamma distribution with parameters α and β. Let V denote the variance of the given normal distribution.

(a) Find the p.d.f. of V.

(b) Show that if $\alpha > 1$, $E(V) = \beta/(\alpha - 1)$.

(c) Show that if $\alpha > 2$, $Var(V) = \beta^2/[(\alpha - 1)^2(\alpha - 2)]$.

26. Suppose that a random sample is to be taken from a normal distribution with a specified value of the mean and an unknown value of the precision W. Suppose also that the prior distribution of W is a gamma distribution and that the coefficient of variation of the posterior distribution of W must be reduced to the value 0.1. Show that this requirement will be satisfied, regardless of the value of the coefficient of variation of the prior distribution, if a sample of size $n = 200$ is taken.

27. Consider a normal distribution with an unknown value of the mean M and an unknown value of the precision R, and suppose that the prior joint distribution of M and R is as specified in Theorem 1 of Sec. 9.6. Find the conditional distribution of R when $M = m$.

28. Consider the conditions specified in Exercise 27. Suppose that the coefficient of variation of the prior distribution of R has the value 0.5. How large a random sample must be taken from the given normal distribution in order that the coefficient of variation of the posterior distribution of R will be reduced to the value 0.1? (*Answer: n = 192.*)

29. Consider the conditions specified in Exercise 27. Suppose that under the conditional posterior distribution of M when $R = 3$, the variance of M must be reduced to the value 0.01. Show that this requirement will be satisfied, regardless of the values of the parameters of the prior distribution, if a random sample of size $n = 34$ is taken from the given normal distribution.

30. The length of time for which a certain man must wait each morning for a bus taking him to work is uniformly distributed on the interval $(0, W)$, where the value of W is unknown and the prior distribution of W is a Pareto distribution with parameters $w_0 > 0$ and $\alpha = 1$. On how many mornings must the man observe his waiting time before he will be able to specify an interval having a length of 0.01 unit such that the probability that the unknown value of $\log W$ lies in this interval is at least 0.95? (*Answer: n = 299.*)

31. Consider the prior distribution of W specified in Exercise 30. On how many mornings must the man observe his waiting time in order that the coefficient of variation of the posterior distribution of W will be reduced to the value 0.01? (*Answer: n = 101.*)

32. Prove Theorem 2 of Sec. 9.7.

33. Consider a uniform distribution on the interval (W_1, W_2), where the values of W_1 and W_2 are unknown, and suppose that the prior joint distribution of W_1 and W_2 is a bilateral bivariate Pareto distribution with parameters $r_1 > 0$, $r_2 > 0$, and $\alpha = 2$. How large a random sample must be taken from the uniform distribution in order that the coefficient of variation of the posterior distribution of the random variable $W_2 - W_1$ will be reduced to the value 0.01? (*Answer: n = 140.*)

34. Suppose that a box contains N balls, of which an unknown number W are red and the rest are blue. Suppose also that the prior distribution of W is a hypergeometric distribution with parameters A, B, and N, where A and B are positive integers such that $A + B \geq N$.

(a) Now suppose that although the exact value of W is unknown, the statistician knows that $r \leq W \leq s$, where r and s are integers such that $0 < r \leq s < N$. Show that there are unique values of A and B such that, under the prior hypergeometric distribution,

$$\Pr\{r \leq W \leq s\} = 1, \qquad \Pr\{W = r\} > 0, \qquad \text{and} \qquad \Pr\{W = s\} > 0.$$

(b) Next, suppose that the statistician knows only that $0 \leq W \leq N$. Show that any prior hypergeometric distribution such that $A \geq N$ and $B \geq N$ will assign positive probability to each integer $0, 1, 2, \ldots, N$.

(c)   Suppose that one ball is selected from the box at random.   What is the probability, under the prior distribution, that it will be red?

(d)   Suppose that n balls $(1 \leq n < N)$ are selected at random from the box without replacement and that x of these balls are red.   Show that the posterior distribution of the number of red balls among the $N - n$ balls remaining in the box is a hypergeometric distribution with parameters $A - x$, $B - (n - x)$, and $N - n$.

35.   Suppose that there are k different types of items in a very large population and let W_i be the unknown proportion of the population that includes items of type i $(i = 1, 2, \ldots, k)$.   Suppose also that the prior distribution of $\mathbf{W} = (W_1, \ldots, W_k)'$ is a Dirichlet distribution with parametric vector $\boldsymbol{\alpha} = (\alpha_1, \ldots, \alpha_k)'$ such that $\alpha_1 + \cdots + \alpha_k = 6$.   How large a random sample of items must be taken in order to be sure that no matter what the values of the individual components of the vector $\boldsymbol{\alpha}$ are and no matter what the observed outcomes are, the posterior variance of each proportion W_i $(i = 1, \ldots, k)$ will be at most 0.005?   (*Answer: n = 43.*)

36.   Consider a bivariate normal distribution with an unknown mean vector $\mathbf{M} = (M_1, M_2)'$ and a precision matrix $\mathbf{r}$ which is known to be

$$\mathbf{r} = \begin{bmatrix} \frac{1}{3} & -\frac{1}{3} \\ -\frac{1}{3} & \frac{4}{3} \end{bmatrix}.$$

Suppose that the prior distribution of $\mathbf{M}$ is a bivariate normal distribution for which the precision matrix is

$$\boldsymbol{\tau} = \begin{bmatrix} 1 & -1 \\ -1 & 6 \end{bmatrix}.$$

How large a random sample must be taken in order that the variance of the posterior distribution of the random variable $M_1 - M_2$ will be reduced to the value 0.01?   (*Answer: n = 297.*)

37.   Suppose that a random $k \times k$ symmetric, positive definite matrix $\mathbf{R}$ has a Wishart distribution with α degrees of freedom $(\alpha > k - 1)$ and precision matrix $\boldsymbol{\tau}$.   Show that the coefficient of variation of the determinant $|\mathbf{R}|$ is

$$\left[\frac{k(2\alpha - k + 3)}{(\alpha - k + 2)(\alpha - k + 1)} \right]^{\frac{1}{2}}.$$

Hint: See Exercise 14 of Chap. 5.

38.   Consider a bivariate normal distribution with a specified mean vector and an unknown precision matrix $\mathbf{R}$.   Suppose that the prior distribution of $\mathbf{R}$ is a Wishart distribution with 3 degrees of freedom and precision matrix $\boldsymbol{\tau}$.   How large a random sample must be taken in order that the coefficient of variation of the posterior distribution of the determinant $|\mathbf{R}|$ will be reduced to the value 0.1?   (*Answer: n = 399.*)

39.   Consider again the conditions described in Exercise 38, and suppose that the elements of the 2×2 matrix $\mathbf{R}$ are

$$\mathbf{R} = \begin{bmatrix} R_{11} & R_{12} \\ R_{21} & R_{22} \end{bmatrix}.$$

How large a random sample must be taken in order that the coefficient of variation of the posterior distribution of the random variable R_{11} will be reduced to the value 0.1?   (*Answer: n = 197.*)

40.   Prove that Eq. (3) of Sec. 9.11 is correct.

41.   Consider a multivariate normal distribution with an unknown value of the mean vector $\mathbf{M}$ and an unknown value of the precision matrix $\mathbf{R}$, and suppose that the prior joint distribution of $\mathbf{M}$ and $\mathbf{R}$ is a multivariate normal-Wishart distribution as specified in Theorem 1 of Sec. 9.10.   What is the conditional distribution of $\mathbf{R}$ when $\mathbf{M} = \mathbf{m}$?

42. Consider a bivariate normal distribution with an unknown value of the mean vector $\mathbf{M}$ and an unknown value of the precision matrix $\mathbf{R}$. Suppose that the prior joint distribution of $\mathbf{M}$ and $\mathbf{R}$ is a bivariate normal-Wishart distribution as specified in Theorem 1 of Sec. 9.10, and suppose that $\alpha = 3$ in this prior distribution. How large a random sample must be taken in order that the coefficient of variation of the posterior distribution of the determinant $|\mathbf{R}|$ will be reduced to the value 0.1? *Hint:* Note that the answer in this exercise must be the same as the answer in Exercise 38.

43. Consider the conditions specified in Exercise 42, and suppose again that $\alpha = 3$ in the prior joint distribution of $\mathbf{M}$ and $\mathbf{R}$. Suppose that a random sample is taken, and let $\mathbf{\mu}^*$ be the location vector and $\mathbf{T}^*$ the precision matrix of the posterior bivariate t distribution of the two-dimensional vector $\mathbf{M}$. Determine the size of a random sample that must be taken in order for the posterior distribution of $\mathbf{M}$ to satisfy the following equation:

$$\Pr[(\mathbf{M} - \mathbf{\mu}^*)'\mathbf{T}^*(\mathbf{M} - \mathbf{\mu}^*) \leq 10] \geq 0.95.$$

Hint: See expression (16) of Sec. 5.6. (*Answer: n = 5.*)

44. Prove Theorem 1 of Sec. 9.13.

45. Prove that if the joint distribution of the random vector $\mathbf{M}$ and the random variable W is a multivariate normal-gamma distribution as specified in Theorem 1 of Sec. 9.13, then the marginal distribution of $\mathbf{M}$ is a multivariate t distribution with 2α degrees of freedom, location vector $\mathbf{\mu}$, and precision matrix $(\alpha/\beta)\tau$.

46. Consider five different normal distributions with unknown means M_1, $\ldots$, M_5 and with a common unknown precision W. Suppose that the prior joint distribution of M_1, $\ldots$, M_5 and W is as follows: For any given value $W = w$, the random variables M_1, $\ldots$, M_5 are independent and have the same normal distribution with mean 3 and precision $2w$. Furthermore, the marginal distribution of W is a gamma distribution with parameters $\alpha = 10$ and $\beta = 5$. Suppose also that a random sample of eight observations is taken from each of the five normal distributions. Let x_{ij} denote the value of the jth observation from the ith distribution, and for $i = 1$, $\ldots$, 5, let $\bar{x}_i$ denote the average of the eight observations from the ith distribution. Finally, let the values μ_1, $\ldots$, μ_5 and c be defined as follows:

$$\mu_i = \frac{3 + 4\bar{x}_i}{5} \qquad i = 1, \ldots, 5,$$

and

$$c = \frac{2.37}{600}\left[50 + 5\sum_{i=1}^{5}\sum_{j=1}^{8}(x_{ij} - \bar{x}_i)^2 + 8\sum_{i=1}^{5}(\bar{x}_i - 3)^2\right].$$

Show that under the posterior joint distribution of M_1, $\ldots$, M_5,

$$\Pr\left[\sum_{i=1}^{5}(M_i - \mu_i)^2 \leq c\right] = 0.95.$$

Hint: See expression (16) of Sec. 5.6.

47. Suppose that $\mathbf{X}_1$, $\ldots$, $\mathbf{X}_n$ is a random sample of symmetric $k \times k$ matrices from a Wishart distribution with m degrees of freedom ($m > k - 1$) and an unknown value of the precision matrix $\mathbf{R}$. Suppose also that the prior distribution of $\mathbf{R}$ is a Wishart distribution with α degrees of freedom and precision matrix τ such that $\alpha > k - 1$ and τ is a $k \times k$ positive definite matrix. Show that the posterior distribution of $\mathbf{R}$ when $\mathbf{X}_i = \mathbf{x}_i$ ($i = 1, \ldots, n$) is a Wishart distribution with $\alpha + mn$ degrees of freedom and precision matrix $\tau + \sum_{i=1}^{n}\mathbf{x}_i$.

limiting posterior
distributions

10.1 IMPROPER PRIOR DISTRIBUTIONS

In certain problems, the prior knowledge that a statistician has about some parameter W may be very slight and vague when compared with the information about W which he expects to acquire from available observations. In such a problem, even though the statistician may be able to find a suitable conjugate family of distributions of W, it may not be easy for him to select the appropriate prior distribution from this family. Because of the vagueness of his prior knowledge in comparison with the information that he will soon have from his observations, it would not be worthwhile for the statistician to expend a great deal of time or effort in determining a specific prior distribution. Instead, he would find it convenient to make use of a standard prior distribution that would be suitable in many situations in which it is desired to represent vague prior information.

Often, the standard prior distribution which is used in one of these problems is an improper distribution in the sense that it is represented by a nonnegative density function whose integral over the whole parameter space Ω is infinite; for any proper probability distribution, this integral must be unity. For example, if Ω is the real line and, because of vagueness, the prior p.d.f. of W is smooth and very widely spread out over the line, then the statistician might find it convenient to assume a uniform, or constant, density over the whole line in order to represent this prior distribution. Such a uniform density is not the p.d.f. of any proper probability distribution on the real line. However, the statistician can often develop a posterior distribution which is a proper distribution by using this uniform density as a prior density and formally carrying out the calculations of Bayes' theorem with some observed values $x_1, \ldots, x_n$. Thus, if $f_n(x_1, \ldots, x_n|w)$ is the likelihood function for these observed values and if

$$0 < \int_{-\infty}^{\infty} f_n(x_1, \ldots, x_n|w) \, dw < \infty, \tag{1}$$

then the posterior p.d.f. $\xi(\cdot \,|x_1, \ldots, x_n)$ of W will be specified by the relation

$$\xi(w|x_1, \ldots, x_n) \propto f(x_1, \ldots, x_n|w). \tag{2}$$

As an illustration of the relation (2), suppose that $X_1, \ldots, X_n$ is a random sample from a normal distribution with an unknown value of the mean W and a specified value of the precision r. If the prior distribution of W is represented by a uniform density over the real line, then it follows from relation (4) of Sec. 9.5 that

$$\xi(w|x_1, \ldots, x_n) \propto \exp\left[-\frac{nr}{2} (w - \bar{x})^2 \right]. \tag{3}$$

Therefore, the posterior distribution of W when $X_i = x_i$ $(i = 1, \ldots, n)$ is a normal distribution with mean $\bar{x}$ and precision nr. Although the prior distribution is improper, the posterior distribution will be a proper normal distribution after just one observation has been made.

Another procedure when the prior knowledge is vague is to use a proper prior distribution of W from some appropriate conjugate family which is indexed by some parameter α, to compute the posterior distribution from the observed values $x_1, \ldots, x_n$, and then to study the limiting posterior distribution as the parameter α approaches some limiting value. Often, this limiting posterior distribution will be a proper probability distribution for W, even though it did not result from any proper prior distribution.

As an illustration of the use of a limiting posterior distribution, consider again the problem of sampling from a normal distribution with an unknown value of the mean W and a specified value of the precision r. A conjugate family of distributions for this problem is described in Theorem 1 of Sec. 9.5. If we let $\tau \to 0$ in the posterior normal distribution of W, as described in that theorem, it can be seen that the limiting result will be a normal distribution with mean $\bar{x}$ and precision nr. This distribution serves as a posterior distribution for W, but it is one that cannot be derived from any proper prior distribution. In this example, the result is the same as the posterior distribution (3), which was obtained by using an improper uniform prior density of W. This agreement could have been anticipated by noting that when the precision $\tau \to 0$ in the prior normal distribution of W, the variance becomes arbitrarily large and hence the distribution becomes spread more and more thinly over the line.

The posterior distribution specified by relation (3) has interesting properties. Let the random variable Z be defined by the equation

$$Z = (nr)^{\frac{1}{2}}(\bar{X} - W). \tag{4}$$

Then it follows from this posterior distribution of W that the conditional distribution of Z for any given value $\bar{x}$ of $\bar{X}$ is a standard normal distribution (i.e., a normal distribution with mean 0 and precision 1). Since this conditional distribution of Z is the same for any given value of $\bar{X}$, we can conclude that the random variables Z and $\bar{X}$ are independent. Furthermore, it follows from the standard sampling theory for normal distributions, as described in Sec. 4.7, that the conditional distribution of Z for any given value w of W is also a standard normal distribution. Therefore, the random variables Z and W are also independent.

Strictly speaking, the two conclusions just stated are mutually incompatible. Under any proper bivariate distribution of $\bar{X}$ and W for which the random variables Z and W are independent, it would not be possible for the random variables Z and $\bar{X}$ also to be independent, unless the random variable Z is equal to a constant with probability 1 (see Exercise 1). The random variable Z is not constant here because of our use of an improper prior distribution. These considerations indicate why the statistician must take particular care when handling and interpreting improper distributions.

Confidence Intervals

We have noted that the conditional distribution of Z for any given value of $\bar{X}$ is the same as the conditional distribution of Z for any given value of W. Because of this property, inferences about W that are based on its posterior distribution will, in general, agree with inferences about W that are based on the standard method of confidence intervals. This method may be described as follows:

Let $u_1(X_1, \ldots, X_n)$ and $u_2(X_1, \ldots, X_n)$ be random variables such that for any given value $w \in \Omega$, the conditional probability satisfies the relation

$$\Pr\{u_1(X_1, \ldots, X_n) \leq w \leq u_2(X_1, \ldots, X_n) | W = w\} \geq \gamma, \quad (5)$$

where γ is a fixed number ($0 < \gamma < 1$). Then for any observed values $x_1, \ldots, x_n$, the following interval can be specified for W:

$$u_1(x_1, \ldots, x_n) \leq W \leq u_2(x_1, \ldots, x_n). \quad (6)$$

This interval is called a *confidence interval* for W, and the number γ is called the *confidence coefficient* of the interval.

For any value of γ ($0 < \gamma < 1$), let k_γ denote the number such that

$$\Phi(k_\gamma) - \Phi(-k_\gamma) = \gamma, \quad (7)$$

where Φ is the standard normal d.f. It follows from the conditional distribution of Z specified earlier that a confidence interval for W with confidence coefficient γ can be defined by the relation

$$\bar{x} - k_\gamma (nr)^{-\frac{1}{2}} \leq W \leq \bar{x} + k_\gamma (nr)^{-\frac{1}{2}}. \quad (8)$$

But it also follows that under the posterior distribution of W, the probability that W lies in the interval in (8) is γ. Thus, according to the standard theory, that interval has confidence γ, and according to the Bayesian or subjective theory, that interval has probability γ. The operational interpretations of these two properties seem to be essentially the same. Therefore, as previously stated, inferences about W based on the two points of view will generally agree.

Further Remarks and References

Improper prior distributions play an important part in the statistical methods discussed by Jeffreys (1961) and Lindley (1965). In the next two sections of this chapter we shall consider posterior distributions which either result from improper prior distributions or are limiting posterior distributions in problems which involve samples from univariate and multivariate normal distributions. Problems which involve samples from other distributions are presented as exercises at the end of the chapter.

We shall show that in many problems in which the posterior distribution of a parameter is obtained in this way, confidence intervals—or, in higher dimensions, confidence sets—for the parameter which have a given confidence coefficient γ will agree with intervals or sets which have posterior probability γ.

Stein (1962a, 1965) has discussed the difficulty of representing vague prior knowledge about a vector **W** of high dimension and other aspects of the use of improper distributions.

10.2 IMPROPER PRIOR DISTRIBUTIONS FOR SAMPLES FROM A NORMAL DISTRIBUTION

Consider the problem treated in Theorem 2 of Sec. 9.5. In this problem, a random sample $X_1, \ldots, X_n$ is to be taken from a normal distribution with a known value of the mean m and an unknown value of the precision W. In the posterior gamma distribution of W as given in that theorem, let $\alpha \to 0$ and $\beta \to 0$. Then the limiting posterior distribution of W is a gamma distribution for which the parameters are $n/2$ and $\frac{1}{2}\Sigma_{i=1}^{n}(x_i - m)^2$.

The same posterior distribution can be obtained from an improper prior distribution. Since the prior distribution of W is a gamma distribution with parameters α and β, then for $w > 0$ the prior p.d.f. ξ of W satisfies the relation

$$\xi(w) \propto w^{\alpha-1}e^{-\beta w}. \tag{1}$$

When we let $\alpha \to 0$ and $\beta \to 0$ in the relation (1), we obtain formally (but only formally, since in that relation we are ignoring the proportionality factor which involves α and β) the following result:

$$\xi(w) \propto \frac{1}{w}. \tag{2}$$

The right side of the relation (2) represents an improper prior density since its integral over Ω is not finite. However, it can be shown that if the improper prior density of W is specified by the relation (2), then the posterior distribution of W will be the proper gamma distribution already found.

Let the random variable Z be defined by the following equation:

$$Z = W\left[\sum_{i=1}^{n}(X_i - m)^2\right]. \tag{3}$$

It follows from the preceding development that the conditional distribution of Z when $X_i = x_i$ $(i = 1, \ldots, n)$ is a χ^2 distribution with n degrees of freedom. Moreover, it is well known and can be easily verified that the conditional distribution of Z when $W = w$ is also a χ^2 distribution with n degrees of freedom. Hence, a confidence interval for W with confidence coefficient γ $(0 < \gamma < 1)$ which is found from the conditional distribution of Z will also be an interval for W whose posterior probability is γ.

As another example involving sampling from a univariate normal distribution, consider the problem treated in Theorem 1 of Sec. 9.6. In this problem, a random sample $X_1, \ldots, X_n$ is to be taken from a distribution for which both the mean M and the precision R are unknown.

In the posterior joint distribution of M and R as given in Theorem 1 of Sec. 9.6, let $\tau \to 0$, let $\alpha \to -\frac{1}{2}$, and let $\beta \to 0$. Then, under the limiting posterior distribution, the conditional distribution of M when $R = r$ is a normal distribution with mean $\bar{x}$ and precision nr, and the marginal distribution of R is a gamma distribution with parameters $(n - 1)/2$ and $\frac{1}{2}\sum_{i=1}^{n}(x_i - \bar{x})^2$. It is assumed that $n \geq 2$. Note that in order to obtain this posterior distribution, the parameter α must violate the condition that $\alpha > 0$, which was assumed in Theorem 1 of Sec. 9.6, and α must now approach the negative number $-\frac{1}{2}$. It also follows from the presentation in Sec. 9.6 that this same posterior joint distribution of M and R can be obtained from an improper prior joint density ξ which is specified over the parameter space Ω by the equation

$$\xi(m, r) = \frac{1}{r}. \tag{4}$$

The joint density specified by Eq. (4) is simply the product of a uniform density in m over the real line and the density $1/r$ over the values $r > 0$. In other words, this joint density is the product of the improper densities which were used when only the value of M or only the value of R was unknown.

Let the random variable T be defined by the equation

$$T = \frac{n^{\frac{1}{2}}(M - \bar{X})}{[\sum_{i=1}^{n}(X_i - \bar{X})^2/(n - 1)]^{\frac{1}{2}}}. \tag{5}$$

We know from the discussion in Sec. 9.6 and from the limiting posterior joint distribution of M and R found above that the conditional distribution of T when $X_i = x_i$ $(i = 1, \ldots, n)$ is a t distribution with $n - 1$ degrees of freedom. But we also know from expression (3) of Sec. 4.12 that the conditional distribution of T when $M = m$ and $R = r$ is a t distribution with $n - 1$ degrees of freedom.

Furthermore, let the random variable U be defined by the equation

$$U = R \left[\sum_{i=1}^{n} (X_i - \bar{X})^2 \right]. \tag{6}$$

Then it follows from the limiting posterior distribution of R that the conditional distribution of U when $X_i = x_i$ $(i = 1, \ldots, n)$ is a χ^2 distribution with $n - 1$ degrees of freedom. As was pointed out in Sec. 4.8, the conditional distribution of U when $M = m$ and $R = r$ is also a χ^2 distribution with $n - 1$ degrees of freedom.

It follows from these remarks that confidence coefficients for the standard confidence intervals for M as found from the conditional distribution of T will agree with the probabilities of the same intervals computed under the limiting posterior distribution. A similar statement can be made for confidence intervals for R as found from the conditional distribution of U. Stone (1963) has described other models which lead to conclusions like these.

Invariant Prior Distributions

Jeffreys (1961) has discussed the use of the prior density specified by Eq. (4), and he has presented arguments for the appropriateness of this density as a representation of vague prior knowledge in scientific work. His arguments are based partly on the following *invariance* properties of this density:

If the statistician has vague prior knowledge about the value of M, then he also has vague prior knowledge about the linear transform $M^* = aM + b$, where a and b are specified constants ($a \neq 0$). Therefore, if it is appropriate to represent the prior knowledge about M by a uniform density over the whole real line, it should also be appropriate to represent the prior knowledge about M^* by a uniform distribution over the whole real line. However, it follows from the theory given in Sec. 3.7 for transformations of random variables that if M has a uniform density, then so also does M^*. Therefore, the uniform density satisfies the desired invariance property.

The density of the precision R specified by Eq. (4) has a similar invariance property. If the statistician has vague prior knowledge about R, then he also has vague prior knowledge about the transform $R^* = R^\alpha$, where α is a specified constant ($\alpha \neq 0$). It can be shown (see Exercise 6a) that if the density of R is proportional to $1/r$ for $r > 0$, then the density of R^* will be proportional to $1/r^*$ for $r^* > 0$. It follows that regardless of whether R^* is the unknown precision, the unknown variance, or the unknown standard deviation, it is appropriate to represent vague prior knowledge about R^* by a density of the form $1/r^*$.

Finally, it should be noted (see Exercise 6b) that if R has a density of the form specified above for $r > 0$, then $S = \log R$ will have a uniform density over the whole real line. Therefore, the joint density of M and R described by Eq. (4) specifies that M and R are independent and both M and $\log R$ have uniform densities over the whole real line.

Invariant distributions are also discussed by Hartigan (1964).

10.3 IMPROPER PRIOR DISTRIBUTIONS FOR SAMPLES FROM A MULTIVARIATE NORMAL DISTRIBUTION

Consider the problem treated in Theorem 1 of Sec. 9.9. In this problem, a random sample $\mathbf{X}_1, \ldots, \mathbf{X}_n$ is to be taken from a k-dimensional multivariate normal distribution with an unknown value of the mean vector $\mathbf{M}$

and a specified precision matrix $\mathbf{r}$. Just as for the univariate distribution, we shall let $\tau \to \mathbf{0}$ in the posterior distribution of $\mathbf{M}$, where $\mathbf{0}$ is the matrix each of whose elements is 0; or, equivalently, we shall take the prior density of $\mathbf{M}$ to be a uniform density over the whole space R^k. Then it can be seen that the posterior distribution of $\mathbf{M}$ becomes a multivariate normal distribution with mean vector $\mathbf{\bar{x}}$ and precision matrix $n\mathbf{r}$.

Let the random vector $\mathbf{Z}$ be defined by the equation $\mathbf{Z} = \mathbf{\bar{X}} - \mathbf{M}$. Then both the conditional distribution of $\mathbf{Z}$ when $\mathbf{X}_i = \mathbf{x}_i$ $(i = 1, \ldots, n)$ and the conditional distribution of $\mathbf{Z}$ when $\mathbf{M} = \mathbf{m}$ will be multivariate normal distributions for which the mean vector is $\mathbf{0}$ and the precision matrix is $n\mathbf{r}$. It follows that if a confidence set for $\mathbf{M}$ constructed from the random vector $\mathbf{Z}$ has confidence coefficient γ, then this set will also have probability γ under the posterior distribution of $\mathbf{M}$.

Now consider the problem treated in Theorem 1 of Sec. 9.10. In this problem, a random sample $\mathbf{X}_1, \ldots, \mathbf{X}_n$ is to be taken from a k-dimensional multivariate normal distribution for which both the mean vector $\mathbf{M}$ and the precision matrix $\mathbf{R}$ are unknown. It is assumed that $n \geq k + 1$. In the posterior joint multivariate normal-Wishart distribution of $\mathbf{M}$ and $\mathbf{R}$, as given in that theorem, let $\nu \to 0$, let $\alpha \to -1$, and let $\tau \to \mathbf{0}$. The result is a joint distribution such that the conditional distribution of $\mathbf{M}$ when $\mathbf{R} = \mathbf{r}$ is a multivariate normal distribution with mean vector $\mathbf{\bar{x}}$ and precision matrix $n\mathbf{r}$, and the marginal distribution of $\mathbf{R}$ is a Wishart distribution with $n - 1$ degrees of freedom and precision matrix $\mathbf{s}$, where $\mathbf{s}$ is defined by Eq. (1) of Sec. 9.10. Note that in order to obtain this posterior distribution, the parameter α must violate the condition that $\alpha > k - 1$, which was assumed in Theorem 1 of Sec. 9.10, and α must now approach the value -1.

Furthermore, it can be seen from the relations (5) and (8) of Sec. 9.10 that this same posterior joint distribution will be obtained if we assume that the improper prior joint density ξ of $\mathbf{M}$ and $\mathbf{R}$ is as follows:

$$\xi(\mathbf{m}, \mathbf{r}) = \frac{1}{|\mathbf{r}|^{(k+1)/2}}. \tag{1}$$

The density specified by Eq. (1) can be interpreted as the product of a uniform density in $\mathbf{m}$ over the space R^k and a density in $\mathbf{r}$ which is proportional to the function on the right side of Eq. (1).

It now follows from the results given in Sec. 9.11 that under the posterior distribution just developed, the marginal distribution of $\mathbf{M}$ will be a multivariate t distribution with $n - k$ degrees of freedom, location vector $\mathbf{\bar{x}}$, and precision matrix $n(n - k)\mathbf{s}^{-1}$. Let the random matrix $\mathbf{S}$ be defined by the equation

$$\mathbf{S} = \sum_{i=1}^{n} (\mathbf{X}_i - \mathbf{\bar{X}})(\mathbf{X}_i - \mathbf{\bar{X}})', \tag{2}$$

and let the random variable Q be defined by the equation

$$Q = \frac{n(n-k)}{k} (\bar{\mathbf{X}} - \mathbf{M})'\mathbf{S}^{-1}(\bar{\mathbf{X}} - \mathbf{M}). \tag{3}$$

Then [see expression (16) of Sec. 5.6] the conditional distribution of Q when $\mathbf{X}_i = \mathbf{x}_i$ $(i = 1, \ldots, n)$ will be an F distribution with k and $n - k$ degrees of freedom.

It is said that a random variable Y has *Hotelling's T^2 distribution* with k and q degrees of freedom $(q \geq k)$ if the random variable $[(q - k + 1)/(kq)]Y$ has an F distribution with k and $q - k + 1$ degrees of freedom. If the random variable Q^* is defined by the equation

$$Q^* = n(n-1)(\bar{\mathbf{X}} - \mathbf{M})'\mathbf{S}^{-1}(\bar{\mathbf{X}} - \mathbf{M}), \tag{4}$$

then it follows from Eq. (3) that the conditional distribution of Q^* when $\mathbf{X}_i = \mathbf{x}_i$ $(i = 1, \ldots, n)$ will be Hotelling's T^2 distribution with k and $n - 1$ degrees of freedom. But it is known [see, e.g., Scheffé (1959), app. 5, or Anderson (1958), chap. 5] that the conditional distribution of Q^* when the values of $\mathbf{M}$ and $\mathbf{R}$ are given is also Hotelling's T^2 distribution with k and $n - 1$ degrees of freedom.

An ellipsoid $\mathcal{E}$ can now be constructed in the space R^k such that $\mathbf{M}$ will lie in this ellipsoid with a specified confidence coefficient γ or, equivalently, with a specified probability γ under the posterior distribution of $\mathbf{M}$. Let γ be a given number $(0 < \gamma < 1)$, and let ϕ_γ be a constant such that if the random variable Q has an F distribution with k and $n - k$ degrees of freedom, then $\Pr(Q \leq \phi_\gamma) = \gamma$. Also, let $\mathbf{x}$ and $\mathbf{s}$ be the observed values of the sample mean vector $\bar{\mathbf{X}}$ and the sample matrix $\mathbf{S}$. Finally, let $\mathcal{E}$ be the set of points $\mathbf{m}$ $(\mathbf{m} \in R^k)$ which satisfy the following relation:

$$(\mathbf{m} - \mathbf{x})'\mathbf{s}^{-1}(\mathbf{m} - \mathbf{x}) \leq \frac{k\phi_\gamma}{n(n-k)}. \tag{5}$$

Since the matrix $\mathbf{s}^{-1}$ is positive definite, the set $\mathcal{E}$ will be an ellipsoid in R^k. It follows from Eqs. (3) and (4) that $\mathcal{E}$ will be a confidence ellipsoid for $\mathbf{M}$ with confidence coefficient γ and also that the posterior probability that $\mathbf{M}$ will lie in $\mathcal{E}$ is γ.

10.4 PRECISE MEASUREMENT

In this section we shall consider further the consequences of using either a proper or an improper uniform prior density for a parameter W when observations which are expected to be very informative are to be made. We shall show that under quite general conditions the posterior distribution of W derived from a uniform prior density over the parameter space

Ω will be a close approximation to the posterior distribution derived from a more carefully specified proper prior distribution. The fact that posterior distributions derived from uniform prior densities are adequate approximations to the actual posterior distributions is a consequence of what Savage calls the *principle of stable estimation* [Edwards, Lindman, and Savage (1963)] or *precise measurement* [Savage (1961), Savage and others (1962)]. Each of these three references contains excellent discussions which pertain not only to the theory of precise measurement but also to the problems of Bayesian statistical inference in general.

One interesting feature of the theory of precise measurement is that specific quantitative results can be obtained which indicate how closely the approximate posterior distribution agrees with the actual posterior distribution. The theorem which will be presented here in this regard is due to Edwards, Lindman, and Savage (1963). This theorem will be valid regardless of whether W is a real-valued parameter or a vector and also regardless of whether W has a discrete prior p.f. on a countable set of values or a prior p.d.f. over R^k ($k \geq 1$). The presentation will be general enough to include all these possibilities.

Let W be a parameter whose possible values are in the parameter space Ω. Also, let X be either a random variable or a random vector whose conditional g.p.d.f. for any given value $W = w$ ($w \in \Omega$) is $f(\cdot | w)$. Let ξ denote the prior g.p.d.f. of W, and, as usual, let $\xi(\cdot | x)$ denote the posterior g.p.d.f. of W resulting from the observed value $X = x$. We shall assume that the prior g.p.d.f. ξ is a bounded function on the set Ω. For the observed value $X = x$, we shall suppose that

$$0 < \int_\Omega f(x|w) \, d\nu(w) < \infty. \tag{1}$$

We can then define the function $\phi(\cdot | x)$ on Ω as follows:

$$\phi(w|x) = \frac{f(x|w)}{\int_\Omega f(x|w') \, d\nu(w')}. \tag{2}$$

The function $\phi(\cdot | x)$ is the posterior g.p.d.f. of W which would result from a uniform prior g.p.d.f., regardless of whether this prior g.p.d.f. is proper or improper. The next theorem provides bounds on the difference between the actual posterior g.p.d.f. $\xi(\cdot | x)$ and the g.p.d.f. $\phi(\cdot | x)$.

Theorem 1 *Let A be any subset of Ω such that*

$$m = \inf_{w \in A} \xi(w) > 0, \tag{3}$$

and let α, β, and γ be three numbers $(0 \leq \alpha < 1, \beta \geq 0,$ *and* $\gamma \geq 0)$ *which satisfy the following three relations:*

$$\int_A \phi(w|x) \, d\nu(w) \geq 1 - \alpha, \tag{4}$$

$$\sup_{w \in A} \xi(w) \leq (1 + \beta)m, \tag{5}$$

and

$$\sup_{w \in A^c} \xi(w) \leq (1 + \gamma)m. \tag{6}$$

Furthermore, let the number ϵ be defined by the equation

$$\epsilon = \max \left\{ \frac{\alpha + \beta}{1 - \alpha}, \frac{\alpha + \beta + \alpha\gamma}{1 + \alpha + \beta + \alpha\gamma} \right\} + \frac{\alpha(2 - \alpha + \gamma)}{1 - \alpha}. \tag{7}$$

Then

$$\int_\Omega |\xi(w|x) - \phi(w|x)| \, d\nu(w) \leq \epsilon. \tag{8}$$

Proof If $f(x|w) \neq 0$, then

$$\frac{\xi(w|x)}{\phi(w|x)} = \frac{\xi(w)}{\int_\Omega \xi(w')\phi(w'|x) \, d\nu(w')}. \tag{9}$$

If $f(x|w) = 0$, then the value of the ratio on the left side of Eq. (9) is indeterminate.   We shall assign to it the value of the right side of Eq. (9) in order that this equation will be valid for every value of $w \in \Omega$.

From relations (3) to (6), we can obtain the following results:

$$\int_\Omega \xi(w)\phi(w|x) \, d\nu(w)$$

$$\leq (1 + \beta)m \int_A \phi(w|x) \, d\nu(w) + (1 + \gamma)m \int_{A^c} \phi(w|x) \, d\nu(w)$$

$$\leq (1 + \beta)m + (1 + \gamma)m\alpha \tag{10}$$

and

$$\int_\Omega \xi(w)\phi(w|x) \, d\nu(w) \geq m(1 - \alpha). \tag{11}$$

Furthermore, by combining the relations (9) to (11) with relations (3) to (6), we can obtain the following inequalities.   For any value $w \in A$,

$$\frac{1}{1 + \alpha + \beta + \alpha\gamma} \leq \frac{\xi(w|x)}{\phi(w|x)} \leq \frac{1 + \beta}{1 - \alpha}. \tag{12}$$

Also, for any value $w \in A^c$,

$$\frac{\xi(w|x)}{\phi(w|x)} \leq \frac{1 + \gamma}{1 - \alpha}. \tag{13}$$

It now follows that

$$\int_\Omega |\xi(w|x) - \phi(w|x)| \, d\nu(w) = \int_\Omega \left| \frac{\xi(w|x)}{\phi(w|x)} - 1 \right| \phi(w|x) \, d\nu(w)$$

$$\leq \max \left\{ \frac{\alpha + \beta}{1 - \alpha}, \frac{\alpha + \beta + \alpha\gamma}{1 + \alpha + \beta + \alpha\gamma} \right\} \int_A \phi(w|x) \, d\nu(w)$$

$$+ \left(\frac{1 + \gamma}{1 - \alpha} + 1 \right) \int_{A^c} \phi(w|x) \, d\nu(w) \leq \epsilon. \blacksquare \quad (14)$$

One consequence of the relation (8) is that for any subset B of Ω, the difference between the probability of B computed from the g.p.d.f. $\phi(\cdot |x)$ and the probability of the same subset B computed from the actual posterior g.p.d.f. $\xi(\cdot |x)$ cannot be greater than ϵ. The value of ϵ which is obtained in Theorem 1 depends on the choice of the set A. An effective choice of the set A will be a set which yields a small value of ϵ, that is, a set A for which α, β, and γ will be small.

A great advantage of Theorem 1 is the simplicity of the three relations (4) to (6). For any set A, the number α specified in the relation (4) can be computed from the g.p.d.f.'s $f(\cdot |w)(w \in \Omega)$ and the observed value x. It is not necessary to consider the prior distribution. The numbers β and γ specified in the relations (5) and (6) can be computed by the statistician from two simple bounds on his prior g.p.d.f. In order that there may exist a set A for which α, β, and γ will be small, the observed value $X = x$ must convey enough information about W to permit the likelihood function $f(x|w)$ to be sharply peaked around some point in Ω. Under these conditions, the statistician knows, without any detailed analysis or specification of his prior g.p.d.f., that the g.p.d.f. $\phi(\cdot |x)$ is a good approximation to his posterior g.p.d.f.

For example, suppose that Ω is the real line and that the prior p.d.f. ξ of W is as sketched in Fig. 10.1. As indicated in that figure, the function ξ is constant on the interval from a to b. Suppose also that when the values $x_1, \ldots, x_n$ of a random sample are observed, the resulting p.d.f. $\phi(\cdot |x_1, \ldots, x_n)$ is as sketched in Fig. 10.1. If the set A is chosen to be the interval from a to b, then the value of α will be equal to the total area of the two shaded regions in Fig. 10.1. Furthermore, it can be seen from the p.d.f. ξ that $\beta = 0$ and $\gamma = 0$. It now follows from Eq. (7) that $\epsilon \leq 3\alpha/(1 - \alpha)$. Therefore, the difference between the probability of any subset of Ω computed from the function $\phi(\cdot |x_1, \ldots, x_n)$ sketched in Fig. 10.1 and the actual posterior probability of that subset cannot be greater than $3\alpha/(1 - \alpha)$.

10.5 CONVERGENCE OF POSTERIOR DISTRIBUTIONS

In the remainder of this chapter we shall consider the limiting properties of posterior distributions as the number of observations in a random

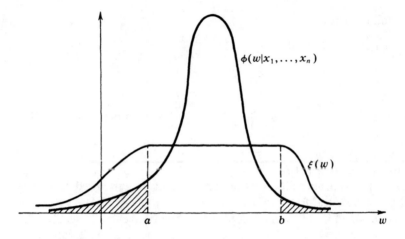

Fig. 10.1 An example of precise measurement.

sample becomes arbitrarily large. We shall show that if $X_1, \ldots, X_n$ is a random sample from a distribution for which the value of some parameter W is unknown and if the value of W is actually w_0, then, under certain conditions, the posterior distribution of W will tend to become more and more concentrated around the value $W = w_0$ as we let $n \to \infty$.

A simple example is that in which the parameter W can have only k different values $w_1, \ldots, w_k$. Suppose that $\Pr(W = w_i) = \xi_i > 0$ for $i = 1, \ldots, k$, and suppose that for any given value $W = w_i$ $(i = 1, \ldots, k)$, the random variables $X_1, \ldots, X_n$ are a random sample from the g.p.d.f. f_i. We shall assume that the g.p.d.f.'s f_i are distinct in the following sense: If S denotes the sample space of any single observation. then

$$\int_S |f_i(x) - f_j(x)| \, d\mu(x) > 0 \qquad \text{for } i \neq j. \tag{1}$$

For any observed values $X_j = x_j$ $(j = 1, \ldots, n)$, let $\xi_i(x_1, \ldots, x_n)$ denote the posterior probability that $W = w_i$ $(i = 1, \ldots, k)$. Thus, this posterior probability is given by the equation

$$\xi_i(x_1, \ldots, x_n) = \frac{\xi_i \Pi_{j=1}^n f_i(x_j)}{\sum_{r=1}^k [\xi_r \Pi_{j=1}^n f_r(x_j)]}. \tag{2}$$

Suppose now that $X_1, \ldots, X_n$ is actually a random sample from the g.p.d.f. f_t, where t is one of the values $1, \ldots, k$. We shall show that with probability 1, the following limiting values must be correct:

$$\lim_{n \to \infty} \xi_t(X_1, \ldots, X_n) = 1 \tag{3}$$

and

$$\lim_{n \to \infty} \xi_i(X_1, \ldots, X_n) = 0 \qquad \text{for } i \neq t. \tag{4}$$

In other words, as $n \to \infty$, all of the posterior probability tends to be concentrated on the correct value w_t of W.

A basic result in problems involving large samples is the strong law of large numbers, which can be stated as follows:

Strong law of large numbers *Let $Y_1, Y_2, \ldots$ be a sequence of independent and identically distributed random variables, for each of which the mean is μ. Then, with probability 1,*

$$\lim_{n \to \infty} \frac{1}{n} \sum_{i=1}^{n} Y_i = \mu. \tag{5}$$

A proof of this result is given in such books as Feller (1957), chap. 10, and (1966), chap. 7; Loève (1963), sec. 16; and Rao (1965), chap. 2. This result will be used in establishing Eqs. (3) and (4).

For any fixed value of i such that $i \neq t$, let μ be defined by the equation

$$\mu = E\left[\log \frac{f_i(X)}{f_t(X)}\right]. \tag{6}$$

Here, X is any single observation and the expectation is computed under the assumption that $W = w_t$. The expectation μ is not necessarily finite. However, since f_i and f_t are distinct g.p.d.f.'s, it follows from Jensen's inequality that

$$\mu < \log E\left[\frac{f_i(X)}{f_t(X)}\right] = \log \int_S \frac{f_i(x)}{f_t(x)} f_t(x)\, d\mu(x)$$

$$= \log \int_S f_i(x)\, d\mu(x) = \log 1 = 0. \tag{7}$$

Therefore, either μ is a finite negative number or μ can be assigned the value $-\infty$. In either case, it follows from the strong law of large numbers that with probability 1,

$$\lim_{n \to \infty} \frac{1}{n} \sum_{j=1}^{n} \log \frac{f_i(X_j)}{f_t(X_j)} = \mu < 0. \tag{8}$$

Hence,

$$\log \lim_{n \to \infty} \prod_{j=1}^{n} \frac{f_i(X_j)}{f_t(X_j)} = \lim_{n \to \infty} \sum_{j=1}^{n} \log \frac{f_i(X_j)}{f_t(X_j)} = -\infty. \tag{9}$$

However, Eq. (9) is equivalent to the following result: If $i \neq t$, then, with probability 1,

$$\lim_{n \to \infty} \prod_{j=1}^{n} \frac{f_i(X_j)}{f_t(X_j)} = 0. \tag{10}$$

Equations (3) and (4) now follow from Eq. (10) and from the expression given in Eq. (2) for the posterior probability $\xi_i(x_1, \ldots, x_n)$.

A limiting result of this type is true for posterior distributions in many problems. Consider the problem treated in Theorem 1 of Sec. 9.5, in which a random sample $X_1, \ldots, X_n$ is taken from a normal distribution with an unknown value of the mean W and a specified value of the precision r. It was shown in that section that the mean μ' of the posterior distribution of W will be a weighted average of the sample mean $\bar{X}$ and the mean of the prior distribution and that the precision of the posterior distribution of W will increase by r units for each observation which is taken.

Now suppose that $X_1, \ldots, X_n$ is, in fact, a random sample from a normal distribution with mean w_0. If we let $n \to \infty$, the relative weight given to $\bar{X}$ in the computation of μ' approaches the value 1. However, by the law of large numbers, $\bar{X} \to w_0$ with probability 1. Therefore, $\mu' \to w_0$ with probability 1. Furthermore, the variance of the posterior distribution of W approaches the value 0. It follows that the posterior distribution of W tends to become more and more concentrated around the value w_0 and that the posterior probability of any open interval containing w_0 must approach the value 1. Other examples are given in Exercises 19 to 21.

Further Remarks and References

The convergence of posterior distributions to the correct value of the parameter as $n \to \infty$ can be demonstrated for more general parameter spaces Ω. However, it becomes necessary to make special assumptions about the prior distribution of the parameter and the family of g.p.d.f.'s $\{f(\cdot \, |w), \ w \in \Omega\}$. This topic has a long history and was studied by Laplace. Some references on this topic and related questions are von Mises (1964), chap. 7; Berk (1966); and the more abstract work of Blackwell and Dubins (1962), Freedman (1963, 1965), Fabius (1964), and Schwartz (1965).

10.6 SUPERCONTINUITY

Now we shall consider a parameter W whose values must lie in an open interval Ω of the real line. The length of Ω may be either finite or infinite.

As usual, we assume that there is a given family of g.p.d.f.'s $\{f(\cdot \mid w),$ $w \in \Omega\}$, but we shall now assume that the observations $X_1, \ldots, X_n$ are, in fact, a random sample from the g.p.d.f. $f(\cdot \mid w_0)$, where w_0 is a specific point in Ω. Under these conditions, we shall study the limiting behavior of the likelihood function computed from the observed values as $n \to \infty$. Specifically, we shall show that under certain regularity conditions, the likelihood function, and hence also the posterior p.d.f. of W, will converge in a special sense to the p.d.f. of a normal distribution.

For any point $w^* \in \Omega$ and any number $\alpha > 0$, we shall define $N(w^*; \alpha)$ as the interval around w^* containing every point in Ω whose distance from w^* is less than α.

Now let g be any real-valued function whose value $g(x, w)$ is specified at every point (x, w) of the product space $S \times \Omega$. Here, as usual, S denotes the sample space of a single observation X. It will be convenient in our development to use the following definition: The function g is *supercontinuous* at the value $w_0 \in \Omega$ if

$$\lim_{\alpha \to 0} E[\sup_{w \in N(w_0; \alpha)} |g(X, w) - g(X, w_0)|] = 0. \tag{1}$$

In Eq. (1) and in the remainder of this development, the expectation is computed with respect to the g.p.d.f. $f(\cdot \mid w_0)$ of each observation.

The meaning of Eq. (1) may be difficult to perceive, and so, before proceeding with the development of the limiting behavior of the likelihood function, we shall first discuss some properties of supercontinuous functions. For $x \in S$ and $\alpha > 0$, we shall let the value $h(x, \alpha)$ be defined as follows:

$$h(x, \alpha) = \sup_{w \in N(w_0; \alpha)} |g(x, w) - g(x, w_0)|. \tag{2}$$

With this definition, we can say that g is supercontinuous at w_0 if

$$\lim_{\alpha \to 0} E[h(X, \alpha)] = 0. \tag{3}$$

In the next two lemmas, we shall show that if the expectation $E[h(X, \alpha)]$ exists for all sufficiently small values of α, then supercontinuity of the function g at w_0 is equivalent to ordinary continuity of the function $g(x, \cdot)$ at w_0 for almost every value of $x \in S$. In regard to these lemmas, it should be kept in mind that when we say that a certain property is true for almost every value of $x \in S$, we mean that the property is true for each value of x with the possible exception of some values which form a subset of S whose probability is 0.

Lemma 1 *If g is supercontinuous at w_0, then $g(x, \cdot)$ is continuous at w_0 for almost every value of $x \in S$.*

Proof Suppose that $x \in S$ is a point such that $g(x, \cdot)$ is not continuous at w_0. Then there is a positive number $\epsilon(x)$ such that $h(x, \alpha) \geq \epsilon(x) > 0$ for every value of $\alpha > 0$. If the set S_0 of such points x has a positive probability, then

$$E[h(X, \alpha)] \geq \int_{S_0} \epsilon(x) f(x|w_0) \, d\mu(x) > 0. \tag{4}$$

Therefore, Eq. (3) would not hold as we let $\alpha \to 0$. Since this is a contradiction, it follows that the set S_0 must have probability 0. In other words, $g(x, \cdot)$ must be continuous at w_0 for almost every value of x in the sample space S. ∎

To prove Lemma 2, we shall use the following theorem, which is known as the Lebesgue dominated convergence theorem [see, e.g., Loève (1963), sec. 7; Halmos (1950), chap. 5; or Kolmogorov and Fomin (1961), sec. 12].

Lebesgue dominated convergence theorem *Let $v_1, v_2, \ldots$ be a sequence of functions on S such that $E[v_n(X)]$ exists for $n = 1, 2, \ldots$ and $\lim_{n \to \infty} v_n(x) = v(x)$ for almost every value of $x \in S$. Suppose that there exists a function M such that $E[M(X)] < \infty$ and $|v_n(x)| \leq M(x)$ for all values of $x \in S$ and $n = 1, 2, \ldots$. Then $E[v(X)]$ exists and $E[v(X)] = \lim_{n \to \infty} E[v_n(X)]$.*

Lemma 2 *If there exists a number $\alpha_0 > 0$ such that $E[h(X, \alpha)] < \infty$ whenever $\alpha \leq \alpha_0$ and if $g(x, \cdot)$ is continuous at w_0 for almost every value of $x \in S$, then g is supercontinuous at w_0.*

Proof At any value of $x \in S$ such that $g(x, \cdot)$ is continuous at w_0, $\lim_{\alpha \to 0} h(x, \alpha) = 0$. Moreover, if $\alpha < \alpha_0$, then $h(x, \alpha) \leq h(x, \alpha_0)$ for all values of $x \in S$. Since $E[h(X, \alpha_0)] < \infty$, it follows from the Lebesgue dominated convergence theorem that

$$\lim_{\alpha \to 0} E[h(X, \alpha)] = E[\lim_{\alpha \to 0} h(X, \alpha)] = 0. \tag{5}$$

Hence, g is supercontinuous at w_0. ∎

Lemmas 1 and 2 now yield the following theorem.

Theorem 1 *Suppose that there exists a number $\alpha_0 > 0$ such that $E[h(X, \alpha)] < \infty$ whenever $\alpha \leq \alpha_0$. Then g is supercontinuous at w_0 if, and only if, $g(x, \cdot)$ is continuous at w_0 for almost every value of $x \in S$.*

To simplify the notation in the remainder of the development, we shall denote the partial derivatives with respect to w of any function v on $S \times \Omega$ by $v'(x, w)$, $v''(x, w)$, and so on. For instance,

$$v''(x, w_1) = \left[\frac{\partial^2 v(x, w)}{\partial w^2}\right]_{w=w_1}. \tag{6}$$

Here, $x \in S$ and $w_1 \in \Omega$ may be any values for which the derivative exists.

As an illustration of the use of supercontinuity, we shall present a condition in the next theorem which permits reversal of the order in which the operations of differentiation and expectation are performed. In other words, differentiation inside the integral sign is permitted. For other results of this type, see Cramér (1946), chap. 7.

Theorem 2 *Suppose that for some value of $\alpha > 0$, the partial derivative $g'(x, w)$ exists for all values of $x \in S$ and all values of $w \in N(w_0; \alpha)$, and suppose that this derivative is supercontinuous at w_0. If $E[g(X, w)]$ exists for all values of $w \in N(w_0; \alpha)$ and if $E[|g'(X, w_0)|] < \infty$, then*

$$E[g'(X, w_0)] = \left\{\frac{\partial E[g(X, w)]}{\partial w}\right\}_{w=w_0}. \tag{7}$$

Proof To prove the correctness of Eq. (7), we must show that

$$E\left[\lim_{w \to w_0} \frac{g(X, w) - g(X, w_0)}{w - w_0}\right] = \lim_{w \to w_0} E\left[\frac{g(X, w) - g(X, w_0)}{w - w_0}\right]. \tag{8}$$

In order to obtain this result by an application of the Lebesgue dominated convergence theorem, we shall show that there exists a function M such that $E[M(X)] < \infty$ and such that, for all values of $w \in N(w_0; \alpha)$ and all values of $x \in S$,

$$\left|\frac{g(x, w) - g(x, w_0)}{w - w_0}\right| \leq M(x). \tag{9}$$

By the mean-value theorem of elementary calculus, there exists a value $w_1(x)$ between w_0 and w such that

$$\frac{g(x, w) - g(x, w_0)}{w - w_0} = g'[x, w_1(x)]. \tag{10}$$

Since $w_1(x) \in N(w_0; \alpha)$,

$$|g'[x, w_1(x)] - g'(x, w_0)| \leq \sup_{w \in N(w_0; \alpha)} |g'(x, w) - g'(x, w_0)|. \tag{11}$$

By hypothesis, $g'(x, w)$ is supercontinuous at w_0. Hence, the right side of the inequality (11), which we shall call $M_1(x)$, has a finite expectation, and the following relation must be satisfied:

$$|g'[x, w_1(x)]| \leq M_1(x) + |g'(x, w_0)|. \tag{12}$$

Since $E[|g'(X, w_0)|] < \infty$ by hypothesis, it can be seen from (10) and (12) that an inequality of the type specified by the relation (9) has been obtained. ∎

10.7 SOLUTIONS OF THE LIKELIHOOD EQUATION

We shall now proceed with our development of the properties of the likelihood function $f(x_1|w) \cdots f(x_n|w)$ when the observations $X_1, \ldots, X_n$ are a large random sample from the g.p.d.f. $f(\cdot|w_0)$. We shall make four assumptions about the family of g.p.d.f.'s $\{f(\cdot|w), w \in \Omega\}$.

Assumption L_1 *The second-order partial derivatives $f''(x|w)$ exist for all values of $x \in S$ and all values of w in some neighborhood $N(w_0; \alpha)$ of w_0, where $\alpha > 0$.*

We know that $\int_S f(x|w) \, d\mu(x) = 1$ for every value of $w \in \Omega$. Therefore, if the integral on the left side of this equation is differentiated with respect to w, the result will be 0. For this reason, the next assumption may be interpreted as stating that we can reverse the order in which the operations of differentiation and integration are performed and still obtain the same result. This property is assumed only for derivatives at the point w_0.

Assumption L_2 $\int_S f'(x|w_0) \, d\mu(x) = 0$ *and* $\int_S f''(x|w_0) \, d\mu(x) = 0.$

For any values of $x \in S$ and $w \in N(w_0; \alpha)$, it is assumed that $f(x|w) > 0$. Let $\lambda(x|w)$ be defined as follows:

$$\lambda(x|w) = \log f(x|w). \tag{1}$$

Assumption L_3 *The functions λ, λ', and λ'' are supercontinuous at w_0, and for all values of $w \in N(w_0; \alpha)$, the expectations $E[\lambda(X|w)]$, $E[\lambda'(X|w)]$, and $E[\lambda''(X|w)]$ are finite.*

For any values of $w \in N(w_0; \alpha)$, let $I(w)$ be defined as follows:

$$I(w) = \int_S [\lambda'(x|w)]^2 f(x|w) \, d\mu(x). \tag{2}$$

Assumption L_4 *The function I is positive and continuous throughout the neighborhood $N(w_0; \alpha)$.*

Now suppose that the number n of observations is made arbitrarily large. It is said that a sequence of statistics $\{\hat{w}_n(X_1, \ldots, X_n); n = 1, 2, \ldots\}$ is a *consistent sequence* if, with probability 1,

$$\lim_{n \to \infty} \hat{w}_n(X_1, \ldots, X_n) = w_0. \tag{3}$$

In the next theorem we shall establish, as a consequence of the four assumptions which have been made, the existence of a consistent sequence of statistics. We shall also show in the next theorem that such a sequence can be constructed from solutions of the following equation, which is called the *likelihood equation*:

$$\sum_{i=1}^{n} \lambda'(X_i|w) = 0. \tag{4}$$

Theorem 1 *Suppose that Assumptions L_1 to L_4 are satisfied. Then, with probability 1, there will exist an integer n_0 such that for each value of $n \geq n_0$, the likelihood equation (4) has a solution $w = \hat{w}_n(X_1, \ldots, X_n)$ and the sequence of statistics $\hat{w}_n(X_1, \ldots, X_n)$ satisfies Eq. (3).*

Proof From Assumption L_3 and Theorem 2 of Sec. 10.6,

$$\left\{ \frac{\partial}{\partial w} E[\lambda'(X|w)] \right\}_{w=w_0} = E[\lambda''(X|w_0)]. \tag{5}$$

However, for all values of $x \in S$,

$$\lambda''(x|w_0) = \frac{f(x|w_0)f''(x|w_0) - [f'(x|w_0)]^2}{f^2(x|w_0)}. \tag{6}$$

Therefore, it follows from Assumptions L_2 and L_4 that

$$E[\lambda''(X|w_0)] = -I(w_0) < 0. \tag{7}$$

From Eq. (5), we can now conclude that the function $E[\lambda'(X|w)]$ is continuous and strictly decreasing at the point $w = w_0$. Furthermore, by Assumption L_2, $E[\lambda'(X|w_0)] = 0$. Hence, for all sufficiently small values of δ ($\delta > 0$), it must be true that $E[\lambda'(X|w_0 - \delta)] > 0$ and $E[\lambda'(X|w_0 + \delta)] < 0$.

For any value of $w \in N(w_0; \alpha)$, it follows from the strong law of large numbers that with probability 1,

$$\lim_{n \to \infty} \frac{1}{n} \sum_{i=1}^{n} \lambda'(X_i|w) = E[\lambda'(X|w)]. \tag{8}$$

Hence, for any value of δ ($\delta > 0$), the following statement must be true:

With probability 1, there will exist an integer n_0 such that for every value of $n \geq n_0$,

$$\frac{1}{n} \sum_{i=1}^{n} \lambda'(X_i|w_0 - \delta) > 0 \quad \text{and} \quad \frac{1}{n} \sum_{i=1}^{n} \lambda'(X_i|w_0 + \delta) < 0. \quad (9)$$

But because of the assumed differentiability of $\lambda'(x|w)$, it follows that for any given values of n and $x_1, \ldots, x_n$, the function $(1/n)\sum_{i=1}^{n}\lambda'(x_i|w)$ is a continuous function of w on the interval where $w_0 - \delta \leq w \leq w_0 + \delta$. Therefore, according to the relation (0), the probability is 1 that Eq. (4) has a solution between $w_0 - \delta$ and $w_0 + \delta$ for any value of $n \geq n_0$. Since δ can be made arbitrarily small, the theorem must be correct.∎

Further Remarks and References

Theorems similar to Theorem 1 that establish the existence of a consistent sequence of solutions of the likelihood equation (4) have been proved under various conditions. Such theorems are presented in the books by Cramér (1946), chap. 33; Wilks (1962), chap. 12; and Rao (1965), chap. 5; and in papers by Huzurbazar (1948), Kulldorff (1957), and Weiss (1963).

For any given sample size n and any values of the observations $X_1, \ldots, X_n$, a *maximum likelihood estimator* is defined to be a value $w = w_n(X_1, \ldots, X_n)$ which maximizes the likelihood function $\Pi_{i=1}^{n} f(X_i|w)$ or, equivalently, which maximizes the logarithm of this function $\sum_{i=1}^{n} \lambda(X_i|w)$. In many standard statistical problems, a maximum likelihood estimator is computed by finding a solution of the likelihood equation (4) and then verifying that this solution actually maximizes the likelihood function. However, unless the statistician has additional information, he cannot be certain that as $n \to \infty$, the sequence of maximum likelihood estimators will actually be the consistent sequence of solutions specified by Theorem 1. Nevertheless, Wald (1949) has proved that under general conditions, the sequence of maximum likelihood estimators will itself be a consistent sequence.

10.8 CONVERGENCE OF SUPERCONTINUOUS FUNCTIONS

In this section we shall prove a theorem which will be helpful when we develop the limiting properties of the likelihood function in the next section. The theorem pertains to the convergence of averages of supercontinuous functions.

Suppose that a function g is supercontinuous at w_0 and that the expectation $E[g(X, w_0)]$ exists. From the strong law of large numbers, it is known that with probability 1,

$$\lim_{n \to \infty} \frac{1}{n} \sum_{i=1}^{n} g(X_i, w_0) = E[g(X, w_0)]. \tag{1}$$

Now let $\{\hat{W}_n = \hat{w}_n(X_1, \ldots, X_n); n = 1, 2, \ldots\}$ be any consistent sequence of statistics, i.e., any sequence which satisfies Eq. (3) of Sec. 10.7. Then it is shown in the next theorem that the following limiting relation is also correct:

$$\lim_{n \to \infty} \frac{1}{n} \sum_{i=1}^{n} g(X_i, \hat{W}_n) = E[g(X, w_0)]. \tag{2}$$

Theorem 1 *Suppose that a function g is supercontinuous at w_0 and that $E[g(X, w_0)]$ exists. If $\{\hat{W}_n; n = 1, 2, \ldots\}$ is any consistent sequence of statistics, then, with probability 1, Eq. (2) must be correct.*

Proof Let $m_0 = E[g(X, w_0)]$, and let δ be any fixed number ($\delta > 0$). We must show that with probability 1, there will exist an integer n_0 such that, for all values of $n \geq n_0$,

$$\left| \frac{1}{n} \sum_{i=1}^{n} g(X_i, \hat{W}_n) - m_0 \right| < \delta. \tag{3}$$

For $x \in S$ and $\alpha > 0$, let the value $h(x, \alpha)$ be as defined in Eq. (2) of Sec. 10.6. If $|\hat{W}_n - w_0| < \alpha$, then

$$\left| \frac{1}{n} \sum_{i=1}^{n} g(X_i, \hat{W}_n) - m_0 \right|$$

$$\leq \frac{1}{n} \sum_{i=1}^{n} |g(X_i, \hat{W}_n) - g(X_i, w_0)| + \left| \frac{1}{n} \sum_{i=1}^{n} g(X_i, w_0) - m_0 \right|$$

$$\leq \frac{1}{n} \sum_{i=1}^{n} h(X_i, \alpha) + \left| \frac{1}{n} \sum_{i=1}^{n} g(X_i, w_0) - m_0 \right|. \tag{4}$$

Since, by hypothesis, g is supercontinuous at w_0, then

$$\lim_{\alpha \to 0} E[h(X, \alpha)] = 0.$$

Therefore, α can be chosen small enough to satisfy the requirement that $E[h(X, \alpha)] < \delta/4$. Also, by hypothesis, $\lim_{n \to \infty} \hat{W}_n = w_0$ with probability 1. Furthermore, in accordance with the strong law of large numbers, the following two equations must be correct with probability 1:

$$\lim_{n \to \infty} \frac{1}{n} \sum_{i=1}^{n} h(X_i, \alpha) = E[h(X, \alpha)] \tag{5}$$

and

$$\lim_{n \to \infty} \frac{1}{n} \sum_{i=1}^{n} g(X_i, w_0) = m_0. \tag{6}$$

Therefore, with probability 1, there will exist an integer n_0 such that, for all values of $n \geq n_0$, the following three relations will be satisfied:

$$|\hat{W}_n - w_0| < \alpha,$$

$$\frac{1}{n} \sum_{i=1}^{n} h(X_i, \alpha) < \frac{\delta}{2}, \tag{7}$$

$$\left| \frac{1}{n} \sum_{i=1}^{n} g(X_i, w_0) - m_0 \right| < \frac{\delta}{2}.$$

The desired result specified in the relation (3) now follows from (4) and (7).■

10.9 LIMITING PROPERTIES OF THE LIKELIHOOD FUNCTION

We shall now suppose that Assumptions L_1 to L_4 are satisfied. In this section we shall show that when the number of observations n is large, the likelihood function and the posterior distribution can be approximated by a certain normal distribution. The mathematical development which we shall present will serve as a formal justification of the following heuristic remarks:

For sufficiently large values of n, we shall let

$$\hat{W}_n = \hat{w}_n(X_1, \ldots, X_n)$$

denote solutions of the likelihood equation which form a consistent sequence as specified in Theorem 1 of Sec. 10.7. To simplify the notation, for any values of the observations $X_1, \ldots, X_n$, we shall let

$$\Lambda_n(w) = \sum_{i=1}^{n} \lambda(X_i | w). \tag{1}$$

When the function Λ_n is expanded in a Taylor series around the value $w = \hat{W}_n$, the following equation is obtained:

$$\Lambda_n(w) = \Lambda_n(\hat{W}_n) + \Lambda_n'(\hat{W}_n)(w - \hat{W}_n)$$
$$+ \tfrac{1}{2}\Lambda_n''(\hat{W}_n)(w - \hat{W}_n)^2 + \cdots . \tag{2}$$

It follows from the definition of $\hat{W}_n$ that $\Lambda'_n(\hat{W}_n) = 0$. Furthermore, it follows from Eq. (7) of Sec. 10.7 and the strong law of large numbers that for large values of n, the following approximation should be appropriate:

$$\frac{1}{n} \Lambda''_n(\hat{W}_n) \approx -I(\hat{W}_n). \tag{3}$$

The leading term on the right side of Eq. (2) does not involve w. Therefore, by ignoring the terms of higher order in Eq. (2), we can write the following approximation to the likelihood function for values of w in a small neighborhood of $\hat{W}_n$:

$$\prod_{i=1}^{n} f(X_i|w) = \exp\left[\Lambda_n(w)\right]$$
$$\propto \exp\left[-\tfrac{1}{2}nI(\hat{W}_n)(w - \hat{W}_n)^2\right]. \tag{4}$$

The interpretation of the relation (4) is that the likelihood function can be approximated by the p.d.f. of a normal distribution for which the mean is $\hat{W}_n$ and the precision is $nI(\hat{W}_n)$. It follows from this approximation that if the prior distribution of the parameter W is represented in a small neighborhood of the point w_0 by a smooth positive density function, then the posterior distribution of W will be highly concentrated in the neighborhood of $\hat{W}_n$, and in this neighborhood it can be approximated by a normal distribution with mean $\hat{W}_n$ and precision $nI(\hat{W}_n)$. In the remainder of this section we shall present a rigorous development of these ideas.

Let $N(w_0; \alpha)$ denote the neighborhood of the value w_0 where Assumptions L_1 to L_4 are assumed to be satisfied. Since $\lim_{n \to \infty} \hat{W}_n = w_0$ with probability 1, it follows that $\hat{W}_n$ will belong to the neighborhood $N(w_0; \alpha)$ for sufficiently large values of n. For any values of the observations $X_1, \ldots, X_n$ and any value of $w \in N(w_0; \alpha)$, let

$$L_n(w) = \left[\prod_{i=1}^{n} f(X_i|w)\right]^{1/n}. \tag{5}$$

By elementary differentiation, it can be verified that

$$L''_n(w) = L_n(w) \left\{\left[\frac{1}{n}\Lambda'_n(w)\right]^2 + \frac{1}{n}\Lambda''_n(w)\right\}. \tag{6}$$

Also, $L'_n(\hat{W}_n) = 0$. Therefore, by expanding the function L_n in a Taylor series around the value $w = \hat{W}_n$ through terms of second order, we obtain the following equation:

$$\frac{L_n(w)}{L_n(\hat{W}_n)} = 1 + \frac{L_n''(\breve{W}_n)}{2L_n(\hat{W}_n)}\,(w - \hat{W}_n)^2,\tag{7}$$

where $\breve{W}_n$ is an appropriate value between w and $\hat{W}_n$.

Now let θ be any fixed number, and suppose that w is specified by the equation

$$w = \hat{W}_n + [nI(\hat{W}_n)]^{-\frac{1}{2}}\theta.\tag{8}$$

Furthermore, let β_n be defined as follows:

$$\beta_n = \frac{L_n''(\breve{W}_n)}{L_n(\hat{W}_n)I(\hat{W}_n)}.\tag{9}$$

Then, by Eqs. (7) to (9),

$$\left[\frac{L_n(w)}{L_n(\hat{W}_n)}\right]^n = \left[1 + \frac{\beta_n\theta^2}{2n}\right]^n.\tag{10}$$

We shall now show that $\lim_{n\to\infty} \beta_n = -1$ with probability 1. By Assumption L_2, the function λ is supercontinuous at w_0. Therefore, it follows from Theorem 1 of Sec. 10.8 that with probability 1,

$$\lim_{n\to\infty}\frac{1}{n}\Lambda_n(\hat{W}_n) = E[\lambda(X|w_0)].\tag{11}$$

Furthermore, by Assumption L_4,

$$\lim_{n\to\infty} I(\hat{W}_n) = I(w_0) > 0.\tag{12}$$

Since $\breve{W}_n$ lies between w and $\hat{W}_n$, it follows from Eq. (8) that $\lim_{n\to\infty} \breve{W}_n = w_0$ with probability 1 for any fixed value of θ. We can again apply Theorem 1 of Sec. 10.8 to obtain a result which is similar to Eq. (11) but in which $\breve{W}_n$ replaces $\hat{W}_n$. Since $(1/n)\Lambda_n(w) = \log L_n(w)$, it now follows that

$$\lim_{n\to\infty}\frac{L_n(\breve{W}_n)}{L_n(\hat{W}_n)} = 1.\tag{13}$$

In addition, according to Assumption L_3, both λ' and λ'' are supercontinuous at w_0. Hence, the following two equations are correct with probability 1:

$$\lim_{n \to \infty} \frac{1}{n} \Lambda_n'(\hat{W}_n) = E[\lambda'(X|w_0)] = 0,$$

$$\lim_{n \to \infty} \frac{1}{n} \Lambda_n''(\hat{W}_n) = E[\lambda''(X|w_0)] = -I(w_0). \tag{14}$$

The desired result that $\lim_{n \to \infty} \beta_n = -1$ now follows from the definition of β_n given in Eq. (9) and from Eqs. (6) and (12) to (14).

The next theorem now follows from Eq. (10).

Theorem 1 *Suppose that Assumptions L_1 to L_4 are satisfied. For any fixed value of θ, let w be specified by Eq. (8). Then, with probability 1,*

$$\lim_{n \to \infty} \frac{\Pi_{i=1}^n f(X_i|w)}{\Pi_{i=1}^n f(X_i|\hat{W}_n)} = \exp\left(-\tfrac{1}{2}\theta^2\right). \tag{15}$$

It now follows from Eqs. (8) and (15) that the relation (4) does provide an approximation to the likelihood function for large values of n and for values of w in a small neighborhood of $\hat{W}_n$.

10.10 NORMAL APPROXIMATION TO THE POSTERIOR DISTRIBUTION

Suppose now that the prior p.d.f. of W is positive and continuous in the neighborhood $N(w_0; \alpha)$. Then, for large values of n, the variation of the prior p.d.f. in the neighborhood of $\hat{W}_n$ will be insignificant when compared with the steepness with which the likelihood function drops off around the value $\hat{W}_n$. We may therefore assume again that the posterior p.d.f. of W will be approximately proportional to the likelihood function. In other words, the posterior p.d.f. will be approximately that of a normal distribution with mean $\hat{W}_n$ and precision $nI(\hat{W}_n)$.

As an example, we shall consider sampling from a Bernoulli distribution for which the parameter W is unknown. For any given value $W = w$ ($0 < w < 1$), the p.f. $f(\cdot|w)$ of any observation X is

$$f(x|w) = w^x(1 - w)^{1-x} \qquad x = 0, 1. \tag{1}$$

By differentiation, we can obtain the following results:

$$\lambda'(x|w) = \frac{x}{w} - \frac{1 - x}{1 - w} \qquad \text{and} \qquad \lambda''(x|w) = -\left[\frac{x}{w^2} + \frac{1 - x}{(1 - w)^2}\right]. \tag{2}$$

It follows that $I(w) = 1/[w(1 - w)]$ for $0 < w < 1$. Furthermore, the statistics $\hat{w}_n(X_1, \ldots, X_n) = \bar{X}$ form a consistent sequence of solutions of the likelihood equation. Hence, it follows that for large values of n, the posterior p.d.f. of W when $X_i = x_i$ $(i = 1, \ldots, n)$ is approximately a normal distribution with mean $\bar{x}$ and precision $n/[\bar{x}(1 - \bar{x})]$. Since it is assumed that the value w_0 of W is neither 0 nor 1, the probability is high that $\bar{X}$ will be neither 0 nor 1 for large values of n.

It should be noted that Theorem 1 of Sec. 10.9 does not cover sampling from a uniform distribution. For that distribution, the p.d.f. $f(x|w)$ is not a continuous function of w.

Further Remarks and References

Asymptotic normality of posterior distributions has been studied by LeCam (1953, 1956, 1966), Lindley (1961b), and Johnson (1967). Anscombe (1964a,b) has discussed transformations of the likelihood function which improve this normal approximation.

10.11 APPROXIMATIONS FOR VECTOR PARAMETERS

Now we shall consider briefly a problem where $\mathbf{W} = (W_1, \ldots, W_k)'$ is a vector parameter whose values lie in a subset Ω of R^k. We shall use the following notation:

$$\lambda(x|\mathbf{w}) = \log f(x|\mathbf{w}),$$

$$\lambda_i(x|\mathbf{w}) = \frac{\partial \lambda(x|\mathbf{w})}{\partial w_i} \qquad i = 1, \ldots, k, \tag{1}$$

$$\lambda_{ij}(x|\mathbf{w}) = \frac{\partial^2 \lambda(x|\mathbf{w})}{\partial w_i \partial w_j} \qquad i, j = 1, \ldots, k.$$

For any value of $\mathbf{w} \in \Omega$, we shall define $\mathbf{I}(\mathbf{w})$ to be the $k \times k$ matrix whose (i, j) element $I_{ij}(\mathbf{w})$ is specified as follows:

$$I_{ij}(\mathbf{w}) = \int_S \lambda_i(x|\mathbf{w})\lambda_j(x|\mathbf{w})f(x|\mathbf{w}) \, d\mu(x) \qquad i, j = 1, \ldots, k. \tag{2}$$

The matrix $\mathbf{I}(\mathbf{w})$ is called the *Fisher information matrix*.

If Assumption L_2 is generalized so that partial derivatives of first and second orders with respect to the components of $\mathbf{w}$ can be taken inside the integral of the g.p.d.f. $f(x|\mathbf{w})$, then the element $I_{ij}(\mathbf{w})$ of the Fisher information matrix can also be written in the following form:

$$I_{ij}(\mathbf{w}) = -\int_S \lambda_{ij}(x|\mathbf{w})f(x|\mathbf{w}) \, d\mu(x) \qquad i, j = 1, \ldots, k. \tag{3}$$

The appropriate generalization of Assumption L_4 then states that the matrix $\mathbf{I}(\mathbf{w})$ is positive definite for each value of $\mathbf{w} \in \Omega$ and that its elements are continuous functions of $\mathbf{w}$.

The proof of Theorem 1 of Sec. 10.7, which established the existence of a consistent sequence of solutions of the likelihood equation, depended heavily on the fact that the parameter w was real-valued, and the extension of that proof to a k-dimensional vector $\mathbf{w}$ is somewhat more complicated. Such extensions have been studied by Chanda (1954) and by Doss (1962, 1963). It is a consequence of these studies and the general results of Wald (1949) that appropriate theorems can be established. In this problem, the set of k likelihood equations is

$$\sum_{i=1}^{n} \lambda_j(X_i|\mathbf{w}) = 0 \qquad j = 1, \ldots, k. \tag{4}$$

Suppose that the observations $X_1, \ldots, X_n$ are actually a random sample from the g.p.d.f. $f(\cdot\,|\mathbf{w}_0)$. Then, under certain conditions which will not be discussed here, there exist solutions $\hat{\mathbf{W}}_n = \hat{\mathbf{w}}_n(X_1, \ldots, X_n)$ of the set of likelihood equations such that $\lim_{n \to \infty} \hat{\mathbf{W}}_n = \mathbf{w}_0$ with probability 1.

Furthermore, Theorem 1 of Sec. 10.9 admits a straightforward generalization to a vector $\mathbf{w}$. For each value of $\mathbf{w} \in \Omega$, we shall let $\mathbf{A}(\mathbf{w})$ be a nonsingular $k \times k$ matrix with continuous elements such that $\mathbf{A}'(\mathbf{w})\mathbf{A}(\mathbf{w}) = \mathbf{I}(\mathbf{w})$. For any fixed vector $\boldsymbol{\theta} = (\theta_1, \ldots, \theta_k)'$, suppose that $\mathbf{w}$ is specified by the equation

$$\mathbf{w} = \hat{\mathbf{W}}_n + n^{-\frac{1}{2}}[\mathbf{A}(\hat{\mathbf{W}}_n)]^{-1}\boldsymbol{\theta}. \tag{5}$$

Under the appropriate generalizations of Assumptions L_1 to L_4, it follows that with probability 1,

$$\lim_{n \to \infty} \frac{\prod_{i=1}^{n} f(X_i|\mathbf{w})}{\prod_{i=1}^{n} f(X_i|\hat{\mathbf{W}}_n)} = \exp\left(-\tfrac{1}{2}\boldsymbol{\theta}'\boldsymbol{\theta}\right). \tag{6}$$

The interpretation of Eqs. (5) and (6) can be stated as follows: Suppose that the prior p.d.f. of $\mathbf{W}$ is positive and continuous in a neighborhood of the value $\mathbf{w}_0$, and suppose that a large number of observations are taken. Then the posterior p.d.f. of $\mathbf{W}$ after the values of $X_1, \ldots, X_n$ and the corresponding value of $\hat{\mathbf{W}}_n$ have been observed will be approximately proportional to the right side of Eq. (6). Hence, from Eq. (5), it can be seen that the posterior p.d.f. of $\mathbf{W}$ is approximately proportional to the following function:

$$\exp\left[-\frac{n}{2}(\mathbf{w} - \hat{\mathbf{W}}_n)'\mathbf{I}(\hat{\mathbf{W}}_n)(\mathbf{w} - \hat{\mathbf{W}}_n)\right]. \tag{7}$$

In other words, the posterior distribution of $\mathbf{W}$ is approximately a multivariate normal distribution for which the mean vector is $\hat{\mathbf{W}}_n$ and the precision matrix is $n\mathbf{I}(\hat{\mathbf{W}}_n)$.

As an example, we shall consider sampling from a population which contains elements of k different types. If W_i is the probability of obtaining an element of type i $(i = 1, \ldots, k)$, then each observation $\mathbf{X} = (X_1, \ldots, X_k)'$ can be regarded as having a multinomial distribution with parameters $n = 1$ and $\mathbf{W} = (W_1, \ldots, W_k)'$, where $W_i > 0$ $(i = 1, \ldots, k)$ and $\Sigma_{i=1}^k W_i = 1$.

Since $\Sigma_{i=1}^k W_i = 1$, the Fisher information matrix $\mathbf{I}(\mathbf{w})$ will be positive definite only if we eliminate one of the components of $\mathbf{W}$, say W_k, and assume that $\mathbf{W} = (W_1, \ldots, W_{k-1})'$. Then, for any point $\mathbf{w} \, \epsilon \, \Omega$,

$$\lambda(\mathbf{X}|\mathbf{w}) = X_1 \log w_1 + \cdots + X_k \log w_k, \tag{8}$$

where $w_k = 1 - \Sigma_{i=1}^{k-1} w_i$. By differentiating, we obtain the following results for $i = 1, \ldots, k - 1$ and $j = 1, \ldots, k - 1$:

$$\lambda_i(\mathbf{X}|\mathbf{w}) = \frac{X_i}{w_i} - \frac{X_k}{w_k},$$

$$\lambda_{ii}(\mathbf{X}|\mathbf{w}) = -\left(\frac{X_i}{w_i{}^2} + \frac{X_k}{w_k{}^2}\right), \tag{9}$$

$$\lambda_{ij}(\mathbf{X}|\mathbf{w}) = -\frac{X_k}{w_k{}^2} \qquad i \neq j.$$

It follows from Eq. (3) that the elements $I_{ij}(\mathbf{w})$ of the $(k - 1) \times (k - 1)$ matrix $\mathbf{I}(\mathbf{w})$ are as follows:

$$I_{ii}(\mathbf{w}) = \frac{1}{w_i} + \frac{1}{w_k},$$

$$I_{ij}(\mathbf{w}) = \frac{1}{w_k} \qquad i \neq j. \tag{10}$$

Suppose now that n observations $\mathbf{X}_1, \ldots, \mathbf{X}_n$ are obtained from the population, and for $i = 1, \ldots, k$, let Y_i be the number of these observations which are of type i. Then a consistent sequence of solutions of the set of likelihood equations (4) is specified by the vector

$$\hat{\mathbf{W}}_n = \hat{\mathbf{w}}_n(\mathbf{X}_1, \ldots, \mathbf{X}_n) = \left(\frac{Y_1}{n}, \ldots, \frac{Y_{k-1}}{n}\right)'. \tag{11}$$

Therefore, for large values of n, the posterior distribution of $\mathbf{W}$ will be approximately a $(k - 1)$-dimensional multivariate normal distribution for which the mean vector is $\hat{\mathbf{W}}_n$ and the precision matrix is $n\mathbf{I}(\hat{\mathbf{W}}_n)$.

The foregoing development leads to the following interesting result. The form of the posterior multivariate normal p.d.f. of $\mathbf{W}$ is specified by expression (7). By performing some elementary algebraic operations, we can obtain the equation

$$n(\mathbf{W} - \hat{\mathbf{W}}_n)'\mathbf{I}(\hat{\mathbf{W}}_n)(\mathbf{W} - \hat{\mathbf{W}}_n) = \sum_{i=1}^{k} \frac{(nW_i - Y_i)^2}{Y_i}. \tag{12}$$

Let Z be the random variable specified by the right side of Eq. (12). It now follows from Exercise 12 of Chap. 5 that for any given values of $Y_1, \ldots, Y_k$, the distribution of Z will be approximately a χ^2 distribution with $k - 1$ degrees of freedom. Moreover, the random variable Z is one version of the standard statistic used in χ^2 tests of goodness of fit. Here, the observed values Y_i are in the denominator, whereas in the more familiar form, the expected values nW_i are in the denominator. It has been shown [Neyman (1949); Jeffreys (1961), chap. 4] that for any given value of $\mathbf{W}$ and for large values of n, the distribution of Z will be approximately a χ^2 distribution with $k - 1$ degrees of freedom. Thus, in this example, inferences about $\mathbf{W}$ based on the posterior distribution of the random variable Z will agree with inferences about $\mathbf{W}$ based on traditional tests of goodness of fit. Of course, from the point of view developed in this section, the statistician has the freedom of working with the full $(k - 1)$-dimensional posterior distribution of $\mathbf{W}$ rather than with just the posterior distribution of the single random variable Z.

The example just given is not an isolated result. It is a well-known and important part of the asymptotic theory of maximum likelihood estimators. [See, e.g., the references given at the end of Sec. 10.7 and also Chernoff (1954).] Under certain conditions, it is shown in this asymptotic theory that for any given value of $\mathbf{W}$ and for large values of n, the distribution of the random vector $n^{\frac{1}{2}}\mathbf{A}(\mathbf{W})(\hat{\mathbf{W}}_n - \mathbf{W})$ will be approximately a multivariate normal distribution whose mean vector is $\mathbf{0}$ and whose precision matrix is the identity matrix. The matrix $\mathbf{A}(\mathbf{W})$ was defined earlier in this section. The appropriate regularity conditions ensure that the random vector $\mathbf{Z} = n^{\frac{1}{2}}\mathbf{A}(\hat{\mathbf{W}}_n)(\hat{\mathbf{W}}_n - \mathbf{W})$ has approximately the same multivariate normal distribution for any given value of $\mathbf{W}$. But the developments of this section have shown that this multivariate normal distribution is also the approximate posterior distribution of $\mathbf{Z}$ for any given values of the observations.

Further Remarks and References

The multinomial example presented in this section has been studied by von Mises in 1919 [see von Mises (1964)]; by Neyman (1929); by Lindley (1965), chap. 7; and by Watson (1966). Lindley (1964) and Bloch and

Watson (1967) have studied the multinomial distribution in more general problems involving contingency tables.

Chernoff (1952, 1956) has studied the asymptotic behavior of the Bayes risk as $n \to \infty$ in certain decision problems.

10.12 POSTERIOR RATIOS

We have seen in the preceding sections that the posterior distributions of parameters in which we are interested will often be approximately normal. Accordingly, we shall assume here that the posterior distribution of a certain k-dimensional vector W is a multivariate normal distribution with mean vector μ and precision matrix τ.

Suppose that the statistician is interested in investigating the likelihood that W has a value in the neighborhood of some given point w_0. Of course, the probability that W lies in any specified region of R^k can be computed from its posterior normal distribution. However, for very small regions, this probability will be close to 0. What is really wanted here is a comparison of the probability of a small neighborhood around w_0 with the probability of any equally small region elsewhere in R^k. In other words, the statistician is interested in learning whether there is any other point in the space R^k which has a neighborhood whose probability is much higher than that of the neighborhood of w_0. Since the probability of a small region around any point w is approximately the value $\xi(w)$ of the p.d.f. of W at the point w multiplied by the k-dimensional volume of the region, the statistician is interested in the ratio $\xi(\mu)/\xi(w_0)$. This is the ratio of the maximum value of the p.d.f. ξ to the value $\xi(w_0)$, and it compares the relative probability of a small region around the mean μ with the probability of a small region around the point w_0.

Since ξ is the p.d.f. of a multivariate normal distribution with mean vector μ and precision matrix τ, it can be seen that for any point $w \in R^k$,

$$\frac{\xi(\mu)}{\xi(w)} = \exp\left[\tfrac{1}{2}(w - \mu)'\tau(w - \mu)\right]. \tag{1}$$

The ratio (1) can be evaluated with the aid of Table 10.1, which gives the values of $\exp(x^2/2)$ for various values of x.

Table 10.1 exhibits some features of the normal p.d.f. that are seldom emphasized. If W has a univariate normal distribution, then it is seen from Table 10.1 that the maximum value of the p.d.f. of W is 7.39 times as large as the value of the p.d.f. at a point two standard deviations away from the mean, is about 90 times as large as the value at a point three standard deviations away from the mean, and is almost 3,000 times as large as the value at a point four standard deviations away from the mean.

One of the disadvantages in studying the ratio of two individual values of the p.d.f. ξ of a parameter W in this way is that the ratio does not necessarily remain the same if W is replaced by a new parameter. For instance, suppose that $V = t(W)$, where t is a one-to-one transformation, and let ξ_V be the p.d.f. of V. Then, unless t is a linear function, it is not necessarily true that the ratios $\xi(w_1)/\xi(w_2)$ and $\xi_V[t(w_1)]/\xi_V[t(w_2)]$

Table 10.1 Values of $\exp(x^2/2)$

x	$\exp(x^2/2)$
0.2	1.02
0.4	1.08
0.6	1.20
0.8	1.38
1.0	1.65
1.2	2.05
1.4	2.66
1.6	3.60
1.8	5.05
2.0	7.39
2.2	11.25
2.4	17.81
2.6	29.37
2.8	50.40
3.0	90.02
3.2	167.34
3.4	323.76
3.6	651.97
3.8	1,366.49
4.0	2,980.96
2.15	10
2.45	20
3.04	100
3.53	500
3.72	1,000

will be equal for arbitrary values w_1 and w_2. However, in many applications we are concerned with the posterior p.d.f. ξ or ξ_V after a large number of observations have been taken. In such a case, much of the posterior probability of W will be concentrated in a small interval. If t is differentiable, then it can be approximated over this interval by a linear function. Hence, for values w_1 and w_2 in this interval, the ratios computed from ξ and ξ_V will be approximately equal.

EXERCISES

1. Let X and Y be random variables with a specified joint distribution, and let $Z = X + Y$. Suppose that the random variables X and Z are independent, and suppose also that Y and Z are independent. Show that there must exist a constant c such that $\Pr\{Z = c\} = 1$.

2. Consider the conditions assumed in Theorem 1 of Sec. 9.2. (a) Show that when we let $\alpha \to 0$ and $\beta \to 0$ in the posterior distribution of W, then the limiting posterior distribution is a proper distribution if at least one of the observed values $x_1, \ldots, x_n$ is 1 and at least one of them is 0. (b) Show that the same posterior distribution is obtained from the improper prior density ξ defined by the equation

$$\xi(w) = \frac{1}{w(1 - w)} \qquad \text{for } 0 < w < 1.$$

3. Suppose that in the prior beta distribution of W specified in Theorem 1 of Sec. 9.2, we assume that $\beta = k\alpha$ for a fixed value $k > 0$ and we then let $\alpha \to 0$. (a) Show that all the moments $E(W^n)$ $(n = 1, 2, \ldots)$ approach the value $1/(k + 1)$. (b) Find a proper prior distribution of W for which these values are its moments, and discuss its appropriateness and usefulness as a prior distribution.

4. (a) Consider the conditions assumed in Theorem 1 of Sec. 9.4. Show that when we let $\alpha \to 0$ and $\beta \to 0$ in the posterior distribution of W, then the limiting posterior distribution is a proper distribution if at least one of the observed values $x_1, \ldots, x_n$ is positive. (b) Show that the same posterior distribution is obtained from the improper prior density ξ defined by the equation $\xi(w) = 1/w$ for $w > 0$.

5. Consider the conditions assumed in Theorem 1 of Sec. 9.7. Show that when we let $w_0 \to 0$ and $\alpha \to 0$ in the posterior distribution of W, then the limiting posterior distribution is the same as the posterior distribution obtained from the improper prior density ξ defined by the equation $\xi(w) = 1/w$ for $w > 0$.

6. Consider a statistical problem involving a parameter W whose value must be positive. Assume that W has an improper prior density ξ such that $\xi(w) \propto 1/w$ for $w > 0$. (a) Show that if a new parameter Y is defined as $Y = W^\alpha$, where $\alpha \neq 0$, then the prior density ξ_Y of Y is of the form $\xi_Y(y) \propto 1/y$ for $y > 0$. (b) Show that if a new parameter Z is defined as $Z = \log W$, then the prior density of Z is a uniform density on the real line.

7. Consider the conditions assumed in Theorem 2 of Sec. 9.7. Show that when we let $r_1 \to \infty$, $r_2 \to -\infty$, and $\alpha \to 0$ in the posterior joint distribution of W_1 and W_2, then the limiting posterior distribution is the same as the posterior distribution obtained from the improper prior joint density ξ which is defined on the parameter space Ω as follows:

$$\xi(w_1, w_2) = \frac{1}{(w_2 - w_1)^2}.$$

8. Consider the conditions assumed in Theorem 1 of Sec. 9.8. (a) Show that when we let $\alpha \to 0$ in the posterior distribution of $\mathbf{W}$, then the limiting posterior distribution is a proper distribution if every component x_i $(i = 1, \ldots, k)$ of the observed vector $\mathbf{x} = (x_1, \ldots, x_k)'$ is positive. (b) Show that the same posterior distribution is obtained from the improper prior density ξ which is defined on the $(k - 1)$-dimensional parameter space Ω as follows:

$$\xi(\mathbf{w}) = \frac{1}{w_1 \cdots w_k}.$$

9. Suppose that in the prior Dirichlet distribution of W as given in Theorem 1 of Sec. 9.8, we assume that $\alpha_i = c_i\alpha$ for fixed values $c_i > 0$ $(i = 1, \ldots, k)$ and then we let $\alpha \to 0$. (a) Show that the limiting value of any moment $E(W_1^{r_1} \cdots W_k^{r_k})$, where $r_1, \ldots, r_k$ are nonnegative integers, is 0 if at least two of the numbers $r_1, \ldots, r_k$ are not zero, and is of the form $c_i/(c_1 + \cdots + c_k)$ if only r_i is not zero. (b) Find a proper prior distribution of W for which these limiting values are its moments, and discuss its appropriateness and usefulness as a prior distribution.

10. Suppose that $X_1, \ldots, X_n$ is a random sample from an exponential distribution with an unknown value of the parameter W, and suppose that W has an improper prior density ξ defined by the equation $\xi(w) = 1/w$ for $w > 0$. Let $Z = W\Sigma_{i=1}^n X_i$. Show that the posterior distribution of Z when $X_i = x_i$ $(i = 1, \ldots, n)$ is the same as the conditional distribution of Z when $W = w$.

11. Consider the conditions assumed in Theorem 1 of Sec. 9.13. (a) Let the random variable Y be defined by the equation

$$Y = W\left[\sum_{i=1}^{n} (\mathbf{X}_i - \bar{\mathbf{X}})'\mathbf{r}(\mathbf{X}_i - \bar{\mathbf{X}}) \right].$$

Show that for any given values of the parameters $\mathbf{M}$ and W, the conditional distribution of Y is a χ^2 distribution with $k(n - 1)$ degrees of freedom. (b) Let the random vector $\mathbf{Z}$ be defined by the equation

$$\mathbf{Z} = \frac{[k(n - 1)]^{\frac{1}{2}}(\bar{\mathbf{X}} - \mathbf{M})}{[\Sigma_{i=1}^n(\mathbf{X}_i - \bar{\mathbf{X}})'\mathbf{r}(\mathbf{X}_i - \bar{\mathbf{X}})]^{\frac{1}{2}}}.$$

Show that for any given values of the parameters $\mathbf{M}$ and W, the conditional distribution of $\mathbf{Z}$ is a multivariate t distribution with $k(n - 1)$ degrees of freedom, location-vector $\mathbf{0}$, and precision matrix $n\mathbf{r}$.

12. (a) Show that when we let $\tau \to 0$, $\alpha \to -k/2$, and $\beta \to 0$ in the posterior joint distribution of $\mathbf{M}$ and W given in Theorem 1 of Sec. 9.13, then, for $n \geq 2$, the limiting posterior distribution is a proper distribution. (b) Show that the same posterior distribution is obtained from the improper prior joint density ξ of $\mathbf{M}$ and W which is defined on the parameter space Ω by the equation $\xi(\mathbf{m}, w) = 1/w$.

13. (a) Let Y be the random variable defined in Exercise 11. Show that under the limiting posterior joint distribution of $\mathbf{M}$ and W found in Exercise 12, the posterior distribution of Y when $X_i = x_i$ $(i = 1, \ldots, n)$ is the same as its conditional distribution when the values of $\mathbf{M}$ and W are given. (b) Show this same result for the random vector $\mathbf{Z}$ defined in Exercise 11.

14. Suppose that $X_1, \ldots, X_{n_1}$ is a random sample of n_1 observations $(n_1 \geq 2)$ from a normal distribution for which both the mean M_1 and the precision R_1 are unknown, and suppose that $Y_1, \ldots, Y_{n_2}$ is a random sample of n_2 observations $(n_2 \geq 2)$ from a different normal distribution for which both the mean M_2 and the precision R_2 are unknown. Let the random variables $\bar{X}$, $\bar{Y}$, S^2, and T^2 be defined as follows:

$$\bar{X} = \frac{1}{n_1} \sum_{i=1}^{n_1} X_i, \qquad\qquad \bar{Y} = \frac{1}{n_2} \sum_{j=1}^{n_2} Y_j,$$

$$S^2 = \frac{1}{n_1 - 1} \sum_{i=1}^{n_1} (X_i - \bar{X})^2, \qquad T^2 = \frac{1}{n_2 - 1} \sum_{j=1}^{n_2} (Y_j - \bar{Y})^2.$$

Finally, suppose that M_1, M_2, R_1, and R_2 have an improper prior joint density ξ such that, for $-\infty < m_i < \infty$ and $r_i > 0$ $(i = 1, 2)$,

$$\xi(m_1, m_2, r_1, r_2) = \frac{1}{r_1 r_2}.$$

(a) Let the random variable Q be defined as $Q = (R_1 S^2)/(R_2 T^2)$. Show that the posterior distribution of Q when $X_i = x_i$ and $Y_j = y_j$ $(i = 1, \ldots, n_1; j = 1, \ldots, n_2)$ is an F distribution with $n_1 - 1$ and $n_2 - 1$ degrees of freedom. (b) Show that this F distribution is also the conditional distribution of Q for any given values of M_i and R_i $(i = 1, 2)$.

15. Suppose that Z_1 is a random variable having a standardized t distribution with d_1 degrees of freedom and that Z_2 is an independent random variable having a standardized t distribution with d_2 degrees of freedom. Let α be a given number such that $0 < \alpha < \pi/2$, and let the random variable Z be defined as

$$Z = (\cos \alpha) Z_1 - (\sin \alpha) Z_2.$$

It is said that Z has a *Behrens distribution* for which there are d_1 and d_2 degrees of freedom and the angle is α. Also, let the random variable D be defined as follows:

$$D = \frac{(M_1 - M_2) - (\bar{X} - \bar{Y})}{[(S^2/n_1) + (T^2/n_2)]^{\frac{1}{2}}},$$

where all the variables on the right side of this equation are as given in Exercise 14. (a) Show that the posterior distribution of D when $X_i = x_i$ and $Y_j = y_j$ $(i = 1, \ldots, n_1; j = 1, \ldots, n_2)$ is a Behrens distribution for which there are $n_1 - 1$ and $n_2 - 1$ degrees of freedom and the angle α is such that

$$\tan \alpha = \left(\frac{n_1 T^2}{n_2 S^2}\right)^{\frac{1}{2}}.$$

(b) Show that the conditional distribution of D when $M_i = m_i$ and $R_i = r_i$ $(i = 1, 2)$ is not a Behrens distribution, and show that the distribution depends on the given values m_i and r_i $(i = 1, 2)$ only through the ratio r_2/r_1.

Note: The problem of making inferences about the difference $M_1 - M_2$ is known as the Behrens-Fisher problem and has been widely discussed in the statistical literature. See, e.g., Fisher (1956), chap. 4; Linnik (1963); Yao (1965); and the other references given by them.

16. Suppose that $X_1, \ldots, X_n$ is a random sample from a uniform distribution on the interval $(W, W + 1)$. (a) For any given prior p.d.f. of W and any observed values $X_i = x_i$ $(i = 1, \ldots, n)$, find the posterior p.d.f. (b) Show that if the improper prior density of W is uniform over the whole line, then the posterior p.d.f. of W is uniform on the interval $I^* = (a, b)$, where

$$a = \max \{X_1, \ldots, X_n\} - 1 \quad \text{and} \quad b = \min \{X_1, \ldots, X_n\}.$$

(c) How large must the value of n be made in order that the probability will be at least 0.95 that the length of the posterior interval I^* will be less than 0.01?

17. Suppose that two observations X_1 and X_2 are taken from a normal distribution for which both the mean M and the precision R are unknown. Also, suppose that the improper prior joint density of M and R is specified by Eq. (4) of Sec. 10.2. If y_1 is the smaller of the two observations and y_2 is the larger of these observations, show that under the posterior distribution of M,

$$\Pr(y_1 < M < y_2) = \tfrac{1}{2}.$$

18. Consider the conditions assumed in Theorem 1 of Sec. 10.4. Prove that the following relations must be satisfied for any subset B of Ω:

$$\frac{1}{1 + \alpha + \beta + \alpha\gamma} \left[\int_B \phi(w|x) \, d\nu(w) - \alpha \right] \leq \int_B \xi(w|x) \, d\nu(w)$$

$$\leq \frac{1 + \beta}{1 - \alpha} \left[\int_B \phi(w|x) \, d\nu(w) + \frac{\alpha + \alpha\gamma}{1 + \beta} \right].$$

19. Suppose that $X_1, \ldots, X_n$ is a random sample from a Bernoulli distribution with parameter w_0. Suppose also that the statistician does not know the value of w_0 and that he assumes that the prior distribution of the parameter W is a beta distribution. Let I be any open interval of the real line which contains the value w_0, and let $\xi_n(I)$ denote the probability of I computed from the posterior distribution of W. Prove that $\lim_{n \to \infty} \xi_n(I) = 1$ with probability 1.

20. Suppose that $X_1, \ldots, X_n$ is a random sample from a Poisson distribution with mean w_0. Suppose also that the statistician does not know the value of w_0 and that he assumes that the prior distribution of W is a gamma distribution. Let I be any open interval of the real line which contains the value w_0, and let $\xi_n(I)$ denote the probability of I computed from the posterior distribution of W. Prove that $\lim_{n \to \infty} \xi_n(I) = 1$ with probability 1.

21. Suppose that $X_1, \ldots, X_n$ is a random sample from a normal distribution with mean m_0 and precision r_0. Suppose also that the statistician does not know the values of m_0 and r_0 and that he assumes that the prior joint distribution of the mean M and the precision R is a normal-gamma distribution as specified in Theorem 1 of Sec. 9.6. Let I be any open interval of the real line which contains the value m_0, and let $\xi_n(I)$ denote the probability of I computed under the posterior marginal distribution of M. Prove that $\lim_{n \to \infty} \xi_n(I) = 1$ with probability 1.

22. For random samples $X_1, \ldots, X_n$ from each of the following distributions, find a consistent sequence of solutions $\hat{w}_n(X_1, \ldots, X_n)$ of the likelihood equation [Eq. (4) of Sec. 10.7], and find the value of $I(w)$ as defined in Eq. (2) of Sec. 10.7.

(a) A Bernoulli distribution with an unknown value of the parameter W $(0 < W < 1)$

(b) A Poisson distribution with an unknown value of the mean W $(W > 0)$

(c) A normal distribution with an unknown value of the mean W $(-\infty < W < \infty)$ and a specified value of the variance σ^2

(d) A normal distribution with a specified value of the mean μ and an unknown value of the variance W $(W > 0)$

(e) A normal distribution with a specified value of the mean μ and an unknown value of the standard deviation W $(W > 0)$

(f) A normal distribution with a specified value of the mean μ and an unknown value of the precision W $(W > 0)$

23. Suppose that $X_1, \ldots, X_n$ is a random sample from a normal distribution for which both the mean W_1 and the precision W_2 are unknown $(-\infty < W_1 < \infty$ and $W_2 > 0)$. If $\mathbf{W} = (W_1, W_2)$, find a consistent sequence of solutions $\hat{\mathbf{w}}_n(X_1, \ldots, X_n)$ of the likelihood equations [Eq. (4) of Sec. 10.11], and find the Fisher information matrix $\mathbf{I}(\mathbf{w})$, as defined by Eq. (2) of Sec. 10.11.

estimation, testing hypotheses, and linear statistical models

11.1 ESTIMATION

Generally speaking, an estimation problem is a statistical decision problem in which the decision made by the statistician is his estimate of the value of some parameter $W = (W_1, \ldots, W_k)'$ whose values belong to a subset Ω of R^k ($k \geq 1$). Since the statistician must estimate the value of W, the decision space D typically coincides with the set Ω. In our discussion of estimation theory, we shall assume, for simplicity, that $\Omega = D = R^k$, although the probability that W lies in certain regions of R^k may be 0.

The statistician's decision $\mathbf{d} = (d_1, \ldots, d_k)' \in R^k$ is his estimate of the value $\mathbf{w} = (w_1, \ldots, w_k)'$ of W, and the loss $L(\mathbf{w}, \mathbf{d})$ which he incurs will reflect the discrepancy between the value $\mathbf{w}$ and his estimate $\mathbf{d}$. For this reason, the loss function L in an estimation problem is often assumed to have the following form:

$$L(\mathbf{w}, \mathbf{d}) = \gamma(\mathbf{w})\Lambda(\mathbf{w} - \mathbf{d}). \tag{1}$$

Here, Λ is a nonnegative function of the error vector $\mathbf{w} - \mathbf{d}$ such that $\Lambda(\mathbf{0}) = 0$ and γ is a nonnegative weighting function which indicates the relative seriousness of a given error vector for different values of the parameter **W**. If the loss $L(\mathbf{w}, \mathbf{d})$ depends only on the error vector $\mathbf{w} - \mathbf{d}$, then the function γ may be taken to be constant on the space R^k. In some problems, the statistician is not required to estimate all the components of the vector **W**. Suppose that he must estimate the first j components of **W** $(1 \leq j < k)$ but that he is not required to estimate the remaining $k - j$ components of **W**. In this situation, the last $k - j$ components of **W** are called *nuisance parameters*. The loss function in an estimation problem involving nuisance parameters can still be represented by a function of the form specified in Eq. (1). In such a problem, the function Λ would involve only the first j components of the error vector $\mathbf{w} - \mathbf{d}$, although the weighting function γ may involve any number of components of **w**. Furthermore, we may still retain the assumption that $D = R^k$. Under this assumption, the statistician is required to estimate all k components of **W**, but the loss function which is used makes it clear that the estimates $d_{j+1}, \ldots, d_k$ of the nuisance parameters are irrelevant.

Consider now an arbitrary estimation problem in which the loss function L has the form specified in Eq. (1), and suppose that after all the observations have been obtained, the posterior p.d.f. of **W** is ξ. A Bayes decision $\mathbf{d}^*$—or, in this context, a Bayes estimate $\mathbf{d}^*$—will be a point $\mathbf{d} \in R^k$ which minimizes the following value of the risk $\rho(\xi, \mathbf{d})$:

$$\rho(\xi, \mathbf{d}) = \int_{R^k} \gamma(\mathbf{w})\Lambda(\mathbf{w} - \mathbf{d})\xi(\mathbf{w})\,d\nu(\mathbf{w}). \tag{2}$$

It should be noted that the same integral must be minimized in a different estimation problem in which the loss function L^* has the simple form $L^*(\mathbf{w}, \mathbf{d}) = \Lambda(\mathbf{w} - \mathbf{d})$ and the posterior p.d.f. ξ^* of **W** satisfies the relation $\xi^*(\mathbf{w}) \propto \gamma(\mathbf{w})\xi(\mathbf{w})$. In other words, the same Bayes decisions will be obtained regardless of whether the nonnegative function γ is a factor of the loss function or a factor of the p.d.f. of **W**. This result is evident from the integral in Eq. (2). For this reason, in discussions of the general theory of estimation, it may be assumed that the function γ in Eq. (1) is constant.

When the parameter W is one-dimensional and its values are in R^1, the loss function in an estimation problem can often be expressed as follows:

$$L(w, d) = a|w - d|^b, \tag{3}$$

where $a > 0$ and $b > 0$. We shall now consider this loss function in detail for the values $b = 1$ and $b = 2$.

11.2 QUADRATIC LOSS

By far the most widely studied loss function in problems involving the estimation of a real parameter W is the quadratic loss function L specified as follows:

$$L(w, d) = a(w - d)^2. \tag{1}$$

The loss function (1) lends itself to mathematical manipulation. Furthermore, the following very crude but routine considerations indicate why it is an acceptable approximation in a wide variety of situations.

For the reasons presented in Sec. 11.1, we shall suppose that the loss $L(w, d)$ depends only on the difference $w - d$. Let $L(w, d) = \Lambda(w - d)$, where Λ is a nonnegative function that can be differentiated at least twice and $\Lambda(0) = 0$. If Λ is expanded in a Taylor series through terms of second order, we obtain

$$L(w, d) = \Lambda(w - d) \approx a_0 + a_1(w - d) + a_2(w - d)^2. \tag{2}$$

If the statistician has enough information about the value of W to be able to choose an estimate d which, with high probability, will be close to w, then the terms of higher order in Eq. (2) will be relatively small and can be ignored. The fact that $\Lambda(0) = 0$ implies that $a_0 = 0$, and the fact that Λ is nonnegative implies that $a_1 = 0$ and that $a_2 \geq 0$. Thus, the form in Eq. (2) reduces to that in Eq. (1).

When the loss function is specified by Eq. (1), the Bayes decision d^* against any given distribution of W will be the number d which minimizes the following value of the risk:

$$E[(W - d)^2] = E(W^2) - 2dE(W) + d^2. \tag{3}$$

Here we are assuming that $E(W^2) < \infty$. An elementary calculation shows that the quadratic function of d in Eq. (3) is minimized when $d = E(W)$. Furthermore, when this value of d is chosen, the minimum value of the risk in Eq. (3) is

$$E\{[W - E(W)]^2\} = \mathrm{Var}(W). \tag{4}$$

Now suppose that X is an observation, which may possibly be a random vector, whose conditional g.p.d.f. when $W = w$ is $f(\cdot \mid w)$. As usual, let ξ denote the prior p.d.f. of W and let $\xi(\cdot \mid x)$ denote the posterior p.d.f. of W when $X = x$. The Bayes estimator δ^* and the Bayes risk $\rho^*(\xi)$ for the quadratic loss function specified by Eq. (1) can now be easily described. By an appropriate choice of unit, we can assume that $a = 1$ in that equation. For any observed value $X = x$, the Bayes decision is $\delta^*(x) = E(W \mid x)$, where $E(W \mid x)$ is the mean of the posterior distribution of W. Furthermore, after the value x has been observed and the estimate $\delta^*(x)$ has been chosen, the risk is $\mathrm{Var}(W \mid x)$, the variance of the posterior distribution of W. Hence, the Bayes risk is specified by the equation

$$\rho^*(\xi) = E[\mathrm{Var}(W \mid X)]. \tag{5}$$

The expectation in Eq. (5) is computed with respect to the marginal g.p.d.f. g of X, which is specified as follows:

$$g(x) = \int_{-\infty}^{\infty} f(x|w)\xi(w) \, d\nu(w). \tag{6}$$

In order to compute the Bayes risk $E[\mathrm{Var}(W|X)]$, it may be helpful to use the relation which was presented in Eq. (5) of Sec. 3.8.

We have assumed that all the expectations which appear in this computation exist. If W is not confined to a bounded interval, then it might happen, for certain observed values x, that the posterior distribution of W has no mean. For such a distribution, there is no estimate for which the expected loss will be finite. Moreover, even if the mean of the posterior distribution exists, the variance $\mathrm{Var}(W|x)$ may not exist. Finally, even if $\mathrm{Var}(W|x)$ is finite for each value of x, the Bayes risk specified by Eq. (5) may not be finite.

As an example in which each expectation exists and is easily computed, we shall suppose that $X_1, \ldots, X_n$ is a random sample from a Poisson distribution for which the value of the mean W is unknown. Suppose that the prior distribution of W is a gamma distribution with parameters α and β. It is known from Theorem 1 of Sec. 9.4 that the posterior distribution of W when $X_i = x_i$ ($i = 1, \ldots, n$) is a gamma distribution with parameters $\alpha + \Sigma_{i=1}^{n} x_i$ and $\beta + n$. It is assumed that the loss function is specified by Eq. (1) with $a = 1$. From the expression for the mean of a gamma distribution, as given in Eq. (4) of Sec. 4.8, it now follows that the Bayes estimator δ^* is defined as

$$\delta^*(X_1, \ldots, X_n) = \frac{\alpha + \Sigma_{i=1}^{n} X_i}{\beta + n}. \tag{7}$$

Furthermore, for any values of $X_1, \ldots, X_n$, the variance of the posterior distribution is

$$\mathrm{Var}(W|X_1, \ldots, X_n) = \frac{\alpha + \Sigma_{i=1}^{n} X_i}{(\beta + n)^2}. \tag{8}$$

Since $E(X_i|W) = W$ for $i = 1, \ldots, n$, it follows that

$$E(X_i) = E[E(X_i|W)] = E(W) = \frac{\alpha}{\beta}. \tag{9}$$

Hence, from Eqs. (5), (8), and (9), we can obtain the following value for the Bayes risk:

$$\rho^*(\xi) = \frac{\alpha}{\beta(\beta + n)}. \tag{10}$$

Suppose now that the sampling cost is c per observation ($c > 0$) and that the statistician can choose the number of observations in the sample. For a sample of n observations, it follows from Eq. (10) that the total risk will be

$$\frac{\alpha}{\beta(\beta + n)} + cn. \tag{11}$$

This total risk will be minimized when

$$n = \left(\frac{\alpha}{c\beta}\right)^{\frac{1}{2}} - \beta. \tag{12}$$

Of course, the actual number n must be a nonnegative integer. If the value specified by Eq. (12) is negative, then the optimal sample size is $n = 0$. In such a case, no observations should be made, and the statistician should estimate W from the prior distribution. If the value specified by Eq. (12) is positive but not an integer, then the optimal sample size is either the integer just below this value or the one just above it.

As was seen in Chap. 10, and in particular in Exercise 4 of that chapter, vague prior knowledge can be represented by letting $\alpha \rightarrow 0$ and $\beta \rightarrow 0$ in the prior distribution of W. When α and β approach these limiting values, the optimal value of n as specified by Eq. (12) becomes indeterminate because it involves the ratio α/β. In this problem, α/β is the mean of the prior distribution of W, and therefore it would be the statistician's estimate of W if no observations were available. Hence, if the statistician has vague prior knowledge about W, it follows from Eq. (12) that the optimal number of observations is approximately $(\hat{w}/c)^{\frac{1}{2}}$, where $\hat{w}$ is the statistician's best estimate of W before any observations have been made. It should be remarked that since the statistician's prior knowledge is vague, he may have difficulty in assigning a precise value to $\hat{w}$. A more satisfactory result can be obtained by modifying the loss function in this problem as follows.

We shall now suppose that the loss function L, instead of being specified by Eq. (1), is specified by the following equation:

$$L(w, d) = \frac{(w - d)^2}{w}. \tag{13}$$

It is again assumed that the prior distribution of W is a gamma distribution with parameters α and β. Then (see Exercise 1) for any values of the observations $X_1, \ldots, X_n$ such that $\alpha + \Sigma_{i=1}^{n}X_i > 1$, the Bayes estimate $\delta^*(X_1, \ldots, X_n)$ is specified by the equation

$$\delta^*(X_1, \ldots, X_n) = \frac{\alpha + \Sigma_{i=1}^{n}X_i - 1}{\beta + n}. \tag{14}$$

Furthermore, the value of the Bayes risk $\rho^*(\xi)$ is $1/(\beta + n)$. It follows that if the sampling cost is c per observation, then the total risk is minimized when $n = c^{-\frac{1}{2}} - \beta$. Hence, if the cost c is small and the prior information about W is vague, then the optimal number of observations is $c^{-\frac{1}{2}}$.

11.3 LOSS PROPORTIONAL TO THE ABSOLUTE VALUE OF THE ERROR

In the preceding section we assumed that the loss in estimating a real-valued parameter W was proportional to the square of the estimation error.   In this section we shall assume that the loss is proportional to the absolute value of the estimation error.   Hence, for $w \, \epsilon \, R^1$ and $d \, \epsilon \, R^1$, the loss $L(w, d)$ is

$$L(w, d) = a|w - d|. \tag{1}$$

When the loss function is specified by Eq. (1), a Bayes decision d^* against any given distribution of W is a number d that minimizes the expectation $E(|W - d|)$.

It is said that a number m is a *median* of the distribution of W if $\Pr(W \geq m) \geq \frac{1}{2}$ and $\Pr(W \leq m) \geq \frac{1}{2}$.   Every distribution has at least one median, but it need not have a unique median.   Figure 11.1 illustrates both a p.f. and a p.d.f. for which the median is not unique.   In either illustration, any number m such that $w_1 \leq m \leq w_2$ is a median.

The next theorem demonstrates that for the loss function specified by Eq. (1), any median of the distribution of W will be a Bayes estimate.

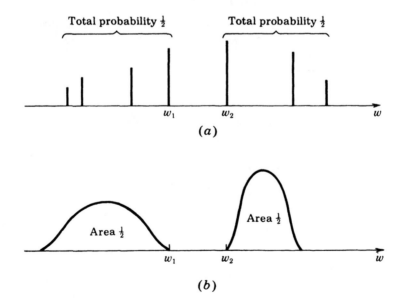

Fig. 11.1 (*a*) A p.f. with a nonunique median. (*b*) A p.d.f. with a nonunique median.

Theorem 1 *Suppose that $E(|W|) < \infty$. Then a number d^* satisfies the equation*

$$E(|W - d^*|) = \inf_{-\infty < d < \infty} E(|W - d|) \tag{2}$$

if, and only if, d^* is a median of the distribution of W.

Proof Suppose that d^* is a median of the distribution of W, and let d be any other number such that $d > d^*$. Then

$$|w - d| - |w - d^*| = \begin{cases} d^* - d & \text{if } w \geq d, \\ d + d^* - 2w & \text{if } d^* < w < d, \\ d - d^* & \text{if } w \leq d^*. \end{cases} \tag{3}$$

Since $(d + d^* - 2w) > (d^* - d)$ when $d^* < w < d$, the following result can be obtained:

$$\begin{aligned} E(|W - d| - |W - d^*|) &\geq (d^* - d)\Pr(W \geq d) \\ &+ (d^* - d)\Pr(d^* < W < d) + (d - d^*)\Pr(W \leq d^*) \\ &= (d - d^*)[\Pr(W \leq d^*) - \Pr(W > d^*)] \geq 0. \end{aligned} \tag{4}$$

The final inequality in the relation (4) follows from the fact that d^* is a median of the distribution of W. The first inequality in the relation (4) will be an equality if, and only if, $\Pr(d^* < W < d) = 0$. The final inequality will be an equality if, and only if,

$$\Pr(W \leq d^*) = \Pr(W > d^*) = \tfrac{1}{2}.$$

Together, these conditions imply that d is also a median. Hence, $E(|W - d|) \geq E(|W - d^*|)$, and there can be equality if, and only if, d is also a median.

A similar proof can also be given when d is any number such that $d < d^*$. ∎

If a vector of observations $\mathbf{X}$ is available, then the statistician can construct a Bayes estimator δ^* by choosing the estimate $\delta^*(\mathbf{x})$ for any observed value $\mathbf{x}$ of $\mathbf{X}$ to be a median of the posterior distribution of W.

As an example, we shall suppose that the sequence $X_1, \ldots, X_n$ is a random sample from a normal distribution with an unknown value of the mean W and a specified precision r. Also, we shall suppose that the value of W must be estimated when the loss function is specified by Eq. (1). By an appropriate choice of unit, we can assume that $a = 1$ in that equation. If the prior distribution of W is a normal distribution with mean μ and precision τ, then it follows from Theorem 1 of Sec. 9.5 that for any values of $X_1, \ldots, X_n$, the posterior distribution of W will be a normal distribution with mean $(\tau\mu + nr\bar{X})/(\tau + nr)$ and precision $\tau + nr$. Since the unique median of a normal distribution coincides with its mean, the

Bayes estimator δ^* is defined as follows:

$$\delta^*(X_1, \ldots, X_n) = \frac{\tau\mu + nr\bar{X}}{\tau + nr}. \tag{5}$$

Also, the Bayes risk $\rho^*(\xi)$ is determined from the following relation:

$$\rho^*(\xi) = E[|W - \delta^*(X_1, \ldots, X_n)|]. \tag{6}$$

In accordance with the above comments, the posterior distribution of $[W - \delta^*(X_1, \ldots, X_n)]$ when $X_i = x_i$ $(i = 1, \ldots, n)$ must be a normal distribution with mean 0 and precision $\tau + nr$, regardless of the values $x_1, \ldots, x_n$. Now, if Y is a random variable having a normal distribution for which the mean is 0 and the precision is p, then it can be shown that

$$E(|Y|) = \left(\frac{2}{\pi p}\right)^{\frac{1}{2}}. \tag{7}$$

Hence, for any given values of $X_1, \ldots, X_n$, the expectation on the right side of Eq. (6) must have the value $\{2/[\pi(\tau + nr)]\}^{\frac{1}{2}}$. It follows from Eq. (6) that

$$\rho^*(\xi) = \left[\frac{2}{\pi(\tau + nr)}\right]^{\frac{1}{2}}. \tag{8}$$

We shall now suppose further that the sampling cost is c per observation $(c > 0)$ and that the statistician can choose the number of observations in the sample. For a sample of n observations, it follows from Eq. (8) that the total risk will be

$$\left[\frac{2}{\pi(\tau + nr)}\right]^{\frac{1}{2}} + cn. \tag{9}$$

The total risk is minimized when

$$n = \frac{1}{(2\pi rc^2)^{\frac{1}{3}}} - \frac{\tau}{r}. \tag{10}$$

Of course, the actual number n must be a nonnegative integer. If we let $\tau \to 0$, the limiting value in Eq. (10) will be the optimal number of observations for this problem when the prior information about W is vague.

11.4 ESTIMATION OF A VECTOR

We shall now consider the problem of estimating a vector

$$\mathbf{W} = (W_1, \ldots, W_k)',$$

where $k \geq 2$. A standard loss function for such a problem is the quadratic loss function L defined as follows for any points $\mathbf{w} \in R^k$ and $\mathbf{d} \in R^k$:

$$L(\mathbf{w}, \mathbf{d}) = (\mathbf{w} - \mathbf{d})'\mathbf{A}(\mathbf{w} - \mathbf{d}). \tag{1}$$

Here $\mathbf{A}$ is a symmetric $k \times k$ nonnegative definite matrix. If $\mathbf{A}$ is actually positive definite, then any nonzero error vector $\mathbf{w} - \mathbf{d}$ leads to a positive loss. On the other hand, if $\mathbf{A}$ is not positive definite, there will be nonzero error vectors which result in no loss. Such vectors exist, for example, if the statistician is interested in estimating only some of the components of $\mathbf{W}$ because the values of the other components are irrelevant to him and can be regarded as nuisance parameters.

We shall suppose that the mean vector and the covariance matrix of $\mathbf{W}$ exist, and we shall let $E(\mathbf{W}) = \mathbf{\mu}$ and $\mathrm{Cov}(\mathbf{W}) = \mathbf{\Sigma}$. A Bayes estimate against the distribution of $\mathbf{W}$ is a point $\mathbf{d} \in R^k$ such that the following expectation is minimized:

$$
\begin{aligned}
E[(\mathbf{W} - \mathbf{d})'\mathbf{A}(\mathbf{W} - \mathbf{d})] \\
= E\{[(\mathbf{W} - \mathbf{\mu})' + (\mathbf{\mu} - \mathbf{d})']\mathbf{A}[(\mathbf{W} - \mathbf{\mu}) + (\mathbf{\mu} - \mathbf{d})]\} \\
= E[(\mathbf{W} - \mathbf{\mu})'\mathbf{A}(\mathbf{W} - \mathbf{\mu})] + (\mathbf{\mu} - \mathbf{d})'\mathbf{A}(\mathbf{\mu} - \mathbf{d}). \tag{2}
\end{aligned}
$$

In the final expression of Eq. (2), the expectation does not involve the decision $\mathbf{d}$. Also, since $\mathbf{A}$ is a nonnegative definite matrix, the second term in that expression is nonnegative for all values of $\mathbf{d}$. Hence, a point $\mathbf{d} \in R^k$ will be a Bayes estimate if, and only if,

$$(\mathbf{\mu} - \mathbf{d})'\mathbf{A}(\mathbf{\mu} - \mathbf{d}) = 0. \tag{3}$$

In particular, the value $\mathbf{d} = \mathbf{\mu}$ is a Bayes estimate of $\mathbf{W}$. Furthermore, if $\mathbf{A}$ is a positive definite matrix, this value of $\mathbf{d}$ is the unique Bayes estimate. If $\mathbf{A}$ is not a positive definite matrix, there will be other values of $\mathbf{d}$ which satisfy Eq. (3). It should be emphasized, however, that the mean vector $\mathbf{\mu} = E(\mathbf{W})$ is always a Bayes estimate for any arbitrary symmetric, nonnegative definite matrix $\mathbf{A}$.

Moreover, it can be shown (see Exercise 11) that

$$E[(\mathbf{W} - \mathbf{\mu})'\mathbf{A}(\mathbf{W} - \mathbf{\mu})] = \mathrm{tr}\,(\mathbf{A}\mathbf{\Sigma}), \tag{4}$$

where $\mathbf{\Sigma}$ is the $k \times k$ covariance matrix of $\mathbf{W}$. Hence, it follows from Eq. (2) that the expected loss for any Bayes estimate $\mathbf{d}$ will be $\mathrm{tr}\,(\mathbf{A}\mathbf{\Sigma})$.

If X is an observation whose conditional g.p.d.f. is $f(\cdot\,|\mathbf{w})$ when $\mathbf{W} = \mathbf{w}$, then it can be seen from the preceding discussion that a Bayes decision function δ^* is defined as $\delta^*(X) = E(\mathbf{W}|X)$ and the Bayes risk $\rho^*(\xi)$ against the prior g.p.d.f. ξ is

$$\rho^*(\xi) = \mathrm{tr}\,\{\mathbf{A}E[\mathrm{Cov}(\mathbf{W}|X)]\}. \tag{5}$$

Here $\mathrm{Cov}(\mathbf{W}|X)$ is the covariance matrix of the posterior distribution of $\mathbf{W}$ when X is given.

For example, suppose that $X_1, \ldots, X_n$ is a random sample from a normal distribution for which both the mean M and the precision R are unknown and must be estimated. Suppose also that when $M = m$ and $R = r$ and the estimated values of M and R are $\hat{m}$ and $\hat{r}$, the loss is specified as follows:

$$L(m, r, \hat{m}, \hat{r}) = a_1(m - \hat{m})^2 + a_2(r - \hat{r})^2 + 2a_3(m - \hat{m})(r - \hat{r}). \quad (6)$$

The requirement that the matrix $\mathbf{A}$ in Eq. (1) be nonnegative definite makes it necessary to assume in this example that $a_1 \geq 0$ and $a_1 a_2 \geq a_3^2$ in Eq. (6).

Now suppose that the prior joint distribution of M and R is a normal-gamma distribution as specified in Theorem 1 of Sec. 9.6. Then the posterior marginal distribution of M will be a t distribution whose location parameter is μ', as specified in that theorem. In order to be certain that the variance of this posterior distribution will be finite, it must be assumed that $n > 2$. Since the p.d.f. of any t distribution is symmetric with respect to the location parameter, it follows that the mean of the posterior distribution of M is μ'. Furthermore, it can be seen from the same theorem that the posterior distribution of R will be a gamma distribution whose mean is $(2\alpha + n)/(2\beta')$, where the value of β' is specified in that theorem. It now follows from the discussion in this section that the Bayes estimators of M and R will be the means of the posterior distributions which we have just described.

Now suppose that we wish to represent vague prior information about M and R. In accordance with the discussion in Sec. 10.2, we can let $\tau \to 0$, $\alpha \to -\frac{1}{2}$, and $\beta \to 0$ in the posterior distribution which we have just found. For any values of the observations $X_1, \ldots, X_n$, it can then be seen from Theorem 1 of Sec. 9.6 that the limiting value $\hat{M}$ of the mean of the posterior distribution of M and the limiting value $\hat{R}$ of the mean of the posterior distribution of R will be specified as follows:

$$\hat{M} = \bar{X} \quad \text{and} \quad \hat{R} = \frac{n - 1}{\sum_{i=1}^n (X_i - \bar{X})^2}. \quad (7)$$

Therefore, the estimators in Eq. (7) can be regarded as approximations to the Bayes estimators of M and R when the prior information is vague and the loss function is of the form (6).

Maximum Likelihood Estimators

We have shown that the mean and the median of the posterior distribution of a one-dimensional parameter W are Bayes estimates in certain decision problems. Also, for a vector parameter $\mathbf{W}$, the mean vector of the posterior distribution is a Bayes estimate when the loss function is quadratic. Since there is no standard definition of a median of a multivariate distribution, there are no analogous standard results for a vector parameter.

Another estimate which can be used for either a one-dimensional parameter or a vector parameter, especially when the loss function has not been specified explicitly, is the value of the parameter at which the maximum of the posterior g.p.d.f. is attained. For any value x of the observation or vector of observations X, let $\xi(\cdot \,|x)$ denote the posterior g.p.d.f. of W on the parameter space Ω. Also, let $\hat{w}(x)$ be a value of w which satisfies the relation

$$\xi[\hat{w}(x)|x] = \sup_{w \, \epsilon \, \Omega} \xi(w|x). \tag{8}$$

If such a value of w exists for any value of x, then it is said that the estimator $\hat{w}(X)$ is a *generalized maximum likelihood estimator* of W.

It should be emphasized that the generalized maximum likelihood estimate $\hat{\mathbf{w}}(x) = [\hat{w}_1(x), \ldots, \hat{w}_k(x)]'$ of a vector parameter $\mathbf{W} = (W_1, \ldots, W_k)'$ has been defined as the point $\mathbf{w} = (w_1, \ldots, w_k)'$ in Ω which maximizes the posterior joint g.p.d.f. of $W_1, \ldots, W_k$. If an estimate $w_1^*(x)$ of W_1 is determined by finding the value of w_1 which maximizes the posterior *marginal* g.p.d.f. of W_1, it will not necessarily be true that $\hat{w}_1(x) = w_1^*(x)$. Either one of these two estimators could have been defined as the generalized maximum likelihood estimator of W_1. After a large number of observations have been taken, the difference between them will typically be small.

Although a generalized maximum likelihood estimator may not be a Bayes estimator for any standard loss function, it can be seen from the development in Chap. 10 that it will be a reasonable estimator whenever the prior distribution and the likelihood function satisfy certain regularity conditions and a large sample is taken. In particular, when the prior g.p.d.f. ξ is constant over the parameter space Ω, a generalized maximum likelihood estimator is the same as a maximum likelihood estimator, which was defined in Sec. 10.7 and has the large-sample properties discussed in Chap. 10. If the prior g.p.d.f. ξ is not constant over Ω, then the value of the generalized maximum likelihood estimator will be the value of w which maximizes the product of the likelihood function and ξ. If the effect of the prior g.p.d.f. ξ becomes negligible when a large number of observations are taken, then the generalized maximum likelihood estimator will also have these same large-sample properties.

Further Remarks and References

Estimation problems are discussed by Blackwell and Girshick (1954), chap. 11, and Raiffa and Schlaifer (1961), chap. 6. More general loss functions in both one dimension and higher dimensions are considered by DeGroot and Rao (1963, 1966). Sacks (1963) and Farrell (1964a) have considered the use of improper prior distributions in estimation problems. Interesting results in regard to the inadmissibility of some standard estimators are given by Stein (1956, 1964) and James and Stein (1961). Evans (1964, 1965) has considered estimating the parameters of normal and multivariate normal distributions. Two books of interest are Deutsch (1965) and Good (1965). The latter deals with both estimation and Bayesian methods in general.

11.5 PROBLEMS OF TESTING HYPOTHESES

We shall now consider a decision problem in which the decision space D contains exactly two decisions $D = \{d_1, d_2\}$. We shall suppose that decision d_1 is appropriate if the parameter W lies in a certain subset Ω_1 of the parameter space Ω and that decision d_2 is appropriate if W lies in the complementary subset $\Omega_2 = \Omega_1{}^c$. However, there may be some points in either Ω_1 or Ω_2 for which the decisions d_1 and d_2 are equally appropriate. We shall simplify our notation by writing $L_i(w)$ instead of $L(w, d_i)$ to denote the loss when $W = w$ and decision d_i is chosen ($i = 1, 2$). In accordance with these comments, we shall assume that the loss function is of the following form:

$$
\begin{aligned}
L_1(w) &= 0 \quad &&\text{for } w \, \epsilon \, \Omega_1, \\
L_1(w) &\geq 0 \quad &&\text{for } w \, \epsilon \, \Omega_2, \\
L_2(w) &= 0 \quad &&\text{for } w \, \epsilon \, \Omega_2, \\
L_2(w) &\geq 0 \quad &&\text{for } w \, \epsilon \, \Omega_1.
\end{aligned}
\tag{1}
$$

Here, $\Omega_1 \cap \Omega_2 = \emptyset$ and $\Omega_1 \cup \Omega_2 = \Omega$. The "indifference set" Ω_0, for which d_1 and d_2 are equally appropriate, is described as

$$
\Omega_0 = \{w: L_1(w) = L_2(w) = 0\}.
$$

In some problems, $\Omega_0 = \emptyset$. In other problems, Ω_0 includes only the points of Ω which lie on the boundary between Ω_1 and Ω_2 and the probability of this set is 0 under the prior distribution of W. In still other problems, however, the set Ω_0 may be larger and may have positive probability.

Problems involving only two decisions are known as *problems of testing hypotheses*. Suppose that a given hypothesis H_1 specifies that the value of W lies in the set Ω_1, and suppose that the alternative hypothesis H_2 specifies that the value of W lies in the set Ω_2. If the probability of the indifference set is 0, then choosing the decision d_1 in this context is tantamount to accepting the hypothesis H_1 and rejecting the hypothesis H_2. Likewise, choosing the decision d_2 is tantamount to rejecting H_1 and accepting H_2. The standard reference for the general theory of testing hypotheses is Lehmann (1959).

A hypothesis H_i ($i = 1, 2$) is called *simple* if the set Ω_i contains just one point, and H_i is called *composite* if Ω_i contains more than one point.

The theory of testing a simple hypothesis H_1 against a simple alternative H_2 was presented in detail in Sec. 8.11. In such a test, there are exactly two points in Ω and two points in D. In general, any test of hypotheses, whether H_i is simple or composite ($i = 1, 2$), can be handled by the methods we have been considering. However, special types of prior distributions are often needed because of the difference between the dimensions of the sets Ω_1 and Ω_2. In some problems, for instance, the parameter space Ω is the k-dimensional space R^k but the set Ω_1 contains only a single point w_1. Such a situation occurs when the statistician has reason to believe, perhaps because of tradition or because of the support of some physical theory, that the parameter W may have the specific value w_1, but he recognizes and must investigate the possibility that W has some other value in Ω. In a situation of this type, the hypothesis H_1 is referred to as the *null hypothesis*. The prior distribution of W is usually such that $\Pr(W = w_1) = p > 0$, and the remaining probability $1 - p$ is spread over $\Omega_2 = \Omega - \{w_1\}$ in accordance with some p.d.f. ξ_2. Thus, for any subset $B \subset \Omega_2$,

$$\Pr(W \in B) = (1 - p) \int_B \xi_2(w) \, d\nu(w). \tag{2}$$

Suppose now that the statistician observes the value of a random variable or random vector X for which the family of conditional g.p.d.f.'s is $\{f(\cdot \mid w), w \in \Omega\}$. Then the posterior probability that $W = w_1$ when $X = x$ must satisfy the following equation:

$$\Pr(W = w_1 \mid X = x) = \frac{pf(x \mid w_1)}{pf(x \mid w_1) + (1 - p) \int_\Omega f(x \mid w) \xi_2(w) \, d\nu(w)}. \tag{3}$$

Also, the posterior conditional p.d.f. $\xi_2(\cdot \mid x)$ of W over the set $\Omega - \{w_1\}$, when it is assumed that $W \neq w_1$, is

$$\xi_2(w \mid x) = \frac{f(x \mid w) \xi_2(w)}{\int_\Omega f(x \mid w') \xi_2(w') \, d\nu(w')} \qquad \text{for } w \neq w_1. \tag{4}$$

For simplicity of notation and computation, the integrals in Eqs. (3) and (4) are defined over the whole set Ω instead of over the subset $\Omega_2 = \Omega - \{w_1\}$ from which the point w_1 is deleted. Since ξ_2 is assumed to be a p.d.f. over Ω_2, the addition or deletion of one point does not affect the value of the integral.

11.6 TESTING A SIMPLE HYPOTHESIS ABOUT THE MEAN OF A NORMAL DISTRIBUTION

As an example of the foregoing development, we shall suppose that $X_1, \ldots, X_n$ is a random sample from a normal distribution for which the value of the mean W is unknown and the value of the precision is 1. We shall let w_1 be a fixed number, and we shall suppose that the prior distribution of W satisfies the following two conditions: (1) $\Pr(W = w_1) = p > 0$. (2) The conditional distribution of W when $W \neq w_1$ is a normal distribution over the real line Ω, exclusive of the single point w_1, for which the mean is μ and the precision is τ. We shall determine the posterior distribution of W when $X_i = x_i$ $(i = 1, \ldots, n)$.

In this problem, the likelihood function is specified by the following equation:

$$f(x_1, \ldots, x_n|w) = (2\pi)^{-n/2} \exp\left[-\tfrac{1}{2} \sum_{i=1}^{n} (x_i - w)^2\right]. \tag{1}$$

Also, the p.d.f. ξ_2 of W when $W \neq w_1$ is

$$\xi_2(w) = \left(\frac{\tau}{2\pi}\right)^{\frac{1}{2}} \exp\left[-\frac{\tau}{2}(w - \mu)^2\right]. \tag{2}$$

Therefore, we can obtain the following equation:

$$\int_{-\infty}^{\infty} f(x_1, \ldots, x_n|w)\xi_2(w)\,dw$$
$$= \left(\frac{1}{2\pi}\right)^{n/2} \left(\frac{\tau}{\tau + n}\right)^{\frac{1}{2}} \exp\left\{-\frac{1}{2}\left[\sum_{i=1}^{n} x_i{}^2 + \tau\mu^2 - \frac{(\tau\mu + n\bar{x})^2}{\tau + n}\right]\right\}. \tag{3}$$

If we now apply Eq. (3) of Sec. 11.5 and simplify the resulting ratio, we can obtain the equation

$$\frac{\Pr(W = w_1|X_1 = x_1, \ldots, X_n = x_n)}{\Pr(W \neq w_1|X_1 = x_1, \ldots, X_n = x_n)}$$
$$= \frac{p}{1 - p}\left(\frac{\tau + n}{\tau}\right)^{\frac{1}{2}} \exp\left\{\frac{n}{2}\left[\frac{\tau}{\tau + n}(\bar{x} - \mu)^2 - (\bar{x} - w_1)^2\right]\right\}. \tag{4}$$

Furthermore, by Theorem 1 of Sec. 9.5, the posterior conditional distribution of W when it is assumed that $W \neq w_1$ will be a normal distribution over the real line, exclusive of the point w_1, for which the mean is $(\tau\mu + n\bar{x})/(\tau + n)$ and the precision is $\tau + n$.

The ratio in Eq. (4) has some interesting features. For very large values of $|\bar{x}|$, the ratio will be close to 0. In other words, there are values of $\bar{x}$ for which the posterior probability that $W \neq w_1$ can be made arbitrarily close to 1. On the other hand, elementary differentiation will show that the ratio in Eq. (4) is maximized when $\bar{x}$ has the following value:

$$\bar{x} = w_1 + \frac{\tau}{n}(w_1 - \mu).\tag{5}$$

For this value of $\bar{x}$, the maximum value of the ratio is

$$\frac{p}{1-p}\left(\frac{\tau+n}{\tau}\right)^{\frac{1}{2}}\exp\left[\frac{\tau}{2}(w_1 - \mu)^2\right].\tag{6}$$

It follows from the values (5) and (6) that for any fixed number of observations n, no values of the observations can increase the probability that $W = w_1$ by more than a limited amount. However, the factor multiplying $p/(1 - p)$ in (6) is greater than 1. Therefore, if a value of $\bar{x}$ close to the value given in Eq. (5) is observed, the posterior probability that $W = w_1$ will be greater than the prior probability p.

The maximum value (6) of the ratio will be smallest when $\mu = w_1$. In other words, it is most difficult to increase the probability that $W = w_1$ when the distribution of W under the alternative hypothesis H_2 is centered at the value w_1. This property is studied further in Exercise 17. It can be seen from the value (6) that as the number of observations n is increased, stronger evidence favoring the hypothesis that $W = w_1$ can be developed.

Finally, it follows from Eq. (4) that the posterior probabilities are somewhat sensitive to the value of τ used in the prior p.d.f. ξ_2. Suppose that the statistician assigns the value p to the prior probability that $W = w_1$ but is vague as to how the remaining probability $1 - p$ is distributed over the real line. He cannot get a simple approximate representation of this vague prior knowledge by letting $\tau \rightarrow 0$ in the prior p.d.f. ξ_2, because the ratio in Eq. (4) grows large without bound as $\tau \rightarrow 0$. This condition brings out the following important fact: If the probability $1 - p$ that $W \neq w_1$ is spread out widely over the real line because τ is small, then even a relatively large value of $\bar{x}$ increases the probability that $W = w_1$. It is traditional in statistical practice to regard an observed value of $\bar{x}$ more than three standard deviations, or $3n^{-\frac{1}{2}}$ units, away from w_1 as strong evidence against the null hypothesis H_1. However, we have now seen that such an observation may actually be evidence in favor of H_1. This concept is discussed in Savage and others (1962) and in Edwards, Lindman, and Savage (1963).

A typical loss function for the decision problem being considered might be of the following form:

$$\begin{aligned} L_1(w) &= a(w - w_1)^2 &&\text{for } -\infty < w < \infty, \\ L_2(w) &= 0 &&\text{for } w \neq w_1, \\ L_2(w_1) &= b. \end{aligned}\tag{7}$$

Here $a > 0$ and $b > 0$. In other words, the loss resulting from the decision d_1 is proportional to the square of the difference between w_1 and the actual value of W. Also, the loss resulting from the decision d_2 is 0 if d_2 is the appropriate decision, and the loss from this decision is b if d_2 is not the appropriate decision. When the distribution of W is specified by the probability p and the p.d.f. ξ_2, the risk $\rho(p,\ \xi_2,\ d_i)$ resulting from the decision d_i $(i\ =\ 1,\ 2)$ will be

$$\rho(p,\ \xi_2,\ d_1)\ =\ (1\ -\ p)aE[(W\ -\ w_1)^2|W\ \neq\ w_1]$$

$$=\ (1\ -\ p)a\left[\frac{1}{\tau}\ +\ (\mu\ -\ w_1)^2\right] \tag{8}$$

or

$$\rho(p,\ \xi_2,\ d_2)\ =\ pb. \tag{9}$$

Thus, the Bayes decision against any distribution of W having the specified form can be found by determining which of these two risks is smaller. Furthermore, by using the posterior distribution of W which we have derived, it is possible to derive a Bayes decision function based on the observations $X_1,\ \ldots,\ X_n$.

11.7 TESTING HYPOTHESES ABOUT THE MEAN OF A NORMAL DISTRIBUTION WHEN THE PRECISION IS UNKNOWN

We shall now consider a normal distribution for which both the mean M and the precision R are unknown. We shall suppose that it is desired to test the null hypothesis H_1 that $M\ =\ m_1$ against the alternative hypothesis H_2 that $M\ \neq\ m_1$. In this problem, the parameter space Ω is the set of all points $(m,\ r)$ such that $-\infty\ <\ m\ <\ \infty$ and $r\ >\ 0$. The subset Ω_1 specified by the hypothesis H_1 is the line containing the points such that $m\ =\ m_1$. The subset Ω_2 specified by the hypothesis H_2 contains all the other points in Ω, as illustrated in Fig. 11.2.

Since Ω_1 is a line in the two-dimensional space Ω, its probability would be 0 under any joint p.d.f. of M and R over Ω. Therefore, in order

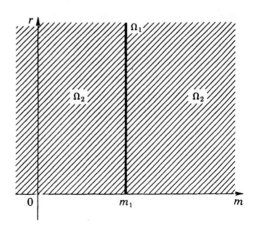

Fig. 11.2   The parameter space Ω.

that a prior distribution will be useful in this problem, the probability p that is assigned to the set Ω_1 must be positive. The distribution of this probability over Ω_1 can then be described in terms of a conditional p.d.f. ξ_1 of the precision R over the line. The remaining probability $1 - p$ is distributed over Ω_2 according to some joint p.d.f. ξ_2 of M and R. In other words, the prior joint distribution of M and R is specified by the following three conditions:

1. $\Pr(M = m_1) = p$.
2. The conditional p.d.f. of R when $M = m_1$ is ξ_1.
3. The conditional joint p.d.f. of M and R when $M \neq m_1$ is ξ_2.

If $X_1, \ldots, X_n$ is a random sample from the normal distribution with mean M and precision R, then the posterior probabilities of Ω_1 and Ω_2 when $X_i = x_i$ $(i = 1, \ldots, n)$ satisfy the following relation:

$$\frac{\Pr(M = m_1 | X_1 = x_1, \ldots, X_n = x_n)}{\Pr(M \neq m_1 | X_1 = x_1, \ldots, X_n = x_n)}$$

$$= \frac{p \int_0^\infty r^{n/2} \exp\left[-(r/2)\Sigma_{i=1}^n(x_i - m_1)^2\right]\xi_1(r)\, dr}{(1 - p) \int_0^\infty \int_{-\infty}^\infty r^{n/2} \exp\left[-(r/2)\Sigma_{i=1}^n(x_i - m)^2\right]\xi_2(m, r)\, dm\, dr}.$$

$$(1)$$

In order to carry the analysis further, we shall now assume that when $M = m_1$, the conditional distribution of R is a gamma distribution with parameters α and β. Furthermore, we shall assume that when $M \neq m_1$, the conditional joint distribution of M and R is a joint normal-gamma distribution as specified in Theorem 1 of Sec. 9.6. This joint normal-gamma distribution is described as follows: When $R = r$, the conditional distribution of M is a normal distribution with mean μ and precision τr, and the distribution of R is a gamma distribution with parameters α and β.

Both the conditional distribution of R when $M = m_1$ and the conditional distribution of R when $M \neq m_1$ are gamma distributions. For simplicity, we shall assume here that these two gamma distributions have the same parameters α and β. This assumption is appropriate if the statistician will assign the same prior distribution to R regardless of whether or not the null hypothesis is correct. If we now perform the integrations indicated in Eq. (1) and simplify the resulting ratio, we obtain the following equation (see Exercise 18):

$$\frac{\Pr(M = m_1 | X_1 = x_1, \ldots, X_n = x_n)}{\Pr(M \neq m_1 | X_1 = x_1, \ldots, X_n = x_n)}$$

$$= \frac{p}{1 - p}\left(\frac{\tau + n}{\tau}\right)^{\frac{1}{2}}\left\{\frac{2\beta + \Sigma_{i=1}^n(x_i - \bar{x})^2 + [\tau n/(\tau + n)](\bar{x} - \mu)^2}{2\beta + \Sigma_{i=1}^n(x_i - \bar{x})^2 + n(\bar{x} - m_1)^2}\right\}^{\alpha + (n/2)}.$$

$$(2)$$

If the prior distribution of M under the alternative hypothesis H_2 is centered at the value m_1, that is, if $\mu = m_1$, then the ratio in Eq. (2) can be simplified further. In particular, if we let $\beta \to 0$ in order to represent vague prior information about R, then the value of the ratio in Eq. (2) will depend on the values $x_1, \ldots, x_n$ of the observations only through the statistic $n(\bar{x} - m_1)^2 / \sum_{i=1}^{n}(x_i - \bar{x})^2$. This statistic is the one which is commonly calculated in performing traditional statistical tests of the hypothesis that $M = m_1$ based on the t distribution.

Now let us consider the posterior distribution of R. In this posterior distribution, as in the prior distribution, both the conditional distribution of R when $M = m_1$ and the conditional distribution of R when $M \neq m_1$ will be gamma distributions. The posterior gamma distribution of R when $M = m_1$ can be found from Theorem 2 of Sec. 9.5. The posterior gamma distribution of R when $M \neq m_1$ can be found from Theorem 1 of Sec. 9.6. We assumed that in the prior distribution of R, the two gamma distributions had the same parameters α and β. It can be seen from the two theorems just mentioned that in the posterior distribution, the two gamma distributions will not have the same parameters.

In this problem, $L_i(m, r)$ denotes the loss which is incurred when $M = m$, $R = r$, and the hypothesis H_i is accepted ($i = 1, 2$). This loss might typically be represented as follows:

$$
\begin{aligned}
L_1(m, r) &= ar(m - m_1)^2 &&\text{for } -\infty < m < \infty, \\
L_2(m, r) &= 0 &&\text{for } m \neq m_1, \\
L_2(m_1, r) &= b.
\end{aligned}
\tag{3}
$$

Here $a > 0$ and $b > 0$. Suppose that $\Pr(M = m_1) = p$, and suppose also that when $M \neq m_1$, the joint distribution of M and R is a normal-gamma distribution as described earlier in this section. Then the risk ρ_1 when decision d_1 is chosen can be computed as follows:

$$
\begin{aligned}
\rho_1 &= (1 - p)aE[R(M - m_1)^2 | M \neq m_1] \\
&= (1 - p)aE\{RE[(M - m_1)^2|R]|M \neq m_1\} \\
&= (1 - p)aE\left\{ R\left[\frac{1}{\tau R} + (\mu - m_1)^2 \right] \,\middle|\, M \neq m_1 \right\} \\
&= (1 - p)a\left[\frac{1}{\tau} + \frac{\alpha}{\beta}(\mu - m_1)^2 \right].
\end{aligned}
\tag{4}
$$

Furthermore, the risk ρ_2 when decision d_2 is chosen is $\rho_2 = pb$. Hence, the Bayes decision is either d_1 or d_2, according to whether ρ_1 or ρ_2 is smaller.

Prior distributions of the type described in this section have been studied by Jeffreys (1961), chaps. 5 and 6, and Lindley (1961b).

11.8 DECIDING WHETHER A PARAMETER IS SMALLER OR LARGER THAN A SPECIFIED VALUE

It is possible to treat by more standard techniques a problem of testing hypotheses in which each of the sets Ω_1 and Ω_2 is large enough to have a positive probability under a prior p.d.f. on the parameter space Ω. A special problem of this type will now be discussed.

We shall suppose that a statistician must decide whether the value of a real-valued parameter W is smaller or larger than a given constant w_0. Also, we shall assume that the losses $L_i(w)$ resulting from the two possible decisions d_i ($i = 1, 2$) are of the following forms:

$$
L_1(w) = \begin{cases} 0 & \text{for } w \leq w_0, \\ w - w_0 & \text{for } w > w_0, \end{cases}
$$

$$
L_2(w) = \begin{cases} w_0 - w & \text{for } w \leq w_0, \\ 0 & \text{for } w > w_0. \end{cases}
$$

$$(1)$$

In other words, decision d_1 is correct if $W < w_0$ and decision d_2 is correct if $W > w_0$. The loss which results from choosing the wrong decision when $W = w$ is $|w - w_0|$.

A loss function of the type (1) is appropriate in fairly general circumstances, as the following argument indicates. In a problem of testing hypotheses about whether a real parameter W is small or large, suppose that the loss $L_1(w)$ from deciding that W is small is a nondecreasing function of w and that the loss $L_2(w)$ from deciding that W is large is a nonincreasing function of w. Suppose also that $L_1(w) = L_2(w)$ at some point w_0 and that the difference $L_1(w) - L_2(w)$ is increasing at the point w_0. Then, under moderate conditions, the difference $L_1(w) - L_2(w)$ can be approximated in a neighborhood around w_0 by an increasing linear function of the form

$$
L_1(w) - L_2(w) = a(w - w_0). \tag{2}
$$

Here $a > 0$, and by a suitable choice of unit it may be assumed that $a = 1$. The loss functions L_1 and L_2 specified by Eq. (1) satisfy Eq. (2) with $a = 1$, as do any loss functions L_1^* and L_2^* of the following form:

$$
L_i^*(w) = L_i(w) + \lambda(w) \qquad i = 1, 2, \tag{3}
$$

where λ is an arbitrary function of w.

As shown in Sec. 8.3, the Bayes decision against any specified distribution of W will be the same for any choice of the function λ. It will be convenient to choose λ so that $\lambda(w) = -L_1(w)$, where $L_1(w)$ is given by Eq. (1). With this choice of λ, we must consider the decision problem in which the loss functions L_i^* are the following linear functions:

$$L_1^*(w) = 0,$$
$$L_2^*(w) = w_0 - w. \tag{4}$$

Now consider any distribution of W such that the mean $E(W) = \mu$ exists. Then the risk from the decision d_1 is $E[L_1^*(W)] = 0$ and the risk from the decision d_2 is $E[L_2^*(W)] = w_0 - \mu$. Thus, the Bayes decision is d_1 if $\mu < w_0$ and is d_2 if $\mu > w_0$.

Let ξ denote the prior distribution of W, and suppose that before the statistician chooses his decision, he can observe the random variable or random vector X for which the conditional g.p.d.f. is $f(\cdot \mid w)$ when $W = w$. More generally, the distribution of X may involve other parameters as well as W, and the prior joint distribution of all the parameters must be specified. Let $E(W \mid x)$ denote the mean of the posterior distribution of W when $X = x$. We shall assume that $E(W \mid x)$ exists for all values of x. Then it follows from the above comments that a Bayes decision function δ^* has the following form:

$$\delta^*(x) = \begin{cases} d_1 & \text{for } E(W \mid x) \leq w_0, \\ d_2 & \text{for } E(W \mid x) > w_0. \end{cases} \tag{5}$$

Also, the risk $r(x)$ from the Bayes decision when $X = x$ is

$$r(x) = \begin{cases} 0 & \text{for } E(W \mid x) \leq w_0, \\ w_0 - E(W \mid x) & \text{for } E(W \mid x) > w_0. \end{cases} \tag{6}$$

Now let G denote the d.f. of the random variable $E(W \mid X)$ when the prior d.f. of W is ξ. In other words, before the observation X is taken, the statistician does not know what the mean $E(W \mid X)$ of the posterior distribution of W will be. However, he can compute its distribution function G from the g.p.d.f.'s $f(\cdot \mid w)$ and the prior d.f. of W. It now follows from Eq. (6) that the Bayes risk $r^*(\xi)$ against the prior d.f. ξ is given by the following equation:

$$r^*(\xi) = E[r(X)] = -\int_{w_0}^{\infty} (w - w_0) \, dG(w). \tag{7}$$

We shall now consider again the original decision problem in which the loss functions are specified by Eq. (1). Let $\rho^*(\xi)$ denote the Bayes risk in this problem. Since we had chosen the function λ so that $\lambda(w) = -L_1(w)$, it follows from Eq. (3) that $\rho^*(\xi) = r^*(\xi) + E[L_1(W)]$. Hence, from Eqs. (7) and (1), it can be seen that the Bayes risk $\rho^*(\xi)$ is specified as follows:

$$\rho^*(\xi) = \int_{w_0}^{\infty} (w - w_0) \, d\xi(w) - \int_{w_0}^{\infty} (w - w_0) \, dG(w). \tag{8}$$

Both integrals in Eq. (8) have the same form. In the first integral, the integration is carried out with respect to the distribution of W; in the second, the integration is carried out with respect to the distribution of $E(W|X)$. Since the loss functions in Eq. (1) are nonnegative, $\rho^*(\xi) \geq 0$. Hence, for any value of w_0, any d.f. ξ, and any observation X, the first integral in Eq. (8) must be at least as large as the second integral.

The integral which appears in Eq. (8) can be studied further. For any d.f. F on the real line for which there is a finite mean, let T_F denote another function on the real line defined as follows:

$$T_F(s) = \int_s^\infty (x - s)\, dF(x) \qquad -\infty < s < \infty. \tag{9}$$

The function T_F can be regarded as a transform of the d.f. F. It can be shown (see Exercise 21) that T_F is a nonnegative convex function which is strictly decreasing at any value s such that $T_F(s) > 0$. Furthermore, if μ is the mean of the d.f. F, then T_F satisfies the following properties:

$$T_F(s) \geq \mu - s \qquad \text{for } -\infty < s < \infty,$$
$$\lim_{s \to -\infty} [T_F(s) - (\mu - s)] = 0, \tag{10}$$
$$\lim_{s \to \infty} T_F(s) = 0.$$

A function T_F that satisfies these properties is sketched in Fig. 11.3. If Y is any random variable whose d.f. is F, then it can be seen from Eq. (9) that $T_F(s) = 0$ at a given value of s if, and only if, $\Pr(Y > s) = 0$. Also, $T_F(s) = \mu - s$ at a given value of s if, and only if,

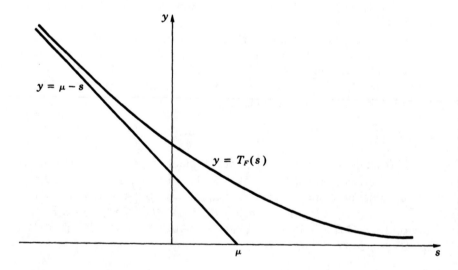

Fig. 11.3 The function T_F.

$\Pr(Y < s) = 0$. Furthermore, if F is the d.f. of a discrete distribution, then the function T_F will be convex but will be composed of linear segments.

Equation (8) can be rewritten as follows:

$$\rho^*(\xi) = T_\xi(w_0) - T_G(w_0). \tag{11}$$

We have already noted that it must be true that $\rho^*(\xi) \geq 0$. Two extreme situations may be identified further. Suppose first that the observation X is irrelevant to the value of W or, equivalently, that no observations are taken. Then the posterior distribution of W must be the same as the prior distribution. Therefore, the mean $E(W|X)$ of the posterior distribution must be equal to μ, the mean of the prior distribution. In other words, the d.f. G represents the distribution that assigns probability 1 to the value μ. It follows that $T_G(w_0) = 0$ if $\mu \leq w_0$ and that $T_G(w_0) = \mu - w_0$ if $\mu > w_0$. When these values are substituted into Eq. (11), we obtain the Bayes risk $\rho^*(\xi)$ when no observations are taken.

Suppose next that the observation X is fully informative about the value of W and that after the value of X has been observed, the statistician will know the value of W. Then $E(W|X) = W$ with probability 1, and it follows that the d.f. G will be the same as the prior d.f. ξ of W. Hence, in this case, $T_\xi(w_0) = T_G(w_0)$ and $\rho^*(\xi) = 0$.

11.9 DECIDING WHETHER THE MEAN OF A NORMAL DISTRIBUTION IS SMALLER OR LARGER THAN A SPECIFIED VALUE

We shall now apply the results which have been developed in Sec. 11.8 to the problem of deciding whether the mean W of a normal distribution which has specified precision is smaller or larger than the value w_0. As usual, we shall let φ denote the p.d.f. of a standard normal distribution (i.e., a normal distribution with mean 0 and unit precision), and we shall let Φ denote the d.f. of a standard normal distribution.

When $F = \Phi$, the transform T_F defined by Eq. (9) of Sec. 11.8 is denoted simply by Ψ. Hence, for $-\infty < s < \infty$,

$$\Psi(s) = \int_s^\infty (x - s)\varphi(x)\,dx = \varphi(s) - s[1 - \Phi(s)]. \tag{1}$$

The function Ψ is a strictly decreasing and strictly convex function on the real line which has the general properties of any function of the form T_F, as specified by the relations (10) of Sec. 11.8. The derivative Ψ' of Ψ is found, from Eq. (1), to be as follows for $-\infty < s < \infty$:

$$\Psi'(s) = \Phi(s) - 1. \tag{2}$$

Furthermore, if F is the d.f. of a normal distribution for which the mean is μ and the precision is τ, then it can be shown (see Exercise 22) that

$$T_F(s) = \tau^{-\frac{1}{2}}\Psi[\tau^{\frac{1}{2}}(s - \mu)]. \tag{3}$$

Now suppose that in the decision problem we have been considering, in which the loss function is specified by Eq. (1) of Sec. 11.8, the statistician can observe the values of a random sample $X_1, \ldots, X_n$ from a normal distribution with mean W and specified precision r. Suppose also that the prior distribution of W is a normal distribution with mean μ and precision τ. Then it follows from Theorem 1 of Sec. 9.5 that the mean $E(W|X_1, \ldots, X_n)$ of the posterior distribution of W is given by the equation

$$E(W|X_1, \ldots, X_n) = \frac{\tau\mu + nr\bar{X}}{\tau + nr}. \tag{4}$$

Since the conditional distribution of $\bar{X}$ when $W = w$ is a normal distribution with mean w and precision nr, then (see Exercise 23) the marginal distribution of $\bar{X}$ is a normal distribution with mean μ and precision $nr\tau/(\tau + nr)$. Hence, the random variable $E(W|X_1, \ldots, X_n)$ defined by Eq. (4) has a normal distribution with mean μ and precision $\tau(\tau + nr)/nr$. It now follows from Eqs. (8) and (9) of Sec. 11.8 and from Eq. (3) that the Bayes risk $\rho^*(\xi)$ against the specified normal prior distribution ξ of W is given by the equation

$$\rho^*(\xi) = \tau^{-\frac{1}{2}}\Psi[\tau^{\frac{1}{2}}(w_0 - \mu)] - \tau_n^{-\frac{1}{2}}\Psi[\tau_n^{\frac{1}{2}}(w_0 - \mu)], \tag{5}$$

where

$$\tau_n = \frac{\tau(\tau + nr)}{nr}. \tag{6}$$

Suppose now that the cost per observation is $c > 0$ and that the number of observations n must be chosen to minimize the total risk $\rho_T = \rho^*(\xi) + cn$. For the purposes of the computation, we shall ignore the fact that n must be an integer. It follows from Eqs. (1), (2), and (5) that the derivative ρ_T' of the total risk with respect to n satisfies the relation

$$\rho_T' = \tfrac{1}{2}\tau_n^{-\frac{1}{2}}\varphi[\tau_n^{\frac{1}{2}}(w_0 - \mu)]\frac{d\tau_n}{dn} + c. \tag{7}$$

By Eq. (6),

$$\frac{d\tau_n}{dn} = \frac{-\tau^2}{rn^2}. \tag{8}$$

Hence, the equation $\rho'_T = 0$ can be written in the following form:

$$\frac{\tau^2 \varphi[\tau_n{}^{\frac{1}{2}}(w_0 - \mu)]}{2r\tau_n{}^{\frac{1}{2}}n^2} = c. \tag{9}$$

For small values of the cost c, the optimal number of observations n will be large. It is seen from Eq. (6) that $\tau_n \to \tau$ as $n \to \infty$. Therefore, when c is small, the optimal sample size n can be computed approximately from the following equation:

$$n = \left\{ \frac{\tau^{\frac{1}{2}}\varphi[\tau^{\frac{1}{2}}(w_0 - \mu)]}{2cr} \right\}^{\frac{1}{2}}. \tag{10}$$

Further Remarks and References

Raiffa and Schlaifer (1961), chap. 5, and Schlaifer (1961), chaps. 20 and 21, present a thorough discussion of the problem just considered and related problems. They give a table of values of the function Ψ, as well as charts of the optimal sample size and the Bayes risk. A previous study related to this problem had been carried out by Grundy, Healy, and Rees (1956).

Bracken and Schleifer (1964) give tables of the p.d.f. of the t distribution and also give tables of the optimal sample size and the Bayes risk for decision problems of the type we have considered in this section, with the additional complication that the precision of the normal distribution is also unknown.

The theory of testing hypotheses can be generalized to cover problems in which the decision space D contains a finite number of decisions rather than only two decisions. A decision problem of this type is called a *multidecision problem*. The problem of deciding whether the mean of a normal distribution is smaller or larger than the value w_0 could be generalized to a multidecision problem in the following way:

Let $w_1 < \cdots < w_k$ be k given values ($k \geq 1$). Suppose that the statistician must decide in which one of the $k + 1$ intervals $(-\infty, w_1]$, $(w_1, w_2]$, . . . , (w_k, ∞) the value of the mean W lies. If the loss resulting from each of the $k + 1$ decisions is chosen appropriately, this problem is one of a class of problems known as *monotone statistical decision problems*. These problems have been studied by Karlin and Rubin (1956a,b). Somewhat more general problems have also been studied by Karlin (1955b; 1957a,b; 1958). The properties of multidecision problems will not be discussed further here.

11.10 LINEAR MODELS

In the remainder of this chapter, we shall discuss some problems of multiple linear regression and the analysis of variance. The statistical models used in these problems are called *linear models* because it is assumed that the mean of each of the observations is a specified linear combination of a fixed set of parameters.

We shall let $\mathbf{M} = (M_1, \ldots, M_k)'$ denote a vector parameter whose values lie in R^k ($k \geq 1$), and we shall let W denote another parameter whose value must be positive. We shall assume that the statistician can observe values $Y_1, \ldots, Y_n$, each of which has a normal distribution whose mean is a linear combination of the k components of $\mathbf{M}$ and whose precision is W. Specifically, it is assumed that the conditional joint distribution of the observations $Y_1, \ldots, Y_n$, when

$$\mathbf{M} = \mathbf{m} = (m_1, \ldots, m_k)'$$

and $W = w$, satisfies the following two properties: (1) $Y_1, \ldots, Y_n$ are independent. (2) For $i = 1, \ldots, n$, the distribution of Y_i is a normal distribution with mean $\sum_{j=1}^{k} x_{ij} m_j$ and precision w. The statistician will know the numbers x_{ij} for $i = 1, \ldots, n$ and $j = 1, \ldots, k$. In some problems, the statistician can actually choose these numbers, at least from some restricted set of values, as part of his experimental plan. In other problems, the values x_{ij} are beyond the statistician's control and he observes these values together with the values of $Y_1, \ldots, Y_n$. It is assumed that in such problems, the observed values x_{ij} do not depend on the values of the parameters $\mathbf{M}$ and W. Therefore, throughout the discussion which will be presented here, the values x_{ij} can be regarded as being fixed.

Let $\mathbf{x}$ denote the $n \times k$ matrix of values x_{ij}; that is, let

$$\mathbf{x} = \begin{bmatrix} x_{11} & \cdots & x_{1k} \\ \cdots & \cdots & \cdots \\ x_{n1} & \cdots & x_{nk} \end{bmatrix}. \tag{1}$$

Also, let $\mathbf{I}$ denote the $n \times n$ identity matrix. Then the conditional joint distribution of $Y_1, \ldots, Y_n$ which we have just described can be summarized as follows: When $\mathbf{M} = \mathbf{m}$ and $W = w$, the conditional distribution of the n-dimensional random vector $\mathbf{Y} = (Y_1, \ldots, Y_n)'$ is a multivariate normal distribution with mean vector $\mathbf{xm}$ and precision matrix $w\mathbf{I}$. Therefore, the likelihood function $f(\mathbf{y}|\mathbf{m}, w)$ will have the following form for any point $\mathbf{y} = (y_1, \ldots, y_n) \in R^n$:

$$f(\mathbf{y}|\mathbf{m}, w) \propto w^{n/2} \exp\left[-\frac{w}{2} (\mathbf{y} - \mathbf{xm})'(\mathbf{y} - \mathbf{xm}) \right]. \tag{2}$$

The problem of making inferences about the values of $\mathbf{M}$ and W on the basis of the observed value $\mathbf{Y} = \mathbf{y}$ is called a problem of *multiple linear regression*. A *least-squares estimate* of $\mathbf{M}$ is defined to be a vector

$m = \hat{m}$ which minimizes the quadratic form $(y - xm)'(y - xm)$ that appears in the exponent of the likelihood function. By differentiation, it can be found that $\hat{m}$ will minimize this quadratic form if, and only if, it satisfies the following equation:

$$x'x\hat{m} = x'y. \tag{3}$$

In the theory of least squares, the system of k equations represented by Eq. (3) is called the system of *normal equations*. It follows from the context in which Eq. (3) was derived that for any vector y and any matrix x, there must always exist at least one solution $\hat{m}$ of this equation. For a more rigorous argument which establishes the existence of a solution of Eq. (3), see Rao (1965), pp. 27 and 181. Of course, if the $k \times k$ matrix $x'x$ is nonsingular, it follows that $\hat{m} = (x'x)^{-1}x'y$, and this vector is the only solution of Eq. (3). In general, however, $x'x$ may be singular and the solution $\hat{m}$ of Eq. (3) may not be unique. It can be verified that if $\hat{m}$ satisfies Eq. (3), then for all values of m, x, and y,

$$(y - xm)'(y - xm) = (m - \hat{m})'x'x(m - \hat{m}) + y'y - \hat{m}'x'x\hat{m}. \tag{4}$$

Suppose now that the prior joint distribution of M and W is a multivariate normal-gamma distribution of the type used in Theorem 1 of Sec. 9.13. Specifically, suppose that the conditional distribution of M when $W = w$ is a k-dimensional multivariate normal distribution with mean vector μ and precision matrix $w\tau$ and that the marginal distribution of W is a gamma distribution with parameters α and β. Thus, the prior joint p.d.f. ξ of M and W is as follows for $m \in R^k$ and $w > 0$:

$$\xi(m, w) \propto w^{k/2} \exp\left[-\frac{w}{2}(m - \mu)'\tau(m - \mu) \right] w^{\alpha-1} \exp(-\beta w). \tag{5}$$

The posterior joint p.d.f. $\xi(\cdot \,|y)$ of M and W when $Y = y$ will be proportional to the product of the right sides of the relations (2) and (5). Let the k-dimensional vector m_1 be defined by the equation

$$m_1 = (\tau + x'x)^{-1}(\tau\mu + x'y). \tag{6}$$

Since the matrix τ is positive definite, the matrix $\tau + x'x$ is also positive definite and therefore can be inverted in Eq. (6), even though the matrix $x'x$ itself may be singular. If $\hat{m}$ is any vector that satisfies Eq. (3), then the following identity can be obtained:

$$\begin{aligned}
(m - \mu)'\tau(m - \mu) &+ (m - \hat{m})'x'x(m - \hat{m}) \\
&= (m - m_1)'(\tau + x'x)(m - m_1) + \mu'\tau\mu \\
&\quad + \hat{m}'x'x\hat{m} - m_1'(\tau + x'x)m_1. \tag{7}
\end{aligned}$$

By using the identities (4) and (7), we can now obtain the following form for the posterior joint p.d.f. of $\mathbf{M}$ and W:

$$\xi(\mathbf{m}, w|\mathbf{y}) \propto \left\{ w^{k/2} \exp\left[-\frac{w}{2}(\mathbf{m} - \mathbf{m}_1)'(\tau + \mathbf{x}'\mathbf{x})(\mathbf{m} - \mathbf{m}_1)\right]\right\}$$
$$\times (w^{\alpha+(n/2)-1}e^{-\beta_1 w}), \quad (8)$$

where

$$\beta_1 = \beta + \tfrac{1}{2}[(\mathbf{y} - \mathbf{x}\mathbf{m}_1)'\mathbf{y} + (\mathbf{\mu} - \mathbf{m}_1)'\tau\mathbf{\mu}]. \quad (9)$$

The posterior p.d.f. specified by Eq. (8) shows that the conditional distribution of $\mathbf{M}$ when $W = w$ is a multivariate normal distribution with mean vector $\mathbf{m}_1$ and precision matrix $w(\tau + \mathbf{x}'\mathbf{x})$ and that the marginal distribution of W is a gamma distribution with parameters $\alpha + (n/2)$ and β_1. It follows from Exercise 45 of Chap. 9 that the posterior marginal distribution of $\mathbf{M}$ will be a multivariate t distribution for which there are $2\alpha + n$ degrees of freedom, the location vector is $\mathbf{m}_1$, and the precision matrix is $[(2\alpha + n)/(2\beta_1)](\tau + \mathbf{x}'\mathbf{x})$.

If $n > k$ and the $k \times k$ matrix $\mathbf{x}'\mathbf{x}$ is nonsingular, we can let $\tau \rightarrow 0$, $\alpha \rightarrow -k/2$, and $\beta \rightarrow 0$ in the posterior distribution of $\mathbf{M}$ and W specified by Eqs. (8) and (9). As can be seen from Eq. (5), this same limiting posterior distribution will be obtained from an improper prior joint density function ξ of the following form:

$$\xi(\mathbf{m}, w) = \frac{1}{w} \quad \text{for } \mathbf{m} \in R^k \text{ and } w > 0. \quad (10)$$

The use of an improper prior density function of this form was discussed in Chap. 10.

Since we are now assuming that the matrix $\mathbf{x}'\mathbf{x}$ is nonsingular, it follows from Eq. (3) that $\hat{\mathbf{m}}$ is specified by the equation

$$\hat{\mathbf{m}} = (\mathbf{x}'\mathbf{x})^{-1}\mathbf{x}'\mathbf{y}. \quad (11)$$

As $\tau \rightarrow 0$ and $\beta \rightarrow 0$, it can be seen from Eqs. (6) and (9) that $\mathbf{m}_1 \rightarrow \hat{\mathbf{m}}$ and $\beta_1 \rightarrow \tfrac{1}{2}(\mathbf{y} - \mathbf{x}\hat{\mathbf{m}})'\mathbf{y}$. Since $(\mathbf{y} - \mathbf{x}\hat{\mathbf{m}})'\mathbf{x}\hat{\mathbf{m}} = 0$, the limiting value of β_1 can be rewritten as $(n - k)\hat{s}^2/2$, where

$$\hat{s}^2 = \frac{1}{n - k}(\mathbf{y} - \mathbf{x}\hat{\mathbf{m}})'(\mathbf{y} - \mathbf{x}\hat{\mathbf{m}}). \quad (12)$$

The limiting posterior joint distribution of $\mathbf{M}$ and W can now be obtained from these limiting values. In particular, the marginal distribution of $\mathbf{M}$ will be a multivariate t distribution for which there are $n - k$ degrees of freedom, the location vector is $\hat{\mathbf{m}}$, as specified by Eq. (11), and the precision matrix is $(1/\hat{s}^2)\mathbf{x}'\mathbf{x}$. The marginal distribution of W will be a gamma distribution for which the parameters are $(n - k)/2$ and $(n - k)\hat{s}^2/2$.

If the values of $\mathbf{M}$ and W must be estimated when there is a quadratic loss function of the type specified in Sec. 11.4, then it follows from the discussion in that section that $\mathbf{M}$ and W should be estimated by the means of their posterior distributions. Therefore, the Bayes estimate of $\mathbf{M}$ will be $\hat{\mathbf{m}}$, which is also the standard least-squares estimate of $\mathbf{M}$. Furthermore, the Bayes estimate of W will be $1/\hat{s}^2$. Here, $\hat{s}^2$ is defined by Eq. (12), and in the theory of least squares, it is the estimate of the variance $1/W$ which is typically used [see, e.g., Rao (1965), p. 185].

11.11 TESTING HYPOTHESES IN LINEAR MODELS

In many problems for which a linear model of the type just described is appropriate, we are interested in studying the hypothesis that the parameter $\mathbf{M}$ lies in a certain subset S_0 of R^k. Furthermore, the subset S_0 can often be described as the set of all vectors $\mathbf{m} \in R^k$ which satisfy linear constraints of the following form:

$$S_0 = \{\mathbf{m}: \mathbf{Am} = \mathbf{b}\}. \tag{1}$$

Here it is assumed that $\mathbf{A}$ is a $p \times k$ matrix $(p < k)$ whose rank is p and that $\mathbf{b}$ is a $p \times 1$ vector. Hence, there is at least one vector $\mathbf{m}$ which satisfies the equation $\mathbf{Am} = \mathbf{b}$, and the set S_0 will not be empty. As examples, S_0 might be the set of all vectors $\mathbf{m} = (m_1, \ldots, m_k)'$ such that $m_1 = 0$ or it might be the set of all vectors $\mathbf{m} = (m_1, \ldots, m_k)'$ such that $m_1 = m_2 = \cdots = m_k$.

In accordance with the discussion in Sec. 11.10, let us suppose that $\mathbf{M}$ has a k-dimensional multivariate t distribution. Since the dimension of the set S_0 is less than k, it follows that $\Pr\{\mathbf{M} \in S_0\} = 0$. However, as discussed in Sec. 10.12, we can assume that $\mathbf{M}$ has a p.d.f. over the space R^k, and we can compare the maximum value of this p.d.f. for values of $\mathbf{M}$ in the set S_0 with the maximum value over the whole space R^k. In this approach, the probability that $\mathbf{M}$ is in a small region around the most likely point in S_0 would be compared with the maximum probability that $\mathbf{M}$ is in a small region of the same size anywhere else in R^k.

Let ξ_M denote the p.d.f. of $\mathbf{M}$, and let the ratio γ be defined as follows:

$$\gamma = \frac{\sup_{\mathbf{m} \in R^k} \xi_M(\mathbf{m})}{\sup_{\mathbf{m} \in S_0} \xi_M(\mathbf{m})}. \tag{2}$$

The value of γ must always be at least 1. However, a very large value of γ indicates that there are values of $\mathbf{M}$ outside S_0 which are much more likely than are any values of $\mathbf{M}$ in S_0. A value of γ close to 1 indicates that no values of $\mathbf{M}$ are much more likely than are certain values of $\mathbf{M}$ in

S_0. If $\gamma = 1$, then the maximum value of ξ_M among all points in R^k is actually attained at a point in the subset S_0. We shall now derive the values of these maxima.

Suppose that $\mathbf{M}$ has a k-dimensional multivariate t distribution with n degrees of freedom, location vector $\mathbf{\mu}$, and precision matrix $\mathbf{T}$. For $\mathbf{m} \in R^k$, let $Q(\mathbf{m})$ be defined as follows:

$$Q(\mathbf{m}) = (\mathbf{m} - \mathbf{\mu})'\mathbf{T}(\mathbf{m} - \mathbf{\mu}). \tag{3}$$

For any subset S_0 of R^k, it follows from Eq. (9) of Sec. 5.6 that the maximum value of the p.d.f. of $\mathbf{M}$ among all points $\mathbf{m} \in S_0$ will be attained at those points of S_0 for which the quadratic form $Q(\mathbf{m})$ is minimized. In the next theorem, we shall derive the minimum value of $Q(\mathbf{m})$ when S_0 is specified by Eq. (1). In the proof of the theorem, we shall make use of the *Schwarz inequality*, which can be described as follows [see, e.g., Rao (1965), p. 42]:

Schwarz inequality *If $\mathbf{y}_1$ and $\mathbf{y}_2$ are any two vectors in R^k such that $\mathbf{y}_1 \neq 0$ and $\mathbf{y}_2 \neq 0$, then*

$$(\mathbf{y}_1'\mathbf{y}_2)^2 \leq (\mathbf{y}_1'\mathbf{y}_1)(\mathbf{y}_2'\mathbf{y}_2). \tag{4}$$

Furthermore, there is equality in the relation (4) if, and only if, there is a number λ such that $\mathbf{y}_1 = \lambda\mathbf{y}_2$.

Theorem 1 *If the set S_0 is specified by Eq. (1) and if the function Q is specified by Eq. (3), then*

$$\inf_{\mathbf{m} \in S_0} Q(\mathbf{m}) = (\mathbf{b} - \mathbf{A}\mathbf{\mu})'(\mathbf{A}\mathbf{T}^{-1}\mathbf{A}')^{-1}(\mathbf{b} - \mathbf{A}\mathbf{\mu}). \tag{5}$$

Furthermore, this minimum value of $Q(\mathbf{m})$ is attained at a unique point $\mathbf{m}_0 \in S_0$, which is specified as follows:

$$\mathbf{m}_0 = \mathbf{\mu} + \mathbf{T}^{-1}\mathbf{A}'(\mathbf{A}\mathbf{T}^{-1}\mathbf{A}')^{-1}(\mathbf{b} - \mathbf{A}\mathbf{\mu}). \tag{6}$$

Proof For any vector $\mathbf{m} \in R^k$, let $\mathbf{x} = \mathbf{m} - \mathbf{\mu}$. When we make this substitution in Eqs. (1) and (3), we see that we must now find the minimum value of the quadratic form $q(\mathbf{x}) = \mathbf{x}'\mathbf{T}\mathbf{x}$ among all vectors $\mathbf{x}$ such that $\mathbf{A}\mathbf{x} = \mathbf{c}$, where $\mathbf{c} = \mathbf{b} - \mathbf{A}\mathbf{\mu}$.

Let the vector $\mathbf{x}_0$ be defined as follows:

$$\mathbf{x}_0 = \mathbf{T}^{-1}\mathbf{A}'(\mathbf{A}\mathbf{T}^{-1}\mathbf{A}')^{-1}\mathbf{c}. \tag{7}$$

Since the $p \times k$ matrix $\mathbf{A}$ has rank p, it follows that the $p \times p$ matrix $\mathbf{A}\mathbf{T}^{-1}\mathbf{A}'$ will also have rank p. Therefore, the inverse of this matrix, which appears in Eq. (7), must exist. It can be seen that $\mathbf{A}\mathbf{x}_0 = \mathbf{c}$ and that

$$q(\mathbf{x}_0) = \mathbf{c}'(\mathbf{A}\mathbf{T}^{-1}\mathbf{A}')^{-1}\mathbf{c}. \tag{8}$$

We shall now show that $q(\mathbf{x}_0) < q(\mathbf{x})$ for any other vector $\mathbf{x}$ such that $\mathbf{A}\mathbf{x} = \mathbf{c}$. This demonstration will complete the proof of the theorem, since Eqs. (5) and (6) can be obtained from the substitutions which have been made.

Suppose first that $\mathbf{c} = \mathbf{0}$. Then $\mathbf{x}_0 = \mathbf{0}$ and $q(\mathbf{x}_0) = 0$. Furthermore, since the matrix $\mathbf{T}$ is positive definite, it follows that $q(\mathbf{x}) = 0$ only when $\mathbf{x} = \mathbf{0}$. Hence, the theorem is true in this case.

Suppose now that $\mathbf{c} \neq \mathbf{0}$, and let $\mathbf{x}$ be any vector such that $\mathbf{A}\mathbf{x} = \mathbf{c}$. Let $\mathbf{U}$ be a $k \times k$ matrix such that $\mathbf{U}'\mathbf{U} = \mathbf{T}$. Also, let $\mathbf{y} = \mathbf{U}\mathbf{x}$ and $\mathbf{y}_0 = \mathbf{U}\mathbf{x}_0$. It follows from the Schwarz inequality that $(\mathbf{y}'\mathbf{y}_0)^2 \leq (\mathbf{y}'\mathbf{y})(\mathbf{y}_0'\mathbf{y}_0)$. However, it can be verified that $\mathbf{y}'\mathbf{y}_0 = q(\mathbf{x}_0)$, $\mathbf{y}'\mathbf{y} = q(\mathbf{x})$, and $\mathbf{y}_0'\mathbf{y}_0 = q(\mathbf{x}_0)$. Therefore, we can conclude that $q(\mathbf{x}_0) \leq q(\mathbf{x})$. Furthermore, if $q(\mathbf{x}_0) = q(\mathbf{x})$, then $\mathbf{y} = \lambda\mathbf{y}_0$ for some number λ. It then follows from the substitution made above that $\mathbf{x} = \lambda\mathbf{x}_0$. Since $\mathbf{A}\mathbf{x} = \mathbf{c}$ and $\mathbf{A}\mathbf{x}_0 = \mathbf{c}$ and we are assuming that $\mathbf{c} \neq \mathbf{0}$, it follows that $\lambda = 1$, which means that $\mathbf{x} = \mathbf{x}_0$. Hence, if $\mathbf{x} \neq \mathbf{x}_0$, then $q(\mathbf{x}_0) < q(\mathbf{x})$.∎

For other results related to Theorem 1, see Rao (1965), p. 48.

It can be seen from Eq. (3) that the minimum value of $Q(\mathbf{m})$ among all points $\mathbf{m}$ in R^k is 0 and that this value is attained at the point $\mathbf{m} = \mathbf{\mu}$. Hence, it follows from the p.d.f. of the multivariate t distribution of $\mathbf{M}$ and from Theorem 1 that the ratio γ specified by Eq. (2) has the following value:

$$\gamma = \left[1 + \frac{1}{n}(\mathbf{b} - \mathbf{A}\mathbf{\mu})'(\mathbf{A}\mathbf{T}^{-1}\mathbf{A}')^{-1}(\mathbf{b} - \mathbf{A}\mathbf{\mu}) \right]^{(n+k)/2}. \tag{9}$$

Hence, $\gamma = 1$ if, and only if, $\mathbf{\mu} \in S_0$; and the value of γ increases as the vector $\mathbf{\mu}$ moves away from the set S_0.

It follows from Eq. (9) that as $n \to \infty$,

$$\gamma \to \exp\left[\tfrac{1}{2}(\mathbf{b} - \mathbf{A}\mathbf{\mu})'(\mathbf{A}\mathbf{T}^{-1}\mathbf{A}')^{-1}(\mathbf{b} - \mathbf{A}\mathbf{\mu})\right]. \tag{10}$$

This result is a restatement of the fact that as the number of degrees of freedom becomes arbitrarily large, the multivariate t distribution converges to a multivariate normal distribution.

When computing the limit in Eq. (10), we assumed that the location vector $\mathbf{\mu}$ and the precision matrix $\mathbf{T}$ remained fixed as $n \to \infty$. However, in problems of multiple linear regression, as the number of observations and the number of degrees of freedom are increased in the posterior multivariate t distribution of $\mathbf{M}$, both $\mathbf{\mu}$ and $\mathbf{T}$ will also vary. Another approach in this situation, for any large but fixed number of degrees of freedom n, is to approximate the multivariate t distribution of $\mathbf{M}$ by a multivariate normal distribution with mean vector $\mathbf{\mu}$ and precision matrix $\mathbf{T}$. The p.d.f. of this multivariate normal distribution involves the same quadratic function $Q(\mathbf{m})$ as the p.d.f. of the multivariate t distribution. Therefore, it can be shown that the value of γ which will be obtained from the multivariate normal distribution will be equal to the limiting value specified in Eq. (10). Thus, for large values of n, this limiting value will be a suitable approximation to the value of γ, and Table 10.1 given in Sec. 10.12 will be applicable here.

11.12 INVESTIGATING THE HYPOTHESIS THAT CERTAIN REGRESSION COEFFICIENTS VANISH

In problems of multiple linear regression, the components of the vector $\mathbf{M} = (M_1, \ldots, M_k)'$ are called *regression coefficients*, and in many such problems, the statistician must decide whether or not the values of certain regression coefficients are close to 0. As an example of the use of the results derived in Sec. 11.11, we shall now investigate the hypothesis that $M_{q+1} = \cdots = M_k = 0$, where q is a fixed number ($1 \leq q \leq k - 1$).

In this problem, we must minimize the function $Q(\mathbf{m})$, as specified by Eq. (3) of Sec. 11.11, among all vectors $\mathbf{m} \, \epsilon \, R^k$ such that

$$m_{q+1} = \cdots = m_k = 0.$$

We shall suppose that the vectors $\mathbf{m}$ and $\mathbf{\mu}$ and the matrices $\mathbf{T}$ and $\mathbf{T}^{-1}$ are partitioned as follows:

$$\mathbf{m} = \begin{bmatrix} \mathbf{m}_1 \\ \mathbf{m}_2 \end{bmatrix}, \quad \mathbf{\mu} = \begin{bmatrix} \mathbf{\mu}_1 \\ \mathbf{\mu}_2 \end{bmatrix}, \quad \mathbf{T} = \begin{bmatrix} \mathbf{T}_{11} & \mathbf{T}_{12} \\ \mathbf{T}_{21} & \mathbf{T}_{22} \end{bmatrix}, \quad \mathbf{T}^{-1} = \begin{bmatrix} \mathbf{V}_{11} & \mathbf{V}_{12} \\ \mathbf{V}_{21} & \mathbf{V}_{22} \end{bmatrix}. \quad (1)$$

Here, $\mathbf{m}_1$ and $\mathbf{\mu}_1$ are q-dimensional vectors and $\mathbf{T}_{11}$ and $\mathbf{V}_{11}$ are $q \times q$ matrices.

It is required to minimize $Q(\mathbf{m})$ among all vectors $\mathbf{m}$ such that $\mathbf{m}_2 = \mathbf{0}$, and this minimum value is specified by Eq. (5) of Sec. 11.11. In this problem, $\mathbf{b} - \mathbf{A}\mathbf{\mu} = -\mathbf{\mu}_2$ and $\mathbf{A}\mathbf{T}^{-1}\mathbf{A}' = \mathbf{V}_{22}$. However, by Eq. (14) of Sec. 5.4, we can write

$$\mathbf{V}_{22}^{-1} = \mathbf{T}_{22} - \mathbf{T}_{21}\mathbf{T}_{11}^{-1}\mathbf{T}_{12}. \quad (2)$$

Therefore, the minimum value of $Q(\mathbf{m})$ is

$$\mathbf{\mu}_2'(\mathbf{T}_{22} - \mathbf{T}_{21}\mathbf{T}_{11}^{-1}\mathbf{T}_{12})\mathbf{\mu}_2. \quad (3)$$

Furthermore, by Eq. (6) of Sec. 11.11, this minimum value is attained at the point $\mathbf{m}$ such that $\mathbf{m}_2 = \mathbf{0}$ and $\mathbf{m}_1$ is specified as follows:

$$\mathbf{m}_1 = \mathbf{\mu}_1 - \mathbf{V}_{12}\mathbf{V}_{22}^{-1}\mathbf{\mu}_2. \quad (4)$$

However, by Eq. (16) of Sec. 5.4,

$$V_{12}V_{22}^{-1} = -T_{11}^{-1}T_{12}. \tag{5}$$

Therefore, the minimum value of $Q(\mathbf{m})$ must be attained at the point $\mathbf{m}$ such that $\mathbf{m}_2 = \mathbf{0}$ and

$$\mathbf{m}_1 = \boldsymbol{\mu}_1 + T_{11}^{-1}T_{12}\boldsymbol{\mu}_2. \tag{6}$$

We shall suppose now that the posterior distribution of $\mathbf{M}$ is the multivariate t distribution derived at the end of Sec. 11.10 for which there are $n - k$ degrees of freedom, the location vector is $\hat{\mathbf{m}} = (\mathbf{x}'\mathbf{x})^{-1}\mathbf{x}'\mathbf{y}$, and the precision matrix is $(1/\hat{s}^2)\mathbf{x}'\mathbf{x}$. We shall also suppose that the k-dimensional vector $\mathbf{x}'\mathbf{y}$ and the $k \times k$ matrix $(\mathbf{x}'\mathbf{x})^{-1}$ are partitioned as follows:

$$\mathbf{x}'\mathbf{y} = \begin{bmatrix} \mathbf{u}_1 \\ \mathbf{u}_2 \end{bmatrix}, \qquad (\mathbf{x}'\mathbf{x})^{-1} = \begin{bmatrix} \mathbf{z}_{11} & \mathbf{z}_{12} \\ \mathbf{z}_{21} & \mathbf{z}_{22} \end{bmatrix}. \tag{7}$$

Here, $\mathbf{u}_1$ is a q-dimensional vector and $\mathbf{z}_{11}$ is a $q \times q$ matrix.

It now follows from Eqs. (2) and (3) that the minimum value Q^* of $Q(\mathbf{m})$ can be written in the following form:

$$Q^* = \frac{1}{\hat{s}^2} (\mathbf{z}_{21}\mathbf{u}_1 + \mathbf{z}_{22}\mathbf{u}_2)'\mathbf{z}_{22}^{-1}(\mathbf{z}_{21}\mathbf{u}_1 + \mathbf{z}_{22}\mathbf{u}_2). \tag{8}$$

Hence, by Eq. (10) of Sec. 11.11, the ratio γ can be approximated by the value $e^{Q^*/2}$. A large value of this ratio indicates that the value of the posterior p.d.f. of $\mathbf{M}$ is much greater at the point $\mathbf{M} = \hat{\mathbf{m}}$ than it is at any point $\mathbf{m}$ for which the last $k - q$ coefficients are 0.

11.13 ONE-WAY ANALYSIS OF VARIANCE

As another example, suppose that we must minimize the function $Q(\mathbf{m})$, as specified by Eq. (3) of Sec. 11.11, among all vectors $\mathbf{m} = (m_1, \ldots, m_k)'$ such that $m_1 = m_2 = \cdots = m_k$. Equations (5) and (6) can be used for finding the magnitude and the location of the minimum of $Q(\mathbf{m})$. However, in this example it is simpler to determine this magnitude and location directly by differentiation instead of by performing the matrix inversions required by the general equations.

If m denotes the common value of the k components of the vector $\mathbf{m}$, then

$$Q(\mathbf{m}) = \sum_{i=1}^{k} \sum_{j=1}^{k} \tau_{ij}(m - \mu_i)(m - \mu_j). \tag{1}$$

Here μ_i denotes the ith component of the location vector $\boldsymbol{\mu}$ and τ_{ij} denotes the (i, j)th component of the precision matrix $\mathbf{T}$. If Q' denotes the derivative of $Q(\mathbf{m})$ with respect to m, we obtain the following equation by differentiating Eq. (1):

$$Q' = 2m \sum_{i=1}^{k} \sum_{j=1}^{k} \tau_{ij} - \sum_{i=1}^{k} \sum_{j=1}^{k} (\mu_i + \mu_j)\tau_{ij}. \tag{2}$$

Let $\mathbf{e}$ denote the k-dimensional vector all components of which have the common value 1. Then it follows from Eq. (2) that $Q' = 0$ when

$$m = \frac{\mathbf{e}'T\mathbf{\mu}}{\mathbf{e}'T\mathbf{e}}. \tag{3}$$

The minimum value Q^* of $Q(\mathbf{m})$ can now be obtained by substituting this value of m into Eq. (1). The result is

$$Q^* = \mathbf{\mu}'T\mathbf{\mu} - \frac{(\mathbf{e}'T\mathbf{\mu})^2}{\mathbf{e}'T\mathbf{e}}. \tag{4}$$

As an application of these results, we shall consider a one-way analysis of variance. In this problem, there are $n = n_1 + \cdots + n_k$ observations Y_{ij} ($j = 1, \ldots, n_i$; $i = 1, \ldots, k$) with the following joint distribution: For any given values $\mathbf{m}$ and w of $\mathbf{M}$ and W, Y_{ij} has a normal distribution with mean m_i and precision w and all n observations are independent. In other words, for $i = 1, \ldots, k$, the sequence Y_{i1}, $\ldots$, Y_{in_i} is a random sample containing n_i observations from a normal distribution with mean m_i and precision w. If the observation vector $\mathbf{Y}$ is regarded as a single long n-dimensional vector, then it can be seen that in this problem the $n \times k$ matrix $\mathbf{x}$ in Eq. (1) of Sec. 11.10 will have the following form:

$$\mathbf{x} = \begin{bmatrix} 1 & 0 & \cdots & 0 \\ \cdots\cdots\cdots\cdots \\ 1 & 0 & \cdots & 0 \\ \hline 0 & 1 & \cdots & 0 \\ \cdots\cdots\cdots\cdots \\ 0 & 1 & \cdots & 0 \\ \hline \cdots\cdots\cdots\cdots \\ \cdots\cdots\cdots\cdots \\ \cdots\cdots\cdots\cdots \\ \hline 0 & 0 & \cdots & 1 \\ \cdots\cdots\cdots\cdots \\ 0 & 0 & \cdots & 1 \end{bmatrix} \begin{matrix} \left.\vphantom{\begin{matrix}1\\1\\1\end{matrix}}\right\} n_1 \\ \left.\vphantom{\begin{matrix}1\\1\\1\end{matrix}}\right\} n_2 \\ \cdot \\ \cdot \\ \left.\vphantom{\begin{matrix}1\\1\end{matrix}}\right\} n_k \end{matrix} \tag{5}$$

Also, $\mathbf{x}'\mathbf{x}$ will be a diagonal matrix the diagonal elements of which are $n_1, \ldots, n_k$.

It follows that if $\hat{\mathbf{m}}$ and $\hat{s}^2$ are defined by Eqs. (11) and (12) of Sec. 11.10, then

$$\hat{\mathbf{m}} = \bar{\mathbf{y}} = (\bar{y}_1, \ldots, \bar{y}_k)', \tag{6}$$

where $\bar{y}_i = (1/n_i)\sum_{j=1}^{n_i} y_{ij}$ for $i = 1, \ldots, k$, and

$$\hat{s}^2 = \frac{1}{n - k} \sum_{i=1}^{k} \sum_{j=1}^{n_i} (y_{ij} - \bar{y}_i)^2. \tag{7}$$

Therefore, if $\mathbf{M}$ and W have the improper prior joint density function specified by Eq. (10) of Sec. 11.10, then it follows from the discussion in that section that the posterior distribution of $\mathbf{M}$ will be a multivariate t distribution for which there are $n - k$ degrees of freedom, the location vector is $\bar{\mathbf{y}}$, and the precision matrix is a diagonal matrix whose ith diagonal element is $n_i/\hat{s}^2$.

From Eqs. (3) and (4), it can be seen that under this posterior distribution, the minimum value Q^* of $Q(\mathbf{m})$ among all vectors $\mathbf{m}$ for which $m_1 = \cdots = m_k$ is

$$Q^* = \frac{\sum_{i=1}^{k} n_i \bar{y}_i^2 - n\bar{\bar{y}}^2}{\hat{s}^2} = \frac{\sum_{i=1}^{k} n_i (\bar{y}_i - \bar{\bar{y}})^2}{\hat{s}^2} \tag{8}$$

where

$$\bar{\bar{y}} = \frac{1}{n} \sum_{i=1}^{k} \sum_{j=1}^{n_i} y_{ij}. \tag{9}$$

This minimum value Q^* is attained at the vector $\mathbf{m}$, all components of which have the common value $\bar{\bar{y}}$. Thus, for large values of n, the ratio of the posterior density of $\mathbf{M}$ at the point $\mathbf{M} = \bar{\bar{y}}$ to the maximum density among all points whose components are equal is approximately $e^{Q^*/2}$, where Q^* is specified by Eq. (8).

Higher-way analyses of variance can be studied in a similar fashion. In general, however, the details become somewhat more complicated because of the greater number of parameters and the greater number of subscripts needed to describe them.

Further Remarks and References

Problems of linear regression and analysis of variance have been studied by Raiffa and Schlaifer (1961), chap. 13, and by Tiao and Zellner (1964a,b), Zellner and Tiao (1964), Box and Tiao (1964), Box and Draper (1965), Tiao and Tan (1965), Zellner and Chetty (1965), Hill (1965), and Duncan (1965). The use of improper prior distributions in such problems has been discussed by Stone and Springer (1965).

EXERCISES

1. Suppose that $X_1, \ldots, X_n$ is a random sample from a Poisson distribution for which the value of the mean W is unknown. Suppose also that the prior distribution of W is a gamma distribution with parameters α and β ($\alpha > 1, \beta > 0$). Finally, suppose that it is desired to estimate the value of W and that the loss function L is specified by Eq. (13) of Sec. 11.2. Prove that the Bayes estimator of W is specified by Eq. (14) of Sec. 11.2, and prove that the Bayes risk is $1/(\beta + n)$.

2. Suppose that $X_1, \ldots, X_n$ is a random sample from a Bernoulli distribution for which the parameter W is unknown, and suppose that the prior distribution of W is a beta distribution with parameters α and β. If the value of W must be estimated when the loss L is specified by the equation $L(w, d) = (w - d)^2$ and if the sampling cost is c per observation, show that the optimal number of observations n is specified by the following equation:

$$n = \left[\frac{\alpha\beta}{c(\alpha + \beta)(\alpha + \beta + 1)} \right]^{\frac{1}{2}} - (\alpha + \beta).$$

3. Suppose that $X_1, \ldots, X_n$ is a random sample from a normal distribution with an unknown value of the mean W and a specified value of the precision r. Suppose also that the prior distribution of W is a normal distribution with mean μ and precision τ. If the value of W must be estimated when the loss L is specified by the equation $L(w, d) = (w - d)^2$ and if the sampling cost is c per observation, show that the optimal number of observations n is specified by the following equation:

$$n = \left(\frac{1}{cr} \right)^{\frac{1}{2}} - \frac{\tau}{r}.$$

4. Suppose that $X_1, \ldots, X_n$ is a random sample from a uniform distribution on the interval $(0, W)$, where the value of W is unknown and the prior distribution of W is a Pareto distribution with parameters w_0 and α ($\alpha > 2$). If the value of W must be estimated when the loss L is specified by the equation $L(w, d) = (w - d)^2$ and if the sampling cost is c per observation, show that the optimal number of observations n is specified by the following equation:

$$n = \left[\frac{2\alpha w_0^2}{c(\alpha - 2)} \right]^{\frac{1}{2}} - (\alpha - 1).$$

5. Let W be a random variable whose d.f. is G, and suppose that $E(|W|^\alpha) < \infty$, where $\alpha > 1$. Prove that the expectation $E(|W - d|^\alpha)$ is minimized when d is the unique number such that

$$\int_{w<d} (d - w)^{\alpha-1} \, dG(w) = \int_{w>d} (w - d)^{\alpha-1} \, dG(w).$$

6. Let W be a random variable with g.p.d.f. ξ, and suppose that ξ is symmetric with respect to the value v, so that $\xi(v + y) = \xi(v - y)$ for all values of y ($-\infty < y < \infty$). Suppose that Λ is a nonnegative convex function on the real line that is symmetric with respect to the value 0, and suppose also that for all values of d ($-\infty < d < \infty$),

$$R(d) = \int_{-\infty}^{\infty} \Lambda(w - d)\xi(w) \, d\nu(w) < \infty.$$

Prove that R is a convex function which is also symmetric with respect to the value v and that $R(d)$ is minimized at the value $d = v$.

7. Suppose that W is a parameter whose value must be estimated when the loss function L is specified by the following equation:

$$L(w, d) = \left(\frac{w - d}{d}\right)^2.$$

It is assumed that $E(W^2) < \infty$. If $E(W) \neq 0$, show that a Bayes estimate d is specified by the equation

$$d = \frac{E(W^2)}{E(W)}.$$

If $E(W) = 0$ and $E(W^2) > 0$, show that there is no Bayes estimate. Finally, show that in either case, the Bayes risk ρ^* is

$$\rho^* = \frac{\text{Var}(W)}{E(W^2)}.$$

8. Suppose that W is a parameter whose value must be estimated when the loss function L is specified as follows:

$$L(w, d) = \begin{cases} k_1(w - d) & \text{for } d \leq w, \\ k_2(d - w) & \text{for } d \geq w. \end{cases}$$

Here k_1 and k_2 are positive constants and it is assumed that $E(|W|) < \infty$. Show that a number d will be a Bayes estimate if, and only if, the following relations are satisfied:

$$\Pr(W \leq d) \geq \frac{k_1}{k_1 + k_2} \quad \text{and} \quad \Pr(W \geq d) \geq \frac{k_2}{k_1 + k_2}.$$

9. Let $X_1, \ldots, X_n$ be a random sample from a Bernoulli distribution for which the parameter W is unknown. Suppose that the value of W must be estimated when the loss function L is

$$L(w, d) = \frac{(w - d)^2}{w(1 - w)}.$$

Suppose also that the prior distribution of W is the uniform distribution on the interval $(0, 1)$. Show that the Bayes estimator δ^* is specified by the equation $\delta^*(X_1, \ldots, X_n) = \bar{X}$ and show that the Bayes risk is $1/n$.

10. Let $X_1, \ldots, X_n$ be a random sample from an exponential distribution for which the parameter W is unknown. Suppose that the value of $1/W$ must be estimated when the loss function L is

$$L(w, d) = \left(\frac{1}{w} - d\right)^2.$$

Suppose also that the prior distribution of W is a gamma distribution with parameters α and β such that $\alpha > 2$. (a) If the number of observations n is fixed, find a Bayes decision function and compute the Bayes risk. (b) If the cost per observation is c, show that the optimal number of observations n is specified by the following equation:

$$n = \frac{\beta}{[c(\alpha - 1)(\alpha - 2)]^{\frac{1}{2}}} - (\alpha - 1).$$

11. Suppose that $\mathbf{W} = (W_1, \ldots, W_k)'$ is a random vector for which the mean vector is $\mathbf{\mu}$ and the covariance matrix is $\mathbf{\Sigma}$. Show that for any fixed $k \times k$

matrix $\mathbf{A}$,

$$E[(\mathbf{W} - \mathbf{\mu})'\mathbf{A}(\mathbf{W} - \mathbf{\mu})] = \text{tr} \ (\mathbf{A}\Sigma).$$

12. Suppose that the random vector $\mathbf{X} = (X_1, \ldots, X_k)'$ has a multinomial distribution for which the parameters are n and $\mathbf{W} = (W_1, \ldots, W_k)'$. Suppose that the value of $\mathbf{W}$ must be estimated when the loss function L is

$$L(\mathbf{w}, \mathbf{d}) = \sum_{i=1}^{k} (w_i - d_i)^2.$$

Finally, suppose that the prior distribution of $\mathbf{W}$ is a Dirichlet distribution for which the parametric vector is $\mathbf{\alpha} = (\alpha_1, \ldots, \alpha_k)'$. Find the Bayes estimator, and show that the value of the Bayes risk is

$$\frac{\alpha_0^2 - \sum_{i=1}^{k}\alpha_i^2}{\alpha_0(\alpha_0 + 1)(\alpha_0 + n)},$$

where $\alpha_0 = \sum_{i=1}^{k}\alpha_i$.

13. Let $X_1, \ldots, X_n$ be a random sample from a normal distribution for which both the mean M and the precision R are unknown. Suppose that the value of M must be estimated and that when $M = m$, $R = r$, and the value of the estimate is d, the loss is

$$L(m, r, d) = (m - d)^2.$$

Suppose also that the prior joint distribution of M and R is a normal-gamma distribution as specified in Theorem 1 of Sec. 9.6, and assume that $\alpha > 1$ in this prior distribution. For any fixed number of observations n, find a Bayes decision function and show that the Bayes risk is $\beta/[(\alpha - 1)(\tau + n)]$.

14. Suppose that $X_1, \ldots, X_n$ is a random sample from a Bernoulli distribution for which the parameter W is unknown, and suppose also that the prior distribution of W is a beta distribution with parameters α and β. Show that when

$$1 - \alpha < \sum_{i=1}^{n} X_i < n + \beta - 1,$$

a generalized maximum likelihood estimator $\hat{w}(X_1, \ldots, X_n)$ of W is specified by the equation

$$\hat{w}(X_1, \ldots, X_n) = \frac{\sum_{i=1}^{n}X_i + \alpha - 1}{n + \alpha + \beta - 2}.$$

15. Suppose that $X_1, \ldots, X_n$ is a random sample from a Poisson distribution for which the mean W is unknown, and suppose also that the prior distribution of W is a gamma distribution with parameters α and β. Show that when $\alpha + \sum_{i=1}^{n}X_i > 1$, a generalized maximum likelihood estimator $\hat{w}(X_1, \ldots, X_n)$ of W is specified by the equation

$$\hat{w}(X_1, \ldots, X_n) = \frac{\sum_{i=1}^{n}X_i + \alpha - 1}{n + \beta}.$$

16. Suppose that $X_1, \ldots, X_n$ is a random sample from a normal distribution for which both the mean M and the precision R are unknown. Suppose also that the prior joint distribution of M and R is a normal-gamma distribution as specified in

Theorem 1 of Sec. 9.6. If μ' and β' are as defined in that theorem, show that generalized maximum likelihood estimators $\hat{m}(X_1, \ldots, X_n)$ and $\mathit{f}(X_1, \ldots, X_n)$ are specified by the equations $\hat{m}(X_1, \ldots, X_n) = \mu'$ and

$$\mathit{f}(X_1, \ldots, X_n) = \frac{\alpha + (n-1)/2}{\beta'}.$$

17. Consider the problem presented in Sec. 11.6. Prove that the posterior probability that $W = w_1$ will be greater than the prior probability p if, and only if, the magnitude of the difference between the observed value of $\bar{x}$ and the value specified in Eq. (5) of Sec. 11.6 is less than the following number:

$$\frac{1}{n}\left\{(\tau + n)\left[\tau(w_1 - \mu)^2 + \log\frac{\tau + n}{\tau}\right]\right\}^{\frac{1}{2}}.$$

18. Verify that Eq. (2) of Sec. 11.7 is correct.

19. Let $X_1, \ldots, X_n$ be a random sample from a Bernoulli distribution for which the value of the parameter W is unknown. Suppose that under the prior distribution of W, $\Pr(W = \frac{1}{2}) = p > 0$ and the remaining probability $1 - p$ is uniformly distributed over the interval $0 < W < 1$. Find the posterior probability that $W = \frac{1}{2}$ when $\Sigma_{i=1}^n X_i = y$, and show that this posterior probability is greater than the prior probability p if, and only if, the following inequality is satisfied:

$$\binom{n}{y}\left(\frac{1}{2}\right)^n > \frac{1}{n+1}.$$

20. Suppose that the random vector $\mathbf{X} = (X_1, \ldots, X_k)'$ has a multinomial distribution with parameters n and $\mathbf{W} = (W_1, \ldots, W_k)'$, where n is a specified positive integer and the values of the components of the vector $\mathbf{W}$ are unknown. Suppose also that the prior distribution of $\mathbf{W}$ satisfies the following three conditions: (a) $\Pr(W_1 = \frac{1}{2}) = p > 0$. (b) When $W_1 = \frac{1}{2}$, the joint distribution of $W_2, \ldots, W_k$ is a uniform distribution on the set of points $(w_2, \ldots, w_k)$ such that $w_i > 0$ $(i = 2, \ldots, k)$ and $\Sigma_{i=2}^k w_i = \frac{1}{2}$. (c) When $W_1 \neq \frac{1}{2}$, the joint distribution of $W_1, \ldots, W_k$ is a uniform distribution on the set of points $(w_1, \ldots, w_k)$ such that $w_i > 0$ $(i = 1, \ldots, k)$, $\Sigma_{i=1}^k w_i = 1$, and $w_1 \neq \frac{1}{2}$. Find the posterior distribution of the vector $\mathbf{W}$ when $X_i = x_i$ $(i = 1, \ldots, k)$, and show that

$$\frac{\Pr(W_1 = \frac{1}{2}|X_1 = x_1, \ldots, X_k = x_k)}{\Pr(W_1 \neq \frac{1}{2}|X_1 = x_1, \ldots, X_k = x_k)} = \frac{p}{1-p}\frac{n+k-1}{2^n(k-1)}\binom{n+k-2}{x_1}.$$

21. Let F be a distribution function on the real line for which the mean μ exists, and let the function T_F on the real line be defined by Eq. (9) of Sec. 11.8. (a) Prove that T_F is a nonnegative convex function which is strictly decreasing at any point s such that $T_F(s) > 0$. (b) Prove that T_F satisfies the relations (10) of Sec. 11.8.

22. If F is the d.f. of a normal distribution for which the mean is μ and the precision is τ, prove that the function T_F satisfies Eq. (3) of Sec. 11.9.

23. Suppose that the joint distribution of two random variables X and Y is as follows: The conditional distribution of X when $Y = y$ $(-\infty < y < \infty)$ is a normal distribution with mean y and variance σ_1^2, and the marginal distribution of Y is a normal distribution with mean μ and variance σ_2^2. Show that the marginal distribution of X is a normal distribution with mean μ and variance $\sigma_1^2 + \sigma_2^2$.

24. The mileage Y that can be obtained from a certain gasoline is a random variable which depends on the amount x of a certain chemical in the gasoline. For any given value of x, the mileage Y has a normal distribution for which the mean is $m_1 + m_2 x + m_3 x^2$ and the precision is w. Here $m_1, m_2, m_3,$ and w are the unknown

values of the parameters $\mathbf{M} = (M_1, M_2, M_3)'$ and W, which have an improper prior joint density of the form specified by Eq. (10) of Sec. 11.10. The values of ten observations, where x and y are measured in appropriate units, are shown in the table.

Amount x	Mileage y
0.10	10.98
0.20	11.14
0.30	13.17
0.40	13.34
0.50	14.39
0.60	14.63
0.70	15.66
0.80	13.78
0.90	15.43
1.00	18.37

(a) Find the posterior distribution of $\mathbf{M}$, and discuss the likelihood that $m_3 = 0$.
(b) Find the value of the Bayes estimator of W when the squared-error loss function is used.

25. The mean yield strengths of three different metallic materials are being studied. The yield strength Y_i of the ith material is normally distributed with mean m_i and precision w ($i = 1, 2, 3$). Here m_1, m_2, m_3, and w are the unknown values of the parameters $\mathbf{M} = (M_1, M_2, M_3)'$ and W, which have an improper prior joint density of the form specified by Eq. (10) of Sec. 11.10. When the yield strengths were measured (in appropriate units) for seven different pieces of material 1, for ten different pieces of material 2, and for five different pieces of material 3, the following values were obtained:

Material 1: 4.52, 3.63, 3.99, 5.00, 6.39, 3.22, 4.90
Material 2: 4.66, 7.04, 6.27, 4.19, 4.81, 6.65, 5.56, 4.60, 6.19, 6.16
Material 3: 6.27, 4.04, 2.87, 4.76, 4.38

(a) Find the posterior distribution of $\mathbf{M}$, and discuss the likelihood that $m_1 = m_2 = m_3$.
(b) Find the value of the Bayes estimator of W when the squared-error loss function is used.

sequential decisions

sequential
sampling

12.1 GAINS FROM SEQUENTIAL SAMPLING

In this chapter we shall consider statistical problems in which the statistician can take his observations X_1, X_2, . . . one at a time from some distribution involving a parameter W whose value is unknown. After each observation X_n, he can evaluate the information he has obtained about W from the observations X_1, . . . , X_n which have been taken up to that time, and he can decide whether to terminate the sampling process or to take another observation X_{n+1}. A sample obtained in this way is called a *sequential sample.*

Consider a statistical decision problem in which there is a fixed cost per observation, and suppose that the statistician can take observations sequentially and can decide after each observation whether he wants to choose a decision $d \, \epsilon \, D$ without further sampling or wants to defer the choice of a decision $d \, \epsilon \, D$ and buy another observation. Then he will typically be able to find a sampling procedure and a decision function which will have a smaller total risk than that of any procedure in which the statistician must commit himself to a fixed total number of observations before any observations are taken. In some problems, however, the cost per observation may be higher when the observations are taken one at a time than when all the observations are taken at one time. In

some other problems—we shall present some nontrivial examples later—
the statistician gains no advantage by being able to sample sequentially,
even when the cost per observation remains the same.

As an example of the gains that can be obtained by sequential sampling, consider a problem in which the statistician must decide whether
or not to purchase a certain lot of manufactured items. Suppose that
he can observe whether or not each item in a random sample of predetermined size is defective and that he must pay a certain sampling cost c
per item. Also, suppose that after the statistician has considered this
cost together with the loss function and his prior distribution, he decides to
take a random sample of 50 items from the lot and to buy the lot if, and
only if, the total number of defective items observed in the random sample
is not greater than 8.   If, however, the observations in the random sample
can be taken and paid for sequentially (i.e., one at a time), it is always
possible for the statistician to follow a procedure which will lead to the
same decision as the above procedure but which has a lower sampling cost.
Specifically, suppose that observations are taken one at a time and that
sampling will be stopped as soon as either the number of defective items
that have been observed exceeds 8 or the number of nondefective items
that have been observed is 42.   Except on rare occasions, one of these
events will occur before 50 observations have been taken.   If the first
event occurs, the statistician will decide not to buy the lot.   If the second
event occurs, he will decide to buy the lot.   In either case, the decision
will always agree with the decision based on all 50 observations.

To further illustrate the gains that may result from sequential
sampling, we shall consider the following statistical decision problem in
some detail.   Suppose that $\Omega = \{w_1, w_2\}$ has just two points and that
$D = \{d_1, d_2\}$ also has just two points.   Suppose further that the loss
function L is specified as follows:

$$L(w_1, d_1) = L(w_2, d_2) = 0,$$
$$L(w_1, d_2) = L(w_2, d_1) = b > 0.$$

Moreover, suppose that each observation X is a discrete random
variable for which the value of the p.f. $f_i(x) = \Pr(X = x | W = w_i)$ is as
follows:

$$f_1(1) = 1 - \alpha, \qquad f_1(2) = 0, \qquad f_1(3) = \alpha;$$
$$f_2(1) = 0, \qquad f_2(2) = 1 - \alpha, \qquad f_2(3) = \alpha. \tag{1}$$

Here α is a given number such that $0 < \alpha < 1$.   In other words, the
observed value of X can be only 1, 2, or 3.   Also, the probability that
$X = 1$ is positive only if $W = w_1$, the probability that $X = 2$ is positive
only if $W = w_2$, and the probability that $X = 3$ is the same under both
values of W.

Finally, suppose that the cost per observation is c, and let the prior distribution of W be specified as follows:

$$\Pr(W = w_1) = \xi = 1 - \Pr(W = w_2).$$

Because of the symmetry in the problem, we can assume without loss of generality that $\xi \leq \frac{1}{2}$.

We shall begin by computing the risk of the procedure which involves taking a fixed number of observations n. It follows from the relations (1) that after an observation X has been taken, the posterior probability that $W = w_1$ will be 1 if $X = 1$, will be 0 if $X = 2$, and will be ξ (that is, the same as the prior probability) if $X = 3$. Thus, the decision problem being considered here is quite special in the following way: After an observation has been taken, either the value of W becomes known or else the distribution of W remains just as it was before the observation was taken. It follows from the loss function in this problem that if the Bayes decision is chosen when $\Pr(W = w_1) = \xi$, then the expected loss will be ξb. Also, if the Bayes decision is chosen either when $\Pr(W = w_1) = 0$ or when $\Pr(W = w_1) = 1$, then the expected loss will be 0. Hence, if the Bayes decision is chosen after n observations $X_1, \ldots, X_n$ have been taken, the expected loss will be ξb if $X_i = 3$ for every observation $(i = 1, \ldots, n)$ and will be 0 if at least one of the observations is not equal to 3.

Regardless of whether $W = w_1$ or $W = w_2$, the probability that $X_i = 3$ for every observation $(i = 1, \ldots, n)$ is α^n. It follows that the total risk $\rho(n)$, including the sampling cost, for the optimal procedure when exactly n observations must be taken is

$$\rho(n) = \xi b \alpha^n + cn. \tag{2}$$

We shall assume that $\rho(1) < \rho(0)$ and that it is therefore worthwhile to take at least one observation. The optimal value n^* of n can then be approximated by regarding n as a continuous variable in Eq. (2) and differentiating $\rho(n)$ in order to find the minimizing value. This differentiation yields the following results:

$$n^* = \left[\log \frac{\xi b \log (1/\alpha)}{c} \right] \frac{1}{\log (1/\alpha)} \tag{3}$$

and

$$\rho(n^*) = \frac{c}{\log (1/\alpha)} \left[1 + \log \frac{\xi b \log (1/\alpha)}{c} \right]. \tag{4}$$

In the remainder of this discussion, we shall assume that n^* is the positive integer which minimizes the value of $\rho(n)$ specified by Eq. (2).

If the statistician can take the n^* observations sequentially, he can reduce the total risk because he can then stop sampling as soon as the value of one of the observations X_i is different from 3. In other words, he will have to take all n^* observations only when $X_i = 3$ for $i = 1, 2,$ $\ldots, n^* - 1$. Under such a sequential procedure, the posterior distribution when sampling terminates will be the same as it was for the procedure with the fixed sample size, and therefore the expected loss will also be the same. However, the number N of observations that will be taken is now a random variable. It follows from the relations (1) that the distribution of N will be the same whether $W = w_1$ or $W = w_2$. Hence,

$$E(N) = E(N|W = w_1) = E(N|W = w_2) < n^*. \tag{5}$$

The value of $E(N)$ can be obtained as follows:

$$E(N) = E(N|W = w_1) = \sum_{j=1}^{n^*} j \Pr(N = j|W = w_1)$$

$$= \sum_{j=1}^{n^*-1} j\alpha^{j-1}(1 - \alpha) + n^*\alpha^{n^*-1} = \frac{1 - \alpha^{n^*}}{1 - \alpha}. \tag{6}$$

The total risk $\bar{\rho}$ from the sequential procedure satisfies the following relation:

$$\bar{\rho} = \xi b\alpha^{n^*} + cE(N) < \rho(n^*). \tag{7}$$

From Eq. (6) and the approximate value of n^* given in Eq. (3), we can now obtain the following result:

$$\bar{\rho} = \frac{c}{1 - \alpha} + \frac{c}{\log (1/\alpha)} \left[1 - \frac{c}{\xi b(1 - \alpha)} \right]. \tag{8}$$

The previous assumption that $\rho(1) < \rho(0)$ is equivalent to the assumption that the expression inside the brackets in Eq. (8) is positive.

Finally, we shall consider the sequential sampling procedure in which the statistician continues taking observations, without any upper bound on the number that can be taken, until he obtains an observation X_i whose value is different from 3. If this procedure is adopted, the statistician can always, when he terminates sampling, choose a decision whose expected loss is 0. Therefore, the total risk ρ^* of this procedure is simply $cE(N)$, where N is again the random number of observations that are taken. For this procedure,

$$E(N) = \sum_{j=1}^{\infty} j\alpha^{j-1}(1 - \alpha) = \frac{1}{1 - \alpha}. \tag{9}$$

Hence,

$$\rho^* = \frac{c}{1 - \alpha}. \tag{10}$$

A comparison of Eqs. (10) and (8) shows that this sequential sampling procedure has a smaller total risk than the other procedures we have considered. However, it should be noted that this procedure may require more—indeed, many more—than n^* observations. For this reason, the suitability of the procedure depends heavily on the assumption that the cost per observation is constant. Under this assumption, it can be demonstrated that this procedure is actually optimal among all possible sampling and decision procedures. In other words, this procedure has a smaller total risk than that of any other procedure. We have assumed that it is worthwhile to take at least one observation X_1. If $X_1 \neq 3$, then the statistician has learned the value of W and the optimal procedure clearly is to stop sampling and to choose the appropriate decision in D. If $X_1 = 3$, the posterior distribution of W is the same as its prior distribution. Hence, when the statistician decides whether or not to take another observation, his position is essentially the same as that at the beginning of the sampling process. Thus, W has the same distribution, and unlimited observations are still available to the statistician at a cost of c per observation. The statistician has spent c units to obtain the first observation. This cannot be regained, regardless of whether or not sampling is continued, and therefore it is irrelevant to his current planning. The statistician must now compare the risk from immediately choosing a decision in D without sampling with the total risk that can be attained through further sampling. Again, the previous assumption that it is worthwhile to take at least one observation indicates that sampling should be continued. The iteration of this argument indefinitely leads to the conclusion that the optimal procedure for the statistician is to continue sampling so long as every observed value is 3 and to stop as soon as one observed value is not 3.

Since there is no limit on the number of observations that might be taken under this procedure, the statistician might spend a large amount of money on sampling costs in a situation in which the most he could lose from an incorrect decision is a relatively small amount b. Of course, the probability that he will need many observations is small. Nevertheless, if he is unlucky and the value of each of a large number of observations is 3, the best thing he can do is to forget about the amount he has already spent on sampling, to continue taking observations, and to hope that his luck will change.

The assumption that there is a constant cost c per observation in this problem is closely related to the idea that the statistician's resources are so vast that at any stage in the procedure he can ignore any amount of money, no matter how large, that he has already spent on sampling. In a practical situation, however, his resources may be limited. When necessary, he can take this factor into account by assuming that the cost of the observations varies. If the statistician is permitted to spend only the amount nc on sampling, a reasonable modification is to assume that each of the first n observations costs c units and that the cost of each subsequent observation is infinite.

Further Remarks and References

The analysis of statistical problems with sequential sampling was largely developed by Wald (1947). He later incorporated the possibility of sequential sampling into his general theory of statistical decisions [Wald (1950)]. Blackwell and Girshick (1954) give an extensive treatment of problems of this type. A wide class of sequential decision problems, called *dynamic-programming problems*, has been characterized and studied by Bellman (1957a). Jackson (1960) has compiled an extensive bibliography on sequential methods in statistics. Various special topics in this area are discussed in the monograph by Wetherill (1966).

12.2 SEQUENTIAL DECISION PROCEDURES

In this section we shall begin the development of the general theory of sequential statistical decision problems. Consider a decision problem specified by a parameter W whose values are in the parameter space Ω, a decision space D, and a loss function L. We shall suppose that before the statistician chooses a decision in D, he will be permitted to observe sequentially the values of a sequence of random variables $X_1, X_2, \ldots$. We shall suppose also that for any given value $W = w$, these observations are independent and identically distributed. It is then said that the observations are a *sequential random sample*. We shall suppose further that the conditional g.p.d.f. of each observation X_i when $W = w$ is $f(\cdot \,|w)$ and that the cost of observing the value of X_i, in its turn, is c_i.

A *sequential decision function*, or a *sequential decision procedure*, has two components. One component may be called a *sampling plan*, or a *stopping rule*. The statistician first specifies whether a decision in D should be chosen without any observations or whether at least one observation should be taken. If at least one observation is to be taken, the statistician specifies, for every possible set of observed values $X_1 = x_1$, $\ldots, X_n = x_n$ $(n \geq 1)$, whether sampling should stop and a decision in D chosen without further observations or whether another value X_{n+1} should be observed.

The second component of a sequential decision procedure may be called a *decision rule*. If no observations are to be taken, the statistician specifies the decision $d_0 \in D$ that is to be chosen. If at least one observation is to be taken, the statistician specifies the decision $\delta_n(x_1, \ldots, x_n) \in D$ that is to be chosen for each possible set of observed values $X_1 = x_1$, $\ldots$, $X_n = x_n$ after which sampling might be terminated.

As in the nonsequential decision problems previously considered, we shall let S denote the sample space of any particular observation X_i. For $n = 1, 2, \ldots$, we shall let $S^n = S \times S \times \cdots \times S$ (with n factors) be the sample space of the n observations $X_1, \ldots, X_n$, and we shall let S^∞ be the sample space of the infinite sequence of observations $X_1, X_2, \ldots$ (see Sec. 2.2). A sampling plan in which at least one observation is to be taken can be characterized by a sequence of subsets $B_n \in S^n$ ($n = 1, 2, \ldots$) which have the following interpretation: Sampling is terminated after the values $X_1 = x_1, \ldots, X_n = x_n$ have been observed if $(x_1, \ldots, x_n) \in B_n$. Another value X_{n+1} is observed if $(x_1, \ldots, x_n)$ $\notin B_n$. If there is some value r for which $B_r = S^r$ or, more generally, if $\Pr[(X_1, \ldots, X_n) \notin B_n \text{ for } n = 1, \ldots, r] = 0$, then sampling must stop after at most r observations have been taken. The specification of the sets B_n for any value of n such that $n > r$ then becomes irrelevant. Nevertheless, it is convenient to assume that the sets B_n will be defined for all values of n.

Each stopping set B_n can be regarded not only as a subset of S^n but also as a subset of S^r, for any value of r such that $r > n$, and as a subset of S^∞. When B_n is regarded as a subset of S^r, where $r > n$, B_n is a cylinder set. In other words, if $(x_1, \ldots, x_r) \in B_n$ and if $(y_1, \ldots, y_r)$ is any other point in S^r such that $y_i = x_i$ for $i = 1, \ldots, n$, then $(y_1, \ldots, y_r) \in B_n$, regardless of the values of the final $r - n$ components. At any particular time in our discussion, the space S^r, of which B_n is regarded as a subset, will be clearly specified by the context.

Suppose that at least one observation is to be taken with a given sampling plan, and let N denote the random total number of observations which will be taken before sampling is terminated. We shall let $\{N = n\}$ denote the set of points $(x_1, \ldots, x_n) \in S^n$ for which $N = n$. In other words, suppose that the values $X_1 = x_1$, $X_2 = x_2, \ldots$, $X_n = x_n$ are observed in sequence. Then sampling will be terminated after the value x_n has been observed (and not before) if, and only if, $(x_1, \ldots, x_n) \in \{N = n\}$. Hence, $\{N = 1\} = B_1$, and for $n > 1$,

$$\{N = n\} = (B_1 \cup \cdots \cup B_{n-1})^c B_n.$$

Similarly, we shall let $\{N \leq n\} = \bigcup_{i=1}^n \{N = n\}$ denote the subset of S^n for which $N \leq n$. The events $\{N = n\}$ and the event $\{N \leq n\}$ involve only the observations $X_1, \ldots, X_n$. Hence, these events are subsets of

S^n. As pointed out above, they can also be regarded as subsets of S^r for $r > n$. Furthermore, the event $\{N > n\} = \{N \leq n\}^c$ involves only the observations $X_1, \ldots, X_n$, and it also can be regarded as a subset of S^r for any value of r such that $r \geq n$.

For any prior g.p.d.f. ξ of W, we shall let $f_n(\cdot \mid \xi)$ denote the marginal joint g.p.d.f. of the observations $X_1, \ldots, X_n$. In other words, for $(x_1, \ldots, x_n) \in S^n$,

$$f_n(x_1, \ldots, x_n \mid \xi) = \int_\Omega f(x_1 \mid w) \, \cdots \, f(x_n \mid w) \xi(w) \, d\nu(w). \tag{1}$$

Furthermore, we shall let $F_n(\cdot \mid \xi)$ denote the marginal joint d.f. of $X_1, \ldots, X_n$. Hence, for any event $A \subset S^n$,

$$\Pr[(X_1, \ldots, X_n) \in A] = \int_A dF_n(x_1, \ldots, x_n \mid \xi). \tag{2}$$

We can now write the following equation:

$$\Pr\{N \leq n\} = \int_{\{N \leq n\}} dF_n(x_1, \ldots, x_n \mid \xi)$$

$$= \int_{\{N=1\}} dF_1(x_1 \mid \xi) + \int_{\{N=2\}} dF_2(x_1, x_2 \mid \xi)$$

$$+ \cdots + \int_{\{N=n\}} dF_n(x_1, \ldots, x_n \mid \xi). \tag{3}$$

The decision rule of a sequential decision procedure is characterized by a decision $d_0 \in D$ and a sequence of functions $\delta_1, \delta_2, \ldots$ with the following property: For any point $(x_1, \ldots, x_n) \in S^n$, the function δ_n specifies a decision $\delta_n(x_1, \ldots, x_n) \in D$. If the sampling plan specifies that an immediate decision in D is to be selected without any sampling, then the decision $d_0 \in D$ is chosen. If, on the other hand, the sampling plan specifies that at least one observation is to be taken and if the observed values $X_1 = x_1, \ldots, X_n = x_n$ satisfy the condition that $(x_1, \ldots, x_n) \in \{N = n\}$, then sampling is terminated and the decision $\delta_n(x_1, \ldots, x_n) \in D$ is chosen. The values of the function δ_n need only be specified on the subset $\{N = n\} \subset S^n$. However, in order that we may be able to consider the decision rule without explicitly describing the sampling plan, we shall find it convenient to assume that each function δ_n is a decision function whose values are specified on the entire space S^n.

Under a sequential decision procedure, as we have already remarked, the total number of observations N that are taken before a decision in D is chosen is a random variable. Of course, a procedure involving a fixed number of observations n can always be obtained by adopting a sampling plan in which $\{N = j\} = \emptyset$, the empty set, for $j = 1, \ldots, n-1$, and in which $\{N = n\} = S^n$. In general, we shall only consider sampling plans for which the probability is 1 that sampling will eventually be terminated. In other words, we shall assume that under the prior distribution of W, the following relation is satisfied:

$$\Pr(N < \infty) = \lim_{n \to \infty} \Pr(N \leq n) = 1. \tag{4}$$

It is reasonable to place such a restriction on any sequential decision procedure that might be used, since the cost of taking observations without ever terminating can be regarded as infinite. However, although Eq. (4) is satisfied, it need not be assumed that there is some finite upper bound n such that $\Pr(N \leq n) = 1$.

12.3 THE RISK OF A SEQUENTIAL DECISION PROCEDURE

The total risk $\rho(\xi, \delta)$ of a sequential decision procedure δ in which at least one observation is to be taken is

$$\rho(\xi, \delta) = E\{L[W, \delta_N(X_1, \ldots, X_N)] + c_1 + \cdots + c_N\}$$

$$= \sum_{n=1}^{\infty} \int_{\{N=n\}} \int_{\Omega} L[w, \delta_n(x_1, \ldots, x_n)]$$

$$\xi(w|x_1, \ldots, x_n) \, d\nu(w) \, dF_n(x_1, \ldots, x_n|\xi)$$

$$+ \sum_{n=1}^{\infty} (c_1 + \cdots + c_n) \Pr\{N = n\}. \tag{1}$$

Here $\xi(\cdot|x_1, \ldots, x_n)$ is the posterior g.p.d.f. of W after the values $X_1 = x_1, \ldots, X_n = x_n$ have been observed.

Alternatively, we could compute the expectation in Eq. (1) by reversing the order of integration. The following result would then be obtained:

$$\rho(\xi, \delta) = \int_{\Omega} \left\{ \sum_{n=1}^{\infty} \int_{\{N=n\}} L[w, \delta_n(x_1, \ldots, x_n)] \right.$$

$$\left. \prod_{i=1}^{n} f(x_i|w) \prod_{i=1}^{n} d\mu(x_i) \right\} \xi(w) \, d\nu(w)$$

$$+ \sum_{n=1}^{\infty} (c_1 + \cdots + c_n) \Pr\{N = n\}. \tag{2}$$

In our development of the theory of sequential statistical decision problems, we shall have little need to refer to any specific value $\xi(w|x_1, \ldots, x_n)$ of the posterior g.p.d.f. of W. However, we shall often have to refer to the entire posterior distribution as represented by its g.p.d.f. Therefore, we shall denote this posterior g.p.d.f. simply by

$\xi(x_1, \ldots, x_n)$. Specifically, if ξ is the prior g.p.d.f. of W, then the posterior g.p.d.f. of W when $X_1 = x_1, \ldots, X_n = x_n$ is $\xi(x_1, \ldots, x_n)$. For any g.p.d.f. ϕ of W, let $\rho_0(\phi)$ be defined as follows:

$$\rho_0(\phi) = \inf_{d \, \epsilon \, D} \int_\Omega L(w, d)\phi(w) \, d\nu(w). \tag{3}$$

In other words, $\rho_0(\phi)$ is the minimum risk from an immediate decision without any further observations when the g.p.d.f. of W is ϕ. We shall continue to assume that against each possible g.p.d.f. ϕ which might arise during the sampling process, there is a Bayes decision in D which actually yields the minimum risk $\rho_0(\phi)$.

A *Bayes sequential decision procedure*, or an *optimal sequential decision procedure*, is a procedure δ for which the risk $\rho(\xi, \delta)$ is minimized. When a statistician is searching for a Bayes sequential decision procedure, he need consider only procedures whose decision functions δ_n ($n = 1, 2, \ldots$) always specify Bayes decisions in D. In other words, if sampling is to be terminated after the values $x_1, \ldots, x_n$ have been observed, then the decision $\delta_n(x_1, \ldots, x_n)$ that is chosen should be Bayes against the posterior g.p.d.f. $\xi(x_1, \ldots, x_n)$ of W. Suppose that a given sequential decision procedure δ will not always lead to a Bayes decision when sampling is terminated. Then a procedure δ^*, which has the same stopping rule as δ but which does always lead to a Bayes decision, will have the property that $\rho(\xi, \delta^*) \leq \rho(\xi, \delta)$.

In accordance with the foregoing comments, we shall assume that the following statement is true for any sequential decision procedure to be considered. Whenever a decision in D is chosen after sampling has been terminated, that decision is a Bayes decision against the posterior distribution of W. Hence, in our discussion of any procedure, we shall not explicitly mention the decision rule. For any such procedure δ which specifies that at least one observation is to be taken, we now have the following relation:

$$\rho(\xi, \delta) = E\{\rho_0[\xi(X_1, \ldots, X_N)]\} + c_1 + \cdots + c_N\}. \tag{4}$$

Furthermore, for the procedure δ_0 which specifies that an immediate decision in D should be chosen without any observations, we have the relation $\rho(\xi, \delta_0) = \rho_0(\xi)$.

In the remainder of this chapter, we shall further restrict ourselves to the important special class of problems in which $c_i = c > 0$ for $i = 1, 2, \ldots$, that is, in which the cost per observation is constant. However, much of what we shall say about such problems in the subsequent sections of this chapter can be adapted without difficulty to more general sampling costs.

12.4 BACKWARD INDUCTION

It is said that a sequential decision procedure δ is *bounded* if there is a positive integer n such that $\Pr(N \leq n) = 1$. In this section and the next few sections, we shall consider problems in which there is a fixed upper bound n on the number of observations that can be taken. In other words, the statistician is restricted to bounded sequential decision procedures in which the number of observations is at most n.

When the statistician employs the technique of backward induction, which will be presented in the next section for finding the optimal bounded sequential decision procedure, he begins by considering the final stage of observation and he then works backward to the first stage of observation. Informally, the reason for using this technique is easily described. In order for the statistician to decide whether he should choose a decision in D without any observation at all or he should observe a value X_1, he must know how he will use X_1 if it is observed. The first question which he must consider is: Will he stop sampling after X_1 has been observed or will he continue and observe X_2? The answer to this question depends, in turn, on the benefit to be gained by observing X_2. This answer leads to the next question: If X_2 is observed, will sampling be terminated then or will it be continued with the observation of X_3? Continuing in this way, the statistician is ultimately led to the following question: If the values of $X_1, \ldots, X_{n-1}$ have been observed, should sampling be terminated without another observation or should the final observation X_n be taken?

Usually, it is not difficult to answer the last question. If X_n is observed, then the fixed limit on the number of observations makes it compulsory to terminate the sampling and to choose a decision in D. Hence, if sampling has not been stopped earlier, the statistician must determine at this final stage whether to choose a decision in D based on the values of $X_1, \ldots, X_{n-1}$ or to take exactly one more observation and then choose a decision in D. The optimal decision will usually depend on the values of $X_1, \ldots, X_{n-1}$ that have been observed.

After the statistician has determined, for each set of values of the observations $X_1, \ldots, X_{n-1}$, whether it is worthwhile to take the final observation X_n, he can begin to work backward. Knowing the values of the observations $X_1, \ldots, X_{n-2}$ at the next to last stage, he can now decide whether it is worthwhile to take the next observation X_{n-1}. Since he knows, for each possible value of X_{n-1}, the optimal procedure from that point, he can compare the risk from making a decision without further observations with the risk if X_{n-1} is observed. By working backward in this way to the first stage, the statistician determines, for each possible value of X_1, the optimal continuation throughout the remaining stages. He can therefore evaluate the risk from observing X_1 and continuing in an optimal fashion thereafter and can compare this risk with the risk from the immediate choice of a decision in D without any observations. These comparisons, at the first and subsequent stages, determine the optimal sequential decision procedure.

The construction of the optimal procedure by backward induction illustrates what Bellman (1957a) has called the *principle of optimality*, which can be stated as follows: The optimal sequential decision procedure must satisfy the requirement that if, at any stage of the procedure, the values $X_1 = x_1, \ldots, X_j = x_j$ $(j < n)$ have been observed, then the continuation of the procedure must be the optimal sequential decision procedure for the problem where the prior distribution of W is $\xi(x_1, \ldots, x_j)$ and the maximum number of observations that can be taken is $n - j$.

12.5 OPTIMAL BOUNDED SEQUENTIAL DECISION PROCEDURES

We shall now present a mathematical development based on the technique of backward induction which leads to the construction of an optimal sequential decision procedure. We are assuming that there is a fixed cost c per observation. If ϕ is any g.p.d.f. of W and if the value $\rho_0(\phi)$ is defined by Eq. (3) of Sec. 12.3, then the expectation $E\{\rho_0[\phi(X)]\}$ is given by the following equation:

$$E\{\rho_0[\phi(X)]\} = \int_S \rho_0[\phi(x)] \, dF_1(x|\phi). \tag{1}$$

Suppose now that the values $X_1 = x_1, \ldots, X_{n-1} = x_{n-1}$ have been observed and the statistician must decide whether to choose a decision in D without another observation or to observe X_n. If $\xi_{n-1} = \xi(x_1, \ldots, x_{n-1})$ denotes the posterior g.p.d.f. of W, then the risk from choosing a decision in D without another observation is $\rho_0(\xi_{n-1})$. If the value $X_n = x$ is observed and a decision in D is then chosen, the risk will be $\rho_0[\xi_{n-1}(x)]$. Hence, the expected total risk from observing X_n and then choosing a decision in D is $E\{\rho_0[\xi_{n-1}(X)]\} + c$. Here, the cost c of observing X_n has been added to the expected loss from the Bayes decision. Hence, when the values $X_1 = x_1, \ldots, X_{n-1} = x_{n-1}$ have been observed, the final choice in the optimal procedure will be as follows: If

$$\rho_0(\xi_{n-1}) < E\{\rho_0[\xi_{n-1}(X)]\} + c, \tag{2}$$

then sampling should be terminated and X_n should not be observed. If the inequality in relation (2) is reversed, then X_n should be observed. If the two sides of relation (2) are equal, then the risk is the same whether sampling is terminated or continued, and we shall assume for convenience that sampling is terminated in this case.

Let $\rho_1(\phi)$ denote the risk from the optimal procedure in which not more than one observation is taken and the g.p.d.f. of W is ϕ. The value of $\rho_1(\phi)$ is specified by the equation

$$\rho_1(\phi) = \min \{\rho_0(\phi), E[\rho_0(\phi(X))] + c\}. \tag{3}$$

In particular, it follows from the above discussion that $\rho_1(\xi_{n-1})$ is the risk from the optimal continuation after the values $X_1 = x_1, \ldots,$ $X_{n-1} = x_{n-1}$ have been observed.

Now we shall move back one stage. Suppose that the values $X_1 = x_1, \ldots, X_{n-2} = x_{n-2}$ have been observed and that $\xi_{n-2} = \xi(x_1, \ldots, x_{n-2})$ is the posterior g.p.d.f. of W. The risk from a decision in D without any further observations is $\rho_0(\xi_{n-2})$. On the other hand, if the value of X_{n-1} is observed and this value is x, the posterior g.p.d.f. becomes $\xi_{n-2}(x)$ and the risk from the optimal continuation at that stage is $\rho_1[\xi_{n-2}(x)]$. When the cost c of observing the value of X_{n-1} is added, the total risk from observing this value and then continuing in an optimal fashion becomes $E\{\rho_1[\xi_{n-2}(X)]\} + c$. Hence, after the values $X_1 = x_1, \ldots, X_{n-2} = x_{n-2}$ have been observed, the optimal procedure can be described as follows: If

$$\rho_0(\xi_{n-2}) \leq E\{\rho_1[\xi_{n-2}(X)]\} + c, \tag{4}$$

then sampling should be terminated and X_{n-1} should not be observed. If the relation (4) is not satisfied, then X_{n-1} should be observed.

Now let $\rho_2(\phi)$ be the risk from the optimal procedure when sampling is terminated after not more than two observations and the g.p.d.f. of W is ϕ. Then $\rho_2(\phi)$ satisfies the relation

$$\rho_2(\phi) = \min \{\rho_0(\phi), E[\rho_1(\phi(X))] + c\}. \tag{5}$$

In particular, it follows from the above discussion that $\rho_2(\xi_{n-2})$ is the risk from the optimal continuation after the values $X_1 = x_1, \ldots, X_{n-2} = x_{n-2}$ have been observed.

In general, for any g.p.d.f. ϕ of W, let $\rho_0(\phi)$ be defined by Eq. (3) of Sec. 12.3 and let $\rho_1(\phi), \rho_2(\phi), \ldots, \rho_n(\phi)$ be defined recursively by the following relation:

$$\rho_{j+1}(\phi) = \min \{\rho_0(\phi), E[\rho_j(\phi(X))] + c\} \quad j = 0, 1, \ldots, n - 1. \tag{6}$$

It is assumed that each expectation in Eq. (6) exists.

On the basis of the results developed in the foregoing discussion, we can state the following theorem.

Theorem 1 *If the prior g.p.d.f. of W is ξ, then $\rho_n(\xi)$ is the total risk from the optimal sequential decision procedure in which not more than n observa-*

tions can be taken. Furthermore, for $j = 1, \ldots, n - 1$, after the values $X_1 = x_1, \ldots, X_{n-j} = x_{n-j}$ have been observed and the posterior g.p.d.f. of W is ξ_{n-j}, the risk from the optimal continuation is $\rho_j(\xi_{n-j})$.

It should be emphasized that the risk $\rho_j(\xi_{n-j})$ does not include the cost $(n - j)c$ of taking the first $n - j$ observations. As mentioned earlier, this cost is irrelevant when the statistician has to decide whether or not further observations should be taken.

We can also state the following theorem, which describes the optimal procedure.

Theorem 2 *Among all sequential decision procedures in which not more than n observations can be taken, the following procedure is optimal: If $\rho_0(\xi) \leq \rho_n(\xi)$, a decision in D is chosen immediately without any observations. Otherwise, X_1 is observed. Furthermore, for $j = 1, \ldots, n - 1$, suppose that the values $X_1 = x_1, \ldots, X_j = x_j$ have been observed and that the posterior g.p.d.f. of W is ξ_j. If $\rho_0(\xi_j) \leq \rho_{n-j}(\xi_j)$, a decision in D is chosen without further observations. Otherwise, X_{j+1} is observed. If sampling has not been terminated earlier, it must be terminated after X_n has been observed.*

The optimal procedure described in Theorem 2 can be summarized as follows: Suppose that at a given stage of the sampling process, the statistician is not permitted to take more than j further observations and that the current g.p.d.f. of W at that stage is ϕ. Then another observation should be taken if, and only if, $\rho_j(\phi) < \rho_0(\phi)$.

The next theorem establishes a simple, but basic, property of the sequence of functions $\rho_0, \rho_1, \ldots$.

Theorem 3 *For any g.p.d.f. ϕ of W,*

$$\rho_n(\phi) \geq \rho_{n+1}(\phi) \qquad for\ n = 0, 1, 2, \ldots. \tag{7}$$

Proof The inequality (7) follows from an induction argument based on Eq. (6). Alternatively, this inequality also follows from the fact that $\rho_n(\phi)$ is the minimum risk among all procedures which are terminated after not more than n observations have been taken and $\rho_{n+1}(\phi)$ is the minimum risk among the larger class of all procedures which are terminated after not more than $n + 1$ observations have been taken. ∎

12.6 ILLUSTRATIVE EXAMPLES

Despite the simple theoretical nature of the optimal procedure just developed, the computation of the function ρ_n for a value of n larger than

4 or 5 is, in many problems, extremely difficult and time-consuming. We shall now present four examples whose solutions are of varying levels of computational difficulty. Each example will provide further insight into the properties of optimal sequential decision procedures.

EXAMPLE 1 Suppose that a sequential random sample $X_1, X_2, \ldots$ can be taken from the Bernoulli distribution for which the value of the parameter W is unknown. Suppose also that either $W = \frac{1}{3}$ or $W = \frac{2}{3}$ and that the statistician must decide which value of W is correct. Therefore, it is assumed that the decision space $D = \{d_1, d_2\}$ and that the loss function is as specified in Table 12.1. Suppose further that each observation costs 1 unit and that the prior distribution of W is specified by the probability $\xi = \Pr(W = \frac{1}{3}) = 1 - \Pr(W = \frac{2}{3})$. We shall compute the values of $\rho_0(\xi)$, $\rho_1(\xi)$, and $\rho_2(\xi)$.

Table 12.1

	d_1	d_2
$W = \frac{1}{3}$	0	20
$W = \frac{2}{3}$	20	0

It is seen from Table 12.1 that d_2 is a Bayes decision if $0 \leq \xi \leq \frac{1}{2}$ and that d_1 is a Bayes decision if $\frac{1}{2} \leq \xi \leq 1$. Hence, it follows that

$$\rho_0(\xi) = \begin{cases} 20\xi & \text{for } 0 \leq \xi \leq \frac{1}{2}, \\ 20(1 - \xi) & \text{for } \frac{1}{2} \leq \xi \leq 1. \end{cases} \tag{1}$$

As Eq. (1) indicates, $\rho_0(\xi) = \rho_0(1 - \xi)$ for $0 \leq \xi \leq 1$. Furthermore, from the symmetry of the problem, it follows that $\rho_j(\xi) = \rho_j(1 - \xi)$ for $0 \leq \xi \leq 1$ and for every value of j ($j = 1, 2, \ldots$). Hence, we shall compute $\rho_1(\xi)$ and $\rho_2(\xi)$ only for values of ξ such that $0 \leq \xi \leq \frac{1}{2}$.

For $x = 0, 1$, let $\xi(x)$ denote the posterior probability that $W = \frac{1}{3}$ when the value of a single observation X is x. By Bayes' theorem,

$$\xi(1) = \frac{\xi}{\xi + 2(1 - \xi)} \quad \text{and} \quad \xi(0) = \frac{2\xi}{2\xi + (1 - \xi)}. \tag{2}$$

From Eqs. (1) and (2), we can now obtain the following results:

$$\begin{aligned} \rho_0[\xi(1)] &= 20\xi(1) & \text{for } 0 \leq \xi \leq \frac{1}{2}; \\ \rho_0[\xi(0)] &= \begin{cases} 20\xi(0) & \text{for } 0 \leq \xi \leq \frac{1}{3}, \\ 20[1 - \xi(0)] & \text{for } \frac{1}{3} < \xi \leq \frac{1}{2}. \end{cases} \end{aligned} \tag{3}$$

The marginal distribution of any observation X is

$$\Pr(X = 1) = \tfrac{1}{3}\xi + \tfrac{2}{3}(1 - \xi) = 1 - \Pr(X = 0). \tag{4}$$

Hence, an elementary computation yields the equation

$$E\{\rho_0[\xi(X)]\} = \rho_0[\xi(1)]\Pr(X = 1) + \rho_0[\xi(0)]\Pr(X = 0)$$
$$= \begin{cases} 20\xi & \text{for } 0 \leq \xi \leq \tfrac{1}{3}, \\ \tfrac{20}{3} & \text{for } \tfrac{1}{3} < \xi \leq \tfrac{1}{2}. \end{cases} \tag{5}$$

Since $c = 1$, it now follows that

$$\rho_1(\xi) = \min \{\rho_0(\xi), E[\rho_0(\xi(X))] + 1\}$$
$$= \begin{cases} 20\xi & \text{for } 0 \leq \xi \leq \tfrac{23}{60}, \\ \tfrac{23}{3} & \text{for } \tfrac{23}{60} < \xi \leq \tfrac{1}{2}. \end{cases} \tag{6}$$

In order to find ρ_2, we make use of the following information. From Eq. (2) it can be shown that $\xi(1) \leq \tfrac{23}{60}$ if, and only if, $\xi \leq \tfrac{46}{83}$; that $\xi(0) \leq \tfrac{23}{60}$ if, and only if, $\xi \leq \tfrac{23}{97}$; and that $\xi(0) \geq \tfrac{37}{60}$ if, and only if, $\xi \geq \tfrac{37}{83}$. A computation which is based on Eqs. (4) and (6) and which makes use of the symmetry of the function ρ_1 then yields the following result:

$$E\{\rho_1[\xi(X)]\} = \begin{cases} 20\xi & \text{for } 0 \leq \xi \leq \tfrac{23}{97}, \\ \dfrac{83\xi + 23}{9} & \text{for } \tfrac{23}{97} < \xi \leq \tfrac{37}{83}, \\ \tfrac{20}{3} & \text{for } \tfrac{37}{83} < \xi \leq \tfrac{1}{2}. \end{cases} \tag{7}$$

Hence,

$$\rho_2(\xi) = \min \{\rho_0(\xi), E[\rho_1(\xi(X))] + 1\}$$
$$= \begin{cases} 20\xi & \text{for } 0 \leq \xi \leq \tfrac{23}{97}, \\ \dfrac{83\xi + 32}{9} & \text{for } \tfrac{23}{97} < \xi \leq \tfrac{37}{83}, \\ \tfrac{23}{3} & \text{for } \tfrac{37}{83} < \xi \leq \tfrac{1}{2}. \end{cases} \tag{8}$$

The functions ρ_0, ρ_1, and ρ_2 are sketched in Fig. 12.1. The computation of ρ_n for a larger value of n is more difficult because the number of linear segments defining the function increases as n becomes larger. Furthermore, if any of the symmetry that has been assumed in this example is relinquished, the computation requires still more detail. However, when there is no upper bound on the number of observations that can be taken, this particular example can be solved by the methods to be presented in Secs. 12.14 to 12.16.

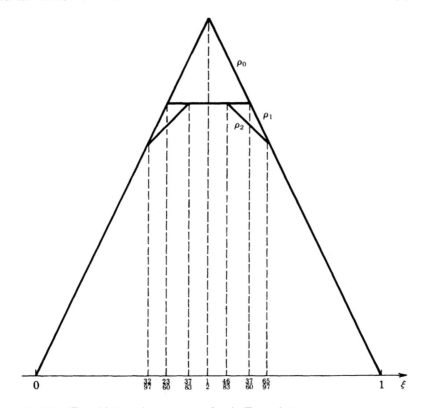

Fig. 12.1 The risk functions ρ_0, ρ_1, and ρ_2 in Example 1.

EXAMPLE 2 Suppose that the conditions are the same as in Example 1 except for the values of the loss function, which are now reduced to those given in Table 12.2.

Table 12.2

	d_1	d_2
$W = \frac{1}{3}$	0	10
$W = \frac{2}{3}$	10	0

Since all the symmetric features of Example 1 are retained here, it is again true that $\rho_j(\xi) = \rho_j(1 - \xi)$ for $0 \leq \xi \leq 1$ and for $j = 0, 1, 2, \ldots$. In this example, $\rho_0(\xi) = 10\xi$ for $0 \leq \xi \leq \frac{1}{2}$. Furthermore, by performing a computation like that in Example 1, we can obtain the following result:

$$E\{\rho_0[\xi(X)]\} = \begin{cases} 10\xi & \text{for } 0 \le \xi \le \frac{1}{3}, \\ \frac{10}{3} & \text{for } \frac{1}{3} < \xi \le \frac{1}{2}. \end{cases} \tag{9}$$

Therefore, by Eq. (3) of Sec. 12.5,

$$\rho_1(\xi) = \begin{cases} 10\xi & \text{for } 0 \le \xi \le \frac{13}{30}, \\ \frac{13}{3} & \text{for } \frac{13}{30} < \xi \le \frac{1}{2}. \end{cases} \tag{10}$$

From Eq. (2) it can be shown that $\xi(1) \le \frac{13}{30}$ if, and only if, $\xi \le \frac{26}{43}$; that $\xi(0) \le \frac{13}{30}$ if, and only if, $\xi \le \frac{13}{47}$; and that $\xi(0) \ge \frac{17}{30}$ if, and only if, $\xi \ge \frac{17}{43}$. Hence, it can be shown that

$$E\{\rho_1[\xi(X)]\} = \begin{cases} 10\xi & \text{for } 0 \le \xi \le \frac{13}{47}, \\ \dfrac{43\xi + 13}{9} & \text{for } \frac{13}{47} < \xi \le \frac{17}{43}, \\ \frac{10}{3} & \text{for } \frac{17}{43} < \xi \le \frac{1}{2}. \end{cases} \tag{11}$$

By Eq. (5) of Sec. 12.5,

$$\rho_2(\xi) = \begin{cases} 10\xi & \text{for } 0 \le \xi \le \frac{13}{30}, \\ \frac{13}{3} & \text{for } \frac{13}{30} < \xi \le \frac{1}{2}. \end{cases} \tag{12}$$

From these results, it can be seen that $\rho_1(\xi) = \rho_2(\xi)$ for all values of ξ $(0 \le \xi \le 1)$. This is, of course, a highly special result. Moreover, it now follows from Eq. (6) of Sec. 12.5 that for $0 \le \xi \le 1$,

$$\begin{aligned} \rho_3(\xi) &= \min \{\rho_0(\xi), E[\rho_2(\xi(X))] + 1\} \\ &= \min \{\rho_0(\xi), E[\rho_1(\xi(X))] + 1\} \\ &= \rho_2(\xi). \end{aligned} \tag{13}$$

Also, for $n = 2, 3, \ldots$, it follows by induction that $\rho_n(\xi) = \rho_1(\xi)$.

The result we have just obtained has the following interpretation. For any prior probability ξ, no matter how large the upper bound on the number of available observations may be, the statistician should not take more than one observation. If $\frac{13}{30} < \xi < \frac{17}{30}$, the statistician should take the observation and then choose a decision in D. Otherwise, he should immediately choose a decision in D without taking any observations.

EXAMPLE 3 Let $X_1, X_2, \ldots$ be a sequential random sample from a normal distribution with unknown mean W and specified precision r. Suppose that the value of W must be estimated when the loss function is specified by the equation $L(w, d) = (w - d)^2$. Suppose also that there is a fixed cost c per observation and that the prior distribution ξ of W is a normal distribution with mean μ and precision τ.

From Sec. 11.2 it can be seen that the risk from choosing a decision without taking any observations is $\rho_0(\xi) = 1/\tau$. Furthermore, for any set of observed values $X_1 = x_1, \ldots, X_n = x_n$, the precision of the posterior distribution $\xi(x_1, \ldots, x_n)$ of W will be $\tau + n\tau$. Hence, the risk from choosing a decision at that stage in the procedure is $\rho_0[\xi(x_1, \ldots, x_n)] = 1/(\tau + n\tau)$. The significance of this result is that the risk ρ_0 depends only on the number of observations that have been taken and not on the observed values. Moreover, it follows from an induction argument based on Eq. (6) of Sec. 12.5 that every function ρ_m ($m = 1, 2, \ldots$), as well as ρ_0, has this property. Thus, for each value of m ($m = 0, 1, 2, \ldots$) and each value of n ($n = 1, 2, \ldots$), there is a constant α_{mn} such that, for all possible observed values $(x_1, \ldots, x_n) \,\epsilon\, S^n$,

$$\rho_m[\xi(x_1, \ldots, x_n)] = \alpha_{mn}. \tag{14}$$

Now consider again Theorem 2 of Sec. 12.5, in which we described the optimal sequential decision procedure among those that terminate after not more than n observations have been taken. Suppose that at a given stage of the procedure, the values $X_1 = x_1, \ldots, X_j = x_j$ have been observed, where $1 \leq j \leq n$. Then it follows from the description of the optimal procedure and from Eq. (14) that no further observations should be taken if $\alpha_{0,j} \leq \alpha_{n-j,j}$ and that another observation should be taken if $\alpha_{0,j} > \alpha_{n-j,j}$. Hence, whether or not another observation should be taken depends only on the numbers j and n and not on the observed values $x_1, \ldots, x_j$. In other words, the optimal sequential decision procedure is, in fact, a procedure in which a fixed number of observations will be taken.

The optimal fixed sample number n^* for this example was found approximately in Exercise 3 of Chap. 11. Thus the procedure in which exactly n^* observations are taken is, in fact, optimal among all sequential decision procedures. Of course, if n^* exceeds the maximum number of observations n that the statistician is permitted to take, then he should take exactly n observations. An exact expression for the optimal number of observations in this example is given in Exercise 3.

Before leaving this example, we shall state the following general theorem, which has now been developed for any sequential decision problem.

Theorem 1 *Let ξ be the prior distribution of the parameter W in a sequential decision problem. For each value of n ($n = 1, 2, \ldots$), suppose that the value of $\rho_0[\xi(x_1, \ldots, x_n)]$ is constant for all observed values $(x_1, \ldots, x_n) \,\epsilon\, S^n$. Then the optimal sequential decision procedure is a procedure in which a fixed number of observations will be taken.*

EXAMPLE 4   In contrast to the estimation problem just presented, we shall consider now the problem of testing hypotheses which we previously considered in some detail in Sec. 11.9.   Suppose that a sequential random sample X_1, X_2, . . . can be taken, at a fixed cost c per observation, from a normal distribution with unknown mean W and specified precision r and that the statistician must decide whether or not W is smaller than a given value w_0.   Thus there are two decisions $D = \{d_1, d_2\}$.   In this example, the loss function L is specified by Eq. (1) of Sec. 11.8.

When the statistician is restricted to taking a fixed number of observations which he must specify before he begins sampling, the optimal fixed number n is specified approximately by Eq. (10) of Sec. 11.9.   In this example, the total risk can be reduced through sequential sampling. However, as we shall now show, even the computation of ρ_2 is quite difficult, and it does not seem possible to develop an explicit expression for the optimal sequential decision procedure.

Suppose that the prior distribution ξ of W is a normal distribution with mean μ and precision τ.   From Eq. (11) of Sec. 11.8 and from the discussion following that equation, it can be seen that the risk $\rho_0(\xi)$ from choosing a decision in D without any observations is

$$\rho_0(\xi) = \begin{cases} T_\xi(w_0) & \text{for } \mu \leq w_0, \\ T_\xi(w_0) - (\mu - w_0) & \text{for } \mu > w_0. \end{cases} \tag{15}$$

By Eq. (3) of Sec. 11.9, we can rewrite $\rho_0(\xi)$ as follows:

$$\rho_0(\xi) = \tau^{-\frac{1}{2}}\Psi(\tau^{\frac{1}{2}}|w_0 - \mu|). \tag{16}$$

To obtain Eq. (16), we have also used the following identity (see Exercise 8) for $-\infty < x < \infty$:

$$\Psi(-x) = \Psi(x) + x. \tag{17}$$

The risk $E\{\rho_0[\xi(X)]\}$ from taking one observation and then choosing a decision in D is equal to the value $\rho^*(\xi)$ obtained from Eq. (5) of Sec. 11.9 when $n = 1$.   Since $E\{\rho_0[\xi(X)]\}$, as well as every function $\rho_n(\xi)$, is a symmetric function of μ with respect to the point $\mu = w_0$, we can write the following equation:

$$E\{\rho_0[\xi(X)]\} = \rho_0(\xi) - \tau_1^{-\frac{1}{2}}\Psi(\tau_1^{\frac{1}{2}}|w_0 - \mu|), \tag{18}$$

where $\tau_1 = \tau(\tau + r)/r$.   It now follows from Eqs. (16) and (18) that if

$$c < \tau_1^{-\frac{1}{2}}\Psi(\tau_1^{\frac{1}{2}}|w_0 - \mu|), \tag{19}$$

then $\rho_1(\xi) = E\{\rho_0[\xi(X)]\} + c$.   Otherwise, $\rho_1(\xi) = \rho_0(\xi)$.

The computation of $E\{\rho_1[\xi(X)]\}$ is quite difficult, and we shall not discuss this example further at this time.   We shall return to it in Sec.

12.11. It has been studied by Chernoff (1961a; 1965a,b), Breakwell and Chernoff (1962), Lindley (1961a), Bather (1962), and Lindley and Barnett (1965). Effective approximations to the optimal procedure have been developed.

12.7 UNBOUNDED SEQUENTIAL DECISION PROCEDURES

We shall now remove the restriction that not more than a certain maximum number of observations can be taken. In other words, we shall now consider decision problems in which the statistician may choose any sequential decision procedure from the class Δ of all procedures that will ultimately terminate, that is, all procedures that satisfy Eq. (4) of Sec. 12.2.

In a sequential decision problem of this type, where there is no given upper bound on the number of observations, the technique of backward induction is not directly applicable in the construction of an optimal sequential decision procedure. In fact, before attempting to construct an optimal sequential decision procedure, we must first establish that there exists an optimal procedure in the class Δ. We can then consider the properties of this optimal procedure and discuss techniques for its construction. It should be emphasized that in some problems, the statistician is unrestricted in his choice of a decision procedure, but the optimal sequential decision procedure in the class Δ is really a bounded procedure. However, it is sometimes difficult to establish even this property.

We shall continue to use much of the notation developed in Secs. 12.2 to 12.5, and we shall consider a general sequential decision problem that is the same as the general problem treated in those sections. Thus we shall suppose that each of the observations $X_1, X_2, \ldots$ costs c units and that $\rho_0[\xi(x_1, \ldots, x_n)]$ is the risk from a Bayes decision in D when the posterior g.p.d.f. of the parameter W is $\xi(x_1, \ldots, x_n)$. Since we are assuming here that the loss function L is nonnegative, it follows that $\rho_0(\xi) \geq 0$ for every g.p.d.f. ξ. In establishing the existence of an optimal sequential decision procedure, the statistician must assume the existence of a suitable lower bound on the function ρ_0. However, it is sufficient that ρ_0 be bounded below by an appropriate function, and the assumption made here that ρ_0 is nonnegative could be somewhat weakened.

The Equivalence between Sequential Decision Procedures and Stopping Variables

We are assuming that the statistician always chooses a Bayes decision in D when sampling is terminated. Therefore, any sequential decision procedure δ is completely determined by the choice of the stopping rule. For any infinite sequence of values $x_1, x_2, \ldots$, we shall let $\delta(x_1, x_2, \ldots)$ denote the total number of observations that will be taken before sampling is terminated when the values $X_1 = x_1, X_2 = x_2, \ldots$ are observed sequentially. Thus, if $\delta(x_1, x_2, \ldots) = n$, for some positive integer n, then sampling will terminate after the nth observation and not before.

The decision to continue or to stop sampling after the first n observations have been obtained can obviously depend only on the values of these n observations and not on the values of any future observations. Therefore, if $\delta(x_1, x_2, \ldots) = n$ for some infinite sequence $x_1, x_2, \ldots$, and if $y_1, y_2, \ldots$ is any other infinite sequence such that $x_i = y_i$ for $i = 1, \ldots, n$, then it must also be true that $\delta(y_1, y_2, \ldots) = n$. In particular, the decision to take the first observation or to choose a decision in D without any observations obviously must be made before any observations have been taken. Therefore, either $\delta(x_1, x_2, \ldots) = 0$ for all sequences $x_1, x_2, \ldots$ or there is no sequence $x_1, x_2, \ldots$ for which this equation is correct.

A function δ which is defined on the space S^∞ of all infinite sequences and which satisfies these properties is called a *stopping variable*. If δ is any sequential decision procedure in the class Δ, then, with probability 1, sampling will eventually be terminated. Therefore, the values $\delta(x_1, x_2, \ldots)$ of the stopping variable will be finite for every infinite sequence $x_1, x_2, \ldots$, with the possible exception of some sequences which form an event whose probability is 0. Conversely, any stopping variable δ whose values are finite, except for an event whose probability is 0, determines a sequential decision procedure in the class Δ. Thus there is an equivalence between sequential decision procedures and stopping variables. Because of this equivalence, we can use the same symbol δ to denote both the procedure and the corresponding stopping variable.

Furthermore, the stopping variable δ can be interpreted as the random number of observations which will be required by a given procedure. Hence, we shall use the same symbol δ to denote both a sequential decision procedure and the random number of observations which will be required by that procedure. For any procedure $\delta \in \Delta$ and any prior distribution ξ of W, the total risk $\rho(\xi, \delta)$ can now be expressed in the following form:

$$\rho(\xi, \delta) = E\{\rho_0[\xi(X_1, \ldots, X_\delta)] + c\delta\}. \tag{1}$$

After the values $X_1 = x_1, \ldots, X_n = x_n$ have been observed in accordance with the procedure δ, then the expected risk from continuing with the procedure δ, including the cost nc of the first n observations, can be denoted by $E[\rho(\xi, \delta)|x_1, \ldots, x_n]$. We have already remarked that the events $\{\delta = n\}$ and $\{\delta > n\}$ can be regarded as subsets of the space S^n and also as subsets of the space S^r, for $r > n$, and of the space S^∞. If $(x_1, \ldots, x_n) \in \{\delta = n\}$, then

$$E[\rho(\xi, \delta)|x_1, \ldots, x_n] = \rho_0[\xi(x_1, \ldots, x_n)] + nc. \tag{2}$$

In the next two sections, we shall prove that there exists an optimal decision procedure $\delta^* \in \Delta$, that is, a procedure δ^* such that

$$\rho(\xi, \delta^*) = \inf_{\delta \in \Delta} \rho(\xi, \delta). \tag{3}$$

12.8 REGULAR SEQUENTIAL DECISION PROCEDURES

A decision procedure $\delta \in \Delta$ is called *regular* if $\rho(\xi, \delta) \leq \rho_0(\xi)$ and if, for every point $(x_1, \ldots, x_n) \in \{\delta > n\}$, the following relation is satisfied for $n = 1, 2, \ldots$:

$$E[\rho(\xi, \delta)|x_1, \ldots, x_n] < \rho_0[\xi(x_1, \ldots, x_n)] + nc. \tag{1}$$

In other words, if δ specifies that at least one observation should be taken, then the risk from δ is smaller than the risk from choosing a decision in D without any observations. Furthermore, whenever δ specifies that another observation should be taken, the expected total risk from continuing must be smaller than the risk from stopping. These comments suggest that if a procedure is not regular, then there may be stages at which the procedure specifies that sampling should be continued although the total risk would be reduced by stopping. For this reason, as we shall prove in the next theorem, the statistician need consider only regular decision procedures.

Theorem 1 *If $\delta \in \Delta$ is any decision procedure which is not regular, then there exists a regular decision procedure $\delta' \in \Delta$ such that $\rho(\xi, \delta') \leq \rho(\xi, \delta)$.*

Proof If $\rho(\xi, \delta) \geq \rho_0(\xi)$, then we can define δ' to be the regular procedure in which a decision in D is chosen without any observations.

Suppose, therefore, that δ specifies that at least one observation should be taken and that $\rho(\xi, \delta) < \rho_0(\xi)$. Let δ' be the procedure which specifies that sampling should be stopped as soon as the observed values $X_1 = x_1, \ldots, X_n = x_n$ do not satisfy the relation (1).

If $(x_1, \ldots, x_n) \in \{\delta = n\}$, then both sides of the relation (1) are equal. Hence, $(x_1, \ldots, x_n) \in \{\delta' \leq n\}$. In other words, the procedure δ' never requires more observations than the procedure δ. Since $\delta \in \Delta$, it follows that $\delta' \in \Delta$. Furthermore, from the definition of the procedure

δ' and from the basic properties of expectation, we can obtain the following relation:

$$
\begin{aligned}
\rho(\xi, \delta') &= \sum_{n=1}^{\infty} \int_{\{\delta'=n\}} \{\rho_0[\xi(x_1, \ldots, x_n)] + nc\} \\
&\qquad\qquad\qquad\qquad\qquad\qquad \times dF_n(x_1, \ldots, x_n|\xi) \\
&\leq \sum_{n=1}^{\infty} \int_{\{\delta'=n\}} E[\rho(\xi, \delta)|x_1, \ldots, x_n] \, dF_n(x_1, \ldots, x_n|\xi) \\
&\qquad - \sum_{n=1}^{\infty} E[\rho(\xi, \delta)|\delta' = n]\mathrm{Pr}(\delta' = n) \\
&= \rho(\xi, \delta).
\end{aligned}
\tag{2}
$$

Hence, $\rho(\xi, \delta') \leq \rho(\xi, \delta)$.

It remains to prove that δ' is a regular procedure. If $(x_1, \ldots, x_n) \in \{\delta' > n\}$, then relation (1) is satisfied, and furthermore, a development similar to that given in relation (2) shows that

$$
E[\rho(\xi, \delta')|x_1, \ldots, x_n] \leq E[\rho(\xi, \delta)|x_1, \ldots, x_n].
\tag{3}
$$

It follows that δ' is a regular procedure.∎

12.9 EXISTENCE OF AN OPTIMAL PROCEDURE

If $\delta_1, \ldots, \delta_k$ are any k sequential decision procedures in the class Δ, we shall let max $\{\delta_1, \ldots, \delta_k\}$ denote the sequential decision procedure in Δ which specifies that another observation should be taken at a given stage if at least one of the procedures $\delta_1, \ldots, \delta_k$ specifies that another observation should be taken at that stage. Thus, under the procedure max $\{\delta_1, \ldots, \delta_k\}$, sampling is continued just as long as it would be continued under at least one of the procedures $\delta_1, \ldots, \delta_k$ and sampling is terminated as soon as the observed values $X_1 = x_1, \ldots, X_n = x_n$ are such that $(x_1, \ldots, x_n) \in \{\delta_i \leq n\}$ for $i = 1, \ldots, k$. Finally, if $\delta_1, \delta_2, \ldots$ is an infinite sequence of decision procedures in Δ, we shall define the decision procedure sup $\{\delta_1, \delta_2, \ldots\}$ in the same way. We can now state the following lemmas.

Lemma 1 *Let $\delta_1 \in \Delta$ and $\delta_2 \in \Delta$ be regular decision procedures, and let $\delta = $ max $\{\delta_1, \delta_2\}$. Then δ is regular, and $\rho(\xi, \delta) \leq \rho(\xi, \delta_i)$ for $i = 1, 2$.*

Proof The procedure δ agrees with the procedure δ_1, except that there may be observed values $X_1 = x_1, \ldots, X_n = x_n$ for which sampling would be terminated under δ_1 but would be continued under δ in accordance with the procedure δ_2. At such a point $(x_1, \ldots, x_n)$, the following relation must be satisfied:

$$
\begin{aligned}
E[\rho(\xi, \delta)|x_1, \ldots, x_n] &= E[\rho(\xi, \delta_2)|x_1, \ldots, x_n] \\
&< \rho_0[\xi(x_1, \ldots, x_n)] + nc \\
&= E[\rho(\xi, \delta_1)|x_1, \ldots, x_n].
\end{aligned}
\tag{1}
$$

Here the first equality is obtained from the fact that the procedures δ and δ_2 agree after the values $x_1, \ldots, x_n$ have been observed. The inequality in relation (1) follows from the fact that the regular decision procedure δ_2 specifies that sampling is to be continued. The final equality in relation (1) simply states that sampling is terminated under the procedure δ_1. It follows that $\rho(\xi, \delta) \leq \rho(\xi, \delta_1)$. The same type of argument shows also that $\rho(\xi, \delta) \leq \rho(\xi, \delta_2)$.∎

Lemma 2 *Let $\delta_i \in \Delta$ $(i = 1, 2, \ldots)$ be a sequence of regular decision procedures, and let $\gamma_n = \max \{\delta_1, \ldots, \delta_n\}$ for $n = 1, 2, \ldots$. Then, for $n = 1, 2, \ldots$, the procedure γ_n is regular, and both of the following relations must be satisfied:*

$$
\rho(\xi, \gamma_n) \leq \rho(\xi, \delta_i) \qquad i = 1, \ldots, n,
\tag{2}
$$

and

$$
\rho(\xi, \gamma_n) \geq \rho(\xi, \gamma_{n+1}).
\tag{3}
$$

Proof Since $\gamma_1 = \delta_1$ and $\gamma_2 = \max \{\delta_1, \delta_2\}$, it follows from Lemma 1 that the procedures γ_1 and γ_2 are regular, that the relation (2) is satisfied when $n = 1$ and $n = 2$, and that the relation (3) is satisfied when $n = 1$. Furthermore, since γ_3 can be written in the form $\gamma_3 = \max \{\gamma_2, \delta_3\}$, it again follows from Lemma 1 and from the properties of γ_2 just proved that γ_3 is regular, that the relation (2) is satisfied when $n = 3$, and that the relation (3) is satisfied when $n = 2$. In general, the lemma is proved by an induction argument.∎

Since each procedure δ_i $(i = 1, 2, \ldots)$ belongs to the class Δ, it can be seen that each procedure γ_i $(i = 1, 2, \ldots)$ will also belong to the class Δ. Thus, $\Pr(\gamma_i < \infty) = 1$ for $i = 1, 2, \ldots$. In the next lemma, it is shown that the procedure sup $\{\delta_1, \delta_2, \ldots\}$ also belongs to the class Δ.

Lemma 3 *Let the decision procedures δ_i and γ_i $(i = 1, 2, \ldots)$ be as defined in Lemma 2, and let $\gamma = \sup \{\delta_1, \delta_2, \ldots\}$. Then $\Pr(\gamma < \infty) = 1$.*

Proof For any decision procedure or stopping variable δ and any positive integer m, $E(\delta) \geq m \Pr(\delta \geq m)$. Hence, it can be seen from relation (3) and from Eq. (1) of Sec. 12.7 that for any positive integers i and m,

$$
\rho(\xi, \gamma_1) \geq \rho(\xi, \gamma_i) \geq cm\Pr(\gamma_i \geq m).
\tag{4}
$$

Therefore,

$$\Pr(\gamma_i \geq m) \leq \frac{\rho(\xi, \gamma_1)}{cm}. \tag{5}$$

For any positive integer m, it follows from the definitions of the procedures γ and γ_i that $\{\gamma_i \geq m\} \subset \{\gamma_{i+1} \geq m\}$ for $i = 1, 2, \ldots$ and that $\{\gamma \geq m\} = \bigcup_{i=1}^{\infty} \{\gamma_i \geq m\}$. Hence,

$$\Pr(\gamma \geq m) = \lim_{i \to \infty} \Pr(\gamma_i \geq m) \leq \frac{\rho(\xi, \gamma_1)}{cm}. \tag{6}$$

We can now obtain the following relation:

$$\lim_{m \to \infty} \Pr(\gamma \geq m) = 0. \tag{7}$$

Equation (7) is equivalent to the statement that $\Pr(\gamma < \infty) = 1.$ ∎

We shall now prove that the risk from the procedure γ is at least as small as the risk from any procedure δ_i in the original sequence. We shall need to use the following result, which is known as the *Fatou-Lebesgue theorem*. For a proof of this theorem, see Loève (1963), p. 125, or Krickeberg (1965), p. 43.

Fatou-Lebesgue theorem *Let $g_1, g_2, \ldots$ be a sequence of nonnegative functions each of which is defined on the space S^∞. Let g be another function whose value at any point $(x_1, x_2, \ldots) \in S^\infty$ is specified as follows:*

$$g(x_1, x_2, \ldots) = \liminf_{n \to \infty} g_n(x_1, x_2, \ldots). \tag{8}$$

Then

$$E(g) \leq \liminf_{n \to \infty} E(g_n). \tag{9}$$

Lemma 4 *Let the decision procedures δ_i and γ_i $(i = 1, 2, \ldots)$ and the decision procedure γ be as defined in Lemmas 2 and 3. Then, for $i = 1, 2, \ldots,$*

$$\rho(\xi, \gamma) \leq \rho(\xi, \gamma_i) \leq \rho(\xi, \delta_i). \tag{10}$$

Proof For any procedure $\delta \in \Delta$, both the corresponding stopping variable δ and the risk $\rho_0[\xi(X_1, \ldots, X_\delta)]$ can be regarded as functions on the space S^∞. From the definition of the procedure γ, it is seen that γ can be represented as $\gamma = \sup \{\gamma_1, \gamma_2, \ldots\}$. Furthermore, by Lemma 3, the following relation will be satisfied with probability 1:

$$\gamma = \lim_{n \to \infty} \gamma_n < \infty. \tag{11}$$

Hence, with probability 1, the following relation will also be satisfied:

$$\rho_0[\xi(X_1, \ldots, X_\gamma)] + c\gamma = \lim_{n \to \infty} \{\rho_0[\xi(X_1, \ldots, X_{\gamma_n})] + c\gamma_n\}.$$

(12)

It now follows from Eq. (1) of Sec. 12.7 and from the Fatou-Lebesgue theorem that

$$\rho(\xi, \gamma) \leq \liminf_{n \to \infty} \rho(\xi, \gamma_n).$$

(13)

From relations (13), (2), and (3), it can be seen that relation (10) is correct.■

The main theorem follows from the foregoing development.

Theorem 1 *There exists an optimal sequential decision procedure in the class Δ.*

Proof Let $\delta_1, \delta_2, \ldots$ be a sequence of procedures in the class Δ such that

$$\lim_{i \to \infty} \rho(\xi, \delta_i) = \inf_{\delta \in \Delta} \rho(\xi, \delta).$$

(14)

It can be assumed that each of the procedures δ_i $(i = 1, 2, \ldots)$ is regular. Let $\delta^* = \sup \{\delta_1, \delta_2, \ldots\}$. Then it follows from Lemmas 3 and 4 that $\delta^* \in \Delta$ and that

$$\rho(\xi, \delta^*) \leq \rho(\xi, \delta_i) \qquad \text{for } i = 1, 2, \ldots.$$

(15)

Therefore, by Eq. (14),

$$\rho(\xi, \delta^*) \leq \inf_{\delta \in \Delta} \rho(\xi, \delta).$$

(16)

Since $\delta^* \in \Delta$, we can conclude that there must be equality in the relation (16) and that the procedure δ^* is optimal.■

Further Remarks and References

The general theory of optimal sequential decision procedures was developed by Wald (1950). It has also been discussed and extended by Arrow, Blackwell, and Girshick (1949), Wald and Wolfowitz (1950), and Le Cam (1955). The proof of the existence of an optimal procedure, as given here, essentially follows Chow and Robbins (1963, 1967b).

12.10 APPROXIMATING AN OPTIMAL PROCEDURE BY BOUNDED PROCEDURES

Now that we have established the existence of an optimal sequential decision procedure δ^* in the class Δ, we shall consider the problem of actually constructing or approximating an optimal procedure. Since the method of backward induction provides a technique for constructing the optimal bounded sequential decision procedure among those that must be terminated after not more than n observations have been taken, it is natural to find out whether the optimal procedure $\delta^* \epsilon \Delta$ can be approximated by these optimal bounded procedures.

For a given sequential decision procedure $\delta \epsilon \Delta$ and any positive integer n, let $\delta' \epsilon \Delta$ be the bounded procedure defined as follows: The procedure δ' agrees with the procedure δ through the first n observations. Under δ', however, sampling will always be terminated after the nth observation, if it has not terminated earlier, regardless of whether or not the procedure δ specifies that further observations should be taken. It is said that the procedure δ' is obtained by *truncating* the procedure δ after n observations. Of course, if δ itself is always terminated after not more than n observations, then the procedures δ' and δ are identical.

The next theorem and its corollary give conditions under which the risk of the procedure obtained by truncating the optimal procedure δ^* after n observations will be close to the Bayes risk $\rho(\xi, \delta^*)$.

Theorem 1 *For $n = 1, 2, \ldots$, let the value Q_n be defined as follows:*

$$Q_n = \int_{\{\delta^* > n\}} \rho_0[\xi(x_1, \ldots, x_n)]\, dF_n(x_1, \ldots, x_n|\xi). \tag{1}$$

Also, for $n = 1, 2, \ldots$, let δ_n be the procedure obtained by truncating the optimal procedure δ^* after n observations. If $\lim\limits_{n \to \infty} Q_n = 0$, then

$$\lim_{n \to \infty} \rho(\xi, \delta_n) = \rho(\xi, \delta^*). \tag{2}$$

Proof For any procedure $\delta \epsilon \Delta$ and any positive integer j, let $T_j(\delta)$ be defined as follows:

$$T_j(\delta) = \int_{\{\delta = j\}} \{\rho_0[\xi(x_1, \ldots, x_j)] + jc\}\, dF_j(x_1, \ldots, x_j|\xi). \tag{3}$$

If the optimal procedure δ^* specifies that no observations should be taken, then the theorem is trivial. Therefore, we shall assume that δ^* specifies that at least one observation is to be taken. Then

$$\rho(\xi, \delta^*) = \sum_{j=1}^{\infty} T_j(\delta^*) = \lim_{n \to \infty} \sum_{j=1}^{n} T_j(\delta^*). \tag{4}$$

Now consider the procedure δ_n, where n is a fixed positive integer. For this procedure the risk is

$$\rho(\xi, \delta_n) = \sum_{j=1}^{n} T_j(\delta_n). \tag{5}$$

The sum in Eq. (5) includes only n terms because the procedure δ_n must be terminated after not more than n observations have been taken. Furthermore, it can be seen from the definition of δ_n that the following relations must be satisfied:

$$\{\delta_n = j\} = \{\delta^* = j\} \qquad \text{for } j = 1, \ldots, n - 1, \tag{6}$$

and

$$\{\delta_n = n\} = \{\delta^* \geq n\} = \{\delta^* = n\} \cup \{\delta^* > n\}. \tag{7}$$

Hence, $T_j(\delta_n) = T_j(\delta^*)$, for $j = 1, \ldots, n - 1$, and

$$T_n(\delta_n) = T_n(\delta^*) + Q_n + V_n, \tag{8}$$

where Q_n is defined by Eq. (1) and $V_n = cn \Pr(\delta^* > n)$.

From Eq. (5) and these remarks, we can write the following relation:

$$\rho(\xi, \delta_n) = \left[\sum_{j=1}^{n} T_j(\delta^*) \right] + Q_n + V_n. \tag{9}$$

The optimal procedure δ^* has a finite risk, which includes a finite expected sampling cost. Hence, $E(\delta^*) < \infty$, and it is implied that $V_n \to 0$ as $n \to \infty$. Since it is assumed that $Q_n \to 0$, Eq. (2) can be obtained from Eqs. (4) and (9).∎

Corollary 1 *Let the procedures δ^* and δ_n ($n = 1, 2, \ldots$) be as defined in Theorem 1. Then Eq. (2) will be satisfied under either of the following two conditions:*

1. *There exists a number K such that, for all values of $(x_1, \ldots, x_n) \in \{\delta^* > n\}$ and for $n = 1, 2, \ldots,$*

$$\rho_0[\xi(x_1, \ldots, x_n)] < nK. \tag{10}$$

2. $\lim_{n \to \infty} E\{\rho_0[\xi(X_1, \ldots, X_n)]\} = 0.$

Proof Suppose that condition 1 is correct. Then, from the definition (1) of Q_n, we can obtain the following result:

$$\lim_{n \to \infty} Q_n \leq \lim_{n \to \infty} nK \Pr(\delta^* > n) = 0. \tag{11}$$

In this last relation, the limiting value must be 0 because $E(\delta^*) < \infty$. Equation (2) can then be obtained in accordance with Theorem 1.

The integrand in Eq. (1) is nonnegative. Therefore, if the region of integration $\{\delta^* > n\}$ in that equation is replaced by the whole sample space S^n, we obtain the following relation:

$$Q_n \leq E\{\rho_0[\xi(X_1, \ldots, X_n)]\}. \tag{12}$$

If condition 2 is correct, then Eq. (2) can again be obtained from Theorem 1 and the relation (12).■

In any particular decision problem, it is usually easy to check condition 2 of Corollary 1. Furthermore, condition 1 can often be verified, in spite of the fact that the optimal procedure δ^* and the set $\{\delta^* > n\}$ are unknown, because in many problems relation (10) will be satisfied for all values $(x_1, \ldots, x_n)$ in the space S^n. These conditions do not require that the loss function L be bounded. In both Example 3 of Sec. 12.6, which involved estimation with a quadratic loss function, and Example 4 of that section, which involved testing hypotheses with a linear loss function, the loss function L is unbounded but both conditions 1 and 2 of Corollary 1 are satisfied.

The next corollary points out that under the prescribed conditions, δ^* can be approximated by the bounded procedure which is optimal among all those that must be terminated after not more than n observations have been taken. As in Sec. 12.5, we shall let $\rho_n(\xi)$ denote the total risk from the optimal bounded procedure, and we shall let $\rho^*(\xi)$ denote the total risk $\rho(\xi, \delta^*)$ from the optimal procedure δ^*. Thus, $\rho^*(\xi)$ is the Bayes risk against the prior distribution ξ.

Corollary 2   *If either condition 1 or condition 2 of Corollary 1 is satisfied, then*

$$\lim_{n \to \infty} \rho_n(\xi) = \rho^*(\xi). \tag{13}$$

Proof   As before, let δ_n be the procedure obtained by truncating δ^* after n observations.   Since δ_n is terminated after not more than n observations have been taken, it follows from the optimal properties of δ^* and $\rho_n(\xi)$ that the following relation must be satisfied for $n = 1, 2, \ldots$:

$$\rho^*(\xi) \leq \rho_n(\xi) \leq \rho(\xi, \delta_n). \tag{14}$$

Equation (13) can now be obtained from Corollary 1.■

It follows from Corollary 2 that when the value of n is sufficiently large, the risk $\rho^*(\xi)$ of the optimal procedure δ^* can, under certain conditions, be approximated by the risk $\rho_n(\xi)$ of the optimal bounded procedure. Of course, the optimal procedure δ^* may itself be a bounded procedure. Suppose, for example, that there exists a value of n such that $\rho_0[\xi(x_1, \ldots, x_n)] \leq c$ for all points $(x_1, \ldots, x_n) \in S^n$. In other words, after the nth observation has been taken, the risk from choosing a decision in D cannot be greater than the cost of the next observation. Hence, the Bayes procedure never requires that more than n observations should be taken.

Further Remarks and References

Some papers dealing with sequential tests of hypotheses and related problems are Sobel and Wald (1949), Sobel (1953), Mikhalevich (1956), Kiefer and Weiss (1957), Wetherill (1961), Moriguti and Robbins (1962), Lechner (1962), Cheng (1963), Anscombe (1963), Ray (1963), Ghosh (1964), Whittle (1964), and Chernoff and Ray (1965). Other references on this topic were given at the end of Sec. 12.6. Asymptotic studies of Bayes procedures have been made by Schwarz (1962), Kiefer and Sacks (1963), Bickel and Yahav (1967), and Lorden (1967).

12.11 REGIONS FOR CONTINUING OR TERMINATING SAMPLING

The optimal sequential decision procedure δ^*, as developed in the preceding section, can be characterized as follows: Let ϕ denote the distribution of W at a given stage of the sampling process. At the beginning of the process, $\phi = \xi$. Another observation should be taken at the given stage if, and only if, there exists some sequential decision procedure δ whose risk $\rho(\phi, \delta)$ is smaller than the risk $\rho_0(\phi)$ from choosing a decision immediately. Observations should be terminated if, and only if, $\rho(\phi, \delta) \geq \rho_0(\phi)$ for every procedure δ.

At any stage of the sampling process in a specific problem, the statistician typically is able to compute the risk $\rho(\phi, \delta)$ for only a few special sequential decision procedures δ. For example, he may be able to compute the risk only for procedures in which a fixed number of observations are to be taken and then a decision in D is to be chosen. In a more restrictive situation, he may be able to compute the risk $\rho(\phi, \delta)$ only for the single procedure δ in which exactly one observation will be taken before a decision in D is chosen. In a given problem, we shall let Δ_0 denote the class of procedures δ for which the statistician can compute the risk $\rho(\phi, \delta)$. Suppose that the statistician follows the procedure in which he takes another observation if, and only if, there is a procedure δ in the class Δ_0 such that $\rho(\phi, \delta) < \rho_0(\phi)$. Then he knows that he will take another observation only when it is optimal to do so.

On the other hand, he may terminate sampling at a stage at which it would have been optimal to continue sampling. Hence, this procedure, which we shall designate as δ_0, provides an "inner bound" for the continuation region of the optimal procedure δ^*. More precisely, the first observation will be taken under the procedure δ_0 only if that observation would also be taken under the optimal procedure δ^*. Also, $\{\delta_0 > n\} \subset \{\delta^* > n\}$ for $n = 1, 2, \ldots$.

To illustrate these ideas, we shall again consider the problem described in Sec. 11.9 and in Example 4 of Sec. 12.6.

EXAMPLE Observations are to be taken from a normal distribution with unknown mean W and specified precision r. The statistician must decide whether $W \leq w_0$ or $W > w_0$ when the loss function is specified by Eq. (1) of Sec. 11.8 and the cost of each observation is c units.

Let δ denote the procedure which specifies that exactly one observation should be taken before a decision in D is chosen. When the distribution ξ of W is a normal distribution with mean μ and precision τ, the risk $\rho(\xi, \delta)$ from the procedure δ may be found by adding the sampling cost c to the value given by Eq. (18) of Sec. 12.6. Also, the risk $\rho_0(\xi)$ from an immediate decision without any observation is given by Eq. (16) of Sec. 12.6. Hence, $\rho(\xi, \delta) < \rho_0(\xi)$ if, and only if, the inequality (19) of Sec. 12.6 is satisfied. Thus, at any stage of the sampling process for which the mean μ and the precision τ of the distribution of W at that stage satisfy the inequality (19) of Sec. 12.6, it is optimal to take another observation. These points are sketched in Fig. 12.2 as the region in the μ, τ-plane where sampling is to be continued. The boundary of this region is the set of points (μ, τ) for which the two sides of the inequality (19) of Sec. 12.6 are equal. The boundary will be symmetric with respect to the line $\mu = w_0$, and the convexity of the upper half of the boundary can be derived from a careful study of the function Ψ. Since $\Psi(0) = (2\pi)^{-\frac{1}{2}}$, it follows from the inequality (19) of Sec. 12.6 that the boundary intersects the line $\mu = w_0$ when τ satisfies the following equation:

$$\tau + \frac{\tau^2}{r} = \frac{1}{2\pi c^2}. \tag{1}$$

This value of τ is designated as τ' in Fig. 12.2.

If $\rho_0(\xi) \leq c$, then it clearly is optimal to terminate sampling, since the cost of the next observation exceeds the risk from an immediate decision. The set of points (μ, τ) for which $\rho_0(\xi) \leq c$ is sketched in Fig. 12.2 as the region where sampling is to be terminated. It can be shown from Eq. (16) of Sec. 12.6 that the boundary of this region is also convex and symmetric and that it intersects the line $\mu = w_0$ when $\tau = 1/(2\pi c^2)$. This value of τ is designated as τ'' in Fig. 12.2. The optimal procedure is

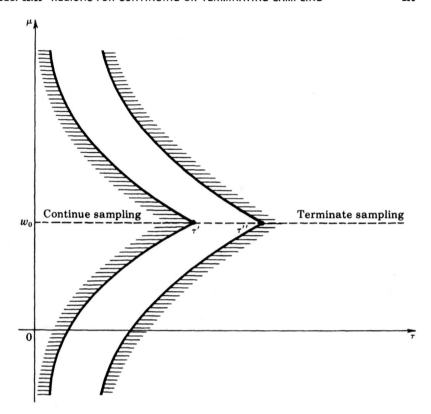

Fig. 12.2   Bounds for the optimal sequential decision procedure.

unknown only for the points (μ, τ) which lie between the two boundaries in Fig. 12.2.

If the regions in the μ, τ-plane in which sampling should be continued or terminated under the optimal procedure were known exactly, then this procedure could be described as follows.   Let μ_0 and τ_0 be the mean and precision of the prior distribution of W, and in general, let μ_n and τ_n be the mean and precision of the posterior distribution of W at any given stage of the sampling process.   If the point (μ_0, τ_0) lies in the region where sampling is to be continued, then the first observation should be taken.   Otherwise, no observations should be taken.   In general, if the point (μ_n, τ_n) lies in the region where sampling is to be continued, then another observation should be taken.   Otherwise, no further observations should be taken.

For each additional observation, the change in the mean μ_n depends on the observed value but the precision always increases by r units. Thus, if we start at some point (μ_0, τ_0) and we suppose that further observations are taken, then the vertical distance between successive points (μ_n, τ_n) will fluctuate while the horizontal distance between the points will always be r units. Suppose that n^* is the smallest value of n such that $\tau_0 + nr \geq \tau''$. Then it can be seen from Fig. 12.2 that sampling must be terminated after not more than n^* observations have been taken. Thus, for any prior precision τ_0, this requirement provides an upper bound on the number of observations that will be taken under the optimal procedure.

12.12 THE FUNCTIONAL EQUATION

We shall now consider some general methods for bounding and approximating the risk $\rho^*(\xi)$ of the optimal procedure in any sequential decision problem. Suppose that at a given stage of the procedure, the distribution of W is ξ and an observation X is taken. Then the posterior distribution of W will be $\xi(X)$, and the risk of the optimal continuation from that point is, by definition, $\rho^*[\xi(X)]$. When the cost c of taking the observation X is included, the total risk of taking an observation and then continuing in an optimal fashion is $E\{\rho^*[\xi(X)]\} + c$, where the expectation is computed with respect to the marginal distribution of X which is induced by the distribution ξ of W. The statistician's alternatives are either to take the observation X or to choose a decision without any further observations. The risk of this latter alternative is $\rho_0(\xi)$. Since the optimal alternative is the one involving the smaller risk, it can be seen that for any distribution ξ of W, the following relation must be satisfied:

$$\rho^*(\xi) = \min \{\rho_0(\xi), E[\rho^*(\xi(X))] + c\}. \tag{1}$$

This relation is the basic functional equation which must be satisfied by the Bayes risk ρ^*. It reflects the principle of optimality described in Sec. 12.4, which may be restated here as follows: Under an optimal procedure, the continuation after an observation has been taken must itself be an optimal procedure when the posterior distribution of W is regarded as a new prior distribution.

Let Ξ^* denote the set of all distributions ϕ of W such that $\rho^*(\phi) = \rho_0(\phi)$. The optimal sequential decision procedure can be characterized as follows: The first observation should be taken if, and only if, the prior distribution ξ of W does not belong to the set Ξ^*. At any subsequent stage of the procedure, sampling is continued if, and only if, the posterior distribution ϕ of W at that stage does not belong to the set Ξ^*. Sampling is terminated as soon as the distribution of W falls into the set Ξ^*.

The function ρ_0 can be computed by the statistician. Hence, if the function ρ^* is also known to him, then the set Ξ^* and the optimal procedure δ^* can be constructed. For instance, if the function ρ^* could be found directly from Eq. (1), then δ^* would be known. A great deal of work has been done in regard to the existence and uniqueness of solutions to functional equations of this type [see, e.g., Bellman (1957a), chap. 4]. The appropriate results will be derived here. In a given problem, it is often impossible to solve Eq. (1) directly and neatly for ρ^*. Nevertheless, general methods of approximating the solution ρ^* do exist, and they also will be described in detail.

We shall let Ξ denote a fixed set of distributions ϕ of W that includes the prior distribution and all the distributions which can possibly arise during the sampling process. In the example of Sec. 12.11, in which the prior distribution of W is a normal distribution with mean μ and precision τ, it might be convenient to take the set Ξ to be the set of all normal distributions whose precision is at least τ. In a decision problem in which W can have only k values, it might be convenient to take the set Ξ to be the set of all distributions over these k values. In general, an appropriate set Ξ can be constructed and described conveniently in any sequential decision problem for which there exists a conjugate family of distributions of W of the type discussed in Chap. 9.

We shall say that a sequential decision problem is *stable* if the following equation is satisfied for every distribution $\phi \in \Xi$:

$$\lim_{n \to \infty} E\{\rho_0[\phi(X_1, \ldots, X_n)]\} = 0. \tag{2}$$

For any prior distribution ϕ of W, let $\rho_n(\phi)$ denote the risk from the procedure that is optimal among all those which must be terminated after not more than n observations have been taken, and let $\rho^*(\phi)$ denote the risk from the optimal procedure in the class Δ. It follows from Corollaries 1 and 2 of Sec. 12.10 that in a stable decision problem, the following relation will be satisfied for every distribution $\phi \in \Xi$:

$$\lim_{n \to \infty} \rho_n(\phi) = \rho^*(\phi). \tag{3}$$

Now we shall consider the following functional equation:

$$\gamma(\phi) = \min \{\rho_0(\phi), E[\gamma(\phi(X))] + c\}. \tag{4}$$

The expectation in Eq. (4) is defined by Eq. (1) of Sec. 12.5, in which an arbitrary function γ now replaces ρ_0. We wish to determine the functions γ defined on the set Ξ such that Eq. (4) is satisfied for every distribution $\phi \in \Xi$. We know from Eq. (1) that $\gamma = \rho^*$ is a solution of this equation. It is shown in the next theorem that in a stable decision problem, the Bayes risk ρ^* is the only nonnegative solution of Eq. (4).

Theorem 1 *In a stable decision problem, the Bayes risk ρ^* is the unique nonnegative function which satisfies Eq. (4).*

Proof Since we already know that ρ^* is a solution of Eq. (4), it remains only to show that there cannot be more than one solution. Suppose that both the nonnegative function α and the nonnegative function β satisfy Eq. (4). Then $\alpha(\phi) \leq \rho_0(\phi)$ and $\beta(\phi) \leq \rho_0(\phi)$ for any $\phi \, \epsilon \, \Xi$, and it is seen from Eq. (4) that the following four relations must be satisfied:

1. If $\alpha(\phi) < \rho_0(\phi)$ and $\beta(\phi) < \rho_0(\phi)$, then
   $\alpha(\phi) - \beta(\phi) = E\{\alpha[\phi(X)] - \beta[\phi(X)]\}$.
2. If $\alpha(\phi) < \rho_0(\phi)$ and $\beta(\phi) = \rho_0(\phi)$, then
   $0 < \beta(\phi) - \alpha(\phi) < E\{\beta[\phi(X)] - \alpha[\phi(X)]\}$.
3. If $\alpha(\phi) = \rho_0(\phi)$ and $\beta(\phi) < \rho_0(\phi)$, then
   $0 < \alpha(\phi) - \beta(\phi) < E\{\alpha[\phi(X)] - \beta[\phi(X)]\}$.
4. If $\alpha(\phi) = \rho_0(\phi)$ and $\beta(\phi) = \rho_0(\phi)$, then $\alpha(\phi) - \beta(\phi) = 0$.

In all four of these cases, the following relation exists for every distribution $\phi \, \epsilon \, \Xi$:

$$|\alpha(\phi) - \beta(\phi)| \leq E\{|\alpha[\phi(X)] - \beta[\phi(X)]|\}. \tag{5}$$

If we now replace ϕ in the inequality (5) by $\phi(X_1)$ and take expectations on both sides, we can write the resulting inequality in the following form:

$$E\{|\alpha[\phi(X_1)] - \beta[\phi(X_1)]|\} \leq E\{|\alpha[\phi(X_1, X_2)] - \beta[\phi(X_1, X_2)]|\}. \tag{6}$$

It follows from the inequalities (5) and (6) and from repeated application of this technique that for every distribution $\phi \, \epsilon \, \Xi$ and every positive integer n, the following relation must be satisfied:

$$|\alpha(\phi) - \beta(\phi)| \leq E\{|\alpha[\phi(X_1, \ldots, X_n)] - \beta[\phi(X_1, \ldots, X_n)]|\}. \tag{7}$$

It is assumed that the decision problem is stable and, hence, that Eq. (2) is satisfied. Since $0 \leq \alpha(\phi) \leq \rho_0(\phi)$ and $0 \leq \beta(\phi) \leq \rho_0(\phi)$, it follows that as $n \to \infty$, the right side of relation (7) must converge to 0 for each distribution $\phi \, \epsilon \, \Xi$. Hence, $\alpha(\phi) = \beta(\phi)$ for each distribution $\phi \, \epsilon \, \Xi$ and Eq. (4) must have a unique nonnegative solution.∎

12.13 APPROXIMATIONS AND BOUNDS FOR THE BAYES RISK

We shall define Γ to be the class of all real-valued functions γ on the set Ξ such that $0 \leq \gamma(\phi) \leq \rho_0(\phi)$ for every distribution $\phi \, \epsilon \, \Xi$. It follows from our assumptions in regard to ρ_0 that the expectation $E\{\gamma[\phi(X)]\}$ exists for every function $\gamma \, \epsilon \, \Gamma$ and every distribution $\phi \, \epsilon \, \Xi$.

Next, for any function $\gamma_0 \in \Gamma$, we shall define a sequence of functions $\gamma_1, \gamma_2, \ldots$ recursively by the following relation: For $\phi \in \Xi$ and $n = 0, 1, \ldots,$

$$\gamma_{n+1}(\phi) = \min \{\rho_0(\phi), E[\gamma_n(\phi(X))] + c\}. \tag{1}$$

Since $\gamma_0 \in \Gamma$, it follows from Eq. (1) that $\gamma_n \in \Gamma$ for $n = 1, 2, \ldots$. The next theorem demonstrates the following important result: In a stable decision problem, the sequence $\gamma_1, \gamma_2, \ldots$ generated by any function $\gamma_0 \in \Gamma$ must converge to the Bayes risk ρ^*.

Theorem 1 *For any function $\gamma_0 \in \Gamma$ in a stable decision problem, let the sequence of functions $\gamma_1, \gamma_2, \ldots$ be defined by Eq. (1). Then, for every distribution $\phi \in \Xi$,*

$$\lim_{n \to \infty} \gamma_n(\phi) = \rho^*(\phi). \tag{2}$$

Proof Let m and n be any positive integers. It can be seen from Eq. (1) and a computation similar to the one leading to relation (5) of Sec. 12.12 that for every distribution $\phi \in \Xi$, the following relation must be satisfied:

$$|\gamma_{n+m}(\phi) - \gamma_n(\phi)| \leq E\{|\gamma_{n+m-1}[\phi(X)] - \gamma_{n-1}[\phi(X)]|\}. \tag{3}$$

When this process is applied repeatedly to the right side of relation (3), as was done with relations (6) and (7) of Sec. 12.12, we obtain the following inequality for every distribution $\phi \in \Xi$:

$$|\gamma_{n+m}(\phi) - \gamma_n(\phi)|$$
$$\leq E\{|\gamma_m[\phi(X_1, \ldots, X_n)] - \gamma_0[\phi(X_1, \ldots, X_n)]|\}. \tag{4}$$

Since $\gamma_m \in \Gamma$ for $m = 0, 1, 2, \ldots$, the right side of relation (4) cannot be greater than $2E\{\rho_0[\phi(X_1, \ldots, X_n)]\}$. Furthermore, since the decision problem is stable, it follows from Eq. (2) of Sec. 12.12 that the right side of relation (4) converges to 0. Hence, for any given distribution $\phi \in \Xi$ and any given number $\epsilon > 0$, it follows from relation (4) that for all sufficiently large values of n and all values of m,

$$|\gamma_{n+m}(\phi) - \gamma_n(\phi)| \leq \epsilon. \tag{5}$$

This relation shows that for each distribution $\phi \in \Xi$, the sequence $\{\gamma_n(\phi); n = 1, 2, \ldots\}$ must converge to some limiting value $\gamma^*(\phi)$. Now we shall let $n \to \infty$ on each side of Eq. (1). By the Lebesgue dominated convergence theorem (see Sec. 10.6),

$$\lim_{n \to \infty} E\{\gamma_n[\phi(X)]\} = E\{\gamma^*[\phi(X)]\}. \tag{6}$$

It can be shown from Eqs. (1) and (6) that γ^* is a solution of Eq. (4) of Sec. 12.12, and it follows from Theorem 1 of Sec. 12.12 that $\gamma^* = \rho^*$.■

As indicated in the next theorem, the convergence of the sequence $\gamma_0, \gamma_1, \gamma_2, \ldots$ to the Bayes risk ρ^* will often be monotone.

Theorem 2   *For any function $\gamma_0 \in \Gamma$, let the sequence $\gamma_1, \gamma_2, \ldots$ be defined by Eq. (1). Then the following four properties must be satisfied.*

   1. *If $\gamma_0(\phi) \leq \gamma_1(\phi)$ for every distribution $\phi \in \Xi$, then $\gamma_n(\phi) \leq \gamma_{n+1}(\phi)$ for $n = 1, 2, \ldots$ and for every distribution $\phi \in \Xi$.*

   2. *If $\gamma_0(\phi) \geq \gamma_1(\phi)$ for every distribution $\phi \in \Xi$, then $\gamma_n(\phi) \geq \gamma_{n+1}(\phi)$ for $n = 1, 2, \ldots$ and for every distribution $\phi \in \Xi$.*

   3. *If $\gamma_0(\phi) \leq \rho^*(\phi)$ for every distribution $\phi \in \Xi$, then $\gamma_n(\phi) \leq \rho^*(\phi)$ for $n = 1, 2, \ldots$ and for every distribution $\phi \in \Xi$.*

   4. *If $\gamma_0(\phi) \geq \rho^*(\phi)$ for every distribution $\phi \in \Xi$, then $\gamma_n(\phi) \geq \rho^*(\phi)$ for $n = 1, 2, \ldots$ and for every distribution $\phi \in \Xi$.*

Proof   To show that property 1 is correct, we shall use an induction argument. Suppose that $\gamma_n(\phi) \leq \gamma_{n+1}(\phi)$ for every distribution $\phi \in \Xi$ and for some nonnegative integer n. Then it can be seen from Eq. (1) that the following relation must be satisfied:

$$\gamma_{n+2}(\phi) = \min \{\rho_0(\phi), E[\gamma_{n+1}(\phi(X))] + c\}$$
$$\geq \min \{\rho_0(\phi), E[\gamma_n(\phi(X))] + c\} = \gamma_{n+1}(\phi). \tag{7}$$

The correctness of each of the other three properties of the theorem can be similarly demonstrated.■

If we choose γ_0 to be ρ_0, then it is seen from Eq. (1) and from Eq. (6) of Sec. 12.5 that $\gamma_n = \rho_n$ for $n = 1, 2, \ldots$, and we know that this sequence satisfies the following relations:

$$\gamma_1(\phi) \geq \gamma_2(\phi) \geq \cdots \geq \rho^*(\phi). \tag{8}$$

Similarly, if we assume that $\gamma_0(\phi) = 0$ for every distribution $\phi \in \Xi$, then obviously $\gamma_0(\phi) \leq \gamma_1(\phi)$ and $\gamma_0(\phi) \leq \rho^*(\phi)$ for every distribution $\phi \in \Xi$. Hence, by Theorem 2, the sequence $\gamma_1, \gamma_2, \ldots$ generated by this choice of γ_0 satisfies the following relations:

$$\gamma_1(\phi) \leq \gamma_2(\phi) \leq \cdots \leq \rho^*(\phi). \tag{9}$$

Furthermore, by Theorem 1, if the decision problem is stable, the sequences $\gamma_1, \gamma_2, \ldots$ in both relations (8) and (9) will converge to ρ^*. However, for any initial choice of $\gamma_0 \in \Gamma$ in a specific problem, the computation of

γ_n for values of n greater than 2 or 3 is typically very difficult, and it is necessary to obtain numerical approximations with the aid of a high-speed computing machine.

Other Functional Equations

We shall now show how other approximations and bounds for the Bayes risk can be obtained from functional equations which are related to, but distinct from, Eqs. (1) and (4) of Sec. 12.12. Let γ_0 be any function in Γ, and consider a new sequential decision problem in which the risk ρ_0 is replaced by γ_0. In this new problem, the cost per observation is still c but the risk from making a decision without any further sampling, when the posterior distribution of W is ϕ, is now assumed to be $\gamma_0(\phi)$. Since $0 \leq \gamma_0(\phi) \leq \rho_0(\phi)$ for every distribution $\phi \in \Xi$, essentially all of the discussion which has been given in this chapter relating to the existence and properties of an optimal sequential decision procedure is applicable to this new decision problem. In particular, since we are assuming that the original problem is stable, it follows from Eq. (2) of Sec. 12.12 that the new decision problem is also stable. Hence, by Theorem 1 of Sec. 12.12, the Bayes risk γ^* in the new decision problem is the unique non-negative function which satisfies the following functional equation for every distribution $\phi \in \Xi$:

$$\gamma(\phi) = \min \{\gamma_0(\phi), E[\gamma(\phi(X))] + c\}. \tag{10}$$

Furthermore, since the risk $\gamma_0(\phi)$ incurred when sampling is terminated in the new decision problem is never larger than the risk $\rho_0(\phi)$ incurred when sampling is terminated in the original decision problem, the following relation must be satisfied for every distribution $\phi \in \Xi$:

$$\gamma^*(\phi) \leq \rho^*(\phi). \tag{11}$$

Next, consider a new decision problem in which the risk ρ_0 is replaced by ρ^*. Then, as we have just seen, the Bayes risk in this new problem is the unique solution of the following functional equation:

$$\gamma(\phi) = \min \{\rho^*(\phi), E[\gamma(\phi(X))] + c\} \qquad \phi \in \Xi. \tag{12}$$

However, it is seen from Eq. (1) of Sec. 12.12 that $\rho^*(\phi) \leq E\{\rho^*[\phi(X)]\} + c$ for every distribution $\phi \in \Xi$. Hence, $\gamma = \rho^*$ is the unique solution of Eq. (12). In other words, if the risk ρ_0 from stopping observations in the original problem is replaced by the risk ρ^* from the optimal continuation, then the Bayes risk in this new problem is again ρ^*, which is the Bayes risk in the old problem. Furthermore, it is never necessary to take any observations in the new problem because the Bayes risk ρ^* can be attained by an immediate decision.

Finally, suppose that γ_0 is chosen so that the following relation is satisfied for every distribution $\phi \in \Xi$:

$$\rho^*(\phi) \leq \gamma_0(\phi) \leq \rho_0(\phi). \tag{13}$$

Then the Bayes risk γ^* in the problem in which ρ_0 is replaced by γ_0 must be at least as large as the Bayes risk ρ^* in the problem where ρ_0 is replaced by ρ^*. Since the reverse inequality has already been demonstrated in relation (11), we have proved the following theorem.

Theorem 3 *Suppose that in a stable decision problem, the function γ_0 satisfies the relation (13). If γ^* is the Bayes risk in a new decision problem, which is defined by replacing the function ρ_0 by the function γ_0, then $\gamma^*(\phi) = \rho^*(\phi)$ for every distribution $\phi \in \Xi$. Equivalently, the function ρ^* is the unique nonnegative solution of Eq. (10).*

Further Remarks and References

Some papers dealing with bounds on the optimal sequential decision procedure and its risk are Hoeffding (1960), Amster (1963), Weiss (1964), Ray (1965), Grigelionis and Shiryaev (1965), and Pratt (1966).

12.14   THE SEQUENTIAL PROBABILITY-RATIO TEST

As an illustration of some of the results of the preceding sections, we shall consider a sequential decision problem in which the parameter space is $\Omega = \{w_1, w_2\}$ and the decision space is $D = \{d_1, d_2\}$. Each space contains exactly two points, and the loss function is given by Table 12.3, in which

Table 12.3

	d_1	d_2
w_1	0	λ_1
w_2	λ_2	0

$\lambda_1 > 0$ and $\lambda_2 > 0$. We shall assume, as before, that a sequential random sample $X_1, X_2, \ldots$ can be taken at a cost of c units per observation. For $i = 1, 2$, we shall let f_i denote the conditional g.p.d.f. of any observation X when $W = w_i$. Since the parameter W can have only two possible values, its distribution at any stage of the sampling process is characterized by the number $\xi = \Pr(W = w_1)$.

It can be seen from Table 12.3 that in this problem the risk $\rho_0(\xi)$ from choosing a decision immediately is

$$\rho_0(\xi) = \min \{\lambda_1 \xi, \lambda_2(1 - \xi)\}. \tag{1}$$

We shall let Δ' denote the class of all sequential decision procedures δ which require that at least one observation should be taken, and for $0 \le \xi \le 1$, we shall define $\rho'(\xi)$ as follows:

$$\rho'(\xi) = \inf_{\delta \, \epsilon \, \Delta'} \rho(\xi, \delta). \tag{2}$$

Thus, the Bayes risk $\rho^*(\xi)$ must satisfy the following requirement:

$$\rho^*(\xi) = \min \{\rho_0(\xi), \rho'(\xi)\}. \tag{3}$$

It can be easily demonstrated that ρ' is a concave, continuous function on the interval $0 \le \xi \le 1$ (see, e.g., Exercise 9). Furthermore, since every procedure $\delta \, \epsilon \, \Delta'$ involves a sampling cost of at least c units, $\rho'(0) = \rho'(1) = c$ and $\rho'(\xi) \ge c$ for all values of ξ $(0 \le \xi \le 1)$. The functions ρ_0 and ρ' are sketched in Fig. 12.3.

As before, we shall let Ξ^* be the set of all values of ξ at which it is optimal to terminate sampling. Thus,

$$\Xi^* = \{\xi \colon \rho_0(\xi) \le \rho'(\xi)\}. \tag{4}$$

Suppose first that

$$\rho' \left(\frac{\lambda_2}{\lambda_1 + \lambda_2} \right) < \frac{\lambda_1 \lambda_2}{\lambda_1 + \lambda_2}. \tag{5}$$

This relation is satisfied in Fig. 12.3, and it can be seen that Ξ^* is the union of the intervals $0 \le \xi \le \xi'$ and $\xi'' \le \xi \le 1$, where ξ' and ξ''

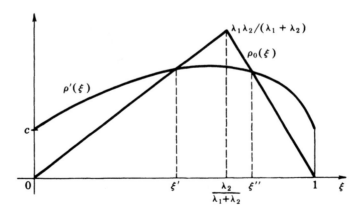

Fig. 12.3 The risks ρ_0 and ρ'.

satisfy the following equations:

$$\lambda_1 \xi' = \rho'(\xi') \qquad \text{and} \qquad \lambda_2(1 - \xi'') = \rho'(\xi''). \tag{6}$$

On the other hand, if the inequality (5) is not satisfied, then it can be seen from Fig. 12.3 that Ξ^* is the entire interval $0 \le \xi \le 1$. In this case, regardless of the prior distribution of W, it is never worthwhile to take any observations.

The set Ξ^* characterizes the Bayes sequential decision procedure, but the difficulty lies in learning enough about the function ρ' to be able to determine the critical values ξ' and ξ'' explicitly. Explicit values can be found in only a few special cases, but useful general approximations will be developed in the next two sections.

Suppose that the prior probability ξ lies in the interval $\xi' < \xi < \xi''$. It then follows from the above discussion that the first observation should be taken. Furthermore, if $\xi(x_1, \ldots, x_n)$ is the posterior probability that $W = w_1$ after the values $X_i = x_i$ ($i = 1, \ldots, n$) have been observed, then the optimal procedure is to continue taking observations whenever the following relation is satisfied:

$$\xi' < \xi(x_1, \ldots, x_n) < \xi''. \tag{7}$$

If the first inequality in relation (7) is not satisfied, no further observations should be taken and the decision d_2 should be chosen. If the second inequality in relation (7) is not satisfied, no further observations should be taken and the decision d_1 should be chosen.

The posterior probability $\xi(x_1, \ldots, x_n)$ can be expressed in the following form:

$$\xi(x_1, \ldots, x_n) = \left[1 + \frac{1 - \xi}{\xi} \frac{f_2(x_1) \cdots f_2(x_n)}{f_1(x_1) \cdots f_1(x_n)} \right]^{-1}. \tag{8}$$

Let constants A and B be defined as follows:

$$A = \frac{\xi(1 - \xi'')}{(1 - \xi)\xi''} \tag{9}$$

and

$$B = \frac{\xi(1 - \xi')}{(1 - \xi)\xi'}. \tag{10}$$

Since $\xi' < \xi < \xi''$, it follows that $A < 1$ and $B > 1$. Furthermore, it can be seen from relation (7) that the optimal procedure is to continue taking observations whenever the following relation is satisfied:

$$A < \frac{f_2(x_1) \cdots f_2(x_n)}{f_1(x_1) \cdots f_1(x_n)} < B. \tag{11}$$

Otherwise, sampling should be stopped.   If the first inequality in relation
(11) is not satisfied, the decision d_1 should be chosen.   If the second
inequality in relation (11) is not satisfied, the decision d_2 should be
chosen.   The problem of finding an optimal sequential decision procedure
has now been reduced to the problem of finding an optimal choice of the
constants A and B.

A sequential decision procedure of the type just described, in which
constants A and B are chosen, with $A < 1 < B$, and sampling is continued
whenever relation (11) is satisfied, is called a *sequential probability-ratio
test*.   Procedures of this type were the first formal methods of sequential
analysis studied by Wald (1947).

Before moving on to a detailed study of the properties of sequential
probability-ratio tests, we shall reconsider Example 2 of Sec. 12.6 in the
light of the results derived in this section.   We have just seen that in a
problem in which both the parameter space Ω and the decision space D
have exactly two points, the optimal sequential decision procedure is
either to choose a decision immediately without any observations or to
use a sequential probability-ratio test.   However, it was found in
Example 2 of Sec. 12.6 that for certain prior distributions, the optimal
procedure was to take exactly one observation and then to terminate
sampling.   It might seem at first that these results are mutually contra-
dictory.   Actually, they are not.   The two results together imply that
in this special example, the limits A and B of the optimal sequential
probability-ratio test are so close together that the likelihood ratio
$f_2(x_1)/f_1(x_1)$ computed from the first observed value x_1 must violate one
of the inequalities in the relation (11).

12.15 CHARACTERISTICS OF SEQUENTIAL PROBABILITY-RATIO TESTS

We shall now turn to the computation of the probabilities of choosing
decision d_1 or decision d_2 and to the computation of the expected sampling
cost for the sequential procedure defined by relation (11) of Sec. 12.14.
Although in many problems exact values for the probabilities of choosing
decisions d_1 and d_2 are not available, simple approximations can be
obtained.   It is seen from relation (11) of Sec. 12.14 that for each sequence
of observed values $X_i = x_i$ ($i = 1, \ldots, n$) which leads to the decision
d_1, the following relation must be satisfied:

$$f_2(x_1) \cdots f_2(x_n) \leq A f_1(x_1) \cdots f_1(x_n). \tag{1}$$

In other words, the value of the joint g.p.d.f. of the observations when
$W = w_2$ cannot be more than A times the value when $W = w_1$.   Since

this statement must be true for each sequence which leads to the decision d_1, we can obtain the following relation:

$$\Pr(d_1|W = w_2) \leq A\Pr(d_1|W = w_1). \tag{2}$$

Similarly, it can be seen from relation (11) of Sec. 12.14 that for each sequence of observed values $X_i = x_i$ $(i = 1, \ldots, n)$ which leads to the decision d_2, the following relation must be satisfied:

$$f_2(x_1) \cdots f_2(x_n) \geq Bf_1(x_1) \cdots f_1(x_n). \tag{3}$$

Hence,

$$\Pr(d_2|W = w_2) \geq B\Pr(d_2|W = w_1). \tag{4}$$

As will be shown a little later in Theorem 1, the probability is 0 that a given sequential probability-ratio test will continue indefinitely without ever leading either to decision d_1 or to decision d_2. Hence, for $i = 1, 2$,

$$\Pr(d_1|W = w_i) + \Pr(d_2|W = w_i) = 1. \tag{5}$$

When this relation is used together with the inequalities (2) and (4), we find that the point determined by the pair of probabilities $[\Pr(d_2|W = w_1),$ $\Pr(d_1|W = w_2)]$ must lie in the shaded region of Fig. 12.4.

If the terminal value of the likelihood ratio in relation (11) of Sec. 12.14 when decision d_1 is chosen is generally close to the lower limit A, then the two sides of the inequality (2) will be approximately equal. If the terminal value of this likelihood ratio when decision d_2 is chosen is generally close to the upper limit B, then the two sides of the inequality

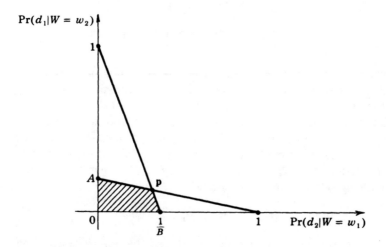

Fig. 12.4 Bounds for $\Pr(d_2|W = w_1)$ and $\Pr(d_1|W = w_2)$.

(4) will also be approximately equal. Hence, as an approximation, we can regard the relations (2) and (4) as equations and can write the following relations:

$$\Pr(d_1|W = w_1) = 1 - \Pr(d_2|W = w_1) \approx \frac{B-1}{B-A},$$
$$\Pr(d_1|W = w_2) = 1 - \Pr(d_2|W = w_2) \approx \frac{A(B-1)}{B-A}. \tag{6}$$

These approximate values of $\Pr(d_2|W = w_1)$ and $\Pr(d_1|W = w_2)$ are the coordinates of the point $\mathbf{p}$ in Fig. 12.4.

For $i = 1, 2, \ldots$, we shall now define the random variable Z_i as follows:

$$Z_i = \log \frac{f_2(X_i)}{f_1(X_i)}. \tag{7}$$

Furthermore, we shall let $a = \log A < 0$ and $b = \log B > 0$. Then the sequential probability-ratio test defined by relation (11) of Sec. 12.14 specifies that sampling should be continued whenever the following relation is satisfied:

$$a < \sum_{i=1}^{n} Z_i < b. \tag{8}$$

When either value $W = w_i$ $(i = 1, 2)$ is given, the random variables $X_1, X_2, \ldots$ are independent and identically distributed. Hence, the random variables $Z_1, Z_2, \ldots$ also have the same properties. The next result [Stein (1946)] indicates, for any given values of a and b, that sampling will terminate with probability 1 and that all moments of the distribution of the random number of observations N will be finite. The assumption that $\Pr(Z_i = 0) < 1$, which is made in the next theorem, simply eliminates the trivial situation in which the g.p.d.f.'s f_1 and f_2 are essentially identical.

Theorem 1 *Let $Z_1, Z_2, \ldots$ be a sequence of independent and identically distributed random variables such that $\Pr(Z_i = 0) < 1$. Consider a sequential procedure which specifies that the value of Z_1 is to be observed and that sampling is to be continued whenever relation (8) is satisfied for the values $Z_1, \ldots, Z_n$ which have been observed at any given stage. Sampling is to be terminated as soon as either inequality in relation (8) is not satisfied. Then $\Pr(N < \infty) = 1$ and $E(N^k) < \infty$ for $k = 1, 2, \ldots$.*

Proof If $\mathrm{Var}(Z_i) = 0$, then $\Pr(Z_i = \alpha) = 1$ for some value of α $(\alpha \neq 0)$ and the result is trivial. Suppose then that $\mathrm{Var}(Z_i) > 0$. Let r be a

fixed positive integer, and define the sequence of random variables
$U_1,\ U_2,\ \ldots$ as follows:

$$U_1 = \sum_{i=1}^{r} Z_i, \qquad U_2 = \sum_{i=r+1}^{2r} Z_i, \qquad U_3 = \sum_{i=2r+1}^{3r} Z_i, \qquad \cdots$$

(9)

Suppose that for some value of i, $|U_i| \geq b - a$. Then, if this value of
U_i is added to any value of $Z_1 + \cdots + Z_{(i-1)r}$ which is in the interval
(a, b), the sum $Z_1 + \cdots + Z_{ir}$ will be outside the interval. In this
case, $N \leq ir$. In other words, if $N > nr$, then $|U_i| < b - a$ for
$i = 1, \ldots, n$. Hence,

$$\Pr(N > nr) \leq \Pr(|U_i| < b - a; i = 1, \ldots, n)$$
$$= [\Pr(|U_1| < b - a)]^n. \qquad (10)$$

Since $\mathrm{Var}(U_1) = r\mathrm{Var}(Z_1)$, we can make $\mathrm{Var}(U_1)$ arbitrarily large by
choosing r sufficiently large. Hence, we can choose r so large that
$\Pr(|U_1| < b - a) = \gamma$, where γ is some number less than 1. It can then
be seen from relation (10) that for $n = 0, 1, 2, \ldots,$

$$\Pr(N > nr) \leq \gamma^n. \qquad (11)$$

It follows that $\lim_{n \to \infty} \Pr(N > n) = 0$ and that $\Pr(N < \infty) = 1$.

Furthermore, it follows from relation (11) that for $t \geq 0$, the following relation must be satisfied:

$$E(e^{tN}) = \sum_{j=1}^{\infty} e^{tj}\Pr(N = j)$$
$$\leq e^{tr}\Pr(0 < N \leq r) + e^{2tr}\Pr(r < N \leq 2r) + \cdots$$
$$\leq e^{tr}\Pr(N > 0) + e^{2tr}\Pr(N > r) + \cdots$$
$$\leq e^{tr} + e^{2tr}\gamma + e^{3tr}\gamma^2 + \cdots = \frac{e^{tr}}{1 - \gamma e^{tr}}. \qquad (12)$$

The final equality in this relation is satisfied for $\gamma e^{tr} < 1$ or, alternatively,
for

$$0 \leq t < \frac{1}{r} \log \frac{1}{\gamma}. \qquad (13)$$

Since $\Pr(N \geq 0) = 1$, it is also true that $E(e^{tN}) < \infty$ for $t < 0$.

Since the moment-generating function $E(e^{tN})$ is finite in a neighborhood of $t = 0$, it follows (see Sec. 3.6) that all the moments $E(N^k)$ of N
must be finite. ■

12.16 APPROXIMATING THE EXPECTED NUMBER OF OBSERVATIONS

We shall now consider the problem of approximating the expected number of observations $E(N)$ required by a sequential probability-ratio test. The next theorem was first given by Wald (1947). The proof we shall present here is similar to one given by Johnson (1959).

Theorem 1 *Suppose that Z_1, Z_2, . . . is a sequence of independent and identically distributed random variables such that $E(Z_i) = m$ ($i = 1$, 2, . . .). For any sequential procedure for which $E(N) < \infty$, the following relation must be satisfied:*

$$E(Z_1 + \cdots + Z_N) = mE(N). \tag{1}$$

Proof Let G denote the common d.f. of each of the random variables Z_i ($i = 1$, 2, . . .). Then, as explained below, the following relation can be derived:

$$
\begin{aligned}
E(Z_1 + \cdots + Z_N) &= \sum_{n=1}^{\infty} \int_{\{N=n\}} (z_1 + \cdots + z_n) \\
&\qquad\qquad \times dG(z_1) \cdots dG(z_n) \\
&= \sum_{n=1}^{\infty} \sum_{i=1}^{n} \int_{\{N=n\}} z_i \, dG(z_1) \cdots dG(z_n) \\
&= \sum_{i=1}^{\infty} \sum_{n=i}^{\infty} \int_{\{N=n\}} z_i \, dG(z_1) \cdots dG(z_n) \\
&= \sum_{i=1}^{\infty} \int_{\{N \geq i\}} z_i \, dG(z_1) \cdots dG(z_i) \\
&= \sum_{i=1}^{\infty} E(Z_i)\mathrm{Pr}(N \geq i) = mE(N). \tag{2}
\end{aligned}
$$

In explaining relation (2), we shall work backward from the final equality. That equality is satisfied because $E(N) = \sum_{i=1}^{\infty}\mathrm{Pr}(N \geq i)$. Next, the penultimate equality is satisfied because the event $\{N \geq i\}$ depends only on the variables $Z_1, \ldots, Z_{i-1}$. Finally, the reversal of the order of summation used to obtain the third equality will be justified if it can be shown that

$$\sum_{i=1}^{\infty} \sum_{n=i}^{\infty} \int_{\{N=n\}} |z_i| \, dG(z_1) \cdots dG(z_n) < \infty. \tag{3}$$

It can be shown that this relation is satisfied by following through the remaining steps of the relation (2) and replacing Z_i by $|Z_i|$.∎

Let Z denote a random variable having the common distribution of each of the random variables $Z_1, Z_2, \ldots$. Then it can be seen from Theorem 1 of Sec. 12.15 and from the theorem we have just proved that for the sequential probability-ratio test defined by relation (8) of Sec. 12.15, the following relation must be satisfied for $i = 1, 2$:

$$E(Z|W = w_i)E(N|W = w_i) = E(Z_1 + \cdots + Z_N|W = w_i)$$
$$= \sum_{j=1}^{2} E(Z_1 + \cdots + Z_N|d_j, W = w_i)\Pr(d_j|W = w_i). \quad (4)$$

If we ignore the difference between the value of the terminating sum $Z_1 + \cdots + Z_N$ and either the boundary value a or the boundary value b, we can approximate the final conditional expectations in Eq. (4) by the following simple values, regardless of whether $W = w_1$ or $W = w_2$:

$$E(Z_1 + \cdots + Z_N|d_1, W = w_i) \approx a,$$
$$E(Z_1 + \cdots + Z_N|d_2, W = w_i) \approx b. \quad (5)$$

Also, we can approximate the probabilities $\Pr(d_j|W = w_i)$ by applying Eqs. (6) of Sec. 12.15. Since $A = e^a$ and $B = e^b$, the results obtained from relations (4) and (5) are

$$E(N|W = w_1) \approx \frac{a(e^b - 1) + b(1 - e^a)}{(e^b - e^a)E(Z|W = w_1)},$$
$$E(N|W = w_2) \approx \frac{a(e^{a+b} - e^a) + b(e^b - e^{a+b})}{(e^b - e^a)E(Z|W = w_2)}. \quad (6)$$

From Jensen's inequality and the definition in Eq. (7) of Sec. 12.15, we can obtain the following relation:

$$E(Z|W = w_1) < \log E\left[\frac{f_2(X)}{f_1(X)}\,\middle|\, W = w_1\right]$$
$$= \log \int_S \frac{f_2(x)}{f_1(x)} f_1(x)\,d\mu(x) = \log 1 = 0. \quad (7)$$

Similarly, $E(Z|W = w_2) > 0$.

The relations (6) of Sec. 12.15 and the relations (6) of this section are actually equalities if the terminating sum $Z_1 + \cdots + Z_N$ must always be exactly equal to a or b. This condition will exist if each of the random variables Z_i can have only three values z, 0, and $-z$ and if both bounds a and b are multiples of z. To illustrate such a case, we shall consider Example 1 of Sec. 12.6.   If, in that example, we regard f_1 as the p.f. of any observation X when $W = \frac{1}{3}$ and we regard f_2 as the p.f. of X when $W = \frac{2}{3}$, then we can write

$$f_1(x) = (\tfrac{1}{3})^x(\tfrac{2}{3})^{1-x} \qquad \text{for } x = 0,\ 1,$$
$$f_2(x) = (\tfrac{2}{3})^x(\tfrac{1}{3})^{1-x} \qquad \text{for } x = 0,\ 1. \tag{8}$$

From Eq. (7) of Sec. 12.15, it follows that the value of Z can be only $\log 2$ or $-\log 2$.   Thus, in Example 1 of Sec. 12.6, the relations (6) of Sec. 12.15 and the relations (6) of this section are exact if a is a negative multiple of $\log 2$ and b is a positive multiple of $\log 2$.

In the decision problem in which the loss function is specified by Table 12.3, the risk $\rho(\xi, \delta)$ of any sequential decision procedure is

$$\rho(\xi, \delta) = \xi\lambda_1 \Pr(d_2|W = w_1) + (1 - \xi)\lambda_2 \Pr(d_1|W = w_2)$$
$$+ c[\xi E(N|W = w_1) + (1 - \xi)E(N|W = w_2)]. \tag{9}$$

For any fixed bounds $a < 0 < b$, the risk of the sequential probability-ratio test defined by relation (8) of Sec. 12.15 can be approximated by substituting in Eq. (9) the expressions in the relations (6) of Sec. 12.15 and the relations (6) of this section.   The optimal sequential probability-ratio test would then be found by determining the values of a and b which minimize Eq. (9).   A comparison of this minimum risk with $\rho_0(\xi)$, which is the risk from an immediate decision without any sampling, would determine whether the optimal procedure is to choose a decision immediately or to use the sequential probability-ratio test based on the minimizing values of a and b.

The procedure for determining the values of a and b that minimize the risk given by Eq. (9) is, in general, quite complicated.   However, when the sampling cost c is small, we can make the following approximations [Chernoff (1959)].   If the sampling cost is small, the optimal procedure will typically involve taking many observations.   In such a case, the bounds a and b of the optimal procedure will be far apart.   In other words, both $-a$ and b will be very large numbers.   From the relations (6) of Sec. 12.15 and the relations (6) of this section, we can then write the following approximations:

$$\Pr(d_2|W = w_1) \approx e^{-b}, \qquad\qquad \Pr(d_1|W = w_2) \approx e^{a},$$
$$E(N|W = w_1) \approx \frac{a}{E(Z|W = w_1)}, \qquad E(N|W = w_2) \approx \frac{b}{E(Z|W = w_2)}. \tag{10}$$

When these approximate values are substituted in Eq. (9) and we let $I_1 = -E(Z|W = w_1)$ and $I_2 = E(Z|W = w_2)$, it is found that the minimizing values of a and b are

$$
a \approx \log c - \log \frac{I_1\lambda_2(1 - \xi)}{\xi},
$$
$$
b \approx \log \frac{1}{c} + \log \frac{I_2\lambda_1\xi}{1 - \xi}. \tag{11}
$$

The values I_1 and I_2 are called *information numbers*. Their importance in statistical theory is discussed by Kullback (1959).

If we exploit the assumption that the cost c is very small, then the values of a and b given in the relations (11) can be approximated simply as

$$
a \approx \log c \quad \text{and} \quad b \approx \log \frac{1}{c}. \tag{12}
$$

From Eqs. (9), (10), and (12), it can be seen that the Bayes risk $\rho^*(\xi)$ from this optimal sequential probability-ratio test is

$$
\rho^*(\xi) \approx [\xi\lambda_1 + (1 - \xi)\lambda_2]c - (c \log c)\left(\frac{\xi}{I_1} + \frac{1 - \xi}{I_2}\right)
$$
$$
\approx (-c \log c)\left(\frac{\xi}{I_1} + \frac{1 - \xi}{I_2}\right). \tag{13}
$$

It is seen from the relations (11) and (12) that the bounds a and b of the optimal procedure depend mainly on the cost c and that they are relatively insensitive to the losses λ_1 and λ_2, the prior probability ξ, and the distribution of the observations. Furthermore, it is seen from the relation (13) that under the optimal procedure, the risk from an incorrect decision, which is of order c, is relatively small in comparison with the expected sampling cost, which is of order $(-c \log c)$.

Further Remarks and References

Optimum properties of the sequential probability-ratio test were developed initially by Wald and Wolfowitz (1948) and by Arrow, Blackwell, and Girshick (1949). Subsequently, extensions and modifications were developed by Dvoretzky, Kiefer, and Wolfowitz (1953), Blasbalg (1957), Savage (1957), Girshick (1958), Anderson and Friedman (1960), Ghosh (1961), Burkholder and Wijsman (1963), and Matthes (1963). These properties are also described by Lehmann (1959), sec. 3.12. Some other modifications and uses of these tests are given by DeGroot and Nadler (1958) and by Anderson (1960, 1964b).

EXERCISES

1. Consider a statistical decision problem in which the parameter space $\Omega = \{w_1, w_2\}$, the decision space $D = \{d_1, d_2\}$, and the loss function L is specified by the accompanying table. Suppose that the statistician can observe the values of a sequential random sample of observations at a cost of c per observation. Suppose also that when $W = w_1$, each observation has a uniform distribution on the interval $(0, 4)$ and that when $W = w_2$, each observation has a uniform distribution on the interval $(3, 7)$. Finally, suppose that the prior distribution of W is

$$\Pr(W = w_1) = \Pr(W = w_2) = \tfrac{1}{2}.$$

Show that the statistician should take at least one observation before choosing a decision in D if $c < \frac{15}{4}$ and that he should choose an immediate decision in D without any observations if $c > \frac{15}{4}$.

Table for Exercise 1

	d_1	d_2
w_1	0	10
w_2	10	0

2. Suppose that in Example 1 of Sec. 12.6, the space D contains three decisions instead of two and that the loss function is as given in the accompanying table. Compute $\rho_0(\xi)$, $\rho_1(\xi)$, and $\rho_2(\xi)$, where $\xi = \Pr(W = \tfrac{1}{3})$.

Table for Exercise 2

	d_1	d_2	d_3
$W = \tfrac{1}{3}$	0	20	8
$W = \tfrac{2}{3}$	20	0	8

3. (a) In Example 3 of Sec. 12.6, suppose that the distribution ξ of W is a normal distribution with mean μ and precision τ. Prove, by induction, that for $n = 1, 2, \ldots$, the risk $\rho_n(\xi)$ is specified as follows: Let $\beta_0 = 0$, $\beta_j = c[\tau + (j - 1)r](\tau + jr)$ for $j = 1, \ldots, n$, and $\beta_{n+1} = \infty$. Then, for $j = 0, 1, \ldots, n$,

$$\rho_n(\xi) = \frac{1}{\tau + jr} + jc \qquad \text{for } \beta_j < r \leq \beta_{j+1}.$$

(b) Suppose that the statistician is not permitted to take more than n observations. Show that, in accordance with the results obtained in part a, the optimal sequential decision procedure is to take exactly j observations, where j is the nonnegative integer such that $\beta_j < r \leq \beta_{j+1}$.

4. Suppose that $X_1, X_2, \ldots$ is a sequential random sample from a Bernoulli distribution with unknown parameter W. Suppose also that the value of W is to be

estimated and that for $0 \leq d \leq 1$ and $0 < w < 1$, the loss function L is specified as follows:

$$L(w, d) = \frac{(w - d)^2}{w(1 - w)}.$$

Suppose further that the prior distribution of W is a uniform distribution on the interval $(0, 1)$ and that the cost per observation is c. Show that the optimal sequential decision procedure is a procedure in which a fixed number of observations will be taken.

    5. Consider a sequential decision problem in which a sequential random sample can be taken from an exponential distribution with unknown parameter W and in which the cost per observation is c. Suppose that the risk $\rho_0(\phi)$ from choosing a decision without further sampling, when W has a distribution ϕ with mean μ and variance σ^2, is $\rho_0(\phi) = \sigma^2/\mu^2$. Show that if the prior distribution of W is a gamma distribution, then the optimal sequential decision procedure is a procedure in which a fixed number of observations will be taken.

    6. Consider a sequential decision problem in which a sequential random sample can be taken from a uniform distribution on the interval $(0, W)$ and in which the cost per observation is c. Suppose that for any distribution ϕ of W which has a finite variance, the risk $\rho_0(\phi)$ is as in Exercise 5. Show that if the prior distribution of W is a Pareto distribution, then the optimal sequential decision procedure is a procedure in which a fixed number of observations will be taken.

    7. Consider a sequential decision problem in which a sequential random sample can be taken from a uniform distribution on the interval (W_1, W_2) and in which the cost per observation is c. Suppose that the risk $\rho_0(\phi)$ from choosing a decision without further sampling, when W_1 and W_2 have the joint distribution ϕ, is $\rho_0(\phi) = \sigma^2/\mu^2$, where $\mu = E(W_2 - W_1)$ and $\sigma^2 = \text{Var}(W_2 - W_1)$. Show that if the prior joint distribution of W_1 and W_2 is a bilateral bivariate Pareto distribution, then the optimal sequential decision procedure is a procedure in which a fixed number of observations will be taken.

    8. Verify Eq. (17) of Sec. 12.6.

    9. Show that in any sequential decision problem, both the risk $\rho_n(\xi)$ of the optimal procedure which is terminated after not more than n observations have been taken $(n = 1, 2, \ldots)$ and the risk $\rho^*(\xi)$ of the optimal procedure δ^* are concave functions of the prior distribution ξ of W (see Theorem 1 of Sec. 8.4).

    10. Suppose that $X_1, X_2, \ldots$ is a sequential random sample from a Poisson distribution with unknown mean W and that the prior distribution ξ of W is a gamma distribution. In each of the following sequential decision problems, prove that Eq. (2) of Sec. 12.10 is satisfied:

        (a) $D = \{d_1, d_2\}$, and the loss function L is as specified by Eq. (1) of Sec. 11.8.

        (b) D is the real line, and the loss function L is $L(w, d) = (w - d)^2$.

    11. Show that the sequential decision problem given in Example 4 of Sec. 12.6 is stable; that is, show that Eq. (2) of Sec. 12.12 is satisfied for every distribution ϕ of W that can arise from the given prior normal distribution.

    12. Show that the sequential decision problems given in Examples 1 and 3 of Sec. 12.6 are stable.

    13. Consider a decision problem in which a sequential random sample can be taken from a Bernoulli distribution with unknown parameter W at a cost of c units per observation. Suppose that W must be estimated when the loss function is $L(w, d) = (w - d)^2$. Show that this decision problem is stable when the prior distribution of W is a beta distribution.

14. In the decision problem given in Exercise 13, let ξ denote the beta distribution with parameters α and β. Show that

$$\rho_1(\xi) = \frac{\alpha + \beta}{\alpha + \beta + 1} \rho_0(\xi) + \min\left\{\frac{1}{\alpha + \beta + 1} \rho_0(\xi), c\right\}.$$

15. Consider the decision problem given in Exercise 13. (a) Show that if the prior distribution of W is a beta distribution, then the optimal sequential decision procedure is a bounded procedure. (b) Find an upper bound for the maximum number of observations that can be taken.

16. In the decision problem given in Exercise 13, suppose that the cost per observation is $c = 0.01$ and that the prior distribution of W is a uniform distribution on the interval $(0, 1)$. (a) Show that the optimal sequential decision procedure requires that not more than three observations should be taken. (b) Determine the optimal procedure.

17. Consider a decision problem in which a sequential random sample can be taken from a Poisson distribution with unknown mean W at a cost of c units per observation. Suppose that W must be estimated when the loss function is

$$L(w, d) = (w - d)^2.$$

Show that this decision problem is stable when the prior distribution of W is a gamma distribution.

18. In the decision problem given in Exercise 17, let ξ denote the gamma distribution with parameters α and β. Show that

$$\rho_1(\xi) = \frac{\beta}{\beta + 1} \rho_0(\xi) + \min\left\{\frac{1}{\beta + 1} \rho_0(\xi), c\right\}.$$

19. Consider the decision problem given in Exercise 17, and let ξ denote the prior gamma distribution of W. For any positive integer k, let $r_k(\xi)$ denote the total risk from the procedure in which exactly k observations must be taken and then a decision d must be chosen. Show that if $\rho_1(\xi) \geq \rho_0(\xi)$, then $r_k(\xi) \geq \rho_0(\xi)$ for $k = 1$, 2, ...

20. Suppose that in a stable decision problem, the function $\gamma \in \Gamma$ satisfies the following relation for every distribution $\phi \in \Xi$:

$$\gamma(\phi) \leq E\{\gamma[\phi(X)]\} + c.$$

Prove that $\gamma(\phi) \leq \rho^*(\phi)$ for every distribution $\phi \in \Xi$.

21. In a given sequential decision problem, suppose that $\gamma_0 \in \Gamma$, and for any positive integer n, let γ_n be defined by Eq. (1) of Sec. 12.13. Show that if ξ is the prior distribution of W, then $\gamma_n(\xi)$ is the Bayes risk among all procedures which must be terminated after not more than n observations have been taken in a new decision problem with the following risk structure: If a decision is chosen immediately without any sampling or if sampling is terminated before all n observations have been taken, the risk from the optimal decision, not including the sampling cost, is $\rho_0(\phi)$, where ϕ is the distribution of W at the stage at which the decision is chosen. If, on the other hand, a decision is not chosen until all n observations have been taken, the risk from the optimal decision is $\gamma_0(\phi)$.

22. Consider a sequential decision problem in which $\Omega = \{w_1, w_2\}$, $D = \{d_1, d_2\}$, and the loss function L is as given in the accompanying table. Suppose that the

value of each observation X can only be either 0 or 1 and that the p.f. f_i of X when $W = w_i$ ($i = 1, 2$) is

$$f_1(1) = \tfrac{1}{4}, \qquad f_1(0) = \tfrac{3}{4};$$
$$f_2(1) = \tfrac{1}{2}, \qquad f_2(0) = \tfrac{1}{2}.$$

Suppose also that the cost c per observation is 1 unit.

(a) Compute the risks ρ_0, ρ_1, and ρ_2 as functions of the prior probability $\xi = \Pr(W = w_1)$.

(b) Assuming that $A = 0.1$ and $B = 5$, find an approximate value of the risk $\rho(\xi, \delta)$ of the sequential probability-ratio test δ in which sampling is continued whenever the relation (11) of Sec. 12.14 is satisfied.

Table for Exercise 22

	d_1	d_2
w_1	0	10
w_2	20	0

23. Consider Example 1 of Sec. 12.6, and suppose that the prior distribution of W is given by the values $\Pr(W = \tfrac{1}{3}) = \Pr(W = \tfrac{2}{3}) = \tfrac{1}{2}$. Utilize the symmetry of the problem to show that the optimal sequential probability-ratio test has symmetric boundaries of the form $a = -h \log 2$ and $b = h \log 2$, where h is a positive integer. Show that the risk of the procedure with these bounds is

$$\frac{20 + 3h(2^h - 1)}{2^h + 1}.$$

Show also that the value $h = 2$ yields the optimal sequential decision procedure. (Be sure to verify the fact that using this sequential procedure is better than choosing a decision without any sampling.)

24. Consider Example 1 of Sec. 12.6 again, and again suppose that the prior distribution of W is as given in Exercise 23. Suppose now, however, that the loss function L is as given in the accompanying table. Show that the optimal sequential decision procedure is a sequential probability-ratio test whose bounds are $a = -3 \log 2$ and $b = 3 \log 2$.

Table for Exercise 24

	d_1	d_2
$W = \tfrac{1}{3}$	0	40
$W = \tfrac{2}{3}$	40	0

25. Consider Example 1 of Sec. 12.6 again, and again suppose that the prior distribution of W is as given in Exercise 23. Suppose, however, that the loss function

is now as given in the accompanying table. Determine the optimal sequential decision procedure. Does this result agree with the result found in Example 2 of Sec. 12.6?

Table for Exercise 25

	d_1	d_2
$W = \frac{1}{3}$	0	10
$W = \frac{2}{3}$	10	0

26. Consider a sequential decision problem in which $\Omega = \{w_1, w_2\}, D = \{d_1, d_2\}$ and the loss function L is as given in the accompanying table. Suppose that each observation X can have only the value 1, 2, or 3 and that the p.f. of X when $W = w_i$ $(i = 1, 2)$ is

$$f_1(1) = \tfrac{1}{4}, \qquad f_1(2) = \tfrac{1}{2}, \qquad f_1(3) = \tfrac{1}{4};$$
$$f_2(1) = \tfrac{1}{4}, \qquad f_2(2) = \tfrac{1}{4}, \qquad f_2(3) = \tfrac{1}{2}.$$

Suppose also that the cost c per observation is 1 unit.

(a) Compute the risks ρ_0, ρ_1, and ρ_2 as functions of the prior probability $\xi = \Pr(W = w_1)$.

(b) Show that the optimal sequential decision procedure when $\xi = \tfrac{1}{2}$ is the sequential probability-ratio test in which $b = -a = 2 \log 2$.

Table for Exercise 26

	d_1	d_2
w_1	0	50
w_2	50	0

27. Consider a sequential decision problem in which $\Omega = \{w_1, w_2, w_3\}$, $D = \{d_1, d_2, d_3\}$, and the loss function L is as given in the accompanying table, where

Table for Exercise 27

	d_1	d_2	d_3
w_1	0	a_{12}	a_{13}
w_2	a_{21}	0	a_{23}
w_3	a_{31}	a_{32}	0

each value $a_{ij} > 0$. Suppose that the cost per observation is c. Also, suppose that each observation X can have only the value 1, 2, or 3 and that the p.f. of X when

$W = w_i$ $(i = 1, 2, 3)$ is

$$f(1|w_1) = \tfrac{1}{2}, \qquad f(2|w_1) = \tfrac{1}{2}, \qquad f(3|w_1) = 0;$$
$$f(1|w_2) = \tfrac{1}{2}, \qquad f(2|w_2) = 0, \qquad f(3|w_2) = \tfrac{1}{2};$$
$$f(1|w_3) = 0, \qquad f(2|w_3) = \tfrac{1}{2}, \qquad f(3|w_3) = \tfrac{1}{2}.$$

Finally, suppose that the prior distribution of W is specified by the values $\Pr(W = w_i) = \xi_i > 0$ $(i = 1, 2, 3)$. Describe the optimal sequential decision procedure.

28. Consider a sequential probability-ratio test in which sampling is continued whenever the relation (11) of Sec. 12.14 is satisfied, where $A < 1 < B$. Suppose that f_i is the p.f. of a Bernoulli distribution with parameter p_i $(i = 1, 2)$, where $0 < p_1 < p_2 < 1$. Show that for appropriate values of r_1, r_2, and s, this test is equivalent to continuing sampling whenever

$$r_1 + sn < \sum_{i=1}^{n} x_i < r_2 + sn.$$

Also, find expressions for r_1, r_2, and s.

29. Suppose that other conditions are the same as in Exercise 28 but that f_i is the p.d.f. of a normal distribution with mean μ_i and precision τ $(i = 1, 2)$, where $\mu_1 < \mu_2$ and $\tau > 0$. Demonstrate the same results as those required in Exercise 28.

30. Suppose that other conditions are the same as in Exercise 28 but that f_i is the p.f. of a Poisson distribution with mean λ_i $(i = 1, 2)$, where $0 < \lambda_1 < \lambda_2$. Demonstrate the same results as those required in Exercise 28.

31. Consider a stable decision problem, and suppose that for each distribution $\phi \in \Xi$ and for each possible value $x_1 \in S$ of a single observation X_1, the following relation must be satisfied:

$$\rho_0(\phi) - E\{\rho_0[\phi(X_1)]\} \geq \rho_0[\phi(x_1)] - E\{\rho_0[\phi(X_1, X_2)]|X_1 = x_1\}.$$

(a) Suppose that $\rho_1(\phi) = \rho_0(\phi)$ for some distribution $\phi \in \Xi$. Show that $\rho_2(\phi) = \rho_0(\phi)$ for that distribution. Then show by induction that

$$\rho_0(\phi) = \rho_n(\phi) = \rho^*(\phi)$$

for $n = 1, 2, \ldots$.

(b) If ϕ denotes the distribution of W at any stage, show that an optimal procedure is to continue sampling *if, and only if,* $\rho_1(\phi) < \rho_0(\phi)$.

32. [From Wald (1950).] Suppose that $X_1, X_2, \ldots$ is a sequential random sample from a uniform distribution on the interval $(W - \tfrac{1}{2}, W + \tfrac{1}{2})$, where the midpoint W of the interval is unknown, and suppose that W has a uniform prior distribution on some specified interval (α, β). Suppose that the value of W must be estimated when the loss function is $L(w, d) = (w - d)^2$ and that the cost per observation is c.

(a) Show that the posterior distribution of W at each stage of the procedure will be a uniform distribution. Show also that when ϕ is a uniform distribution on an interval of length s, the risk $\rho_0(\phi)$ from choosing a decision without further sampling is $\rho_0(\phi) = s^2/12$.

(b) Show that if ϕ is a uniform distribution on an interval of length s, then

$$E\{\rho_0[\phi(X)]\} = \begin{cases} \dfrac{s^2}{12}\left(1 - \dfrac{s}{2}\right) & \text{if } s \leq 1, \\[2mm] \dfrac{1}{12}\left(1 - \dfrac{1}{2s}\right) & \text{if } s > 1. \end{cases}$$

(c) Let $\alpha(s) = E\{\rho_0[\phi(X)]\}$, as given in part b. Show that since $(s^2/12) - \alpha(s)$ is an increasing function of s, then the inequality in Exercise 31 is satisfied for any uniform distribution ϕ and any observed value x_1.

(d) Suppose that the distribution of W at a given stage is a uniform distribution on an interval of length s. Use Exercise 31 to conclude that an optimal procedure at that stage is to take another observation if, and only if, $s > s^*$, where s^* is the unique solution of the following equation:

$$\frac{s^2}{12} - \alpha(s) = c.$$

33. Consider a sequential decision problem in which $\Omega = \{w_1, w_2\}$, $D = \{d_1, d_2\}$, and the loss function L is as given in the accompanying table. Suppose that if $W = w_1$, each observation in a sequential random sample has a uniform distribution

Table for Exercise 33

	d_1	d_2
w_1	0	200
w_2	200	0

on the interval $(0, 1)$ and that if $W = w_2$, each observation has a uniform distribution on the interval $(0, 2)$. Also, suppose that the cost c per observation is 1 unit and that the prior distribution of W is specified by the values $\Pr(W = w_1) = \Pr(W = w_2) = \frac{1}{2}$. Prove that the optimal sequential decision procedure is to terminate sampling as soon as an observed value x lies in the interval $1 \le x \le 2$ and to terminate the procedure after seven observations have been taken if none of these observed values lies in this interval.

13

optimal

stopping

13.1 INTRODUCTION

The problems we considered in Chap. 12 fall under the general heading of problems of *optimal stopping*. In each of these problems, the statistician takes observations sequentially, and at each stage he must decide whether to stop and suffer a specified stopping risk or continue and take the next observation at some specified sampling cost. Each sequential statistical decision problem treated in Chap. 12 was reduced to a problem of optimal stopping by developing the stopping risks and the distribution of the observations from the given specification of the parameter space, decision space, loss function, and prior distribution of the parameters and from the understanding that only Bayes decisions would ever be chosen. Clearly, there are broad classes of problems involving optimal stopping in which the stopping risks, the joint distribution of the observations, and the sampling costs can be assigned directly, without reference to the structure of any underlying statistical decision problem. These problems are characterized as follows: There is a finite or infinite sequence of potential observations having some specified joint distribution. Although the statistician can take these observations sequentially, the procedure is subject to given stopping risks and sampling costs. Hence, his objective is to find a stopping rule which minimizes the expected total risk. In such a situation it is irrelevant, or even inappropriate, to try to specify the components of some underlying statistical decision problem.

Breiman (1964) has written an introductory survey of this topic.

In the next several sections, we shall discuss some special problems of optimal stopping. Then the existence of optimal stopping rules will be demonstrated for a general class of problems which includes the sequential statistical decision problems treated in Chap. 12. In the final sections of the present chapter we shall discuss martingales and Markov processes, which are two important classes of problems.

One new feature of this chapter, in contrast to our development of statistical decision problems, is that it will be more convenient to formulate a problem in terms of the statistician's utility or gain, rather than in terms of his loss or risk. Thus, in the problems to be considered here, an optimal stopping rule is one that maximizes the statistician's expected gain.

We shall begin with some special problems of optimal stopping which have received a great deal of attention in the statistical literature.

13.2 THE STATISTICIAN'S REWARD

The problem to be treated now has been called, among other names, the marriage problem, the secretary problem, and the beauty-contest problem. We shall present it in more neutral terms, but the reader who wishes to do so will have no difficulty in translating the presentation into more congenial settings.

Suppose that an adventurous statistician in an exotic land has won the favor of its ruler, who tells him that he will be permitted to inspect a specified number n ($n \geq 2$) of the most beautiful works of art in the kingdom and that he may then select one of them as a reward. The objects are to be shown to the statistician in a random order, and he is to inspect them sequentially. After having inspected any number r ($1 \leq r \leq n$) of the objects, the statistician will be able to rank them from most preferable (rank 1) to least preferable (rank r). If another object is then inspected, he can insert it into his ranking without otherwise changing the ranking. At any stage, the statistician can either stop the inspection process and accept the object just inspected or continue and inspect another object. Once he has decided not to accept a particular object, he can never go back and select it at a later stage. If he has not stopped and selected an object earlier, then he must accept the nth object as his reward. The basic feature of this problem, and the reason for formulating it in the above terms, is that the *only* relevant information which the statistician obtains about each object is its relative rank among those that have already been inspected.

For $i = 1, \ldots, n$, let $U(i)$ be the utility to the statistician of selecting the object that has rank i among all n objects. We shall assume that $U(1) \geq U(2) \geq \cdots \geq U(n)$. Under any specified stopping rule, let the random variable X denote the rank of the object that is selected. The statistician's problem is to find a stopping rule which maximizes $E[U(X)]$. No sampling cost or inspection cost need be considered in this problem.

For $a = 1, \ldots, r$ and $r = 1, \ldots, n$, we shall let $U^*(a, r)$ denote the expected utility of the optimal continuation when r objects have already been inspected and the rth object has been found to have rank a among the r that have been inspected. Also, we shall let $U_0(a, r)$ denote the expected utility if the rth object is selected and inspection is thus terminated. Since inspection must be terminated with the nth object, if not earlier, then the following relation must be satisfied:

$$U^*(a, n) = U_0(a, n) = U(a) \qquad a = 1, \ldots, n. \tag{1}$$

Now consider the probability that the object which has rank a among the first r objects inspected actually has rank b among all n objects. This event can also be described as follows: In a random sample of r objects from the entire set of n objects, exactly $a - 1$ objects are from the highest ranking $b - 1$ objects, one object has the rank b, and the remaining $r - a$ objects are from the lowest ranking $n - b$ objects. Hence, the probability is

$$\frac{\binom{b-1}{a-1}\binom{n-b}{r-a}}{\binom{n}{r}}. \tag{2}$$

The rank b must lie between the bounds $a \leq b \leq n - r + a$. It follows from the definition of U_0 and the expression (2) that for $a = 1, \ldots, r$ and $r = 1, \ldots, n$,

$$U_0(a, r) = \sum_{b=a}^{n-r+a} U(b) \frac{\binom{b-1}{a-1}\binom{n-b}{r-a}}{\binom{n}{r}}. \tag{3}$$

Since it is assumed that the objects are inspected in a random order, the $(r + 1)$st object to be inspected is equally likely to have any one of the ranks $1, \ldots, r + 1$ among the first $r + 1$ objects. If its rank is b

$(b = 1, \ldots, r + 1)$, then the expected utility of the optimal continuation is $U^*(b, r + 1)$. Since each of the $r + 1$ values of b has probability $1/(r + 1)$, it is evident that after r objects have been inspected, the expected utility of inspecting one more object and then continuing in an optimal way is

$$\frac{1}{r + 1} \sum_{b=1}^{r+1} U^*(b, r + 1). \tag{4}$$

It can be seen from the expression (4) and the definitions of U^* and U_0 that for $a = 1, \ldots, r$ and $r = 1, \ldots, n - 1$, U^* satisfies the following functional equation:

$$U^*(a, r) = \max \left\{ U_0(a, r), \frac{1}{r + 1} \sum_{b=1}^{r+1} U^*(b, r + 1) \right\}. \tag{5}$$

Thus, the values of $U^*(a, r)$ can be found from Eqs. (1), (3), and (5) by starting from the last stage and applying the technique of backward induction. The optimal procedure is to continue inspection if $U^*(a, r) > U_0(a, r)$ and to stop when $U^*(a, r) = U_0(a, r)$.

13.3 CHOICE OF THE UTILITY FUNCTION

We shall now consider in detail a special choice of the utility function U for the statistician's reward. We shall suppose that $U(1) = 1$ and that $U(b) = 0$ for $b = 2, \ldots, n$. This utility function means that the statistician must maximize the probability of selecting the object which has rank 1 among all n objects. It can then be seen from Eq. (3) of Sec. 13.2 that for $r = 1, \ldots, n$, the value of U_0 is

$$\begin{cases} U_0(1, r) = \dfrac{r}{n}, \\ U_0(a, r) = 0 \qquad a = 2, \ldots, r. \end{cases} \tag{1}$$

Since $U_0(a, r) = 0$ for $a > 1$, it is clear that $U^*(a, r) > U_0(a, r)$ for $a > 1$ and $r = 1, \ldots, n - 1$. In other words, the statistician should certainly continue inspection if the rank a of the object he has just inspected is anything other than 1. It follows that for each fixed value of r $(r = 1, \ldots, n)$, the value of $U^*(a, r)$ will be the same for $a = 2, \ldots, r$. If we let $V(r)$ denote this value, it follows from Eq. (5) of Sec. 13.2 that

$$V(r) = \frac{1}{r + 1} \sum_{b=1}^{r+1} U^*(b, r + 1). \tag{2}$$

For $a = 1$, Eq. (5) of Sec. 13.2 can be written in the following form:

$$U^*(1, r) = \max \left\{ \frac{r}{n}, \frac{1}{r+1} U^*(1, r+1) + \frac{r}{r+1} V(r+1) \right\}. \quad (3)$$

Also, Eq. (2) can be written as follows:

$$V(r) = \frac{1}{r+1} U^*(1, r+1) + \frac{r}{r+1} V(r+1). \quad (4)$$

Furthermore, by Eq. (1) of Sec. 13.2, $U^*(1, n) = 1$ and $V(n) = 0$. Hence, Eqs. (3) and (4) can be solved recursively by working backward from the last stage. After the rth object has been inspected, the optimal procedure is of the following form: If the rank a of this object is not 1, then inspection should be continued. If the object has rank 1 among the r objects which have been inspected, then inspection should be stopped if $U^*(1, r) = r/n$ and should be continued if $U^*(1, r) > r/n$. The problem now is to find a useful expression for $U^*(1, r)$.

Lemma 1 *Suppose that $U^*(1, r) = r/n$ for some value of r $(r < n)$. Then $U^*(1, r + 1) = (r + 1)/n$.*

Proof Suppose that $U^*(1, r + 1) > (r + 1)/n$. Then, if r is replaced by $r + 1$ in Eqs. (3) and (4), it can be seen that $U^*(1, r + 1) = V(r + 1)$. Thus, by Eq. (3),

$$U^*(1, r) = \max \left\{ \frac{r}{n}, U^*(1, r+1) \right\} \geq \max \left\{ \frac{r}{n}, \frac{r+1}{n} \right\} = \frac{r+1}{n}. \quad (5)$$

But this contradicts the hypothesis that $U^*(1, r) = r/n$.∎

Suppose that under the optimal procedure, inspection should be stopped if the observed rank of the rth object is 1. Then, by Lemma 1, inspection should be stopped also if the observed rank of any later object is 1. Hence, the problem is to find the smallest value of r for which $U^*(1, r) = r/n$.

Lemma 2 *If $U^*(1, r) = r/n$ for some value $r > 1$, then*

$$V(r - 1) = \frac{r-1}{n} \left(\frac{1}{n-1} + \frac{1}{n-2} + \cdots + \frac{1}{r-1} \right). \quad (6)$$

Proof The lemma is proved by backward induction on r. It is correct for $r = n$ because the following conditions exist: $U^*(1, n) = 1$; $V(n) = 0$; and from Eq. (4), $V(n - 1) = 1/n$. We shall assume then that the lemma is correct for $r = k + 1$, and we shall demonstrate its correctness for $r = k$. From Eq. (4) and the hypothesis that $U^*(1, k) = k/n$, it follows that

$$V(k - 1) = \frac{1}{n} + \frac{k - 1}{k} V(k). \tag{7}$$

Furthermore, it follows from Lemma 1 that $U^*(1, k + 1) = (k + 1)/n$. Hence, by the induction hypothesis, we can obtain the following equation:

$$V(k - 1) = \frac{1}{n} + \frac{k - 1}{k} \left[\frac{k}{n} \left(\frac{1}{n - 1} + \frac{1}{n - 2} + \cdots + \frac{1}{k} \right) \right]$$

$$= \frac{k - 1}{n} \left(\frac{1}{n - 1} + \frac{1}{n - 2} + \cdots + \frac{1}{k} + \frac{1}{k - 1} \right) \cdot \blacksquare \tag{8}$$

From Eqs. (3) and (4),

$$U^*(1, r) = \max \left\{ \frac{r}{n}, V(r) \right\}. \tag{9}$$

We shall let r^* be the smallest value of r for which $U^*(1, r) = r/n$. It can be seen from Eq. (9) and Lemmas 1 and 2 that when $r = r^*$, the following relation must be satisfied:

$$\frac{r}{n} \geq \frac{r}{n} \left(\frac{1}{n - 1} + \frac{1}{n - 2} + \cdots + \frac{1}{r} \right). \tag{10}$$

Furthermore, it follows from Lemma 2 that $V(r^* - 1)$ is equal to the value of the right side of the inequality (10) when $r = r^* - 1$. Hence, from Eq. (9) and the definition of r^*, it is seen that the inequality (10) is not satisfied when $r = r^* - 1$. Finally, when the factor r/n is canceled from both sides of the inequality (10), the truth of the following statement becomes evident. If the inequality (10) is not satisfied for a given value of r, then this relation is not satisfied for any smaller value of r. Hence, r^* is the smallest value of r that will satisfy the relation (10) or, equivalently, that will satisfy the following relation:

$$\frac{1}{n - 1} + \frac{1}{n - 2} + \cdots + \frac{1}{r} \leq 1. \tag{11}$$

The optimal procedure can now be characterized as follows: Let r^* be the smallest positive integer that satisfies the relation (11). The procedure should not be stopped before r^* objects have been inspected. If the (r^*)th object has rank 1, the inspection process should be stopped and that object should be selected. Otherwise, the inspection process should be continued until an object of rank 1 is found.

If the (r^*)th object has rank 1 among those which have been inspected and the statistician therefore selects it, then the probability that he is selecting the object which actually has rank 1 among all n

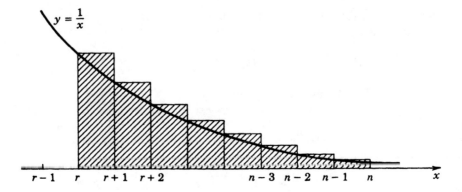

Fig. 13.1 An upper bound for log (n/r).

objects is $U_0(1, r^*) = r^*/n$. If the (r^*)th object does not have rank 1 among those which have been inspected and the statistician therefore continues the inspection process, then the probability that he will select the object which actually has rank 1 among all n objects is

$$V(r^*) = \frac{r^*}{n} \left(\frac{1}{n-1} + \frac{1}{n-2} + \cdots + \frac{1}{r^*} \right). \tag{12}$$

By relation (11), the value of $V(r^*)$ thus computed will be close to r^*/n. Hence, in either event, the probability of success is about r^*/n.

A simple approximation to r^* can be given. In Fig. 13.1, the area under the curve between $x = r$ and $x = n$ is log (n/r) and the sum of the areas of the rectangles is the sum on the left side of the relation (11). In Fig. 13.2, the area under the curve between $x = r - 1$ and $x = n - 1$

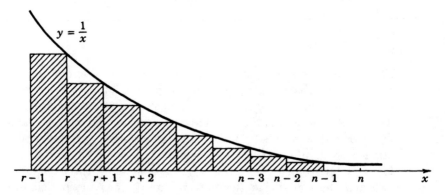

Fig. 13.2 A lower bound for log $[(n - 1)/(r - 1)]$.

is log $[(n - 1)/(r - 1)]$ and the sum of the areas of the rectangles is again the sum on the left side of the relation (11). Hence, the following relation must be satisfied:

$$\log \frac{n}{r} = \int_r^n \frac{dx}{x} < \frac{1}{n-1} + \frac{1}{n-2} + \cdots + \frac{1}{r}$$
$$< \int_{r-1}^{n-1} \frac{dx}{x} = \log \frac{n-1}{r-1}. \quad (13)$$

It follows from the relation (13) that for large values of n, the value of log (n/r^*) is about equal to 1. Hence, $r^* \approx n/e$. Accordingly, the optimal procedure is to inspect approximately the proportion $1/e$ of the objects and to select the first one thereafter which has rank 1. The probability of selecting the object which actually has rank 1 in the whole set of n objects will be approximately $1/e$, even for arbitrarily large values of n.

Further Remarks and References

The development in this section is essentially that given by Lindley (1961a). He also discusses the version presented in Exercise 3, in which the statistician must minimize the expected rank of the object he selects. That version has also been studied by Chow, Moriguti, Robbins, and Samuels (1964). They have shown that as $n \to \infty$, the expected rank under the optimal procedure is

$$\prod_{t=1}^{\infty} \left(1 + \frac{2}{t}\right)^{1/(t+1)} \approx 3.87.$$

Gilbert and Mosteller (1966) have made an extensive study of the problem given in this section and some related problems, including the version given in Exercise 1, in which the statistician must maximize the probability of selecting one of the k objects of highest rank. They also list several references on problems of this type.

13.4 SAMPLING WITHOUT RECALL

The problem discussed in Secs. 13.2 and 13.3 is highly special in that the only information which the statistician gets from inspecting an object is the rank of that object relative to the objects inspected earlier. In the context of this fantasy, it is assumed that at the beginning of the process, the statistician knows nothing about the overall quality of the n works of art he will see. Hence, even though the statistician realizes that the first object inspected by him has extraordinary beauty and value, he must assume that it is just as likely that this object is one of the lowest ranking of the n objects as that the object is one of the highest ranking. In the problems which we shall now discuss, the statistician has much more information. We shall assume that the statistician can assign a numerical utility to each object as soon as he inspects it and, most importantly, that he knows the distribution of these utilities over the entire population of objects from which the n objects can be regarded as a sample.

More formally, we shall suppose that a sequential random sample $X_1, X_2, \ldots$ can be taken from a distribution on the real line for which it is known that the d.f. is F. It is assumed that the mean of this distribution exists. We shall suppose also that there is a given upper bound n ($n \geq 2$) on the number of observations which can be taken. At the jth stage of the sampling process ($j = 1, \ldots, n - 1$), after the values $X_1 = x_1, \ldots, X_j = x_j$ have been observed, the statistician either can stop and accept the value x_j as his gain or can continue sampling and observe X_{j+1}. He must stop and accept the final observed value x_n as his gain if he has not stopped earlier. As usual, we shall let N denote the random number of observations which are taken under a specified stopping rule. The problem is to find a stopping rule that maximizes the expected gain $E(X_N)$. This is a problem of *sampling without recall* because, as in the problem discussed in Secs. 13.2 and 13.3, once the statistician decides to pass up a particular observation without stopping, he cannot return to that observation at a later time and accept its value as his gain.

At any stage of the sampling process, after at least one observation has been taken, the statistician's position is characterized by the value x of the most recent observation taken and by the number j of observations which remain before he will be forced to stop. For $j = 0, 1, \ldots, n - 1$, we shall let $V_j(x)$ denote the expected gain from the optimal procedure under these conditions. Hence, after the values $X_1 = x_1, \ldots, X_i = x_i$ have been observed, the expected gain from the optimal procedure is $V_{n-i}(x_i)$. Since the statistician must stop and accept the final observed value x as his gain when no further observations are available, it follows that $V_0(x) = x$ for $-\infty < x < \infty$.

Furthermore, for $j = 1, \ldots, n$, we shall let v_j denote the maximum expected gain that can be obtained when j observations remain to be taken and at least one of these observations is taken before sampling is stopped. Since the statistician is sampling without recall from a known distribution in this problem, the expected gain v_j from taking at least one further observation and then continuing in an optimal way will be a constant whose value depends only on the number j of observations remaining and does not depend on the values of the observations which have already been taken. After one further observation X has been taken, the expected gain from the optimal continuation over the remaining $j - 1$ stages will be

$V_{j-1}(X)$. Hence, the expected gain v_j satisfies the following equation:

$$v_j = E[V_{j-1}(X)] = \int_{-\infty}^{\infty} V_{j-1}(x) \, dF(x). \tag{1}$$

For $j = 1, \ldots, n - 1$, the expected gain $V_j(x)$ from the optimal procedure must be the maximum of the gain x from stopping and the expected gain v_j from continuing. Thus,

$$V_j(x) = \max \{x, v_j\}. \tag{2}$$

Since $V_0(x) = x$, the functions V_1, V_2, $\ldots$ and the numbers v_1, v_2, $\ldots$ can be determined successively from Eqs. (1) and (2). Since a maximum of n observations can be taken and at least one observation must be taken, the expected gain from the optimal procedure is v_n. If, at the beginning of the sampling process, the statistician has the additional option of accepting a gain x_0 without taking any observations, then the expected gain from the optimal procedure is $V_n(x_0)$.

Now let T_F be the transform of the d.f. F defined by Eq. (9) of Sec. 11.8. This transform exists since we are assuming that the mean of F exists. It can be shown (see Exercise 5) that the transform T_F has the following property:

$$E[\max \{X, s\}] = s + T_F(s) \qquad -\infty < s < \infty. \tag{3}$$

It can be seen from Eqs. (1) and (2) that for $j = 1, 2, \ldots,$

$$v_{j+1} = v_j + T_F(v_j). \tag{4}$$

By Eq. (1), $v_1 = E(X)$. Hence, the sequence $v_1, v_2, \ldots$ can be computed successively from Eq. (4). If not more than n observations can be taken, an optimal procedure is to continue sampling whenever an observed value $x_j < v_{n-j}$ and to stop the process as soon as some observed value $x_j \geq v_{n-j}$. The expected gain from this procedure is v_n.

Some specific examples are presented in Exercises 6 to 9. Problems like these have been discussed by Gilbert and Mosteller (1966), Guttman (1960), and Moser (1956)—and also by Cayley (1875), more than 80 years before the others.

13.5 FURTHER PROBLEMS OF SAMPLING WITH RECALL AND SAMPLING WITHOUT RECALL

We shall now consider the problem of sampling without recall from a known distribution when there is no upper bound on the number of observations which can be taken but there is a fixed cost c per observation. Specifically, we shall let $X_1, X_2, \ldots$ be a sequential random sample from a distribution for which it is known that the d.f. is F, and we shall suppose that if the statistician stops the sampling process after the values $X_1 = x_1, \ldots, X_n = x_n$ have been observed, his gain is $x_n - cn$. Hence, the problem is to find a stopping rule which maximizes $E(X_N - cN)$. Since there is no upper bound on the number of observations which the statistician can take, we cannot be certain that there exists an optimal stopping rule. However, it will be shown in Sec. 13.9 that if the variance of the d.f. F is finite, then the maximum expected gain v^* among all stopping rules will be finite and there will exist a stopping rule whose expected gain is v^*. Accordingly, we shall assume here that such an optimal procedure exists, and we shall construct this procedure and determine the value of v^*.

After the first observation X has been taken, the statistician either can stop the sampling process or can continue to take observations. If he stops, his gain will be the value of X minus the sampling cost c. Suppose that he continues the sampling process. Then, since he is sampling without recall, he is in precisely the same position that he was in at the beginning of the process, but he has already spent the amount c for the first observation. Hence, if he continues sampling, the expected gain from the optimal continuation is again v^* minus the cost c of the observation already taken.

It follows that after the first observation X has been taken, an optimal procedure is to continue sampling if $X < v^*$ and to terminate sampling if $X \geq v^*$. Furthermore, the expected gain from this optimal procedure will be $E[\max \{X, v^*\} - c]$. However, the expected gain from the optimal procedure is assumed to be v^*. Therefore, we have developed the following equation:

$$v^* = E[\max \{X, v^*\}] - c. \tag{1}$$

By Eq. (3) of Sec. 13.4, we can rewrite Eq. (1) as follows:

$$T_F(v^*) = c. \tag{2}$$

Since $c > 0$, it follows from the properties of the function T_F given in Sec. 11.8 that there is a unique number v^* which satisfies Eq. (2).

The optimal stopping rule and its expected gain are characterized by this unique value of v^*. An optimal rule is to continue sampling whenever an observed value $x_j < v^*$ and to stop the process as soon as some observed value $x_j \geq v^*$. The expected gain from this procedure is v^*.

In particular, when the sample is taken from a normal distribution with mean μ and standard deviation σ, we find from Eq. (3) of Sec. 11.9 that

$$v^* = \mu + \sigma\Psi^{-1}\left(\frac{c}{\sigma}\right). \tag{3}$$

Sampling with Recall

We shall now consider the problem of *sampling with recall*. We shall suppose again that a sequential random sample $X_1, X_2, \ldots$ can be taken at a cost of c units per observation from a distribution whose d.f. is F. We shall assume that if the statistician stops sampling after he has observed the values $X_1 = x_1, \ldots, X_n = x_n$, his gain is max $\{x_1, \ldots, x_n\}$ $- cn$. In other words, the statistician can stop sampling at any stage, select any observation which he has taken earlier, and accept as his gain the value of that observation minus the total sampling cost.

The fundamental point to be made here is that since the observations are independent and identically distributed, sampling with recall extends no additional advantage to the statistician over sampling without recall, for the following reason. When the statistician takes observations either with or without recall, he passes up an observation and continues the sampling process whenever the value of the observation is less than the gain he expects from further sampling. Moreover, this expected gain remains fixed from stage to stage, regardless of the number of observations previously taken and regardless of their values. Hence, the statistician cannot acquire any new information from further observations that would induce him to terminate the sampling process and choose for his gain the value of an observation that had been passed up earlier because its value was insufficient. Even when a statistician is permitted to sample with recall, an optimal procedure is to continue the sampling process until some observed value $x_j \geq v^*$, where v^* is the unique number which satisfies Eq. (2). The expected gain from this procedure is v^*.

When the distribution from which the sequential sample is being taken involves unknown parameters with a specified prior distribution, the situation changes. In such a case, the expected gains from sampling with recall and sampling without recall will be different. An observation which is passed up as having an insufficient value at an early stage of the sampling process, when the statistician believes that the distribution is likely to be very favorable for him, may later be regarded as acceptable, after additional observations have indicated to the statistician that the distribution probably is not so favorable as he had previously thought. When a statistician is sampling with recall, he will stop the sampling process and will accept as his gain the value of an observation taken earlier. We shall illustrate these remarks in the next two sections by considering the problem of sampling from a normal distribution with unknown mean.

Further Remarks and References

Problems of the type given in this section have been studied by MacQueen and Miller (1960), Derman and Sacks (1960), and Chow and Robbins

(1961, 1963). Other variations have been studied by McCall (1965) and Elfving (1967).

Problems involving sampling from an unknown distribution, which are similar to those to be treated in the next two sections, were considered by Sakaguchi (1961).

The problem of finding a stopping rule that maximizes $E[(X_1 + \cdots + X_N)/N]$ and related problems have been studied by Chow and Robbins (1965, 1967a), Teicher and Wolfowitz (1966), Dubins and Teicher (1967), and Dvoretzky (1967).

13.6 SAMPLING WITHOUT RECALL FROM A NORMAL DISTRIBUTION WITH UNKNOWN MEAN

Suppose that a sequential random sample $X_1, X_2, \ldots$ can be taken at a cost of c units per observation from a normal distribution with unknown mean W and specified precision. Without loss of generality, we can assume that this precision is 1. As usual, we shall assume that the prior distribution of W is normal. Hence, the posterior distribution at each stage of the sampling process will also be normal. The statistician must find a stopping rule that maximizes his expected gain $E(X_N - cN)$.

In this section we shall assume that an optimal stopping rule exists and that the expected gain of the optimal stopping rule is finite. This assumption will be justified in Sec. 13.9. It will be shown in that section that the assumption is correct for a sequential random sample from a distribution which involves a parameter whose value is unknown if the variance of the common marginal distribution of each observation is finite. It is clear that this requirement is satisfied in the problem of sampling from a normal distribution which is being considered here. If the variance of the prior distribution of W is σ_0^2, then the variance of the marginal distribution of each observation is $\sigma_0^2 + 1$.

Throughout this section and the next one, we shall use the notation $Z \sim \mathfrak{N}(\mu, \tau)$ to indicate that a random variable Z has a normal distribution with mean μ and *precision* τ.

At each stage of the sampling process, the statistician's position is characterized by the triple (r, μ, τ), where r is the value of the most recent observation taken and μ and τ are the mean and the precision of the current posterior distribution of W. Thus, r represents the gain to the statistician, exclusive of the amount he has already spent on sampling, if he terminates the sampling process. As we know, the amount already spent on sampling is irrelevant to his decision as to whether he should terminate the process or continue sampling. If $W \sim \mathfrak{N}(\mu, \tau)$ and another observation is taken whose value is x, then we know from Theorem 1 of

Sec. 9.5 that the posterior distribution of W is $W \sim \mathfrak{N}[\mu(x), \tau + 1]$, where

$$\mu(x) = \frac{\tau\mu + x}{\tau + 1}. \tag{1}$$

Hence, the position (r, μ, τ) is transformed into the new position $[x, \mu(x), \tau + 1]$. Furthermore, it follows from Exercise 23 of Chap. 11 that when $W \sim \mathfrak{N}(\mu, \tau)$, the marginal distribution of the next observation X is $X \sim \mathfrak{N}[\mu, \tau/(\tau + 1)]$.

We shall let $V(r, \mu, \tau)$ be the expected gain, exclusive of the amount already spent on sampling, when the optimal procedure is continued from the position (r, μ, τ). At this stage the statistician either may stop sampling and accept the gain r or may, at a cost c, take another observation and then continue the process in an optimal way. It follows from the above discussion and the principle of optimality that if $X \sim \mathfrak{N}[\mu, \tau/(\tau + 1)]$, then V satisfies the following functional equation:

$$V(r, \mu, \tau) = \max \{r, E[V(X, \mu(X), \tau + 1)] - c\}. \tag{2}$$

Furthermore, for any position (r, μ, τ) and any constant k, the function V satisfies the following invariance relation:

$$V(r - k, \mu - k, \tau) = V(r, \mu, \tau) - k. \tag{3}$$

This relation can be verified by noting that the effect of reducing μ by k is equivalent to the effect of working with new variables $X' = X - k$ and $W' = W - k$. In other words, the effect is the same as if the value of each future observation were reduced by k. If the gain r which results from terminating the process is also reduced by k, then it is obvious that the gain of any procedure starting from the position (r, μ, τ) is likewise reduced. The relation (3) is thus obtained.

It follows from Eq. (3) that at each stage of the sampling process, the statistician's position can be reduced through a translation to one for which $\mu = 0$. Accordingly, we shall let $U(r, \tau) = V(r, 0, \tau)$. Then U satisfies the functional equation in the following lemma.

Lemma 1 *If Y_τ is a random variable such that $Y_\tau \sim \mathfrak{N}[0, (\tau + 1)/\tau]$, then, for any values of r and τ $(\tau > 0)$,*

$$U(r, \tau) = \max \{r, E[U(Y_\tau, \tau + 1)] - c\}. \tag{4}$$

Proof From Eqs. (1) to (3) and from the definition of U, we can write the following relation:

$$U(r, \tau) = \max \left\{ r, E\left[V\left(X, \frac{X}{\tau + 1}, \tau + 1\right)\right] - c \right\}$$
$$= \max \left\{ r, E\left[U\left(\frac{\tau X}{\tau + 1}, \tau + 1\right) + \frac{X}{\tau + 1}\right] - c \right\}. \tag{5}$$

Since $X \sim \mathfrak{N}[0, \tau/(\tau + 1)]$, then $E(X) = 0$. Also, if we let

$$Y_\tau = \tau X/(\tau + 1),$$

then Y_τ has the distribution specified in the lemma. Hence, Eq. (4) follows from the relation (5).■

For any value of $\tau > 0$, let the distribution of the random variable Y_τ be as specified in Lemma 1, and let $\alpha(\tau)$ be defined as follows:

$$\alpha(\tau) = E[U(Y_\tau, \tau + 1)] - c. \tag{6}$$

Although the function α has not yet been determined, the results of Lemma 1 can be interpreted to give the following description of an optimal procedure.

Theorem 1 *An optimal procedure at the position (r, μ, τ) is to terminate the sampling process and accept the gain r if $r - \mu \geq \alpha(\tau)$ and to continue sampling otherwise. If sampling is continued, the expected value under the optimal procedure is $\mu + \alpha(\tau)$.*

The precision of W increases by 1 unit each time an observation is taken. Hence, for any given prior precision $\tau_0 > 0$, the optimal procedure will be known if we can determine the sequence of values $\alpha(\tau_0 + n)$ for $n = 0, 1, \ldots .$ The basic recursion relation for this sequence is given in the following lemma. In this lemma, the function Ψ is defined by Eq. (1) of Sec. 11.9.

Lemma 2 *For any value of $\tau > 0$, let $\tau' = [\tau/(\tau + 1)]^{\frac{1}{2}}$. Then*

$$\alpha(\tau) = \tau'\Psi\left[\frac{-\alpha(\tau + 1)}{\tau'}\right] - c. \tag{7}$$

Proof From Eqs. (6) and (4), it follows that

$$\alpha(\tau) = E[\max \{Y_\tau, \alpha(\tau + 1)\}] - c. \tag{8}$$

If we now apply Eq. (3) of Sec. 13.4 and Eq. (3) of Sec. 11.9, we obtain the following result:

$$E[\max \{Y_\tau, \alpha(\tau + 1)\}] = \alpha(\tau + 1) + \tau'\Psi\left[\frac{\alpha(\tau + 1)}{\tau'}\right]. \tag{9}$$

Finally, from Eq. (17) of Sec. 12.6, we can obtain Eq. (7).■

By inverting Eq. (7), we obtain the following equation:

$$\alpha(\tau + 1) = -\tau'\Psi^{-1}\left[\frac{\alpha(\tau) + c}{\tau'}\right]. \tag{10}$$

Hence, the entire sequence of values $\alpha(\tau_0 + n)$, for $n = 0, 1, \ldots$, can be determined from the value of any single term in the sequence. We shall now derive some further properties of the function α and shall develop some approximations to the values of the terms $\alpha(\tau_0 + n)$.

It is helpful at this point to return to Eqs. (4) and (7) and to derive them again from another point of view. Consider a problem of sampling without recall in which the statistician can observe sequentially the independent random variables $Y_1, Y_2, \ldots$ at a cost of c units per observation. Suppose that the random variables in this sequence are independent but are not identically distributed. Assume that for $n = 1, 2, \ldots$, the distribution of Y_n is

$$Y_n \sim \mathfrak{N}\left(0, \frac{\tau_0 + n}{\tau_0 + n - 1}\right). \tag{11}$$

In this problem, the statistician's position after he has observed the values of $Y_1, \ldots, Y_n$ can again be represented by (r, τ), where r is the value of the most recent observation Y_n and $\tau = \tau_0 + n$ is an index of the number of observations that have been taken. Again, we shall let $U(r, \tau)$ denote the expected gain, exclusive of the amount that has already been spent on sampling, when the optimal procedure is continued from the position (r, τ). If another observation Y_{n+1} is taken, then the position (r, τ) is transformed into the new position $(Y_{n+1}, \tau + 1)$ and Y_{n+1} has the same distribution as the random variable Y_τ in Lemma 1. It follows that the optimal expected gain $U(r, \tau)$ in this problem must also satisfy the functional equation (4). Furthermore, if $\alpha(\tau)$ is specified by Eq. (6), then $\alpha(\tau)$ again is the optimal expected gain among those procedures which specify that at least one more observation should be taken, and $\alpha(\tau)$ again satisfies Eq. (7). This interpretation of the function U as the value of an auxiliary problem of optimal stopping suggests the properties of $\alpha(\tau)$ presented in the next theorem.

Theorem 2   *The function α is a continuous, strictly increasing function of τ ($\tau > 0$). Furthermore, if τ' is defined as in Lemma 2, then*

$$\alpha(\tau) > \tau'\Psi^{-1}\left(\frac{c}{\tau'}\right) \qquad \text{for } \tau > 0, \tag{12}$$

and

$$\lim_{\tau \to \infty} \alpha(\tau) = \Psi^{-1}(c). \tag{13}$$

Proof   Only an intuitive outline of the proof will be presented here, but a rigorous proof can be developed along these same lines.

Suppose that each of the observations Y_n ($n = 1, 2, \ldots$) in the above auxiliary problem of optimal stopping had mean 0 and precision τ.

This problem would then be the one considered in Sec. 13.5.   As pointed out in Eq. (3) of Sec. 13.5, the maximum expected gain $v^*(\tau)$ is

$$v^*(\tau) = \tau^{-\frac{1}{2}}\Psi^{-1}(\tau^{\frac{1}{2}}c).\tag{14}$$

It can be shown (see Exercise 17) that for any fixed cost $c > 0$, v^* is a decreasing function of τ.

In the auxiliary problem of optimal stopping which we are considering, it is seen from relation (11) that the precisions of the observations Y_1, Y_2, . . . are not equal.   In fact, the precision of Y_n decreases as n becomes larger.   Since the precision of Y_1 is $(\tau + 1)/\tau$, the expected gain $\alpha(\tau)$ from the optimal procedure will be larger than the expected gain $v^*[(\tau + 1)/\tau]$ which would be obtained if all future observations Y_2, Y_3, . . . had the same precision as Y_1.   The inequality (12) follows from Eq. (14).

Furthermore, since the precision $(\tau + n)/(\tau + n - 1)$ of the observation Y_n ($n = 1, 2, . . .$) is a decreasing function of τ, the same type of argument shows that the expected gain $\alpha(\tau)$ from the optimal procedure is an increasing function of τ.

The relation (13) can be established as follows: When τ is very large in the auxiliary sampling problem, the precision of each observation Y_n will be just slightly larger than 1.   If each precision were exactly 1, then it again follows from Eq. (3) of Sec. 13.5 that the expected gain from the optimal procedure would be $\Psi^{-1}(c)$.   It is now plausible, and can be proved, that the actual expected gain $\alpha(\tau)$ from the optimal procedure can be made arbitrarily close to $\Psi^{-1}(c)$ by selecting a sufficiently large value for τ.

The relation (13) is, in effect, a statement about the continuity of the function α as $\tau \to \infty$.   It is equally plausible that α is also continuous at each value of τ ($\tau > 0$), but we omit the details here.■

If we let $\alpha_0(\tau)$ denote the right side of the inequality (12), then the function α_0 is a lower bound for the function α.   Since τ' is an increasing function of τ, it follows from Exercise 17 that α_0 is an increasing function of τ.   Furthermore, since $\tau' \to 1$ as $\tau \to \infty$, it can be seen from the definition of α_0 that

$$\lim_{\tau \to \infty} \alpha_0(\tau) = \Psi^{-1}(c).\tag{15}$$

Hence, the lower bound α_0 has some of the same properties that α has. The function α_0 is a continuous, increasing function with the limiting value specified by Eq. (15).   It follows that for large values of τ, $\alpha_0(\tau)$ is a reasonable approximation to $\alpha(\tau)$.   An approximation to the optimal procedure can now be obtained from Theorem 1 by replacing $\alpha(\tau)$ in the description of the optimal procedure given there by the following value:

$$\alpha_0(\tau) = \tau' \Psi^{-1} \left(\frac{c}{\tau'} \right). \tag{16}$$

Since $\alpha_0(\tau) < \alpha(\tau)$ for $\tau > 0$, the procedure based on this approximation may be terminated before the optimal procedure would be terminated, but it would never be continued when the optimal procedure would not be continued.

For further details about this problem and the problem to be considered in the next section, see DeGroot (1968).

13.7 SAMPLING WITH RECALL FROM A NORMAL DISTRIBUTION WITH UNKNOWN MEAN

As in the preceding section, we shall suppose that a sequential random sample can be taken at a cost of c units per observation from a normal distribution with unknown mean W and unit precision. Again we shall suppose that the prior distribution of W is normal. However, we shall now consider the problem of sampling with recall. In this problem, the statistician must find a stopping rule which maximizes his expected gain $E[\max \{X_1, \ldots, X_N\} - cN]$.

At each stage of the sampling process, the statistician's position is again characterized by a triple (r, μ, τ), where r is the maximum value of the observations he has taken and μ and τ are the mean and the precision of the current posterior distribution of W. Thus, r again represents the gain to the statistician, exclusive of the amount already spent on sampling, if he terminates the procedure. If the procedure is continued from this position and the value of the next observation is x, then the position (r, μ, τ) is transformed into the new position $[r(x), \mu(x), \tau + 1]$, where $r(x) = \max \{r, x\}$ and $\mu(x)$ is defined by Eq. (1) of Sec. 13.6.

As before, if $V(r, \mu, \tau)$ denotes the expected gain from the optimal procedure starting from the position (r, μ, τ) and $X \sim \mathfrak{N}[\mu, \tau/(\tau + 1)]$, then V must satisfy the following functional equation:

$$V(r, \mu, \tau) = \max \{r, E(V[r(X), \mu(X), \tau + 1]) - c\}. \tag{1}$$

Furthermore, since V again satisfies the invariance relation (3) of Sec. 13.6, we can again let $U(r, \tau) = V(r, 0, \tau)$. Then U must satisfy the functional equation given in the following lemma.

Lemma 1 *If X is a random variable such that $X \sim \mathfrak{N}[0, \tau/(\tau + 1)]$, then, for any values of r and τ ($\tau > 0$),*

$$U(r, \tau) = \max \left\{ r, E \left(U \left[r(X) - \frac{X}{\tau + 1}, \tau + 1 \right] \right) - c \right\}. \tag{2}$$

Since the proof of this lemma is similar to that of Lemma 1 of Sec. 13.6, it is not given here. However, Eq. (2) is more complicated than Eq. (4) of Sec. 13.6 because the expectation on the right side of Eq. (2) depends on the value of r as well as on τ. Despite this complication, the next lemma shows that it is still optimal to take another observation if, and only if, $r - \mu < \beta(\tau)$ for some appropriate critical value $\beta(\tau)$.

Lemma 2 *Suppose that $U(r, \tau) > r$ for given values of r and τ. Then $U(r_1, \tau) > r_1$ for any value $r_1 < r$.*

Proof Since $U(r, \tau) > r$, there exists a procedure which involves taking at least one observation whose expected gain, starting from the position $(r, 0, \tau)$, is greater than r. Let Z be the maximum of the values which are observed under this procedure before the sampling process is terminated. Also, let F_0 be the distribution function of Z, and let N be the number of observations taken. The expected gain of the procedure will then be $E[\max \{r, Z\} - cN]$, and the statement that this value is greater than r can be represented as follows:

$$T_{F_0}(r) > cE(N). \tag{3}$$

Since T_{F_0} is a decreasing function, the relation (3) must also be satisfied for any value $r_1 < r$. Hence, the same sampling procedure that is worthwhile from the position (r, τ) is also worthwhile from the position (r_1, τ), and it follows that $U(r_1, \tau) > r_1$.∎

Thus, as in the problem of sampling without recall, the function U is again of the following form:

$$\begin{aligned} U(r, \tau) &= r \qquad \text{for } r \geq \beta(\tau), \\ U(r, \tau) &> r \qquad \text{for } r < \beta(\tau). \end{aligned} \tag{4}$$

One important difference between the present problem and the problem of sampling without recall—a difference that complicates the present problem—is that here, for a given value of τ, the function U is not constant for $r < \beta(\tau)$ but is a strictly increasing function of r. There remains the problem of determining the function β.

For any values of r and τ, we shall let

$$U_1(r, \tau) = E\left\{U\left[r(X) - \frac{X}{\tau + 1}, \tau + 1\right]\right\} - c. \tag{5}$$

In this equation, $X \sim \mathfrak{N}[0, (\tau')^2]$, where $\tau' = [\tau/(\tau + 1)]^{\frac{1}{2}}$ and $r(X) = \max \{r, X\}$. Then Eq. (2) can be rewritten as follows:

$$U(r, \tau) = \max \{r, U_1(r, \tau)\}. \tag{6}$$

It is seen from Eq. (6) that $U(r, \tau) \geq r$ for any values of r and τ. There-fore, we can obtain the following relation:

$$U_1(r, \tau) \geq E\left[\max\{r, X\} - \frac{X}{\tau + 1}\right] - c$$

$$= r + \frac{\Psi(\tau'r)}{\tau'} - c. \tag{7}$$

The final equality in Eq. (7) follows from Eq. (3) of Sec. 13.4 and Eq. (3) of Sec. 11.9 and from the fact that $E(X) = 0$.

Now suppose that for a given position (r, τ),

$$\frac{\Psi(\tau'r)}{\tau'} > c. \tag{8}$$

Then, by the relations (6) and (7), it follows that $U(r, \tau) > r$. This inequality states that the expected gain $U(r, \tau)$ from the optimal procedure is greater than the gain r from terminating the sampling process. Hence, when the inequality (8) is satisfied, the optimal procedure must specify that another observation should be taken. It now follows from the relations (4) that $r < \beta(\tau)$ whenever the inequality (8) is satisfied. By solving the inequality (8) for r, we can obtain the following relation for all values of τ $(\tau > 0)$:

$$\beta(\tau) \geq \frac{\Psi^{-1}(\tau'c)}{\tau'}. \tag{9}$$

Thus, the relation (9) provides a lower bound for the function β. We shall now show that, in fact, the equality in relation (9) must be satisfied for all values of c and τ which exceed certain minimal levels.

Lemma 3   *The equality in relation* (9) *is satisfied whenever*

$$\frac{\tau}{\tau + 1} \geq 2(1 - \pi c^2). \tag{10}$$

Proof   The inequality (9) was derived by showing that if the inequality (8) is satisfied, then the optimal procedure specifies that another observation should be taken. To prove the lemma, we must now show that if the inequality (8) is not satisfied and it is assumed that the relation (10) is satisfied, then the optimal procedure specifies that the sampling process should be terminated.

Accordingly, let S^* denote the set of all pairs of values (r, τ) for which the relation (10) is satisfied and for which

$$\frac{\Psi(\tau'r)}{\tau'} \leq c. \tag{11}$$

To prove the lemma, we must show that $U(r, \tau) = r$ for any pair $(r, \tau) \in S^*$.

For any pair $(r, \tau) \in S^*$ and for any number x, let

$$r^* = \max \{r, x\} - \frac{x}{\tau + 1}. \tag{12}$$

We shall show first that $(r^*, \tau + 1) \in S^*$. Since τ satisfies the relation (10), it can be seen that relation (10) will also be satisfied when τ is replaced by $\tau + 1$. It must now be shown that the relation (11) is satisfied by $(r^*, \tau + 1)$.

For any given values of r and τ, the minimum value of r^* among all values of x is $(\tau')^2 r$. Also, for each fixed value of τ, the left side of the relation (11) is a decreasing function of r. If we let $\tau'' = [(\tau + 1)/(\tau + 2)]^{\frac{1}{2}}$, then it is sufficient to show that

$$\frac{\Psi[\tau''(\tau')^2 r]}{\tau''} \leq c. \tag{13}$$

The relation (13) is equivalent to the following relation:

$$r \geq \frac{\Psi^{-1}(\tau'' c)}{\tau''(\tau')^2}. \tag{14}$$

Also, the relation (11) is equivalent to the following relation:

$$r \geq \frac{\Psi^{-1}(\tau' c)}{\tau'}. \tag{15}$$

Hence, it suffices to show that when relation (10) is satisfied, the right side in relation (15) is at least as large as the right side in relation (14). Since $\tau' < \tau''$, it follows from Exercise 17 that

$$\frac{\Psi^{-1}(\tau' c)}{\tau'} > \frac{\Psi^{-1}(\tau'' c)}{\tau''}. \tag{16}$$

Furthermore, since $\Psi^{-1}(x) \leq 0$ if, and only if, $x \geq (2\pi)^{-\frac{1}{2}}$, it follows from the relation (10) that the value of the right side in the relation (16) cannot be positive and, hence, that this value must be at least as large as the right side of the relation (14). These remarks yield the desired inequality.

We have now shown that if the statistician's position (r, τ) belongs to the set S^*, then any position $(r^*, \tau + 1)$ which he might reach by taking another observation, and any subsequent position which he might reach by taking further observations, will also belong to the set S^*. Hence, for the purposes of this lemma, in trying to find a function U which satisfies Eq. (2), we need consider the values of U only on the set S^*.

We shall now show that for $(r, \tau) \in S^*$, the values $U(r, \tau) = r$ satisfy Eq. (2). It follows from relation (5) that for these values of U, equality must be satisfied in the relation (7). Therefore, by the relation (11),

$U_1(r, \tau) \leq r$. It now follows from Eq. (6) that the value $U(r, \tau) = r$ must satisfy Eq. (2).

The equation $U(r, \tau) = r$ indicates that for $(r, \tau) \epsilon S^*$, the optimal procedure is to terminate the sampling process.■

For any given position (r, μ, τ) in the problem of sampling with recall, we can now summarize the results that have been obtained as follows:

Theorem 1 *From any position (r, μ, τ) for which τ satisfies the inequality (10), an optimal procedure is to terminate the sampling process if*

$$r - \mu \geq \frac{\Psi^{-1}(\tau'c)}{\tau'}, \tag{17}$$

and to continue the sampling process otherwise. From any position for which τ does not satisfy the inequality (10), it is still optimal to continue the sampling process when the relation (17) is not satisfied.

If $c \geq 1/\pi^{\frac{1}{2}}$, then the inequality (10) will be satisfied for all values of τ $(\tau > 0)$. In this case, the optimal procedure is completely specified in Theorem 1. If $c > 1/(2\pi)^{\frac{1}{2}}$, then the inequality (10) will be satisfied for all sufficiently large values of τ. However, if $c \leq 1/(2\pi)^{\frac{1}{2}}$, then the inequality (10) will not be satisfied for any values of τ.

Consider now the myopic procedure which looks ahead only one stage from any given position. According to this procedure, the sampling process would be continued from any position if, and only if, it would be optimal to continue from that position when the statistician is permitted to take only one more observation before being forced to terminate the process. This procedure specifies that the process should be terminated at any position (r, μ, τ) if, and only if, the inequality (17) is satisfied. Furthermore, as $\tau \to \infty$ and the statistician gains more precise knowledge about the value of W, the limiting value of the right side in the relation (17) is $\Psi^{-1}(c)$. This limit is the value that characterizes the optimal procedure when W is known (see Sec. 13.5 and Exercise 14). Hence, the procedure characterized by the inequality (17), which is actually optimal when the relation (10) is satisfied, is a good approximation to the optimal procedure whenever τ is sufficiently large.

13.8 EXISTENCE OF OPTIMAL STOPPING RULES

In this section and the next one we shall demonstrate the existence of an optimal stopping rule in a broad class of problems, including those considered in Secs. 13.5 to 13.7 and the sequential statistical decision problems considered in Chap. 12. We shall rely heavily on the development given in Secs. 12.8 and 12.9, where the existence of an optimal sequential decision procedure was established, and we shall emphasize here only the following two topics: (1) the necessity for extending the results of Sec. 12.9 in order to cover more general problems of optimal stopping and (2) the additional conditions which must be imposed in order to accomplish this extension.

Let $X_1, X_2, \ldots$ be a sequence of observations which have a specified joint distribution, and for $n = 1, 2, \ldots$, let $Y_n = y_n(X_1, \ldots, X_n)$ be a random variable whose value depends on the first n observations $X_1, \ldots, X_n$. We shall suppose that if the statistician terminates the sampling process after having observed the values of $X_1, \ldots, X_n$, his gain is Y_n. We are now interested in determining whether there exists a stopping rule which maximizes the expected gain $E(Y_N)$.

For a given stopping rule, the expectation $E(Y_N)$ exists if, and only if, the following relation is satisfied:

$$E(|Y_N|) = \sum_{n=1}^{\infty} E(|Y_n| \mid N = n) \Pr(N = n) < \infty. \tag{1}$$

In particular, it can be seen that $E(Y_N)$ exists if there exists a bound $M < \infty$ such that any one of the following three conditions is satisfied:

$$E(|Y_n| \mid N = n) \leq M \qquad n = 1, 2, \ldots; \tag{2}$$

$$\Pr(|Y_n| \leq M) = 1 \qquad n = 1, 2, \ldots; \tag{3}$$

$$\Pr(N \leq M) = 1, E(|Y_n|) < \infty \qquad n = 1, \ldots, M. \tag{4}$$

The preceding statement that the observations $X_1, X_2, \ldots$ in the sequential sample have some specified joint distribution is general enough to apply both to problems of sampling from a known distribution and to problems of sampling from a distribution involving parameters which themselves have a prior distribution. The basic difference is illustrated by a comparison of the following situations: When $X_1, X_2, \ldots$ is a sequential random sample from a known distribution, the observations are *independent* and are identically distributed. When $X_1, X_2, \ldots$ is a sequential random sample from a distribution involving an unknown parameter W with a given prior distribution, the observations are still identically distributed. However, because of their dependence on W, they are no longer independent. In either case, the joint distribution of the observations is well defined.

We shall make two assumptions pertaining to the sequence $Y_1, Y_2, \ldots$. First, we shall let the random variable Z be defined as $Z = \sup_n Y_n$. In other words, for any sequence of observations $X_1, X_2, \ldots$, the value of Z is the supremum of the values in the sequence

$Y_1, Y_2, \ldots$ We shall assume that

$$E(|Z|) = M < \infty. \tag{5}$$

It follows from the assumption (5) that even if the statistician could observe the entire sequence $X_1, X_2, \ldots$ and then select any one of the values $Y_1, Y_2, \ldots$ as his gain, the expected gain $E(Z)$ would still be finite.

Second, to make certain that the statistician will not wish to continue taking observations indefinitely without ever terminating the sampling process, we shall assume that with probability 1,

$$\lim_{n \to \infty} Y_n = -\infty. \tag{6}$$

In accordance with the discussion in Sec. 12.7, we shall use the same symbol δ to denote a given stopping rule, the corresponding stopping variable, and the random number of observations required by the given stopping rule. We shall let Δ denote the set of stopping rules δ for which $\Pr(\delta < \infty) = 1$. For any procedure $\delta \epsilon \Delta$, we shall let $v(\delta)$ denote the value of the expected gain $E(Y_\delta)$. Also, we shall let

$$v^* = \sup_{\delta \epsilon \Delta} v(\delta). \tag{7}$$

Since $Y_\delta \leq Z$ for any procedure $\delta \epsilon \Delta$, it follows from Eq. (5) that $v^* < \infty$. We shall now show that there exists an optimal stopping rule $\delta^* \epsilon \Delta$, that is, a stopping rule δ^* for which $v(\delta^*) = v^*$. As mentioned above, the development will be similar to that in Secs. 12.8 and 12.9.

Theorem 1 *If the relations* (5) *and* (6) *are satisfied, then there exists a stopping rule* $\delta^* \epsilon \Delta$ *such that* $v(\delta^*) = v^*$.

Proof It is said that a stopping rule $\delta \epsilon \Delta$ is regular if, for every set of observed values $X_1 = x_1, \ldots, X_n = x_n$ for which $(x_1, \ldots, x_n) \epsilon \{\delta > n\}$, the following relation is satisfied ($n = 1, 2, \ldots$):

$$E[v(\delta)|x_1, \ldots, x_n] > y_n(x_1, \ldots, x_n). \tag{8}$$

This definition is completely analogous to that given by relation (1) of Sec. 12.8. As before, we need only consider regular stopping rules.

Let $\delta_1, \delta_2, \ldots$ be a sequence of regular stopping rules in Δ for which

$$\lim_{n \to \infty} v(\delta_n) = v^*. \tag{9}$$

For $n = 1, 2, \ldots$, we shall define $\gamma_n = \max \{\delta_1, \ldots, \delta_n\}$ to be the stopping rule which specifies that another observation should be taken at any stage if, and only if, at least one of the stopping rules $\delta_1, \ldots, \delta_n$

specifies that another observation should be taken at that stage. Similarly, we shall define $\delta^* = \sup\{\delta_1, \delta_2, \ldots\} = \sup\{\gamma_1, \gamma_2, \ldots\}$ to be the stopping rule which specifies that another observation should be taken whenever at least one of the stopping rules $\delta_1, \delta_2, \ldots$ specifies that another observation should be taken. Then it can be seen, as in Lemma 2 of Sec. 12.9, that $\gamma_n \in \Delta$ and that for $i = 1, \ldots, n$ and for $n = 1, 2, \ldots$,

$$v(\delta_i) \leq v(\gamma_n) \leq v(\gamma_{n+1}). \tag{10}$$

It follows from relations (9) and (10) that $v(\gamma_n) \to v^*$ as $n \to \infty$. Hence, the optimality of the stopping rule δ^* will be established if we can show that $\delta^* \in \Delta$ and that

$$\lim_{n \to \infty} v(\gamma_n) = v(\delta^*). \tag{11}$$

As before, for those sequences of observed values $X_1, X_2, \ldots$ for which $\lim_{n \to \infty} \gamma_n = \sup_n \gamma_n < \infty$, we can write the following relations:

$$\delta^* = \lim_{n \to \infty} \gamma_n \quad \text{and} \quad Y_{\delta^*} = \lim_{n \to \infty} Y_{\gamma_n}. \tag{12}$$

On the other hand, if $\lim_{n \to \infty} \gamma_n = \sup_n \gamma_n = \infty$, we shall let $\delta^* = \infty$ and $Y_{\delta^*} = -\infty$. Then it follows from Eq. (6) that the limiting relations (12) are satisfied in this case also.

Since $Y_{\gamma_n} \leq Z$ for $n = 1, 2, \ldots$, it now follows from the relation (5) and the Fatou-Lebesgue theorem that

$$v^* = \lim_{n \to \infty} v(\gamma_n) = \lim_{n \to \infty} E(Y_{\gamma_n})$$
$$\leq E(\limsup_{n \to \infty} Y_{\gamma_n}) = E(Y_{\delta^*}) = v(\delta^*). \tag{13}$$

Hence,

$$v(\delta^*) \geq v^*. \tag{14}$$

We can now establish the condition that $\Pr(\delta^* < \infty) = 1$. If $\Pr(\delta^* = \infty) > 0$, then $\Pr(Y_{\delta^*} = -\infty) > 0$, and it follows that $E(Y_{\delta^*}) = -\infty$. This result is a contradiction of the relation (14). Therefore, $\delta^* \in \Delta$.

It now follows from Eq. (7) that $v(\delta^*) \leq v^*$. This result, together with the relation (14), establishes the optimality of δ^*.■

Theorem 1 is general enough to cover a wide variety of problems of optimal stopping. If we let $Y_n = -\{\rho_0[\xi(X_1, \ldots, X_n)] + cn\}$ for $n = 1, 2, \ldots$, then the theorem includes the sequential statistical decision problems described in Chap. 12. If the statistician is not permitted to take more than n observations in a given problem, then we may let $Y_t = -\infty$ for $t > n$.

We shall now present a simple example in which there is no optimal stopping rule.

EXAMPLE Suppose that the statistician can make a sequence of favorable bets as follows: Each time he bets, his fortune either will be doubled with probability p $(p > \frac{1}{2})$ or will be reduced to 0 with probability $q = 1 - p$. For a given initial fortune x_0, let X_j represent the statistician's fortune after he has made j bets $(j = 1, 2, \ldots)$. Then, for any given finite sequence of fortunes $X_1 = x_1, \ldots, X_n = x_n$, the distribution of X_{n+1} is such that $X_{n+1} = 2x_n$ with probability p and $X_{n+1} = 0$ with probability q. We shall show that there is no stopping rule which maximizes $E(X_N)$.

Suppose that at some stage the statistician's fortune is $x > 0$. Then, since $p > \frac{1}{2}$, his expected fortune after one more bet is $2px > x$. Hence, so long as the statistician's fortune has not been reduced to 0, it is worthwhile for him to make one more bet. Each time he bets, however, there is a probability $q > 0$ that he will lose. Hence, if he continues to bet as long as he has any money, his fortune will certainly be reduced to 0 sooner or later. Thus, there is no optimal stopping rule.

A reasonable procedure in this example is to select some integer n and, if one's fortune has not been reduced to 0 during the first n bets, to terminate betting and to accept the fortune $2^n x_0$. Of course, the rule which calls for stopping after $n + 1$ bets has a greater expected gain.

13.9 EXISTENCE OF OPTIMAL STOPPING RULES FOR PROBLEMS OF SAMPLING WITH RECALL AND SAMPLING WITHOUT RECALL

In this section we shall indicate the conditions needed in order that Theorem 1 of Sec. 13.8 can be used to establish the existence of optimal stopping rules in the problems of sampling with and without recall that were considered in Secs. 13.5 to 13.7. We shall need to use the following result, which is known as the Borel-Cantelli lemma.

Borel-Cantelli lemma *Let $A_1, A_2, \ldots$ be a sequence of events in an arbitrary probability space such that $\sum_{n=1}^{\infty} \Pr(A_n) < \infty$. Furthermore, let $A^* = \bigcap_{j=1}^{\infty} \bigcup_{n=j}^{\infty} A_n$. Then $\Pr(A^*) = 0$.*

Proof The probability $\Pr(A^*)$ must satisfy the following relation:

$$\Pr(A^*) = \lim_{j \to \infty} \Pr(\bigcup_{n=j}^{\infty} A_n) \leq \lim_{j \to \infty} \sum_{n=j}^{\infty} \Pr(A_n). \tag{1}$$

Since $\sum_{n=1}^{\infty} \text{Pr}(A_n) < \infty$, it follows that the final series in the relation (1) must approach 0 as $j \to \infty$. ∎

The event A^* defined in the Borel-Cantelli lemma can be interpreted as the event which occurs if, and only if, an infinite number of the events in the sequence $A_1, A_2, \ldots$ occur. Hence, the Borel-Cantelli lemma states that if $\sum_{n=1}^{\infty} \text{Pr}(A_n) < \infty$, then, with probability 1, only a finite number of the events A_n ($n = 1, 2, \ldots$) can occur.

We shall now consider the problem of sampling with recall. It is assumed in the next lemma that the observations $X_1, X_2, \ldots$ are identically distributed, but they need not be independent. Hence, as explained earlier, the lemma includes both problems of sampling from a known distribution and problems of sampling from a distribution involving unknown parameters.

Lemma 1 *Let $X_1, X_2, \ldots$ be a sequence of identically distributed random variables whose common d.f. is F. Let $c > 0$ be a given number, and let*

$$Y_n = \max \{X_1, \ldots, X_n\} - nc \qquad for\ n = 1, 2, \ldots. \tag{2}$$

Also, let $Z = \sup_n Y_n$. If the mean of the d.f. F exists, then, with probability 1, $Z < \infty$ and $\lim_{n \to \infty} Y_n = -\infty$. If the variance of the d.f. F is finite, then $E(|Z|) < \infty$.

Proof It follows from Eq. (2) and the definition of Z that $Z \geq \sup_n \{X_n - nc\}$. On the other hand, it can be seen that for $n = 1, 2, \ldots$,

$$Y_n \leq \max \{X_1 - c, X_2 - 2c, \ldots, X_n - nc\}. \tag{3}$$

Hence, it is also true that $Z \leq \sup_n \{X_n - nc\}$. It is now clear that Z can be represented as follows:

$$Z = \sup_n \{X_n - nc\}. \tag{4}$$

Suppose now that the mean of F exists, and let X be any random variable whose d.f. is F. Then, for any constant α ($\alpha > 0$),

$$\sum_{n=1}^{\infty} \text{Pr}(X > n\alpha) = \sum_{n=1}^{\infty} [1 - F(n\alpha)]$$

$$\leq \int_0^{\infty} [1 - F(\alpha x)]\, dx = \frac{1}{\alpha} \int_0^{\infty} x\, dF(x) < \infty. \tag{5}$$

The final equality in the relation (5) follows from integration by parts [see, e.g., Feller (1966), p. 148]. Hence, $\sum_{n=1}^{\infty} \text{Pr}(X > n\alpha) < \infty$. In particular, if we let $\alpha = c/2$ and use the fact that each observation X_n has the same distribution as X, then we can write the following relation:

$$\sum_{n=1}^{\infty} \Pr\left(X_n > \frac{nc}{2}\right) < \infty. \tag{6}$$

It follows from the Borel-Cantelli lemma that with probability 1, there can be only a finite number of values of n for which $X_n > nc/2$. Hence, with probability 1, the following condition will be satisfied:

$$\lim_{n \to \infty} (X_n - nc) = -\infty. \tag{7}$$

Therefore, it follows from Eq. (4) that $\Pr(Z < \infty) = 1$.

To show that $Y_n \to -\infty$ with probability 1, we shall define the random variable T_n as follows:

$$T_n = \max \{X_1, \ldots, X_n\} - \frac{nc}{2} \quad \text{for } n = 1, 2, \ldots. \tag{8}$$

Also, we shall let $U = \sup_n T_n$. Then if c is replaced by $c/2$, it follows from the first part of this lemma that $\Pr(U < \infty) = 1$. But, for $n = 1$, $2, \ldots$,

$$Y_n = T_n - \frac{nc}{2} \leq U - \frac{nc}{2}. \tag{9}$$

Hence, $Y_n \to -\infty$ as $n \to \infty$.

Now, let N be a random variable the values of which are positive integers and which is defined by the following relation:

$$Z = X_N - cN. \tag{10}$$

In other words, N is the random index at which the supremum $\sup_n \{X_n - nc\}$ is attained. As a consequence of the results derived in the first part of this lemma, it follows that with probability 1, the value of N will be well defined.

For $k = 1, 2, \ldots$, we shall let A_k be the event defined as follows:

$$A_k = \{Z > 0 \text{ and } ck < X_N \leq c(k + 1)\}. \tag{11}$$

Then $\{Z > 0\} = \bigcup_{k=1}^{\infty} A_k$. Furthermore, since $X_N \leq c(k + 1)$ whenever the event A_k occurs, then it is also true that $Z \leq c(k + 1)$ whenever the event A_k occurs. Therefore, if F_Z denotes the d.f. of Z, the following relation must be satisfied:

$$\int_0^{\infty} z \, dF_Z(z) \leq \sum_{k=1}^{\infty} c(k + 1)\Pr(A_k). \tag{12}$$

Since $X_N - cN > 0$ and $X_N \le c(k + 1)$ whenever the event A_k occurs, then $N < k + 1$ whenever the event A_k occurs. Hence, for $k = 1$, $2, \ldots$,

$$A_k \subset \bigcup_{i=1}^{k} \{ck < X_i \le c(k + 1)\} \tag{13}$$

and

$$\Pr(A_k) \le \sum_{i=1}^{k} \Pr\{ck < X_i \le c(k + 1)\}$$

$$= k\Pr\{ck < X \le c(k + 1)\}. \tag{14}$$

From relations (12) and (14), the following result can now be obtained:

$$\int_0^\infty z \, dF_Z(z) \le c \sum_{k=1}^{\infty} (k + 1)^2 \Pr\{ck < X \le c(k + 1)\}$$

$$\le c \sum_{k=1}^{\infty} \int_{ck}^{c(k+1)} \left(\frac{x}{c} + 1\right)^2 dF(x)$$

$$= c \int_c^\infty \left(\frac{x}{c} + 1\right)^2 dF(x)$$

$$\le cE\left[\left(\frac{X}{c} + 1\right)^2\right]. \tag{15}$$

If the variance of the d.f. F is finite, then the final expectation in the relation (15) is finite, and therefore, so also is the first integral in the relation (15).

Since $Z \ge X_1 - c$, the following relation is also true:

$$\int_{-\infty}^0 z \, dF_Z(z) > -\infty. \tag{16}$$

Hence, if the variance of F is finite, then $E(|Z|) < \infty$. ∎

Theorem 1 of Sec. 13.8 and Lemma 1 of this section can now be combined to establish the existence of those optimal stopping rules whose existence could only be assumed in Secs. 13.5 to 13.7. The results of this development will be summarized in the next two theorems.

Theorem 1 Let X_1, X_2, . . . be a sequential random sample from a distribution for which it is known that the d.f. is F. Let c be a fixed cost per observation, and for $n = 1, 2, \ldots$, let the random variables Y_n and Y'_n be defined as follows:

$$Y_n = \max \{X_1, \ldots, X_n\} - nc \qquad and \qquad Y'_n = X_n - nc. \tag{17}$$

If $E(X_i^2) < \infty$ for $i = 1, 2, \ldots$, then there exists a stopping rule which maximizes both $E(Y_N)$ and $E(Y_N')$. This rule is to terminate the sampling process as soon as some observed value $x_j \geq v^*$, where v^* is the unique solution of Eq. (1) of Sec. 13.5.

Theorem 2 *Let X_1, X_2, $\ldots$ be a sequential random sample from a distribution which involves a parameter W whose value is unknown, and suppose that W has a specified prior distribution. Also, let c be a fixed cost per observation, and for $n = 1, 2, \ldots$, let the random variables Y_n and Y_n' be defined by Eq. (17). If $E(X_i^2) = E[E(X_i^2|W)] < \infty$ for $i = 1, 2, \ldots$, then there exists a stopping rule which maximizes $E(Y_N)$ and a stopping rule which maximizes $E(Y_N')$.*

Further Remarks and References

The existence of optimal stopping rules under various conditions has been studied by Snell (1952); Chow and Robbins (1961, 1963, 1967b); Bramblett (1965), on whose work the proof of Lemma 1 is based; Haggstrom (1966); Yahav (1966), the source of Exercise 18; Chernoff (1967a); and Siegmund (1967).

13.10 MARTINGALES

The concept of a *martingale* provides an abstract representation of the general idea of a sequence of fair games or gambles. Martingales and the related concepts of supermartingales and submartingales have become of increasing importance in many facets of probability theory, and they can play an important part in the study of optimal stopping rules in very general classes of problems.

In this section and the next few sections, we shall first give precise definitions of these concepts and we shall then explore the properties of stopping rules when the statistician's sequence of possible gains forms a martingale, a submartingale, or a supermartingale. We shall also utilize some of these properties in the development of stopping rules for more general problems.

Let X_1, X_2, $\ldots$ be a sequence of observations with a specified joint distribution. In general, these observations need not be either independent or identically distributed. As in Sec. 12.2, we shall let S^n denote the sample space of the first n observations $X_1, \ldots, X_n$ ($n = 1$, $2, \ldots$) and shall let $F_n(x_1, \ldots, x_n)$ denote the joint d.f. of $X_1, \ldots, X_n$ evaluated at any point $(x_1, \ldots, x_n) \in S^n$. As described in Sec. 12.2, any subset $A \subset S^n$ can, and will, also be regarded as a subset of S^r for any positive integer $r > n$.

For $n = 1, 2, \ldots$, we shall let $Y_n = y_n(X_1, \ldots, X_n)$ be a random variable whose value depends on the first n observations $X_1, \ldots, X_n$. We shall again suppose that if the statistician terminates the sampling process after the values of $X_1, \ldots, X_n$ have been observed, his gain is Y_n.

It is said that the sequence $Y_1, Y_2, \ldots$ is a *martingale with respect to the sequence* $X_1, X_2, \ldots$ if, for $n = 1, 2, \ldots$, $E(Y_n)$ exists and, with probability 1,

$$E(Y_{n+1}|X_1, \ldots, X_n) = Y_n. \tag{1}$$

It is clear that a martingale provides a reasonable representation of a sequence of fair gambles because Eq. (1) simply states that at any stage of the sequential process, the statistician's expected gain after another value has been observed is equal to his current gain.

More simply, a sequence of random variables $Y_1, Y_2, \ldots$ is said to be a *martingale*, without reference to any other sequence $X_1, X_2, \ldots$, if for $n = 1, 2, \ldots$, $E(Y_n)$ exists and, with probability 1,

$$E(Y_{n+1}|Y_1, \ldots, Y_n) = Y_n. \tag{2}$$

Two important examples of martingales are the following.

EXAMPLE 1 Suppose that $X_1, X_2, \ldots$ is a sequence of independent random variables for which $E(X_i) = 0$ $(i = 1, 2, \ldots)$, and let $Y_n = \sum_{i=1}^{n} X_i$ for $n = 1, 2, \ldots$. Then the sequence $Y_1, Y_2, \ldots$ is a martingale with respect to the sequence $X_1, X_2, \ldots$.

EXAMPLE 2 Suppose that $X_1, X_2, \ldots$ is a sequence of independent random variables for which $E(X_i) = 1$ $(i = 1, 2, \ldots)$, and let $Y_n = \prod_{i=1}^{n} X_i$ for $n = 1, 2, \ldots$. Then the sequence $Y_1, Y_2, \ldots$ is a martingale with respect to the sequence $X_1, X_2, \ldots$.

In these examples, the random variables $X_1, X_2, \ldots$ must be independent and must have a common expectation, but they need not be identically distributed.

We shall now define the closely related concepts of a submartingale and a supermartingale. It is said that the sequence $Y_1, Y_2, \ldots$ is a *submartingale with respect to the sequence* $X_1, X_2, \ldots$ if, for $n = 1, 2, \ldots$, $E(Y_n)$ exists and, with probability 1,

$$E(Y_{n+1}|X_1, \ldots, X_n) \geq Y_n. \tag{3}$$

It is said that the sequence $Y_1, Y_2, \ldots$ is a *supermartingale with respect to the sequence* $X_1, X_2, \ldots$ if, for $n = 1, 2, \ldots$, $E(Y_n)$ exists and, with probability 1,

$$E(Y_{n+1}|X_1, \ldots, X_n) \leq Y_n. \tag{4}$$

Clearly, a submartingale provides a representation of a sequence of gambles that are either fair or favorable to the statistician because the inequality (3) states that at any stage of the sequential process, his expected gain after another value has been observed is never less than his current gain. Similarly, a supermartingale provides a representation of a sequence of gambles that are either fair or unfavorable to the statistician. It is unfortunate for explanatory purposes in our present context that the term "supermartingale" refers to unfavorable gambles and the term "submartingale" refers to favorable gambles. In older publications, submartingales are called semimartingales and supermartingales are called lower semimartingales. It should be noted that a martingale is both a submartingale and a supermartingale.

To simplify the notation throughout this entire discussion of martingales, we shall let $y_j = y_j(x_1, \ldots, x_j)$ for $j = 1, 2, \ldots$ in all the integrals which will appear. Furthermore, since the argument of the j-dimensional d.f. F_j is always a point $(x_1, \ldots, x_j) \epsilon S^j$, the differential $dF_j(x_1, \ldots, x_j)$ in any of these integrals will be denoted simply by dF_j.

The next theorem exhibits a fundamental property of martingales, submartingales, and supermartingales that will be useful in deriving the results to follow.

Theorem 1 *Let the sequence* $Y_1, Y_2, \ldots$ *be a supermartingale, a submartingale, or a martingale with respect to the sequence* $X_1, X_2, \ldots$. *Let* r *and* n *be positive integers such that* $r > n$, *and let* A_n *be an arbitrary subset of* S^n. *If the sequence is a supermartingale, then*

$$\int_{A_n} y_n \, dF_n \geq \int_{A_n} y_r \, dF_r. \tag{5}$$

If the sequence is a submartingale, the inequality in the relation (5) *is reversed. If the sequence is a martingale, there is equality in the relation* (5).

Proof We shall prove the correctness of the inequality (5) for supermartingales. The proofs of the other parts of the theorem are completely analogous.

We shall assume first that $r = n + 1$, and we shall let S_{n+1} be the sample space of the random variable X_{n+1}. Then we can obtain the following relation:

$$\int_{A_n} y_{n+1} \, dF_{n+1} = \int_{A_n} \int_{S_{n+1}} y_{n+1} \, dF(x_{n+1}|x_1, \ldots, x_n) \, dF_n$$

$$= \int_{A_n} E(Y_{n+1}|x_1, \ldots, x_n) \, dF_n \leq \int_{A_n} y_n \, dF_n. \tag{6}$$

In the first integral in the relation (6), the subset A_n is regarded as a subset of S^{n+1}, and in the second integral, A_n is regarded as a subset of S^n. The expression $F(x_{n+1}|x_1, \ldots, x_n)$ in the second integral denotes the conditional d.f. of X_{n+1} when $X_1 = x_1, \ldots, X_n = x_n$. The final inequality in the relation (6) follows from the definition of a supermartingale.

Since the event A_n can be regarded as a subset of S^r for any integer r ($r \geq n$), the above development demonstrates equally well that

$$\int_{A_n} y_{r+1} \, dF_{r+1} \leq \int_{A_n} y_r \, dF_r \qquad r \geq n. \tag{7}$$

Together, the relations (6) and (7) yield the relation (5).∎

Further Remarks and References

Martingales were first studied by Lévy (1937) and Ville (1939). The name was first used by Ville. The general theory of martingales has mainly been developed by Doob (1953), and many of the results given in the next three sections are based on his work. Loève (1963), Neveu (1965), and Meyer (1966) also present general discussions of martingales at an advanced level.

13.11 STOPPING RULES FOR MARTINGALES

We shall now consider the problem of determining the expected gain $E(Y_N)$ for various stopping rules when the sequence of gains $Y_1, Y_2, \ldots$ is a martingale. It follows from Eq. (1) of Sec. 13.10 that $E(Y_n) = E(Y_1)$ for $n = 1, 2, \ldots$. Hence, it is natural to expect that $E(Y_N) = E(Y_1)$ for a broad class of stopping rules, if not for all stopping rules. Similarly, if $Y_1, Y_2, \ldots$ is a supermartingale, then $E(Y_1) \geq E(Y_2) \geq \cdots$, and we might therefore expect that $E(Y_N) \leq E(Y_1)$. An analogous statement can be made for submartingales. However, these intuitive relations are not completely general for all sequences and for all stopping rules. In this section and the next one, we shall present various sufficient conditions for these relations to be correct. It should be kept in mind that for any stopping rule, the event $\{N > n\}$ can be regarded as a subset of S^n or as a subset of S^r for $r \geq n$. We shall consider only stopping rules that belong to the class Δ of rules for which $\Pr(N < \infty) = 1$.

Theorem 1 *Let the sequence $Y_1, Y_2, \ldots$ be a supermartingale, a submartingale, or a martingale with respect to the sequence $X_1, X_2, \ldots$, and consider a stopping rule for which $E(Y_N)$ exists.*

Suppose that the sequence is a supermartingale and that the following relation is satisfied:

$$\lim_{n \to \infty} \int_{\{N>n\}} y_n \, dF_n \geq 0. \tag{1}$$

Then $E(Y_N) \leq E(Y_1)$.

Suppose that the sequence is a submartingale and that the inequality in the relation (1) is reversed. Then $E(Y_N) \geq E(Y_1)$.

Suppose that the sequence is a martingale and that there is equality in the relation (1). Then $E(Y_N) = E(Y_1)$.

Proof We shall prove the correctness of the theorem for supermartingales. The proofs of the other two parts are completely analogous.

Since it is assumed that $E(Y_N)$ exists, we can write the following relation:

$$\begin{aligned} E(Y_N) &= \lim_{n \to \infty} \sum_{j=1}^{n} \int_{\{N=j\}} y_j \, dF_j \\ &= \lim_{n \to \infty} \sum_{j=1}^{n} \left(\int_{\{N>j-1\}} y_j \, dF_j - \int_{\{N>j\}} y_j \, dF_j \right). \end{aligned} \tag{2}$$

Also, it follows from Theorem 1 of Sec. 13.10 that for $j = 2, 3, \ldots$,

$$\int_{\{N>j-1\}} y_j \, dF_j \leq \int_{\{N>j-1\}} y_{j-1} \, dF_{j-1}. \tag{3}$$

Hence,

$$\begin{aligned} E(Y_N) &\leq \lim_{n \to \infty} \left(\int_{\{N>0\}} y_1 \, dF_1 - \int_{\{N>n\}} y_n \, dF_n \right) \\ &= E(Y_1) - \lim_{n \to \infty} \int_{\{N>n\}} y_n \, dF_n. \end{aligned} \tag{4}$$

By hypothesis, the final limit in the relation (4) is nonnegative. Therefore, $E(Y_N) \leq E(Y_1)$.∎

Since the limit in the relation (1) must be 0 for any bounded stopping rule, the next corollary follows from Theorem 1.

Corollary 1 *Let the sequence $Y_1, Y_2, \ldots$ be a supermartingale, a submartingale, or a martingale with respect to the sequence $X_1, X_2, \ldots$, and consider any stopping rule such that $\Pr(N \leq n) = 1$ for some positive integer n. For a supermartingale, $E(Y_1) \geq E(Y_N) \geq E(Y_n)$. For a submartingale, $E(Y_1) \leq E(Y_N) \leq E(Y_n)$. For a martingale, the conclusion is that $E(Y_1) = E(Y_N) = E(Y_n)$.*

Proof As stated above, the first half of each part of the conclusion follows from Theorem 1. The second half of each part follows from Theorem 1 of Sec. 13.10.   For example, if the sequence is a supermartingale, the following relation must be satisfied:

$$E(Y_N) = \sum_{j=1}^{n} \int_{\{N=j\}} y_j \, dF_j \geq \sum_{j=1}^{n} \int_{\{N=j\}} y_n \, dF_n = E(Y_n). \blacksquare \qquad (5)$$

Theorem 1 is based on the hypothesis that $E(Y_N)$ exists for a given stopping rule.   We shall now present a simple condition that ensures the existence of $E(Y_N)$ for *all* stopping rules in the class Δ.

Theorem 2   *Let the sequence $Y_1, Y_2, \ldots$ be a supermartingale, a submartingale, or a martingale with respect to the sequence $X_1, X_2, \ldots$.   If there exists a bound $M < \infty$ such that $E(|Y_n|) \leq M$ for $n = 1, 2, \ldots$, then $E(Y_N)$ exists for all stopping rules in the class Δ.*

Proof   We shall give the proof for submartingales.   The proof for supermartingales is equivalent, and that for martingales is somewhat simpler.

For any random variable Y, we shall define the random variables Y^+ and Y^- by the following relations:

$$Y^+ = \max\{Y, 0\} \qquad \text{and} \qquad Y^- = \max\{-Y, 0\}. \qquad (6)$$

Then $Y = Y^+ - Y^-$ and $|Y| = Y^+ + Y^- = 2Y^+ - Y$.   For any stopping rule $\delta \, \epsilon \, \Delta$ and any positive integer n, we shall let δ_n denote the stopping rule obtained by truncating the procedure δ after n observations have been taken.   In this proof, we shall again use the same symbol to denote both a given stopping rule and the random number of observations required by that stopping rule.

For any stopping rule $\delta \, \epsilon \, \Delta$ and any positive integer n,

$$E(|Y_{\delta_n}|) = \sum_{j=1}^{n} \int_{\{\delta=j\}} |y_j| \, dF_j + \int_{\{\delta>n\}} |y_n| \, dF_n. \qquad (7)$$

Therefore, if we let $n \to \infty$ in Eq. (7), it can be seen that

$$E(|Y_\delta|) \leq \liminf_{n \to \infty} E(|Y_{\delta_n}|). \qquad (8)$$

To prove the theorem, we shall show that the right side of the relation (8) is finite.   Since the sequence $Y_1, Y_2, \ldots$ is a submartingale with respect to the sequence $X_1, X_2, \ldots$, it follows from Exercise 21 that the sequence $Y_1^+, Y_2^+, \ldots$ is also a submartingale with respect to $X_1, X_2, \ldots$.   Furthermore, since $\Pr(\delta_n \leq n) = 1$, we can apply Corollary 1 and obtain the following relation:

$$E(|Y_{\delta_n}|) = 2E(Y_{\delta_n}^+) - E(Y_{\delta_n}) \leq 2E(Y_n^+) - E(Y_1)$$
$$\leq 2E(|Y_n|) - E(Y_1) \leq 2M - E(Y_1). \qquad (9)$$

Thus the value of the right side in the relation (8) cannot be greater than $2M - E(Y_1)$.∎

13.12 UNIFORMLY INTEGRABLE SEQUENCES OF RANDOM VARIABLES

It is said that the random variables Y_1, Y_2, . . . are *uniformly integrable* if the following relation is satisfied uniformly in n:

$$\lim_{a \to \infty} \int_{\{|Y_n| \geq a\}} |y_n| \, dF_n = 0. \tag{1}$$

In other words, the variables Y_1, Y_2, . . . are uniformly integrable if, for any number $\epsilon > 0$, there exists a sufficiently large number a_0 such that the following condition is satisfied for every value of n ($n = 1, 2, \ldots$):

$$\int_{\{|Y_n| \geq a_0\}} |y_n| \, dF_n < \epsilon. \tag{2}$$

It can be shown (see Exercise 22) that if the random variables Y_1, Y_2, . . . are uniformly integrable, then there is an upper bound $M < \infty$ such that $E(|Y_n|) \leq M$ for $n = 1, 2, \ldots$. Also (see Exercise 23), the following relation will be satisfied for every stopping rule in the class Δ:

$$\lim_{n \to \infty} \int_{\{N > n\}} |y_n| \, dF_n = 0. \tag{3}$$

The next result now follows from Theorems 1 and 2 of Sec. 13.11.

Theorem 1 *Let the sequence Y_1, Y_2, . . . be a supermartingale, a submartingale, or a martingale with respect to the sequence X_1, X_2, . . . , and suppose that the random variables Y_1, Y_2, . . . are uniformly integrable. Then $E(Y_N)$ exists for any stopping rule in the class Δ. For a supermartingale, $E(Y_N) \leq E(Y_1)$. For a submartingale, $E(Y_N) \geq E(Y_1)$. For a martingale, $E(Y_N) = E(Y_1)$.*

A simple special case of a uniformly integrable sequence of random variables is that in which the variables are uniformly bounded, i.e., when there exists an upper bound $M < \infty$ such that $\Pr(|Y_n| \leq M) = 1$ for $n = 1, 2, \ldots$. In fact (see Exercise 24), if there is an upper bound $M < \infty$ such that $E(Y_n^2) \leq M$ for $n = 1, 2, \ldots$, then the variables Y_1, Y_2, . . . are uniformly integrable.

Theorem 1 states that if the random variables Y_1, Y_2, . . . in a martingale are uniformly integrable, then all stopping rules yield the same expected gain $E(Y_N) = E(Y_1)$. In other words, it is impossible for the statistician to find a stopping rule that is better than the one which always requires terminating the process after the value of X_1 has been observed. Moreover, it is impossible for even the most foolish statistician to use a stopping rule for which the expected gain will be smaller than $E(Y_1)$. It follows from Corollary 1 of Sec. 13.11 that the same remarks will be true for *any* martingale if sampling must be terminated after not more than n observations have been taken. However, if the random variables $Y_1, Y_2, \ldots$ in a martingale are not uniformly integrable and there is no upper bound on the number of observations that can be taken, then $E(Y_N)$ may differ from $E(Y_1)$, as shown in the following example.

EXAMPLE Consider a gambler who bets in accordance with the following procedure: On his first bet, the probabilities are equal that he will either win 1 unit or lose 1 unit. If he wins 1 unit on the first bet, then his gain remains fixed at 1 unit throughout the rest of the process. If he loses 1 unit on the first bet, then on his second bet the probabilities are equal that he either wins 2 units or loses 2 units. Thus, his total gain after the two bets is either 1 unit or -3 units. Again, if his total gain after the two bets is 1 unit, then his gain remains fixed at 1 unit throughout the rest of the process. If his total gain is -3 units, then on his third bet the probabilities are equal that he either wins 4 units or loses 4 units. Thus, his total gain after three bets is either 1 unit or -7 units. The process continues in this way. As long as the gambler keeps losing, his possible gain or loss on the next bet doubles. In this way, when he finally wins a bet, his total gain will be exactly 1 unit. Once his total gain does become 1 unit, then it remains fixed at this value throughout the rest of the process.

Formally, this process can be described as follows: Let Y_n denote the gambler's total gain after n stages of the process ($n = 1, 2, \ldots$). Then $\Pr(Y_1 = 1) = \Pr(Y_1 = -1) = \frac{1}{2}$. Also, for $n = 1, 2, \ldots$, the distribution is

$$\Pr(Y_{n+1} = 1 | Y_n = 1) = 1, \tag{4}$$

and

$$\begin{aligned}\Pr(Y_{n+1} = 1 | Y_n &= -2^n + 1) \\ &= \Pr(Y_{n+1} = -2^{n+1} + 1 | Y_n = -2^n + 1) \\ &= \tfrac{1}{2}.\end{aligned} \tag{5}$$

It follows from the relations (4) and (5) that the sequence $Y_1, Y_2, \ldots$ is a martingale.

Consider now the stopping rule under which sampling is terminated as soon as the value $Y_n = 1$ is observed. This is a valid stopping rule because

$$\Pr(N = j) = (\tfrac{1}{2})^j \qquad \text{for } j = 1, 2, \ldots$$

and, hence, $\Pr(N < \infty) = 1$. Under this stopping rule, it can be seen that $\Pr(Y_N = 1) = 1$ and, therefore, $E(Y_N) = 1$. Hence, we have constructed a martingale and a stopping rule for which $E(Y_1) = 0$ but $E(Y_N) = 1$.

This example described the gambling system under which the gambler doubles his bet at each stage and stops as soon as he realizes a positive gain. It is a system under which the gambler cannot lose, provided that he has the unlimited resources of both money and time necessary for him to be able to continue doubling his bet indefinitely. This system is applicable even if the probability p that the gambler will win at any stage is less than $\tfrac{1}{2}$, i.e., even if each bet that he makes is unfavorable for him. When $p < \tfrac{1}{2}$, the sequence of gains $Y_1, Y_2, \ldots$ is a supermartingale.

It was possible to construct this example for the following reason: Although each of the random variables Y_n is bounded and the entire sequence $Y_1, Y_2, \ldots$ is uniformly bounded above by the value 1, nevertheless the sequence is not bounded below and, in fact, is not uniformly integrable.

The expected gain from a martingale or submartingale has been considered by Dubins and Freedman (1966) and by Chow (1967).

13.13 MARTINGALES FORMED FROM SUMS AND PRODUCTS OF RANDOM VARIABLES

We shall now develop some other results based on Theorem 1 of Sec. 13.11. Although the next theorem is quite general, it is particularly useful for studying martingales formed from sums of random variables.

Theorem 1 *Let the sequence $Y_1, Y_2, \ldots$ be a supermartingale, a submartingale, or a martingale with respect to the sequence $X_1, X_2, \ldots$. Suppose that for a given stopping rule, there exists a bound $M < \infty$ such that $E(|Y_1|) < M$ and, for $n = 1, 2, \ldots$ and for all points $(x_1, \ldots, x_n) \, \epsilon \, \{N > n\}$, the following relation is satisfied:*

$$E(|Y_{n+1} - Y_n| \mid x_1, \ldots, x_n) \leq M. \tag{1}$$

Suppose also that $E(N) < \infty$. Then $E(Y_N)$ exists. For a supermartingale, $E(Y_N) \leq E(Y_1)$. For a submartingale, $E(Y_N) \geq E(Y_1)$. For a martingale, $E(Y_N) = E(Y_1)$.

Proof If we let $y_0 = 0$, then

$$E(|Y_N|) = \sum_{j=1}^{\infty} \int_{\{N=j\}} |y_j|\, dF_j$$

$$\leq \sum_{j=1}^{\infty} \int_{\{N=j\}} \sum_{i=1}^{j} |y_i - y_{i-1}|\, dF_j$$

$$= \sum_{j=1}^{\infty} \sum_{i=1}^{j} \int_{\{N=j\}} |y_i - y_{i-1}|\, dF_j. \tag{2}$$

By reversing the order of summation in the final expression in Eq. (2), we can obtain the following relation:

$$E(|Y_N|) \leq \sum_{i=1}^{\infty} \sum_{j=i}^{\infty} \int_{\{N=j\}} |y_i - y_{i-1}|\, dF_j$$

$$= \sum_{i=1}^{\infty} \int_{\{N>i-1\}} |y_i - y_{i-1}|\, dF_i$$

$$= E(|Y_1|) + \sum_{i=2}^{\infty} \int_{\{N>i-1\}} E(|Y_i - Y_{i-1}| \mid x_1, \ldots, x_{i-1})\, dF_{i-1}$$

$$\leq M + M \sum_{i=2}^{\infty} \Pr(N > i - 1)$$

$$= M \sum_{i=0}^{\infty} \Pr(N > i)$$

$$= M E(N) < \infty. \tag{3}$$

Hence, $E(Y_N)$ exists.

Furthermore,

$$\left| \int_{\{N>n\}} y_n\, dF_n \right| \leq \int_{\{N>n\}} |y_n|\, dF_n \leq \sum_{i=1}^{n} \int_{\{N>n\}} |y_i - y_{i-1}|\, dF_n$$

$$= \sum_{j=n+1}^{\infty} \sum_{i=1}^{n} \int_{\{N=j\}} |y_i - y_{i-1}|\, dF_j. \tag{4}$$

Since we have already shown that the double sum in the final expression in the relation (2) is finite, it follows that

$$\lim_{n \to \infty} \sum_{j=n+1}^{\infty} \sum_{i=1}^{n} \int_{\{N=j\}} |y_i - y_{i-1}|\, dF_j = 0. \tag{5}$$

The theorem follows from relations (4) and (5) and Theorem 1 of Sec. 13.11.∎

Corollary 1 *Suppose that the random variables X_1, X_2, . . . are independent and identically distributed and that $E(X_n) = 0$ ($n = 1, 2, . . .$). Let $Y_n = \Sigma_{j=1}^n X_j$ for $n = 1, 2, ...$ Then the sequence Y_1, Y_2, . . . is a martingale with respect to the sequence X_1, X_2, ... If $E(N) < \infty$ for a given stopping rule, then $E(Y_N)$ exists and $E(Y_N) = E(Y_1) = 0$.*

Proof Since the random variables X_1, X_2, . . . are identically distributed with a finite mean, there is a number M such that $E(|X_n|) = M$ ($n = 1, 2, . . .$). Furthermore, for all values of x_1, . . . , x_n ($n = 1, 2, . . .$),

$$E(|Y_{n+1} - Y_n| \mid x_1, \ldots, x_n) = E(|X_{n+1}|) = M. \tag{6}$$

The corollary now follows from Theorem 1.∎

It is of interest to note that Theorem 1 of Sec. 12.16, which we proved earlier by other methods, is a simple consequence of Corollary 1.

Corollary 1 can be given the following interpretation. Suppose that the statistician can make a sequence of independent and identically distributed gambles and that the expected gain from any one of these gambles is 0. Let X_j be the gain from the jth gamble, and let Y_n be the total gain from the first n gambles. Then the expected total gain $E(Y_N)$ must be 0 for any stopping rule under which the expected number of gambles that will be made before gambling is terminated is finite.

The next theorem indicates that the situation is somewhat different for martingales constructed from products of random variables which are independent and identically distributed. We shall consider only nonnegative products. Suppose that X_1, X_2, . . . is a sequence of identically distributed random variables. Then, for the purposes of this theorem, we shall let $\Pr(\cdot \mid f)$ and $E(\cdot \mid f)$ denote probabilities and expectations computed when the g.p.d.f. of each X_n is f.

Theorem 2 *Suppose that X_1, X_2, . . . are independent and identically distributed random variables such that their common g.p.d.f. is f and $f(x) = 0$ for $x < 0$. Suppose also that $E(X_n|f) = 1$ ($n = 1, 2, . . .$). Let $Y_n = \Pi_{j=1}^n X_j$ for $n = 1, 2, ...$ Then the sequence Y_1, Y_2, . . . is a martingale with respect to the sequence X_1, X_2, ... Furthermore, let f^* be the g.p.d.f. defined as follows:*

$$f^*(x) = xf(x) \qquad for \; -\infty < x < \infty. \tag{7}$$

Then, for any stopping rule for which $\Pr(N < \infty|f) = 1$,

$$E(Y_N|f) = \Pr(N < \infty|f^*). \tag{8}$$

Proof The function f^* is a g.p.d.f. It is nonnegative since $f(x) = 0$ for

$x < 0$. Also,

$$\int_{-\infty}^{\infty} f^*(x)\,d\mu(x) = \int_0^{\infty} xf(x)\,d\mu(x) = E(X_n|f) = 1. \tag{9}$$

Hence,

$$
\begin{aligned}
E(Y_N|f) &= \sum_{n=1}^{\infty} \int_{\{N=n\}} \left(\prod_{i=1}^{n} x_i\right) \left[\prod_{i=1}^{n} f(x_i)\right] \prod_{i=1}^{n} d\mu(x_i) \\
&= \sum_{n=1}^{\infty} \int_{\{N=n\}} \left[\prod_{i=1}^{n} f^*(x_i)\right] \prod_{i=1}^{n} d\mu(x_i) \\
&= \sum_{n=1}^{\infty} \Pr(N = n|f^*) = \Pr(N < \infty|f^*). \blacksquare \tag{10}
\end{aligned}
$$

For any martingale Y_1, Y_2, . . . formed from products of non-negative random variables which are independent and identically distributed and for any stopping rule in the class Δ, it follows from Theorem 2 that $E(Y_N) \leq 1$. The next corollary provides a simple condition on the stopping rule which guarantees that $E(Y_N) = 1$.

Corollary 2 *Suppose that the random variables X_1, X_2, . . . satisfy the conditions specified in Theorem 2, and let the sequence Y_1, Y_2, . . . be the martingale defined in Theorem 2. Suppose also that for a given stopping rule in the class Δ, there is a bound $M < \infty$ such that, for $n = 1, 2, \ldots$,*

$$E(Y_n|N > n) \leq M. \tag{11}$$

Then, $E(Y_N) = 1$.

Proof By Theorem 2, we need only show that if the g.p.d.f. f^* is defined by Eq. (7), then

$$\lim_{n \to \infty} \Pr(N > n|f^*) = 0. \tag{12}$$

However, the following relation can be written:

$$
\begin{aligned}
\Pr(N > n|f^*) &= \int_{\{N>n\}} \left[\prod_{i=1}^{n} f^*(x_i)\right] \prod_{i=1}^{n} d\mu(x_i) \\
&= \int_{\{N>n\}} \left(\prod_{i=1}^{n} x_i\right) \left[\prod_{i=1}^{n} f(x_i)\right] \prod_{i=1}^{n} d\mu(x_i) \\
&= E(Y_n|N > n, f)\,\Pr(N > n|f) \leq M\,\Pr(N > n|f).
\end{aligned}
$$

Since, for any stopping rule in the class Δ, $\Pr(N > n|f) \to 0$ as $n \to \infty$, Eq. (12) is established. $\blacksquare$

As another application of Theorem 2, we shall quote the next corollary, which was first developed by Wald (1947) and is known as the *fundamental identity of sequential analysis*. This corollary was originally applied by Wald to the sequential probability-ratio test.

Corollary 3 *Let X_1, X_2, $\ldots$ be a sequence of independent and identically distributed random variables. Suppose that for a given stopping rule in the class Δ, there is a bound $M < \infty$ such that $|x_1 + \cdots + x_n| \leq M$ for all points $(x_1, \ldots, x_n) \in \{N > n\}$ and for $n = 1, 2, \ldots$. Also, let $\psi(t) = E(e^{tX_i})$ be the moment-generating function of X_i. Then, for any value of t such that $\psi(t) < \infty$,*

$$E\{e^{t(X_1 + \cdots + X_N)}[\psi(t)]^{-N}\} = 1. \tag{13}$$

The proof of Corollary 3 is left as Exercise 26.

Further Remarks and References

Martingales that are formed from sums of random variables have been considered by Chow, Robbins, and Teicher (1965), Teicher (1966), Andrews and Blum (1966), and Farrell (1964b; 1966a,b).

Some papers on the fundamental identity of sequential analysis which give references to earlier work on this topic are Bellman (1957b), Bahadur (1958), Ruben (1959), Tweedie (1960), and Miller (1961).

13.14 REGULAR SUPERMARTINGALES

We shall now consider the importance of supermartingales in general problems of optimal stopping. As we shall see, the properties of supermartingales can be utilized to facilitate the determination of optimal stopping rules in a wide variety of problems.

Let the sequence Y_1, Y_2, $\ldots$ be a supermartingale with respect to the sequence X_1, X_2, $\ldots$. In the preceding three sections, we presented conditions under which $E(Y_N) \leq E(Y_1)$ for a given stopping rule. We shall now give the following definition: A supermartingale is *regular* if $E(Y_N) \leq E(Y_1)$ for every stopping rule for which $E(Y_N)$ exists.

It follows from Theorem 1 of Sec. 13.12 that if the random variables Y_1, Y_2, $\ldots$ are uniformly integrable, then the supermartingale is regular. This concept of regularity is related to, but is not the same as, the concept of stability introduced in Chap. 12 for sequential decision problems. Both concepts are introduced to guarantee that even though a stopping rule may not be bounded, its properties can be determined as well-behaved limits of the corresponding properties of bounded procedures.

Consider now an arbitrary problem of optimal stopping which is defined by observations X_n and gains $Y_n = y_n(X_1, \ldots, X_n)$ for $n = 1$, $2, \ldots$, and suppose that there exists an optimal stopping rule under which the expected gain is finite. Suppose also that at a given stage of the procedure, the values $X_1 = x_1, \ldots, X_n = x_n$ have been observed and the following relation is satisfied:

$$E(Y_{n+1}|x_1, \ldots, x_n) > y_n(x_1, \ldots, x_n). \tag{1}$$

Then, as we have noted on several earlier occasions, it is clearly optimal to continue sampling, since the expected gain from taking the next observation and then terminating the procedure is greater than the gain from stopping without further observations.

On the other hand, suppose that the following relation is satisfied:

$$E(Y_{n+1}|x_1, \ldots, x_n) \leq y_n(x_1, \ldots, x_n). \tag{2}$$

In order to decide whether or not it is optimal to terminate the sampling process, the statistician must analyze the process further. Thus, although the expected gain from taking just one more observation is not greater than the gain from terminating the sampling process immediately, there may exist a sequential continuation under which the expected gain will exceed $y_n(x_1, \ldots, x_n)$. However, suppose that for any set of observed values $X_1 = x_1, \ldots, X_n = x_n$ for which the relation (2) is satisfied, the sequence of future gains $Y_{n+1}, Y_{n+2}, \ldots$ is a regular supermartingale with respect to the sequence of future observations $X_{n+1}, X_{n+2}, \ldots$. The meaning of the assumption that the future gains form a super-martingale is as follows: Once the sampling process reaches a stage at which it would be unfavorable to observe exactly one more value, then at any future stage that can be reached through further sampling, it would also be unfavorable to observe one more value. The further assumption that the supermartingale is regular guarantees that the expected gain from any sequential continuation must be unfavorable. Hence, when any procedure specifies continuing the sampling process beyond the stage at which the observed values are $X_1 = x_1, \ldots, X_n = x_n$ and it is known that the expected gain $E(Y_N|x_1, \ldots, x_n)$ exists, then the following relation must be satisfied:

$$E(Y_N|x_1, \ldots, x_n) \leq E(Y_{n+1}|x_1, \ldots, x_n) \leq y_n(x_1, \ldots, x_n). \tag{3}$$

The inequalities (3) show that when any set of observed values $X_1 = x_1$, $\ldots, X_n = x_n$ satisfies the relation (2), it is optimal to terminate the procedure. This property is summarized in the next theorem.

Theorem 1 *Consider a problem of optimal stopping in which an optimal stopping rule exists. Suppose that for any set of observed values $X_1 = x_1$, . . . , $X_n = x_n$ which satisfies the relation (2), the sequence of future gains $Y_{n+1}, Y_{n+2}, \ldots$ is a regular supermartingale with respect to the sequence of future observations $X_{n+1}, X_{n+2}, \ldots .$ Then an optimal procedure after any set of values $X_1 = x_1$, . . . , $X_n = x_n$ has been observed is to continue sampling if relation (1) is satisfied and to terminate the sampling process if relation (2) is satisfied.*

EXAMPLE As a simple illustration of Theorem 1, consider the following problem. A contestant in a quiz program is to be asked a sequence of questions. He is given an initial prize of r dollars when he enters the contest, and he receives an additional prize of s dollars for each question he answers correctly. However, as soon as he does not answer one of the questions correctly, he must forfeit all his prize money and he is eliminated from the contest. He may stop at any time before he has been eliminated and may then accept his accumulated prize money as his gain. Suppose that the probability that he will answer any given question correctly is p $(0 < p < 1)$ and that all questions are independent.

We shall let Y_n be the contestant's gain at the nth stage ($n = 0, 1, 2, \ldots$). Then, if he has answered the first n questions correctly, $Y_n = r + ns$, and if he has answered one of them incorrectly, $Y_n = 0$. If $Y_n = y$ ($y > 0$) at a given stage, then $Y_{n+1} = y + s$ with probability p and $Y_{n+1} = 0$ with probability $1 - p$. Hence,

$$E(Y_{n+1}|Y_n = y) = p(y + s). \tag{4}$$

It can be seen that $E(Y_{n+1}|Y_n = y) \leq y$ if, and only if,

$$y \geq \frac{ps}{1 - p}. \tag{5}$$

The inequality (5) shows that when the contestant's gain reaches the value $ps/(1 - p)$, it is unfavorable for him to try to answer just one more question. Since his gain y can only increase as long as he has not been eliminated, it can be seen that when the inequality (5) is satisfied, the sequence of future gains $Y_{n+1}, Y_{n+2}, \ldots$ is a supermartingale.

Furthermore, when $Y_n = y$, the distribution of Y_{n+j} ($j = 1, 2, \ldots$) is

$$\Pr(Y_{n+j} = y + js) = p^j, \qquad \Pr(Y_{n+j} = 0) = 1 - p^j. \tag{6}$$

It follows that the random variables $Y_{n+1}, Y_{n+2}, \ldots$ are uniformly integrable. Hence, by Theorem 1 of Sec. 13.12, the supermartingale is regular.

According to Theorem 1, the optimal procedure for the contestant is to try to answer questions until his prize money y reaches the level given in the relation (5) and then to stop.

Other examples to which Theorem 1 applies are provided by Exercises 31 and 32 of Chap. 12 and Exercise 14 of this chapter.

The situation described in Theorem 1 is called the *monotone case* by Chow and Robbins (1961) and has been studied in detail by them. Thus the result presented in Theorem 1 can be described as follows: In the monotone case, an optimal procedure for the statistician is the myopic procedure in which his decision at any stage is the same as the decision which would be appropriate at that stage if he were permitted to take at most one more observation before being forced to terminate the sampling process.

13.15 SUPERMARTINGALES AND GENERAL PROBLEMS OF OPTIMAL STOPPING

Supermartingales and regular supermartingales can be used to characterize the optimal stopping rules and their expected gains in completely general problems, as is shown in the papers by Snell (1952), Haggstrom (1966), and Chernoff (1967a). We shall consider a general problem of optimal stopping defined by the sequence of observations $X_1, X_2, \ldots$ and the sequence of gains $Y_1, Y_2, \ldots$ under the assumption that each expectation $E(Y_n)$ exists. We shall let $Z = \sup_n Y_n$, and we shall suppose that $E(|Z|) < \infty$.

Furthermore, for $n = 1, 2, \ldots$, we shall let $U_n = u_n(X_1, \ldots, X_n)$ be a function of the first n observations, and we shall assume that the sequence $U_1, U_2, \ldots$ is a supermartingale with respect to the sequence $X_1, X_2, \ldots$ such that $U_n \geq Y_n$ $(n = 1, 2, \ldots)$ with probability 1. We shall also suppose that if the sequence $T_1, T_2, \ldots$ is any other supermartingale with respect to the sequence $X_1, X_2, \ldots$ such that $T_n \geq Y_n$ $(n = 1, 2, \ldots)$ with probability 1, then it is also true that $T_n \geq U_n$ $(n = 1, 2, \ldots)$ with probability 1. It is said that such a sequence $U_1, U_2, \ldots$ is a *minimal supermartingale*. If, in this definition, the requirement that the sequences $U_1, U_2, \ldots$ and $T_1, T_2, \ldots$ be supermartingales is replaced by the requirement that they be *regular* supermartingales, then the sequence $U_1, U_2, \ldots$ is called a *minimal regular supermartingale*. The existence of minimal supermartingales and minimal regular supermartingales can be established. The next theorem shows that they will satisfy the basic functional equation, which is also satisfied by the expected gain from the optimal stopping rule.

Theorem 1 *If the sequence $U_1, U_2, \ldots$ is either a minimal supermartingale or a minimal regular supermartingale, then for $n = 1, 2, \ldots$, the* *following equation will be satisfied with probability* 1:

$$u_n(X_1, \ldots, X_n)$$
$$= \max \{y_n(X_1, \ldots, X_n), E(U_{n+1}|X_1, \ldots, X_n)\}. \quad (1)$$

Proof We shall give the proof for a minimal regular supermartingale; the proof for a minimal supermartingale is virtually identical. Since the sequence U_1, U_2, . . . is a supermartingale and $U_n \geq Y_n$, then for $n = 1, 2, \ldots$, the following relation will be satisfied with probability 1:

$$u_n(X_1, \ldots, X_n)$$
$$\geq \max \{y_n(X_1, \ldots, X_n), E(U_{n+1}|X_1, \ldots, X_n)\}. \quad (2)$$

Suppose that for some value of n, say $n = m$, the probability is positive that there will be strict inequality in the relation (2). Define the sequence $T_1 = t_1(X)$, $T_2 = t_2(X_1, X_2)$, . . . as follows:

$$\begin{cases} t_j(X_1, \ldots, X_j) = E(Z|X_1, \ldots, X_j) & \text{for } j < m, \\ t_m(X_1, \ldots, X_m) = \max \{y_m(X_1, \ldots, X_m), \\ \qquad\qquad\qquad\qquad\qquad E(U_{m+1}|X_1, \ldots, X_m)\}, \\ t_j(X_1, \ldots, X_j) = u_j(X_1, \ldots, X_j) & \text{for } j > m. \end{cases}$$
$$(3)$$

It can be shown from this definition and Exercise 28 that the sequence T_1, T_2, . . . must be a supermartingale with respect to the sequence X_1, X_2, ... Furthermore, it is a regular supermartingale since the sequence U_1, U_2, . . . is a regular supermartingale and $T_j = U_j$ for $j > m$. Finally, for $n = 1, 2, \ldots$, the probability that $T_n \geq Y_n$ is 1. However, when $n = m$, the probability is positive that there will be strict inequality in the relation (2). Hence, there is positive probability that $T_m < U_m$. But this contradicts the hypothesis that the sequence U_1, U_2, . . . is a minimal regular supermartingale. Hence, for $n = 1, 2, \ldots$, there is equality in the relation (2) with probability 1.∎

If $V_n = v_n(X_1, \ldots, X_n)$ denotes the expected gain under the optimal stopping rule after the values of $X_1, \ldots, X_n$ have been observed, then V_n also satisfies a functional equation like (1). Hence, in those problems where the equation has a unique solution, it follows that $U_n = V_n$ for $n = 1, 2, \ldots$. Indeed, it is shown in the references given at the beginning of this section that the sequence V_1, V_2, . . . of optimal expected gains will be the minimal regular supermartingale under very general conditions.

13.16   MARKOV PROCESSES

In this section we shall define a Markov process and begin the discussion of optimal stopping rules for problems involving such a process. Let

Z_1, Z_2, . . . be a sequence of random variables or, more generally, a sequence of random vectors, each of which takes values in the sample space S. Since the space S may be a fairly general subset of some vector space R^k, it will be convenient to introduce explicitly the σ-field $\mathfrak{A}$ of subsets of S for which probabilities are defined. The sequence Z_1, Z_2, . . . is a Markov process if, at any stage, the distribution of future values Z_{n+1}, Z_{n+2}, . . . of the sequence depends only on the current value Z_n and does not depend on past values Z_1, . . . , Z_{n-1}. More precisely, it is said that the sequence Z_1, Z_2, . . . is a *Markov process* if, for every event $A \in \mathfrak{A}$, for all values z_1, . . . , z_n in S, and for $n = 1, 2, . . .$, the following equation is satisfied:

$$\Pr(Z_{n+1} \in A \,|\, Z_1 = z_1, \ldots, Z_n = z_n) = \Pr(Z_{n+1} \in A \,|\, Z_n = z_n). \quad (1)$$

It is said that a Markov process has *stationary transition probabilities* if the conditional distributions in Eq. (1) remain the same from stage to stage. In other words, for each event $A \in \mathfrak{A}$ and each value $z \in S$, the conditional probability $\Pr(Z_{n+1} \in A \,|\, Z_n = z)$ is the same for $n = 1, 2, ...$ We shall assume that for each value $z \in S$, this conditional distribution can be represented by a g.p.d.f. $f(\cdot \,|z)$ on S. Hence, in a Markov process with stationary transition probabilities, the following relation must be satisfied:

$$\Pr(Z_{n+1} \in A \,|\, Z_1 = z_1, \ldots, Z_n = z_n) = \int_A f(z_{n+1}|z_n) \, d\mu(z_{n+1}). \quad (2)$$

The function f is called the *transition function* of the process.

We shall consider now a problem of optimal stopping in which the observations Z_1, Z_2, . . . constitute a Markov process with stationary transition probabilities. The value of Z_n is called the *state* of the process at the nth stage. We shall assume that the process starts from a given initial state $z_0 \in S$ and, hence, that the g.p.d.f. of Z_1 is $f(\cdot \,|z_0)$. In this context, the sample space S is also called the *state space*. When the state space is finite or countable, the Markov process is called a *Markov chain*.

Suppose that at the nth stage the process is in state z. In other words, suppose that $Z_n = z$. We shall assume that at this stage, as at any stage, the statistician either can terminate the sampling process and receive a reward $b(z)$ or can pay a cost $c(z)$ and continue sampling. Here the reward function b and the cost function c depend only on the state z and do not depend on the stage n of the process.

In addition, there may be a subset $S_0 \subset S$ of states in which the statistician is forced to stop sampling, and there may be a subset $S_1 \subset S$ of states in which he is forced to continue sampling. Thus, if $z \in S_0$, then the statistician must stop sampling and receive $b(z)$; if $z \in S_1$, he must pay $c(z)$ and continue sampling.   In many problems, of course, S_0 or S_1 is empty.

If the statistician decides not to take any observations, his gain is $b(z_0)$.   If he takes at least one observation and terminates the sampling process after the values $Z_1 = z_1, \ldots, Z_n = z_n$ have been observed, his total gain is $b(z_n) - c(z_0) - c(z_1) - \cdots - c(z_{n-1})$.   Among all rules which specify that at least one observation must be taken, the problem is to find a stopping rule which maximizes the expected total gain.   This expectation can be expressed as follows:

$$E[b(Z_N) - c(z_0) - c(Z_1) - \cdots - c(Z_{N-1})]. \tag{3}$$

Most of the sequential problems that have been described in this chapter and in Chap. 12 are of this type.   For example, in the sequential statistical decision problem considered in Secs. 12.7 to 12.9, we could let the initial state ξ_0 be the given prior distribution of the parameter W and, in general, could let ξ_n be the posterior distribution of W after the first n observations $X_1, \ldots, X_n$ have been taken.   When the prior distribution and all possible posterior distributions can be indexed by a finite number of parameters (as, for example, when all the distributions belong to a conjugate family), then all the states ξ_n can be regarded as points in a finite-dimensional state space S.   Since, for any given value of W, the observations $X_1, X_2, \ldots$ are independent and identically distributed, it follows that the transition probabilities from the distribution ξ_n to the posterior distribution ξ_{n+1} are the same at all stages.   Hence, the sequence $\xi_1, \xi_2, \ldots$ forms a Markov process with stationary transition probabilities.

Furthermore, the risk from terminating the sampling process and choosing a decision when the process has reached the state $\xi \, \epsilon \, S$ is $\rho_0(\xi)$.   Hence, the reward function is $b(\xi) = -\rho_0(\xi)$.   Since the cost of taking another observation at any stage is c, the cost function is $c(\xi) = c$, which is a constant.   In this problem, both the set S_0 and the set S_1 are empty.

Next, consider the bounded sequential decision problem described in Secs. 12.4 and 12.5.   This problem is similar to the above problem, but the sampling process must be terminated after not more than n observations have been taken.   Let the state Z of the process be defined as the pair $Z = (j, \xi_j)$, where j is a nonnegative integer specifying the number of observations which have been taken and ξ_j is, as before, the distribution of W at that stage.   The given initial state is $z_0 = (0, \xi_0)$.   On the basis of this definition, the process is a Markov process with stationary transition probabilities.   From any state of the form (j, ξ), the next transition must be, of course, to a state of the form $(j + 1, \xi')$, where ξ' is a posterior distribution which can be obtained from the distribution ξ with a single observation.

The reward and cost functions are again as given above. Since the statistician must terminate the sampling process after not more than n observations have been taken, the set S_0 contains all states of the form (n, ξ) and the set S_1 is empty.

By utilizing the technique of including the stage j as a component of the state of the process, we can formulate many problems within the framework presented here. Thus, in the above examples, suppose that the cost of each observation is not constant but that the cost c_j of the jth observation depends on the value of j. If the stage j is included as a component of the state, the cost function c for the process is still of the appropriate form.

Other examples are listed in Exercise 31.

Further Remarks and References

Markov chains are discussed in most standard texts on probability, as well as in the books of Kemeny and Snell (1960), Chung (1960), and Kemeny, Snell, and Knapp (1966). At a more advanced level are the two volumes by Dynkin (1965) on Markov processes.

Most of the material on Markov processes in the rest of this chapter, including Exercise 33, is based on the paper of Breiman (1964). Kemeny and Snell (1958) discuss some problems of optimal stopping for Markov chains with finite state spaces.

13.17 STATIONARY STOPPING RULES FOR MARKOV PROCESSES

We shall now assume that for any initial state $z_0 \in S$, there exists an optimal stopping rule which maximizes the expected total gain given by the expression (3) of Sec. 13.16 and that this maximum value is finite. Even for a simple Markov chain with a finite state space, however, such an optimal rule need not exist. For example, suppose that there is a state $z \in S$ for which $c(z) < 0$ and for which $|c(z)|$ is very large. Suppose also that as the chain evolves, it will be in the state z relatively often. Then it may be worthwhile for the statistician to continue the sampling process indefinitely and to receive the positive amount $-c(z)$ each time the chain reaches the state z. On the other hand, we have already studied a wide variety of problems in each of which there exists an optimal procedure that yields a finite expected gain. Hence, our assumption is meaningful.

Technically, the statistician's decision to terminate the sampling process or to continue sampling at the nth stage may depend on all the states $z_0, z_1, \ldots, z_n$ of the process up to that time. However, as we have stated before, his decision cannot affect the amount of money already spent on sampling. Furthermore, the reward $b(z_n)$ from terminating the sampling process depends only on the current state z_n. Also, since we are considering a Markov process with stationary transition probabilities, the expected gain from any procedure which specifies that sampling should be continued depends only on the current state z_n. Accordingly, we shall restrict ourselves to stopping rules for which the statistician's decision at any stage depends only on the state of the process at that stage. A stopping rule δ with this property is called a *stationary* stopping rule. Such a rule is characterized by the set S_δ of states for which sampling will be terminated ($S_\delta \subset S$). Since sampling must be terminated in the set S_0 and must be continued in the set S_1, then $S_0 \subset S_\delta$ and $S_1 \subset S_\delta{}^c$ for every stationary stopping rule δ. Furthermore, if the initial state $z_0 \notin S_\delta$, then the stopping set S_δ must satisfy the requirement that sampling will not be continued indefinitely. In order to satisfy this requirement, the probability must be 1 that at least one of the states $Z_1, Z_2, \ldots$ will belong to S_δ.

The next step in the solution of the problem is to eliminate the states in the set S_1. When the statistician decides whether or not to terminate the sampling process at any state $z \in (S_0 \cup S_1)^c$, he takes into consideration the following fact: If he continues sampling and the process moves to a state in S_1, he will not be permitted to terminate sampling until the process has subsequently moved to a state outside the set S_1, and he must pay the sampling cost of each transition until such a state is reached. Hence, as the statistician proceeds from any state $z \in S_1{}^c$, he must compute the probability distribution of the first state in the continuation that will again lie in $S_1{}^c$, and he must compute the expected sampling cost until a state in $S_1{}^c$ is again reached. For simplicity, we shall assume that these computations are carried out for states in S_0, as well as for states in $(S_0 \cup S_1)^c$, even though the statistician is forced to stop in S_0. These values computed by the statistician become the transition function and the cost function in a new reduced problem in which the state space is $S_1{}^c$ and there are no states at which the statistician is forced to continue sampling. An optimal procedure in the reduced problem is also optimal in the original problem, and the expected total gain is the same in both problems.

We are assuming here, of course, that when the process starts from any state in $S_1{}^c$, the probability is 1 that the process will return to $S_1{}^c$ at some future stage. Otherwise, there would be positive probability that the process would enter S_1 and remain there forever and that the statistician would never again be allowed to terminate the sampling process. Furthermore, we assume that the initial state of the process is in $S_1{}^c$. Even if this initial state were in S_1, the statistician's decision problem would not really begin until the process had reached a state in $S_1{}^c$.

The new transition function on $S_1{}^c$ can be computed as follows:

For any starting state $z \in S$, let $f^*(\cdot | z)$ denote the g.p.d.f. on $S_1{}^c$ of the first state in $S_1{}^c$ that will be reached later in the process. If $z \in S_1{}^c$, then $f^*(\cdot | z)$ is the g.p.d.f. of the first state in $S_1{}^c$ that will be reached through further sampling. Then for any state $z \in S$ and any state $z' \in S_1{}^c$, the functions f^* must satisfy the following functional equation:

$$f^*(z'|z) = f(z'|z) + \int_{S_1} f^*(z'|z_1) f(z_1|z) \, d\mu(z_1). \tag{1}$$

This equation can be verified by the following reasoning: The state z' can be reached either on the first transition, for which the density is $f(z'|z)$, or on a later transition. If more than one transition is required, the first transition is to a point $z_1 \in S_1$ and the process eventually enters $S_1{}^c$ at z'. The density of this transition is given by the integral in Eq. (1).

Similarly, for any starting state $z \in S$, let $c^*(z)$ denote the expected sampling cost until some state in $S_1{}^c$ is reached. Then c^* must satisfy the following functional equation:

$$c^*(z) = c(z) + \int_{S_1} c^*(z_1) f(z_1|z) \, d\mu(z_1). \tag{2}$$

Equation (2) can be verified by the following reasoning: The cost of the first transition from z is $c(z)$. If the first transition is to a state $z_1 \in S_1$, the additional expected cost is $c^*(z_1)$.

EXAMPLE Consider a Markov chain with four states and stationary transition probabilities. Let $S = \{1, 2, 3, 4\}$. Suppose that the transition function is given by Table 13.1, where the entry in the ith row and the jth column is $f(j|i) = \Pr(Z_{n+1} = j | Z_n = i)$. Suppose also that S_0 is empty but that $S_1 = \{3, 4\}$. Finally, suppose that the values of the reward function are $b(1) = 20$ and $b(2) = 14$ [the values of $b(3)$ and $b(4)$ are irrelevant] and that the sampling costs are $c(1) = c(2) = 1$ and $c(3) = c(4) = 2$. It is required to determine the optimal stopping rule.

Clearly, if $f^*(\cdot | i)$ denotes the frequency function on $S_1{}^c = \{1, 2\}$ defined above, then $f^*(1|i) = 1 - f^*(2|i)$ for $i = 1, 2, 3, 4$, and it can be seen from Table 13.1 and Eq. (1) that the values $f^*(1|i)$ must satisfy the

Table 13.1

	1	2	3	4
1	$\frac{1}{4}$	$\frac{1}{4}$	$\frac{1}{4}$	$\frac{1}{4}$
2	0	$\frac{1}{3}$	0	$\frac{2}{3}$
3	$\frac{1}{4}$	$\frac{1}{4}$	$\frac{1}{2}$	0
4	$\frac{1}{2}$	0	$\frac{1}{4}$	$\frac{1}{4}$

Table 13.2

	1	2
1	$\frac{7}{12}$	$\frac{5}{12}$
2	$\frac{5}{9}$	$\frac{4}{9}$

following relations:

$$f^*(1|1) = \tfrac{1}{4} + \tfrac{1}{4}f^*(1|3) + \tfrac{1}{4}f^*(1|4),$$
$$f^*(1|2) = \tfrac{2}{3}f^*(1|4),$$
$$f^*(1|3) = \tfrac{1}{4} + \tfrac{1}{2}f^*(1|3), \tag{3}$$
$$f^*(1|4) = \tfrac{1}{2} + \tfrac{1}{4}f^*(1|3) + \tfrac{1}{4}f^*(1|4).$$

By solving these equations, it is found that $f^*(1|1) = \frac{7}{12}$, $f^*(1|2) = \frac{5}{9}$, $f^*(1|3) = \frac{1}{2}$, and $f^*(1|4) = \frac{5}{6}$. Hence, after we have eliminated the states in S_1, the transition probabilities for the reduced state space $S_1{}^c$ are those given in Table 13.2.

From the specified values of $c(i)$ and from Eq. (2), it can be seen that the expected sampling costs $c^*(i)$ must satisfy the following relations:

$$c^*(1) = 1 + \tfrac{1}{4}c^*(3) + \tfrac{1}{4}c^*(4),$$
$$c^*(2) = 1 + \tfrac{2}{3}c^*(4),$$
$$c^*(3) = 2 + \tfrac{1}{2}c^*(3), \tag{4}$$
$$c^*(4) = 2 + \tfrac{1}{4}c^*(3) + \tfrac{1}{4}c^*(4).$$

By solving these equations, the following results are obtained: $c^*(1) = 3$, $c^*(2) = \frac{11}{3}$, and $c^*(3) = c^*(4) = 4$. Hence, the problem is reduced to one involving the two states $\{1, 2\}$, the transition probabilities given in Table 13.2, the rewards $b(1) = 20$ and $b(2) = 14$, and the sampling costs $c^*(1) = 3$ and $c^*(2) = \frac{11}{3}$.

When the process is in state 1, it is clear that the statistician should terminate the sampling process because sampling costs are positive and $b(1)$ is the largest reward attainable. His only remaining decision occurs when the process is in state 2. He must then decide whether to terminate the sampling process or to continue sampling until state 1 is reached. If he stops, his gain is $b(2) = 14$. If he continues, each transition has probability $\frac{5}{9}$ of moving the process to state 1, and it follows from the geometric distribution that the expected number of transitions until state 1 is reached will be $\frac{9}{5}$. Therefore, the expected sampling cost will be $\frac{9}{5}(\frac{11}{3}) = 6.6$. Since his reward when state 1 is reached will be $b(1) = 20$, his expected net gain from continuing the process until state 1 is reached will be $20 - 6.6 = 13.4$.

Hence, when the process is in state 2, it is optimal to terminate the procedure. We have now found that regardless of whether the process begins in state 1 or in state 2, the statistician should collect his reward without any sampling.

13.18 ENTRANC⁻-FEE PROBLEMS

We shall now return to the general development, and as a consequence of the discussion in Sec. 13.17, we shall assume that the set S_1 is empty. The next step is to transform the given problem of optimal stopping into an equivalent one in which the reward function vanishes identically. Let b be the reward function in the given problem, and for each state $z \in S$, let the expectation $E[b(Z_1)|z]$ be defined as follows:

$$E[b(Z_1)|z] = \int_S b(z_1)f(z_1|z)\, d\mu(z_1). \tag{1}$$

It is assumed that the integral in Eq. (1) exists for all states $z \in S$. We shall now consider a transformed problem in which the reward function b' and the cost function c' have the following values: For any state $z \in S$, $b'(z) = 0$ and

$$c'(z) = b(z) - \{E[b(Z_1)|z] - c(z)\}. \tag{2}$$

The cost $c'(z)$ defined by Eq. (2) is the difference between the gain in the original problem from terminating sampling when the process is in state z and the expected gain from taking exactly one more observation and then terminating sampling.

Now consider a stationary stopping rule δ for which the stopping set is S_δ. Suppose that when the stopping rule δ is used for any initial state $z \in S$, the expected gain $V(z)$ in the original problem is finite. If $z \in S_\delta$, then the process is terminated and $V(z) = b(z)$. If $z \in S_\delta{}^c$, then the sampling cost $c(z)$ is paid and the process moves to a new state Z_1 from which the expected gain is $V(Z_1)$. Therefore, the function V must satisfy the following equation:

$$V(z) = \begin{cases} b(z) & \text{for } z \in S_\delta, \\ E[V(Z_1)|z] - c(z) & \text{for } z \in S_\delta{}^c. \end{cases} \tag{3}$$

Similarly, suppose that when the stopping rule δ is used, the expected gain $T(z)$ in the transformed problem is also finite. Then the function T must satisfy the following equation:

$$T(z) = \begin{cases} 0 & \text{for } z \in S_\delta, \\ E[T(Z_1)|z] - c'(z) & \text{for } z \in S_\delta{}^c. \end{cases} \tag{4}$$

By adding the value $b(z)$ to both sides of Eq. (4) and applying Eq. (2), it can be seen that a function $T(z)$ will satisfy Eq. (4) if, and only if, the function $V(z) = T(z) + b(z)$ will satisfy Eq. (3). It follows that for any initial state $z \in S$ and any stopping rule, the expected gain in the transformed problem will be equal to the original expected gain in the original problem minus $b(z)$. Thus, the transformed problem is equivalent to the original problem, and both problems must have the same optimal stopping rules.

A transformed problem of the type being considered here, in which the reward function is identically 0, is called an *entrance-fee problem*. In such a problem, the statistician must pay a cost, or entrance fee, at each stage in order to be permitted to observe the process for another stage, but whenever he terminates the process, he receives nothing. Under these conditions it may be advantageous to take some observations because some of the entrance fees may be negative.

In certain cases, the solution of an entrance-fee problem is easily obtained. Consider an entrance-fee problem in which the cost function is c. Without loss of generality, it can be assumed that the set S_1 is empty. If $c(z) < 0$ for some state $z \in S_0{}^c$, then the statistician should certainly continue sampling when the process is in state z. He gains $-c(z)$ by continuing, and he can always stop after the next transition at no additional cost. On the other hand, if the process is in some state $z \in S_0{}^c$ for which $c(z) \geq 0$ and if it is impossible for the process ever again to reach a state z' for which $c(z') < 0$, then it is clearly optimal to terminate sampling. We can therefore state the following theorem.

Theorem 1 *Consider an entrance-fee problem in which the cost function is c and the set S_1 is empty. Let S^* be the set of all states z for which $c(z) \geq 0$, and suppose that the only possible transition from any state in S^* is either to a state in S^* or to a state in S_0. Then it is optimal to terminate the sampling process as soon as a state in $S^* \cup S_0$ is reached.*

Theorem 1 is analogous to Theorem 1 of Sec. 13.14 in the following respect. It simply states that if the only possible transitions from an unfavorable state are to other unfavorable states, then it is optimal to conform to the myopic rule which specifies that the process should be continued only when the statistician will realize an immediate gain by continuing it. We have already considered some problems in which the myopic rule is optimal. Others are provided by the example in Sec. 13.17 (also see Exercise 32) and Exercise 33.

13.19 THE FUNCTIONAL EQUATION FOR A MARKOV PROCESS

We shall conclude this chapter with a brief discussion of the functional equation characterizing optimal stopping rules. We shall consider an entrance-fee problem in which the set S_1 is empty and which is defined by the cost function c, the transition function f, and the set S_0. For any state $z \, \epsilon \, S$, we shall let $V(z)$ denote the expected total gain from an optimal stopping rule when the initial state of the process is z. Then V satisfies the following functional equation:

$$V(z) = \begin{cases} 0 & \text{for } z \, \epsilon \, S_0, \\ \max \, \{0, \, E[V(Z_1)|z] - c(z)\} & \text{for } z \, \epsilon \, S_0^c. \end{cases} \tag{1}$$

Since we are assuming that an optimal procedure exists, Eq. (1) must have a solution. The next two theorems provide some conditions under which Eq. (1) will have only one solution satisfying certain restrictions.

Suppose that U and V are functions on S and that both functions satisfy Eq. (1). If $z \, \epsilon \, S_0^c$ and either $U(z) > 0$ or $V(z) > 0$, we can use the type of argument that was used in the proof of Theorem 1 of Sec. 12.12 to obtain the following relation:

$$|U(z) - V(z)| \leq E[|U(Z_1) - V(Z_1)| \, | \, z_0 = z]. \tag{2}$$

For all other states $z \, \epsilon \, S$, $U(z) = V(z) = 0$. Hence, the relation (2) must be satisfied for all states $z \, \epsilon \, S$. By iterating this argument, we can obtain the following result for $n = 1, 2, \ldots$ and all states $z \, \epsilon \, S$:

$$|U(z) - V(z)| \leq E[|U(Z_n) - V(Z_n)| \, | \, z_0 = z]. \tag{3}$$

It is clear that any function V which satisfies Eq. (1) is nonnegative on the set S. We shall suppose that the expected total gain from the optimal procedure can be bounded above on S by a nonnegative function H which satisfies the following relation for every state $z \, \epsilon \, S$:

$$\lim_{n \to \infty} E[H(Z_n)|z_0 = z] = 0. \tag{4}$$

This supposition is completely analogous to the assumption made in Eq. (2) of Sec. 12.12 that the sequential statistical decision problems considered in that section were stable.

Theorem 1 *Let H be a nonnegative function on S such that Eq. (4) is satisfied for all states $z \, \epsilon \, S$. Then there exists at most one function V which satisfies Eq. (1) and for which $V(z) \leq H(z)$ for all states $z \, \epsilon \, S$.*

Proof Suppose that U and V are two functions which satisfy Eq. (1) and that both functions are bounded above by H. Then $|U(Z_n) - V(Z_n)| \leq H(Z_n)$ for all values of n. Hence, it follows from the relations (3) and (4) that $U(z) = V(z)$ for all states $z \, \epsilon \, S$.∎

The next theorem is applicable mainly to Markov chains with a finite or countable state space. Since the statistician is forced to terminate sampling when the process is in a state belonging to the set S_0, the problem is unchanged if the transition function is modified on S_0 so that the following requirement is satisfied: Once the process is in any state $z \epsilon S_0$, it remains in that same state at all future stages. A state with this property is called an *absorbing* state. We are therefore saying that the statistician's problem remains unchanged if it is assumed that all the states of S_0 are absorbing. Furthermore, suppose that for some state $z \epsilon S$, it is known that the optimal stopping rule requires the statistician to terminate sampling when the process is in state z. Then the problem remains unchanged if the state z is included in the set S_0 and is assumed to be absorbing, since the statistician will stop the procedure in that state anyway. These remarks help to justify and fulfil the hypotheses in the following theorem.

Theorem 2 *Suppose that all the states in the set S_0 are absorbing and that from any initial state $z \epsilon S$, the probability is 1 that the process will ultimately enter the set S_0. Then there exists at most one function V which satisfies Eq. (1) and which is bounded on the set S.*

Proof Suppose that U and V are two functions which are bounded on S and satisfy Eq. (1). Since $U(z) = V(z) = 0$ for $z \epsilon S_0$, it follows from the hypotheses of the theorem that for any given initial state $z_0 \epsilon S$, the following relation will be satisfied with probability 1:

$$\lim_{n \to \infty} |U(Z_n) - V(Z_n)| = 0. \tag{5}$$

Since U and V are bounded functions, it follows from Eq. (5) that the expectation on the right side of the relation (3) converges to 0 as $n \to \infty$. Hence, $U(z) = V(z)$ for all states $z \epsilon S$.∎

For any Markov chain with a finite state space, any function defined on S is bounded. Hence, under the conditions of Theorem 2, there can be at most one function which satisfies Eq. (1).

EXERCISES

1. Suppose that in the problem considered in Secs. 13.2 and 13.3, the utility function U is

$$U(i) = \begin{cases} 1 & \text{for } i = 1, \ldots, k, \\ 0 & \text{for } i = k + 1, \ldots, n. \end{cases}$$

Here $1 < k < n$. In this problem, the statistician must maximize the probability of selecting an object which has rank k or higher. Show that the function U_0 can be written in the following form: For $1 \leq a \leq k$,

$$U_0(a, r) = \sum_{b=a}^{k} \frac{\binom{k}{b} \binom{n-k}{r-b}}{\binom{n}{r}},$$

and $U_0(a, r) = 0$ for $a > k$.

2. Suppose that $n = 5$ and $k = 2$ in Exercise 1. Show that the optimal procedure is as follows: Never accept the first object. Stop and accept the second object if it has rank 1, but continue the inspection process otherwise. (Alternatively, never accept the second object, since such a decision leads to the same probability of success.) Stop and accept the third object if it has rank 1, but continue the inspection otherwise. Stop and accept the fourth object if it has either rank 1 or rank 2, but otherwise continue the inspection to the last object. Show that the probability of success with this procedure is 0.7.

3. Suppose that in the problem considered in Secs. 13.2 and 13.3, the utility function U is

$$U(i) = n + 1 - i \qquad i = 1, 2, \ldots, n.$$

In this problem, maximizing the expected utility of the rank of the object selected by the statistician is equivalent to minimizing its expected rank. Show that for $a = 1, \ldots, r$ and $r = 1, \ldots, n$, the function U_0 can be written in the following form:

$$U_0(a, r) = (n + 1) \left(1 - \frac{a}{r + 1}\right).$$

4. Suppose that $n = 5$ in Exercise 3. Show that the optimal procedure is as follows: Never accept the first object. Stop and accept the second object if it has rank 1, but continue the inspection process otherwise. Stop and accept the third object if it has rank 1, but continue the inspection otherwise. Stop and accept the fourth object if it has either rank 1 or rank 2, but otherwise inspect the last object. Show that the expected utility with this procedure is $\frac{79}{20}$.

5. Prove that Eq. (3) of Sec. 13.4 is correct for any d.f. F on the real line for which the mean exists.

6. A sequential random sample $X_1, X_2, \ldots$ is to be taken from a uniform distribution on the interval $(0, b)$, and there is a specified upper bound n on the number of observations that can be taken. If the sampling process is terminated after the values $X_1 = x_1, \ldots, X_j = x_j$ have been observed $(j = 1, \ldots, n)$, the statistician's gain is x_j. Suppose that he must find a stopping rule which maximizes his expected gain. Show that the expected gain v_n from the optimal procedure has the following values: $v_1 = b/2$ and, for $n = 1, 2, \ldots$,

$$v_{n+1} = \frac{b}{2} \left[1 + \left(\frac{v_n}{b}\right)^2\right].$$

7. Suppose that a sequential random sample is to be taken from the exponential distribution with parameter β. For the conditions of Exercise 6, show that $v_1 = 1/\beta$ and, for $n = 1, 2, \ldots$,

$$v_{n+1} = v_n + \frac{1}{\beta} \exp\left(-\beta v_n\right).$$

8. Suppose that a sequential random sample is to be taken from the Pareto distribution with parameters x_0 and α $(x_0 > 0, \alpha > 1)$. For the conditions of Exercise

6, show that $v_1 = [\alpha/(\alpha - 1)]x_0$ and, for $n = 1, 2, \ldots,$

$$v_{n+1} = v_n \left[1 + \frac{1}{\alpha - 1} \left(\frac{x_0}{v_n} \right)^\alpha \right].$$

9. Suppose that a sequential random sample is to be taken from the normal distribution with mean μ and standard deviation σ. For the conditions of Exercise 6, show that $v_1 = \mu$, $v_2 = \mu + \sigma/(2\pi)^{\frac{1}{2}}$, and in general for $n = 1, 2, \ldots,$

$$v_{n+1} = \mu + \sigma\Psi\left(\frac{\mu - v_n}{\sigma}\right).$$

10. Suppose that two observations X_1, X_2 can be taken sequentially from the standard normal distribution. If the sampling process is terminated after the value $X_1 = x_1$ has been observed, the statistician's gain is x_1. If the sampling process is not terminated until both values $X_1 = x_1$ and $X_2 = x_2$ have been observed, the gain is $(x_1 + x_2)/2$. Show that the expected gain from the optimal stopping rule is $1/(8\pi)^{\frac{1}{2}}$.

11. Suppose that a sequential random sample is to be taken from the uniform distribution on the interval (a, b) at a cost of c units per observation. Let v^* be the expected gain from the optimal procedure for the problem of sampling without recall described in Sec. 13.5. Show that

$$v^* = \begin{cases} b - [2c(b - a)]^{\frac{1}{2}} & \text{if } c \leq \dfrac{b - a}{2}, \\ \dfrac{b + a}{2} - c & \text{if } c > \dfrac{b - a}{2}. \end{cases}$$

12. Suppose that a sequential random sample is to be taken from the exponential distribution with parameter β. For the conditions of Exercise 11, show that

$$v^* = \begin{cases} -\dfrac{1}{\beta} \log (\beta c) & \text{if } c \leq \dfrac{1}{\beta}, \\ \dfrac{1}{\beta} - c & \text{if } c > \dfrac{1}{\beta}. \end{cases}$$

13. Suppose that a sequential random sample is to be taken from the Pareto distribution with parameters x_0 and α ($x_0 > 0$, $\alpha > 2$). For the conditions of Exercise 11, show that

$$v^* = \begin{cases} \left[\dfrac{x_0{}^\alpha}{(\alpha - 1)c} \right]^{1/(\alpha-1)} & \text{if } c \leq \dfrac{x_0}{\alpha - 1}, \\ \dfrac{\alpha x_0}{\alpha - 1} - c & \text{if } c > \dfrac{x_0}{\alpha - 1}. \end{cases}$$

14. Consider the problem of sampling with recall from a distribution whose d.f. is F at a cost of c units per observation. It is assumed that F has a finite variance. At any stage of the sampling process, let r denote the maximum value among those observations that have been taken. Moreover, let $V(r)$ denote the expected gain (exclusive of the amount already spent on sampling) that results from taking exactly one more observation and then terminating the sampling process. Show that an optimal procedure is to continue the sampling process if, and only if, $r < V(r)$. In other words, show that an optimal procedure for the statistician is the myopic procedure in which his decision at any stage is the same as the decision which would be appropriate at that stage if he were permitted to take at most one more observation before being forced to terminate the sampling process.

15. Let $X_1, X_2, \ldots$ be a sequential random sample from a distribution whose d.f. is F, and let μ denote the mean of this distribution. Suppose that there is no

sampling cost but that the statistician may be forced to terminate the sampling process at any stage without recall, i.e., he will be forced to accept as his gain the value of the observation he has just taken. Specifically, suppose that at a given stage of the sampling process, the statistician has neither decided to terminate the procedure nor been forced to terminate it. Then, regardless of the values that have been observed, there is a fixed probability p $(0 < p < 1)$ that he will be forced to terminate the process after the next observation. He may, of course, terminate the procedure at any stage even though he is not forced to terminate it there. If the procedure is terminated after he has observed the values $X_1 = x_1, \ldots, X_n = x_n$, then his gain is x_n. The problem is to find a stopping rule which maximizes his expected gain $E(X_N)$. Show that an optimal procedure is as follows: If he has not been forced to terminate the procedure earlier, he should terminate it the first time he observes a value $x_j \geq v^*$, where v^* is the unique solution of the following equation:

$$(1 - p)T_F(v^*) = p(v^* - \mu).$$

Show that v^* is also the expected gain from this procedure.

16. Let $X_1, X_2, \ldots$ be a sequential random sample from a distribution whose d.f. is F, and let μ denote the mean of this distribution. Suppose that there is no sampling cost but that the statistician's gain is discounted. In other words, if he terminates the sampling process after having observed the values $X_1 = x_1, \ldots, X_n = x_n$, his gain is $q^n x_n$, where $0 < q < 1$. The problem is to find a stopping rule that maximizes his expected gain $E(q^N X_N)$. Show that an optimal procedure is for him to terminate the sampling process the first time he observes a value $x_j \geq v^*$, where v^* is the unique solution of the following equation:

$$qT_F(v^*) = (1 - q)v^*.$$

Show that v^* is also the expected gain from this procedure.

17. (a) Prove that for any fixed value of α $(-\infty < \alpha < \infty)$, the function $\Psi(\alpha s)/s$ is a decreasing function of s $(s > 0)$.

(b) Prove that for any fixed value of α $(\alpha > 0)$, the function $\Psi^{-1}(\alpha s)/s$ is a decreasing function of s $(s > 0)$.

18. (a) Let W be a parameter which can take only two values w_1 and w_2, and suppose that $\Pr(W = w_1) = \Pr(W = w_2) = \frac{1}{2}$. Also, suppose that for each given value $W = w_i$ $(i = 1, 2)$, $X_1, X_2, \ldots$ is a sequential random sample from the distribution specified as follows:

$$\Pr(X = 0|W = w_1) = \Pr(X = 2|W = w_1) = \frac{1}{2};$$

and

$$\Pr(X = 0|W = w_2) = \frac{1}{2},$$
$$\Pr(X = 4|W = w_2) = \Pr(X = 12|W = w_2) = \frac{1}{4}.$$

Finally, suppose that the cost per observation is 1 unit. For the problem of sampling with recall, find a stopping rule that maximizes the expected gain $E[\max \{X_1, \ldots, X_N\} - N]$.

(b) Show that even when the statistician is sampling without recall, he can use the same stopping rule to maximize the expected gain $E(X_N - N)$.

19. If the sequence $Y_1, Y_2, \ldots$ is a supermartingale with respect to the sequence $X_1, X_2, \ldots$, show that the sequence $-Y_1, -Y_2, \ldots$ is a submartingale with respect to $X_1, X_2, \ldots$.

20. Let the sequence $Y_1, Y_2, \ldots$ be a martingale with respect to the sequence $X_1, X_2, \ldots$. Also, let g be a convex function such that $E[g(Y_n)]$ exists for

$n = 1, 2, \ldots$ Show that the sequence $g(Y_1), g(Y_2), \ldots$ is a submartingale with respect to $X_1, X_2, \ldots$

21. Let $Y_1, Y_2, \ldots$ be a submartingale with respect to the sequence X_1, $X_2, \ldots$ Also, let g satisfy the conditions in Exercise 20, and in addition, suppose that g is nondecreasing. Show that the sequence $g(Y_1), g(Y_2), \ldots$ is a submartingale with respect to $X_1, X_2, \ldots$

22. If the random variables $Y_1, Y_2, \ldots$ are uniformly integrable, prove that there is an upper bound $M < \infty$ such that $E(|Y_n|) < M$ for $n = 1, 2, \ldots$

23. If $Y_1, Y_2, \ldots$ is a sequence of uniformly integrable random variables, prove that Eq. (3) of Sec. 13.12 is satisfied for every stopping rule in the class Δ.

24. If $Y_1, Y_2, \ldots$ is a sequence of random variables such that $E(Y_n^2) \leq M < \infty$ for $n = 1, 2, \ldots$, show that the variables $Y_1, Y_2, \ldots$ are uniformly integrable.

25. Whenever a certain gambler makes a bet, he either wins \$1 with probability $\frac{1}{2}$ or loses \$1 with probability $\frac{1}{2}$. He makes a sequence of independent bets of this type and continues until he has either realized a gain of r dollars or suffered a loss of s dollars, where r and s are fixed positive integers. Show that the probability that he will stop betting because he has gained r dollars is $s/(r + s)$.

26. Prove Corollary 3 of Sec. 13.13. *Hint:* Use Theorem 1 of Sec. 12.15.

27. Let W be a parameter whose values are in the parameter space Ω. Suppose that for any given value $W = w$ $(w \in \Omega)$, the observations $X_1, X_2, \ldots$ form a sequential random sample from the distribution with g.p.d.f. $f(\cdot | w)$. Suppose also that the prior g.p.d.f. of W is ξ. Let w_0 be any fixed point in Ω. For $n = 1, 2, \ldots$, define Y_n to be the value of the posterior g.p.d.f. of W at the point w_0 after the values of $X_1, \ldots, X_n$ have been observed. Show that the sequence $Y_1, Y_2, \ldots$ is a martingale with respect to the sequence $X_1, X_2, \ldots$

28. Let Z and $X_1, X_2, \ldots$ be random variables with some specified joint distribution for which $E(Z)$ and $E(X_i)$ exist $(i = 1, 2, \ldots)$. Let $Y_n = E(Z|X_1, \ldots, X_n)$ for $n = 1, 2, \ldots$ Show that the sequence $Y_1, Y_2, \ldots$ is a martingale with respect to the sequence $X_1, X_2, \ldots$

29. Let $X_1, X_2, \ldots$ be a sequence of independent and identically distributed random variables for which $E(X_i) = 0$ and $\text{Var}(X_i) = \sigma^2 < \infty$ $(i = 1, 2, \ldots)$. Define Y_n as follows, for $n = 1, 2, \ldots$:

$$Y_n = \left(\sum_{i=1}^{n} X_i \right)^2 - n\sigma^2.$$

Show that the sequence $Y_1, Y_2, \ldots$ is a martingale with respect to the sequence $X_1, X_2, \ldots$

30. Suppose that the random variables $X_1, X_2, \ldots$ satisfy the conditions specified in Exercise 29. Suppose also that under a given stopping rule in the class Δ, there is a bound $M < \infty$ such that $|x_1 + \cdots + x_n| \leq M$ for all points $(x_1, \ldots, x_n) \in \{N > n\}$ and for $n = 1, 2, \ldots$ Finally, suppose that $E(N) < \infty$. Prove that

$$\text{Var}(X_1 + \cdots + X_N) = \sigma^2 E(N).$$

31. Formulate the processes generated in each of the following problems of optimal stopping as Markov processes with stationary transition probabilities. Describe the transition function f, the reward function b, the cost function c, and the sets S_0 and S_1 in each problem.

(a) The problem of sampling without recall from a known distribution, which is described in Sec. 13.5

(b) The problem of sampling with recall from a known distribution, which is also described in Sec. 13.5

(c) The problem of sampling without recall from a normal distribution with unknown mean, which is described in Sec. 13.6

(d) The problem of sampling with recall from a normal distribution with unknown mean, which is described in Sec. 13.7

(e) The problems described in Exercises 6 to 9

(f) The problems described in Exercises 15 and 16

(g) The problems described in the example at the end of Sec. 13.8 and the example in Sec. 13.14

32. Find the optimal stopping rule in the example in Sec. 13.17 by transforming the reduced version with the transition probabilities given in Table 13.2 into an entrance-fee problem.

33. A motorist is driving along a straight highway toward his destination, and he is looking for a parking place. As he drives along, he can observe only one parking place at a time, and he notes whether or not it is occupied. Assume that unoccupied places occur independently and that the probability that any given place will be unoccupied is p $(0 < p < 1)$. If a space is unoccupied, he may stop and park there; if it is occupied, he is forced to continue. If he has not parked in a space by the time he reaches his destination on the highway, then he continues driving beyond it, under the same probability conditions, until he finds an unoccupied space. His loss when he parks is proportional to the distance he must walk to his destination. Thus, if he parks in a place which is k parking places from his destination, then his loss is $\alpha|k|$, where $\alpha > 0$. Let

$$k^* = -\frac{\log 2}{\log (1 - p)}.$$

Show that he should continue driving until he is not more than k^* parking places from his destination and then should park in the first space he finds unoccupied.

34. Consider a Markov chain with five states $S = \{1, 2, 3, 4, 5\}$ and stationary transition probabilities given by the accompanying table. Suppose that $S_0 = \{3\}$ and $S_1 = \{4, 5\}$. Also, suppose that the reward function is $b(1) = b(2) = 7$ and $b(3) = 10$ and that the cost function is $c(1) = c(2) = 1$, $c(4) = -1$, and $c(5) = 2$. (a) Show that the optimal stopping rule is to terminate sampling when the process is in either state 2 or state 3. (b) Suppose that the process starts from state 1. Show that under the optimal rule, the expected total gain is $4\frac{3}{6}$.

Table for Exercise 34

	1	2	3	4	5
1	$\frac{1}{4}$	$\frac{1}{4}$	$\frac{1}{4}$	$\frac{1}{4}$	0
2	0	0	$\frac{1}{4}$	$\frac{1}{4}$	$\frac{1}{2}$
3	0	0	1	0	0
4	$\frac{1}{4}$	$\frac{1}{4}$	$\frac{1}{4}$	0	$\frac{1}{4}$
5	$\frac{1}{2}$	$\frac{1}{4}$	$\frac{1}{4}$	0	0

sequential choice
of experiments

14.1 INTRODUCTION

In this chapter we shall consider sequential decision problems in which the statistician may be required to make two choices at any stage. First, he may have to decide whether to continue experimenting or to terminate the process. Second, if he decides to continue, he may have to choose one of two or more feasible experiments that are available at that stage. In other, but equivalent, terms the statistician may have to choose at each stage one random variable from an appropriate class which he wishes to observe at that stage. Through his sequential choice of experiments, the statistician can exercise some control over the distributions of the observations generated during the process and, hence, over the distributions of his rewards and costs. His problem is again to maximize the expected value of some appropriately defined gain function.

Because of the relatively great freedom of choice available to the statistician, the problem of finding an optimal sequential decision procedure is typically difficult. Indeed, even when the question of whether to continue experimenting or to terminate the process is completely removed from the problem by the requirement that the process must be continued for exactly n stages and then terminated, significantly difficult choices still remain. Although every problem of this type has a fixed number of stages n, such a problem is still a sequential decision problem in the following respect: The statistician chooses the experiments sequentially, and his choice of one of the alternatives available at any stage can depend on all observations made in the previous stages.

There is another class of sequential decision problems in which no stopping rule is required. In any problem of this class, there are an infinite number of stages in the process; i.e., the process continues indefinitely without ever being terminated. At each stage, the statistician must choose one of the available alternatives. An important feature of any problem of this class is that the statistician's total gain or total loss is typically the sum of an infinite series of random variables. Therefore, the statistician must be certain that the specification of the problem guarantees the convergence of this series. Most of the processes which will be discussed in this chapter either will have a fixed finite number of stages or else will be continued indefinitely without being terminated.

We shall begin our discussion of these problems by giving some introductory explanations and results for a general class of processes known as Markovian decision processes. We shall then illustrate the broad scope of these processes by discussing their applications to such topics as inventory problems, search techniques, adaptive control processes, betting systems, two-armed-bandit problems, and sufficient experiments in statistical decision problems.

14.2 MARKOVIAN DECISION PROCESSES WITH A FINITE NUMBER OF STAGES

Let S denote the sample space, or state space, of a sequential process with n stages, and suppose that at each stage, the state of the process can be represented as a point $z \in S$. Suppose also that at each stage of the process, the statistician must choose one alternative, or experiment, from a given class A of alternatives. We shall now complete the description of the decision problem by specifying the reward function and transition function as follows.

Suppose that the process is in a state $z \in S$ at some stage and that the statistician chooses an alternative $a \in A$. He then receives a reward or an expected reward $r(z, a)$, and the process moves to a new state in the space S whose g.p.d.f. is given by the transition function $f(\cdot \, | z, a)$. It is assumed that both the transition function f and the reward function r depend only on the current state z of the process and on the alternative a selected by the statistician. Neither of these functions varies from stage to stage, and neither depends on the state of the process at any earlier stages or on the alternatives selected at those stages. However, as mentioned in Sec. 13.16, this is not a severe restriction for the following reason: Even when the transition and reward functions vary from stage to stage, these functions can typically be made to satisfy the above restrictions by redefining the state space S so that the stage of the process is regarded as a component of its state. Processes which meet these requirements are often called *Markovian decision processes*.

Suppose that the process is initially in a given state $z_0 \in S$, that the statistician selects an alternative $a_0 \in A$ and receives the reward $r(z_0, a_0)$, and that the process moves to a new state Z_1 in accordance with the g.p.d.f. $f(\cdot | z_0, a_0)$. The statistician then observes the new state $Z_1 = z_1$, selects an alternative $a_1 \in A$, and receives the reward $r(z_1, a_1)$; and the process moves to a new state Z_2 in accordance with the g.p.d.f. $f(\cdot | z_1, a_1)$. By continuing in this way, the statistician ultimately observes the state $Z_{n-1} = z_{n-1}$ at stage $n - 1$, selects an alternative $a_{n-1} \in A$, and receives the reward $r(z_{n-1}, a_{n-1})$; and the process finally moves to its terminal state Z_n in accordance with the g.p.d.f. $f(\cdot | z_{n-1}, a_{n-1})$. There is no additional difficulty if we assume that the statistician also receives a terminal reward $r_0(z_n)$. His problem is to select a sequence of alternatives $a_0, a_1, \ldots, a_{n-1}$ which maximizes his expected total gain. This gain can be expressed as follows:

$$r(z_0, a_0) + E[r(Z_1, a_1) + \cdots + r(Z_{n-1}, a_{n-1}) + r_0(Z_n)]. \tag{1}$$

An optimal procedure can be readily characterized by backward induction. For any state $z \in S$, let $V_0(z) = r_0(z)$ and let the functions $V_1, V_2, \ldots, V_n$ be defined recursively by the following relation:

$$V_j(z) = \sup_{a \in A} \left\{ r(z, a) + \int_S V_{j-1}(z')f(z'|z, a) \, d\mu(z') \right\}. \tag{2}$$

From this definition, it can be seen that

$$V_1(z) = \sup_{a \in A} \{ r(z, a) + E[r_0(Z_n)|Z_{n-1} = z, a_{n-1} = a] \}. \tag{3}$$

Hence, $V_1(z)$ is the maximum expected gain from the last stage of the process if $Z_{n-1} = z$. In general, for $j = 2, \ldots, n$, it can be seen from Eq. (2) that $V_j(z)$ is the maximum expected gain over the last j stages of the process if $Z_{n-j} = z$. Also, it can be seen that an optimal choice of the alternative a_{n-j} is any value $a_{n-j} = a$ which yields the supremum in Eq. (2), provided that the supremum is actually attained at some point in the space A. As usual, the statistician must work his way back through the sequence $a_n, a_{n-1}, \ldots$ in order to determine the optimal choice of the initial alternative a_0.

14.3 MARKOVIAN DECISION PROCESSES WITH AN INFINITE NUMBER OF STAGES

In a problem in which there is no natural finite terminal stage, it is often appropriate to construct a Markovian decision process with an infinite number of stages in the following manner: The state space S, the alternative space A, the reward function r, and the transition function f will be defined as before.   We shall let β be a given number $(0 < \beta < 1)$ that, as we shall see, is a factor used to discount future rewards in comparison with current rewards.   The process begins in some given state $z_0 \in S$ and is continued indefinitely without ever being terminated.   At the jth stage $(j = 0, 1, 2, \ldots)$, the statistician observes the state z_j of the process, selects an alternative $a_j \in A$, and receives the reward $r(z_j, a_j)$; and the process moves to a new state Z_{j+1} in accordance with the g.p.d.f. $f(\cdot \mid z_j, a_j)$. His problem is to select an infinite sequence of alternatives $a_0, a_1, a_2, \ldots$ which will maximize the expected value of his discounted total gain. This gain can be expressed as follows:

$$r(z_0, a_0) + E\left[\sum_{j=1}^{\infty} \beta^j r(Z_j, a_j)\right]. \tag{1}$$

The expected total gain (1) reflects the role of the discount factor β. At any stage of the process, the statistician regards the promise of receiving a reward of one unit after n more stages have passed as being equivalent to receiving a reward of β^n units at the current stage.   Discount factors are widely used in problems pertaining to economic planning and investment.   A process involving an infinite number of stages and discounted rewards has the following two important features.

The first feature is that in many problems in which there are an infinite number of stages, the statistician's expected total gain from an optimal procedure will nevertheless be finite.   In the simplest case, there will be a bound $M < \infty$ such that $|r(z, a)| \leq M$ for all states $z \in S$ and all alternatives $a \in A$.   In such a case, it can be seen from the value (1) that the total gain from any sequential procedure cannot be greater than $M/(1 - \beta)$.

The second important feature is that the statistician's decision problem is stationary in time in the following sense.   Suppose that at the nth stage, the process is in the state $Z_n = z_n$.   Since the rewards he has received from his choices at stages 0 through $n - 1$ will not be affected by future choices, the statistician must now maximize his expected total gain over the remainder of the process.   This future gain can be expressed as follows:

$$\beta^n \left\{r(z_n, a_n) + E\left[\sum_{j=1}^{\infty} \beta^j r(Z_{n+j}, a_{n+j}) \mid Z_n = z_n\right]\right\}. \tag{2}$$

In this process, the transition function remains the same from stage to stage. Hence, except for the factor β^n, the expected total gain (2) from the nth stage through the remainder of the process is the same as the expected total gain (1) over the entire process when z_n is regarded as the initial state. Therefore, at any stage the decision problem faced by the statistician is the same as the one he faced initially, but there is a different initial state.

We shall assume, for any initial state z ($z \in S$), that there exists an optimal procedure and that the expected total gain from the procedure is finite. It can then be shown that there must be a *stationary* procedure that is optimal for every initial state z. In this context, a stationary procedure is one in which the alternative selected at any stage of the process depends only on the state of the process at that stage and is not otherwise dependent on the stage. In other words, a stationary procedure is characterized by a function δ which assigns, to each state $z \in S$, an alternative $\delta(z) \in A$. When the statistician uses the stationary procedure δ, he chooses the alternative $\delta(z) \in A$ whenever the process is in the state $z \in S$.

Let $V(z)$ denote the expected total gain from the optimal procedure when the initial state of the process is z ($z \in S$). Then, in the usual way, the values $V(z)$ must satisfy the following equation for all points $z \in S$:

$$V(z) = \sup_{a \in A} \left\{ r(z, a) + \beta \int_S V(z')f(z'|z, a) \, d\mu(z') \right\}. \tag{3}$$

Suppose that the function V is a bounded function on the set S. The next two theorems, which were developed by Strauch (1966) and Prescott (1967), show that the function V will be the only bounded solution of Eq. (3) and that the method of successive approximations can be used to obtain a sequence of functions which will converge uniformly to V.

Theorem 1 *For any given value of β $(0 < \beta < 1)$, there is at most one bounded function V which satisfies Eq. (3).*

Proof Suppose that V_1 and V_2 are two bounded functions which satisfy Eq. (3), and let $m = \sup_{z \in S} |V_1(z) - V_2(z)|$. Then $V_1(z) \leq V_2(z) + m$ for $z \in S$. Therefore, for all states $z \in S$,

$$V_1(z) = \sup_{a \in A} \left\{ r(z, a) + \beta \int_S V_1(z')f(z'|z, a) \, d\mu(z') \right\}$$

$$\leq \sup_{a \in A} \left\{ r(z, a) + \beta \int_S [V_2(z') + m]f(z'|z, a) \, d\mu(z') \right\}$$

$$= V_2(z) + \beta m. \tag{4}$$

Similarly, if V_1 and V_2 are interchanged in Eq. (4), it follows that $V_2(z) \leq V_1(z) + \beta m$ for all states $z \epsilon S$. Thus, $|V_1(z) - V_2(z)| \leq \beta m$ for all states $z \epsilon S$. By the definition of m, it must be true that $m = 0$. In other words, $V_1(z) = V_2(z)$ for all states $z \epsilon S$.∎

Theorem 2 *Let g_0 be an arbitrary bounded function on the set S. Also, for $n = 1, 2, \ldots$, let the function g_n be defined recursively by the following relation:*

$$g_n(z) = \sup_{a \epsilon A} \left\{ r(z, a) + \beta \int_S g_{n-1}(z')f(z'|z, a) \, d\mu(z') \right\} \qquad z \epsilon S. \qquad (5)$$

If V is a bounded function which satisfies Eq. (3), then the following relation must be satisfied uniformly for all states $z \epsilon S$:

$$\lim_{n \to \infty} g_n(z) = V(z). \qquad (6)$$

Proof Since both g_0 and V are bounded functions, there exists a number $m < \infty$ such that $|g_0(z) - V(z)| \leq m$ for all states $z \epsilon S$. Hence, for all states $z \epsilon S$,

$$g_1(z) = \sup_{a \epsilon A} \left\{ r(z, a) + \beta \int_S g_0(z')f(z'|z, a) \, d\mu(z') \right\}$$

$$\leq \sup_{a \epsilon A} \left\{ r(z, a) + \beta \int_S [V(z') + m]f(z'|z, a) \, d\mu(z') \right\}$$

$$= V(z) + \beta m. \qquad (7)$$

Similarly, $g_1(z) \geq V(z) - \beta m$ for all states $z \epsilon S$. Hence, $|g_1(z) - V(z)| \leq \beta m$ for all states $z \epsilon S$.

It follows from the iteration of this argument that $|g_n(z) - V(z)| \leq \beta^n m$ for $n = 1, 2, \ldots$. Thus the uniform convergence indicated by Eq. (6) is implied.∎

Further Remarks and References

A wide selection of problems involving Markovian decision processes, at varying levels of generality, are described in the books by Bellman (1957a, 1961), Howard (1960), and Martin (1967). Other references are the papers by Karlin (1955a), Blackwell (1961, 1962, 1964, 1965, 1967), Maitra (1965, 1966), Strauch (1966), Derman (1962, 1963, 1964, 1966), Derman and Veinott (1967), Fisher (1968), Ross (1968), Fisher and Ross (1968), and MacQueen (1966). Some papers which apply primarily to economics and management but are of general interest are Arrow (1962, 1964), Marschak (1963a), and MacQueen (1964a).

Another class of sequential decision problems includes those known as compound decision problems and empirical Bayes procedures. Such procedures were first studied systematically by Robbins. Some references in these areas are Robbins (1956b, 1964), Hannan (1957), Hannan and Robbins (1955), Hannan and Van Ryzin (1965), Johns (1957, 1961), Samuel (1963a,b; 1964; 1965a,b), and Van Ryzin (1966a,b).

14.4 SOME BETTING PROBLEMS

Suppose that on each play of a certain game, a gambling statistician has a fixed probability p of winning $(0 < p < 1)$. We shall assume that he begins with a given fortune $Y_0 > 0$ and that on the first play of the game, he may bet any amount x_1 such that $0 \leq x_1 \leq Y_0$. If he wins on the first play, then he gains the amount x_1 which he has bet. If he loses on the first play, then he loses the amount x_1. For $j = 1, 2, \ldots$, we shall let Y_j be the statistician's fortune after the jth play. On each play of the game, the statistician may bet any amount which does not exceed his current fortune. On the $(j + 1)$st play, for instance, the statistician may bet any amount x_{j+1} such that $0 \leq x_{j+1} \leq Y_j$. On this play, the probability that he will win the amount x_{j+1} will be p and the probability that he will lose the amount x_{j+1} will be $q = 1 - p$. It will be assumed that the outcomes of all plays of the game are independent and that the game will be played n times, where n is a fixed number. We shall also suppose that there is a given utility function U. The statistician's problem is to construct a sequential betting procedure that maximizes the expected utility of his terminal fortune $E[U(Y_n)]$.

For $j = 0, 1, 2, \ldots$ and $y \geq 0$, we shall let $V_j(y)$ denote the maximum value of $E[U(Y_n)]$ when the number of plays remaining is j and the statistician's fortune is y, that is, $Y_{n-j} = y$. Then, for any fortune $y \geq 0$, it follows that $V_0(y) = U(y)$. Furthermore, for $j = 0, 1, 2, \ldots$, the values $V_j(y)$ satisfy the following relation:

$$V_{j+1}(y) = \sup_{0 \leq x \leq y} [pV_j(y + x) + qV_j(y - x)]. \tag{1}$$

By successively applying Eq. (1) for $j = 0, 1, \ldots, n - 1$, the statistician can find $V_n(Y_0)$, which is the value of $E[U(Y_n)]$ under the optimal procedure. The optimal procedure itself is also described by Eq. (1) for the following reason: When the number of plays remaining is $j + 1$ $(j = 0, 1, \ldots, n - 1)$ and the statistician's fortune is y, his optimal bet is an amount x that yields the supremum on the right side of Eq. (1). We shall now consider some specific examples.

EXAMPLE 1 First we shall determine an optimal procedure when the utility function U is *convex*. In other words, we shall suppose that the statistician is a risk taker, as discussed in Sec. 7.6.

For any fixed value y $(y > 0)$, both the function $U(y + x)$ and the function $U(y - x)$ must be convex functions of x on the interval $0 \leq x \leq y$. This condition implies, in turn, that the linear combination $pU(y + x) + qU(y - x)$ is also convex on the interval $0 \leq x \leq y$. Since the maximum value of any convex function on a closed, bounded interval is attained at one of the end points of the interval, the application of Eq. (1) when $j = 0$ yields the following result:

$$V_1(y) = \max \{U(y), pU(2y) + qU(0)\}. \tag{2}$$

In other words, on the final play of the game, the optimal procedure is as follows: If $U(y) \geq pU(2y) + qU(0)$, the statistician should bet nothing. Otherwise, he should bet his entire fortune y.

Moreover, both $U(y)$ and $pU(2y) + qU(0)$ are convex functions of y for $y > 0$. Since the maximum of two convex functions is itself convex, it follows from Eq. (2) that $V_1(y)$ is convex. By repeating the above argument, it can be seen from Eq. (1) when $j = 1$ that on the penultimate play of the game, the optimal procedure again requires that the statistician should bet either nothing or his entire fortune. Furthermore, $V_2(y)$ will be a convex function. By continuing in this way, we learn that on every play of the game, the statistician should bet either nothing or the entire fortune which he has accumulated at that stage. In other words, for $j = 0, 1, 2, \ldots$, the function $V_j(y)$ is convex and satisfies the following equation:

$$V_{j+1}(y) = \max \{V_j(y), pV_j(2y) + qV_j(0)\}. \tag{3}$$

Now we shall suppose that $p \geq \frac{1}{2}$, that is, that each play is favorable, or at least fair, to the statistician. Since $U(y)$ is the utility to the statistician of having a terminal fortune y, we can assume that U is an increasing function. By induction, it follows from Eq. (3) that V_j is an increasing function for each value of j ($j = 0, 1, \ldots$). Since V_j is also a convex function, the following relation must be satisfied for any value of y ($y > 0$):

$$V_j(y) \leq \tfrac{1}{2}V_j(2y) + \tfrac{1}{2}V_j(0) \leq pV_j(2y) + qV_j(0). \tag{4}$$

According to Eq. (3), the optimal procedure therefore requires that the statistician should bet his entire fortune on every play of the game. Although this procedure is optimal, it has the *unfortunate* feature that there is a high probability that the statistician will lose his entire fortune. In other words, at the end of n plays, either his fortune will be $2^n Y_0$ or else he will have nothing. The probability that he will have nothing is $1 - p^n$.

As a special limiting case, suppose that U is a linear function. In this case, we can assume that $U(y) = y$ ($y \geq 0$) and that the statistician must maximize his expected terminal fortune. If $p > \frac{1}{2}$, he should bet his entire fortune on each play. If $p = \frac{1}{2}$, all betting procedures will yield the same expected terminal fortune. Under such conditions, the statistician will be indifferent about how much he bets on each play and about whether he even bets at all.

EXAMPLE 2 Now we shall determine an optimal procedure when the convex utility function U is defined, for $y \geq 0$ and some fixed number α ($\alpha > 1$), as follows:

$$U(y) = y^{\alpha}. \tag{5}$$

Also, we shall suppose that $0 < p < \frac{1}{2}$.
By Eq. (2),

$$V_1(y) = \max \{y^{\alpha}, 2^{\alpha}py^{\alpha}\} = y^{\alpha} \max \{1, 2^{\alpha}p\}. \tag{6}$$

Hence, regardless of the value of the statistician's fortune y at that stage, the optimal procedure on the final play of the game is as follows: If $p > 2^{-\alpha}$, the statistician should bet his entire fortune. On the other hand, if $p \leq 2^{-\alpha}$, he should bet nothing. In either case, V_1 is simply a multiple of U. Therefore, the same analysis can be carried out on the next to the last play and, by induction, on every play. In summary, the optimal procedure can be described as follows: If $p > 2^{-\alpha}$, the statistician should bet his entire fortune on every play. If $p \leq 2^{-\alpha}$, he should not bet at all.

EXAMPLE 3 Finally, we shall determine an optimal procedure when the utility function U is concave. In other words, we shall suppose that the statistician is a risk averter.

As usual, it is assumed that U is an increasing function. Therefore, if $p \leq \frac{1}{2}$, it follows that for any values x and y such that $0 \leq x \leq y$, the following relation must be satisfied:

$$U(y) \geq \tfrac{1}{2}U(y + x) + \tfrac{1}{2}U(y - x) \geq pU(y + x) + qU(y - x). \tag{7}$$

From Eq. (1), it can be seen that $V_1(y) = U(y)$ for $y \geq 0$. Also, it follows, by induction, that $V_j(y) = U(y)$ for all values $y \geq 0$ and for $j = 1, 2, \ldots$. Under these conditions, therefore, the statistician should not bet at all.

On the other hand, suppose that $p > \frac{1}{2}$. Since each play is now favorable to him, the statistician may wish to bet a positive amount on a given play even though he is a risk averter. As a specific example, suppose that $\frac{1}{2} < p < 1$ and that the utility function is defined as follows for $y > 0$:

$$U(y) = \log y. \tag{8}$$

Then the value of $V_1(y)$ can be computed from the following equation:

$$V_1(y) = \sup_{0 \le x \le y} [p \log (y + x) + q \log (y - x)]. \tag{9}$$

By differentiating the right side of Eq. (9), we find that the supremum occurs at the value $x = (p - q)y$. Hence,

$$
\begin{aligned}
V_1(y) &= \log y + p \log p + q \log q + \log 2 \\
&= \log y + \alpha, \tag{10}
\end{aligned}
$$

where $\alpha = p \log p + q \log q + \log 2$. Thus, $V_1(y) = U(y) + \alpha$, and from Eq. (1), it follows that $V_2(y) = V_1(y) + \alpha$. In general, for $y > 0$ and for $j = 1, 2, \ldots$,

$$V_{j+1}(y) = V_j(y) + \alpha. \tag{11}$$

Furthermore, for each value of j and each value of y, the supremum on the right side of Eq. (1) is attained at the value $x = (p - q)y$.

Hence, the optimal procedure can be summarized as follows: On each play of the game, the statistician should bet the fixed proportion $p - q$ of his current fortune. Under this procedure, the expected utility of his terminal fortune will be

$$E(\log Y_n) = \log Y_0 + n(p \log p + q \log q + \log 2). \tag{12}$$

Further Remarks and References

A variety of interesting gambling problems have been studied in a highly abstract way by Dubins and Savage (1965). Some papers dealing with other gambling problems are those by Kelly (1956), Breiman (1961), MacQueen (1961, 1964b), and Ferguson (1965). Other short notes are by Molenaar and van der Velde (1967), Freedman (1967), and Freedman and Purves (1967).

Thorp (1961, 1962) has discovered and developed a favorable betting procedure for the game of blackjack, or twenty-one.

14.5 TWO-ARMED-BANDIT PROBLEMS

Consider two random variables X and Y. Suppose that the distribution of X depends on the value of a parameter W_1 and that the distribution of Y depends on the value of another parameter W_2. We shall let $f_X(\cdot \ | w_1)$ denote the g.p.d.f. of X when $W_1 = w_1$ $(w_1 \, \epsilon \, \Omega_1)$, and we shall let $f_Y(\cdot \ | w_2)$

denote the g.p.d.f. of Y when $W_2 = w_2$ ($w_2 \in \Omega_2$). Also, we shall suppose that the statistician must take a fixed total of n observations. At each of the n stages, he may choose to take either an observation on X or one on Y. His choices can be made sequentially, in the sense that the choice of the random variable at any stage can be based on the actual observed values of the random variables chosen at the earlier stages. It is assumed that the outcomes of the n observations are independent in the following sense: Suppose that the random variable X is chosen for observation at some given stage. Then, regardless of the choices and outcomes of the observations at all preceding stages, the conditional g.p.d.f. of the observation at the given stage, when $W_1 = w_1$, is $f_X(\cdot \,|w_1)$. A similar property is satisfied if the random variable Y is chosen for observation at any stage.

The statistician's problem is to find a sequential procedure that maximizes the expected value of the sum of the n observations. A problem of this type is called a *two-armed-bandit problem* since the statistician's choice is similar to that of a gambler who puts money in a slot machine which is known as a two-armed bandit. On each play, the person must decide whether to put his money in one arm or the other. More serious applications relate to the choice between two different medical treatments, either of which may be prescribed for any given patient.

Let ξ denote the prior joint distribution of the parameters W_1 and W_2. If W_1 and W_2 are independent under this prior distribution, then they will remain independent throughout the sampling process. Each observation on X will lead to a new posterior distribution of W_1, but the observation will not give the statistician any further information about the value of W_2. On the other hand, if W_1 and W_2 are dependent under the prior distribution ξ, then each observation made on either one of the variables will lead, in general, to new posterior distributions of both W_1 and W_2.

For each prior distribution ξ and each positive integer n, we shall let $V_n(\xi)$ denote the maximum expected sum of n observations that can be attained from any possible sequential procedure. It is assumed that this value is finite.

Consider the procedure under which the first observation is made on X and an optimal procedure is then followed over the remaining $n - 1$ observations. Suppose that after the first observation X has been taken, the posterior joint distribution of W_1 and W_2 is $\xi(X)$. Then the expected sum of the remaining $n - 1$ observations is $V_{n-1}[\xi(X)]$. Hence, under this procedure, the expected sum of all n observations is $E\{X + V_{n-1}[\xi(X)]\}$.

Similarly, if the first observation is made on Y and an optimal procedure is adopted thereafter, the expected sum of all n observations is $E\{Y + V_{n-1}[\xi(Y)]\}$. Since the first observation must be made either on

X or on Y, the value of $V_n(\xi)$ must satisfy the following equation:

$$V_n(\xi) = \max \{E[X + V_{n-1}(\xi(X))], E[Y + V_{n-1}(\xi(Y))]\}. \tag{1}$$

The expectations in Eq. (1) are computed with respect to the prior distribution ξ. When we add the initial condition that $V_0(\xi) = 0$ for all joint distributions ξ of W_1 and W_2, Eq. (1) can be solved successively for the functions $V_1, V_2, \ldots, V_n$. We shall now consider some special cases.

14.6 TWO-ARMED-BANDIT PROBLEMS WHEN THE VALUE OF ONE PARAMETER IS KNOWN

In this section we shall suppose that the statistician knows the value of the parameter W_2. Specifically, we shall suppose that $W_2 = w_2$ and that the prior g.p.d.f. of W_1 is ξ. In this case, the statistician gains no information about the unknown value of W_1 by taking observations on Y, and the posterior distribution of W_1 after an observation on Y will be the same as its distribution before the observation was taken. Suppose, therefore, that at some stage of the process, it is optimal for the statistician to take an observation on Y rather than one on X. Then, regardless of the value of that observation, he cannot receive any information from the observation which would cause him to switch to X for the next observation. It follows that the optimal procedure is either to take all observations on Y or else to begin with observations on X and continue with observations on X as long as the distribution of W_1 is favorable or at least seems promising. If the statistician takes the first observation on X, either the observed values will lead him to continue taking observations on X throughout the entire process or else at some stage it will become optimal for him to switch to observations on Y. Once he switches, all further observations should be made on Y.

The optimal choice at each stage of the process will depend only on the current distribution of W_1 and on the number of observations which remain to be taken. When many observations remain to be taken and W_1 has a given g.p.d.f. ξ, the statistician may be willing to take the first few observations on X if there is at least a small chance that the value of W_1 will be favorable. On the other hand, if only a few observations remain to be taken, the statistician may not be willing to experiment with X.

Since the statistician's choice at the first stage has been reduced to either taking an observation on X and then proceeding optimally or taking all n observations on Y, Eq. (1) of Sec. 14.5 can now be replaced by the following equation:

$$V_n(\xi) = \max \{E[X + V_{n-1}(\xi(X))], nE(Y|W = w_2)\}. \tag{1}$$

This problem can be regarded as a problem of optimal stopping because the statistician must only decide when to stop taking observations on X and to switch to Y.

We shall now consider an example. Suppose that both X and Y can have only the values 0 and 1 and that for any given values $W_1 = w_1$ and $W_2 = w_2$ such that $0 \le w_1 \le 1$ and $0 \le w_2 \le 1$, the p.f.'s of X and Y are

$$
\begin{aligned}
f_X(1|w_1) &= w_1, & f_X(0|w_1) &= 1 - w_1; \\
f_Y(1|w_2) &= w_2, & f_Y(0|w_2) &= 1 - w_2.
\end{aligned}
\tag{2}
$$

Suppose also that W_1 can have only the values $\frac{1}{2}$ and 0 and that it is known that $W_2 = \frac{1}{4}$. Let the prior distribution of W_1 be specified as follows:

$$
\xi = \Pr(W_1 = \tfrac{1}{2}) = 1 - \Pr(W_1 = 0).
\tag{3}
$$

Finally, suppose that the total number of observations n that will be taken is fixed.

If, in this special problem, the value of any observation on X is 1, then the statistician knows that $W_1 = \frac{1}{2}$. Since $W_2 = \frac{1}{4}$, the optimal continuation must then be to take all remaining observations on X. For this reason, the optimal procedure will have the following simple description: The statistician takes the first r observations on X. If the value is found to be 1 for at least one of these observations, then all the remaining $n - r$ observations should also be taken on X. On the other hand, if the value is 0 for each of the first r observations on X, then the statistician should switch and take the remaining $n - r$ observations on Y. If we include the possibility that the statistician simply takes all n observations on Y, in which case $r = 0$, then there must be a value of r $(r = 0, 1, \ldots, n)$ for which this procedure is optimal.

Although it would be possible to develop the optimal procedure from Eq. (1), we shall take advantage of the special form of the procedure and adopt the following approach. Let Z denote the sum of the n observations taken. Also, for each value of r $(r = 0, 1, \ldots, n)$, let $T(n, r)$ denote the value of the expectation $E(Z)$ when the procedure described above is followed. The problem then is to find the value of r that maximizes $T(n, r)$.

When $r = 0$, all n observations are made on Y. Since the expected value of each observation is $\frac{1}{4}$, it follows that

$$
T(n, 0) = \frac{n}{4}.
\tag{4}
$$

Now consider any value of r such that $1 \le r \le n$. For any number j $(j = 1, \ldots, r)$, let A_j denote the event that the value is 0 for each of the first $j - 1$ observations on X but is 1 for the jth observation. Also, let B_r denote the event that the value is 0 for each of the r observations on X. It then follows that

$$T(n, r) = \sum_{j=1}^{r} E(Z|A_j) \Pr(A_j) + E(Z|B_r) \Pr(B_r). \tag{5}$$

For $j = 1, \ldots, r$, it can be seen that $\Pr(A_j|W_1 = \frac{1}{2}) = (\frac{1}{2})^j$ and $\Pr(A_j|W_1 = 0) = 0$. Hence, $\Pr(A_j) = \xi/2^j$. Furthermore, after the event A_j has occurred, the statistician knows that exactly one of the first j observations has the value 1 and that $W_1 = \frac{1}{2}$. Since each of the remaining $n - j$ observations is then to be made on X, the expected value of the sum of those observations will be $(n - j)/2$. Hence, the following relation can be obtained:

$$E(Z|A_j) = 1 + \frac{n - j}{2} = \frac{n - j + 2}{2}. \tag{6}$$

Similarly, $\Pr(B_r|W_1 = \frac{1}{2}) = (\frac{1}{2})^r$ and $\Pr(B_r|W_1 = 0) = 1$. Hence, $\Pr(B_r) = (\xi/2^r) + 1 - \xi$. If the event B_r occurs, then each of the first r observations has the value 0 and the remaining $n - r$ observations will be made on Y. The expected value of the sum of those observations will be $(n - r)/4$. Hence, the following relation can be obtained:

$$E(Z|B_r) = \frac{n - r}{4}. \tag{7}$$

From Eq. (5), it can be seen that for $r = 1, \ldots, n$, the value of $T(n, r)$ is

$$T(n, r) = \frac{\xi}{2} \sum_{j=1}^{r} \frac{n - j + 2}{2^j} + \frac{n - r}{4} \left(\frac{\xi}{2^r} + 1 - \xi \right). \tag{8}$$

Next, by applying Eqs. (4) and (8), we can obtain the following relation for $r = 0, 1, \ldots, n - 1$:

$$T(n, r + 1) - T(n, r) = \frac{\xi(n - r + 1)}{2^{r+3}} - \frac{1 - \xi}{4}. \tag{9}$$

The right side of Eq. (9) is a decreasing function of r. Therefore, $T(n, r)$ will be maximized when r is the smallest integer for which the value in Eq. (9) is negative. If the value in Eq. (9) is nonnegative for every value of r ($r = 0, 1, \ldots, n - 1$), then $T(n, r)$ will be maximized when $r = n$.

In particular, it follows that $T(n, r)$ will be maximized at $r = 0$ if, and only if, $T(n, 1) - T(n, 0) \leq 0$. Thus, an optimal procedure specifies that all n observations should be taken on Y if the following relation is satisfied:

$$\xi \leq \frac{2}{n + 3}. \tag{10}$$

In this example, the number ξ can be regarded as the probability that X is the more favorable variable. When $n = 3$, the statistician will take at least one observation on X only if $\xi > \frac{1}{3}$. When $n = 100$, he will take at least one observation on X whenever $\xi > \frac{2}{103}$.

14.7 TWO-ARMED-BANDIT PROBLEMS WHEN THE PARAMETERS ARE DEPENDENT

The development in this section is based on the work of Feldman (1962). We shall suppose again that both X and Y have Bernoulli distributions and the p.f.'s for any given values $W_1 = w_1$ and $W_2 = w_2$ ($0 \leq w_1 \leq 1$ and $0 \leq w_2 \leq 1$) are as specified by Eqs. (2) of Sec. 14.6. Also, we shall let a and b be given numbers such that $0 \leq a < b \leq 1$, and we shall suppose that either $W_1 = a$ and $W_2 = b$ or else $W_1 = b$ and $W_2 = a$. Thus, in this problem, the parameters W_1 and W_2 are highly dependent. The parameter vector (W_1, W_2) can have only the two values (a, b) and (b, a). Therefore, the distribution of (W_1, W_2) can be specified by the single number $\phi = \Pr(W_1 = a) = \Pr(W_2 = b)$.

Suppose that the total number of observations n is fixed, and let Z be the sum of the n observations. Since the value of each observation must be 0 or 1, the sum Z is simply the number of 1s among the observed values. It will be shown that the sequential procedure which maximizes $E(Z)$ is a myopic procedure. Hence, this procedure will have a very simple form.

For any sequential procedure, we shall let M be the total number of observations taken on the random variable for which the probability is a that the observed value will be 1. The statistician does not know with certainty whether this variable is X or Y.

The number M may be considered the number of mistakes made by the statistician during the procedure, because the ideal procedure would be for him to take every observation on the random variable for which there is the higher probability b that the value will be 1.

Lemma 1 *A sequential procedure maximizes $E(Z)$ if, and only if, it minimizes $E(M)$.*

Proof For any given procedure, let N_X and N_Y ($N_X + N_Y = n$) denote the number of observations taken on X and Y, respectively. Furthermore, for $j = 1, \ldots, n$ we shall define random variables A_j, B_j, and C_j as follows: $A_j = 1$ if an observation is taken on X at the jth stage, and $A_j = 0$ otherwise; $B_j = 1$ if an observation is taken on Y at the jth stage, and $B_j = 0$ otherwise; C_j is the value of the observation at the jth stage. Then, for any given values $W_1 = w_1$ and $W_2 = w_2$, the following relation can be obtained:

$$
\begin{aligned}
E(Z) &= E\left(\sum_{j=1}^{n} C_j\right) = \sum_{j=1}^{n} \Pr(C_j = 1) \\
&= \sum_{j=1}^{n} \Pr(C_j = 1 | A_j = 1)\,\Pr(A_j = 1) \\
&\quad + \sum_{j=1}^{n} \Pr(C_j = 1 | B_j = 1)\,\Pr(B_j = 1) \\
&= w_1 \sum_{j=1}^{n} \Pr(A_j - 1) + w_2 \sum_{j=1}^{n} \Pr(B_j = 1) \\
&= w_1 E\left(\sum_{j=1}^{n} A_j\right) + w_2 E\left(\sum_{j=1}^{n} B_j\right) \\
&= w_1 E(N_X) + w_2 E(N_Y).
\end{aligned}
\tag{1}
$$

Since all expectations in Eq. (1) are computed with respect to the given values $W_1 = w_1$ and $W_2 = w_2$, the final result in this equation is more properly written as

$$
\begin{aligned}
E(Z | W_1 = w_1, W_2 = w_2) &= w_1 E(N_X | W_1 = w_1, W_2 = w_2) \\
&\quad + w_2 E(N_Y | W_1 = w_1, W_2 = w_2).
\end{aligned}
\tag{2}
$$

In the particular problem being considered, for any prior probability $\xi = \Pr(W_1 = a)$, the following result will be obtained:

$$
\begin{aligned}
E(Z) &= \xi[aE(N_X | W_1 = a) + bE(N_Y | W_1 = a)] \\
&\quad + (1 - \xi)[bE(N_X | W_2 = a) + aE(N_Y | W_2 = a)] \\
&= \xi\{aE(N_X | W_1 = a) + b[n - E(N_X | W_1 = a)]\} \\
&\quad + (1 - \xi)\{b[n - E(N_Y | W_2 = a)] + aE(N_Y | W_2 = a)\} \\
&= bn - (b - a)[\xi E(N_X | W_1 = a) + (1 - \xi)E(N_Y | W_2 = a)] \\
&= bn - (b - a)E(M).
\end{aligned}
\tag{3}
$$

Since $b - a > 0$, it follows from the relation (3) that a procedure will maximize $E(Z)$ if, and only if, it minimizes $E(M)$.∎

For $j = 1, 2, \ldots, n$ and $0 \leq \phi \leq 1$, we shall consider the situation in which the total number of observations remaining to be taken is j and the distribution of W_1 and W_2 is specified by the probability $\phi = \Pr(W_1 = a)$. In this situation, let $\delta_j{}^X$ denote the procedure which specifies that the first observation should be taken on X and then an optimal procedure should be adopted over the remaining $j - 1$ observations, and let $m_j{}^X(\phi)$ denote the expected number of mistakes during the j observations for which the procedure $\delta_j{}^X$ is used. Similarly, let $\delta_j{}^Y$

denote the procedure which specifies that the first observation should be taken on Y and then an optimal procedure should be adopted over the remaining $j - 1$ observations, and let $m_j{}^Y(\phi)$ denote the expected number of mistakes when the procedure $\delta_j{}^Y$ is used.

Furthermore, let $\delta_j{}^{XY}$ be the procedure which specifies that the first observation should be taken on X, the second observation should be taken on Y, and then an optimal procedure should be adopted over the remaining $j - 2$ observations.   Similarly, let $\delta_j{}^{YX}$ be the procedure which specifies that the first observation should be taken on Y, the second observation should be taken on X, and then an optimal procedure should be adopted over the remaining $j - 2$ observations.   Also, let $m_j{}^{XY}(\phi)$ and $m_j{}^{YX}(\phi)$ be the expected numbers of mistakes for these procedures.

As usual, for any prior probability ϕ, we shall let $\phi(X)$, $\phi(Y)$, $\phi(X, Y)$, or $\phi(Y, X)$ denote the posterior probability when either one observation is taken on the indicated random variable or two observations are taken in the indicated order.   Of course, for any specified values of X and Y, the posterior probabilities $\phi(X, Y)$ and $\phi(Y, X)$ will be equal.

Lemma 2 *For $j = 2, 3, \ldots$ and for $0 \leq \phi \leq 1$, $m_j{}^{XY}(\phi) = m_j{}^{YX}(\phi)$.*

Proof For either the procedure $\delta_j{}^{XY}$ or the procedure $\delta_j{}^{YX}$, the statistician will make exactly one mistake in the first two observations.   Furthermore, the posterior probabilities $\phi(X, Y)$ and $\phi(Y, X)$, after two observations have been taken, will have the same value regardless of the order of the first two observations.   Since the two procedures agree for the remaining $j - 2$ observations, the expected number of mistakes during these remaining observations will be the same for both procedures.∎

Lemma 3 *For each fixed value of t in the interval $0 \leq t \leq 1$, the probabilities $\Pr[\phi(X) \geq t]$ and $\Pr[\phi(Y) \geq t]$ are nondecreasing functions of ϕ $(0 \leq \phi \leq 1)$.*

Proof Let $\alpha = (1 - b)/(1 - a)$, and let $\beta = b/a$.   Then the following relations can be obtained:

$$\phi(X) = \begin{cases} \left(1 + \beta\dfrac{1 - \phi}{\phi}\right)^{-1} & \text{if } X = 1, \\[2ex] \left(1 + \alpha\dfrac{1 - \phi}{\phi}\right)^{-1} & \text{if } X = 0. \end{cases} \tag{4}$$

Furthermore,

$$\begin{aligned} \Pr(X = 1) &= a\phi + b(1 - \phi), \\ \Pr(X = 0) &= (1 - a)\phi + (1 - b)(1 - \phi). \end{aligned} \tag{5}$$

Thus, for any value of t $(0 \leq t \leq 1)$,

$$
\Pr[\phi(X) \geq t] = \begin{cases} 0 & \text{for } \phi < \dfrac{\alpha t}{1 - t + \alpha t}, \\[2ex] (1 - b) + (b - a)\phi & \\[1ex] & \text{for } \dfrac{\alpha t}{1 - t + \alpha t} \leq \phi < \dfrac{\beta t}{1 - t + \beta t}, \\[2ex] 1 & \text{for } \phi \geq \dfrac{\beta t}{1 - t + \beta t}. \end{cases}
$$

$$(6)$$

The right side of Eq. (6) is a nondecreasing function of ϕ.

A similar argument can be used to show that $\Pr[\phi(Y) \geq t]$ is also a nondecreasing function of ϕ. ∎

We are now ready to prove that at every stage the statistician should take an observation on the random variable X or Y for which there is the greater probability that the observed value will be 1. In other words, the optimal procedure is the myopic procedure under which the statistician makes a choice at each stage as if it were the final stage.

Theorem 1 *Let δ^* be a procedure which specifies that an observation should be taken on X at any stage for which $\phi = \Pr(W_1 = a) < \frac{1}{2}$ and that an observation should be taken on Y at any stage for which $\phi > \frac{1}{2}$. Then δ^* is an optimal sequential procedure.*

Proof When $\phi = \frac{1}{2}$, the procedure δ^* is unspecified, and it may be defined arbitrarily. For $j = 1, 2, \ldots, n$ and for $0 \leq \phi \leq 1$, let $m_j(\phi)$ denote the expected number of mistakes when j observations are taken under an optimal procedure. Then, since the first observation under an optimal procedure must be made either on X or on Y, we can write the following relation:

$$
m_j(\phi) = \min \{m_j^X(\phi), m_j^Y(\phi)\}. \tag{7}
$$

Let the difference $\gamma_j(\phi)$ be defined as follows:

$$
\gamma_j(\phi) = m_j^X(\phi) - m_j^Y(\phi). \tag{8}
$$

To prove that δ^* is an optimal procedure, we need only show that $\gamma_j(\phi) \leq 0$ for $\phi < \frac{1}{2}$ and that $\gamma_j(\phi) \geq 0$ for $\phi > \frac{1}{2}$. From the symmetry of the problem, it is clear that $\gamma_j(\frac{1}{2}) = 0$. Hence, the theorem will be proved if we show that $\gamma_j(\phi)$ is a nondecreasing function of ϕ for any fixed value of j $(j = 1, 2, \ldots)$. We shall demonstrate by induction that this condition is satisfied.

Since $m_1^X(\phi) = \phi$ and $m_1^Y(\phi) = 1 - \phi$, it follows that

$$\gamma_1(\phi) = 2\phi - 1,$$

which is a nondecreasing function. Now we shall suppose that $\gamma_j(\phi)$ is nondecreasing for $j = 1, \ldots, k$, and we shall prove that $\gamma_{k+1}(\phi)$ is also nondecreasing. Consider the procedures δ_{k+1}^{XY} and δ_{k+1}^{YX}, which were defined earlier. By the induction hypothesis, δ^* is optimal when the number of observations remaining is k or fewer. Hence, after the first two observations have been taken, the procedure δ^* must be the optimal continuation over the remaining $k - 1$ observations. We shall next define the following differences:

$$\begin{aligned}
\gamma_{k+1}^X(\phi) &= m_{k+1}^{XY}(\phi) - m_{k+1}^X(\phi), \\
\gamma_{k+1}^Y(\phi) &= m_{k+1}^{YX}(\phi) - m_{k+1}^Y(\phi).
\end{aligned} \tag{9}$$

From Lemma 2 and Eq. (8), it follows that

$$\gamma_{k+1}(\phi) = \gamma_{k+1}^Y(\phi) - \gamma_{k+1}^X(\phi). \tag{10}$$

Now let $V_{k+1}^X(\phi)$ be the difference between the number of mistakes during the last $k + 1$ observations when the procedure δ_{k+1}^{XY} is used and the number when the procedure δ_{k+1}^X is used. Then

$$E[V_{k+1}^X(\phi)] = \gamma_{k+1}^X(\phi). \tag{11}$$

Also, if the random variable $V_{k+1}^Y(\phi)$ is defined in a similar way, then

$$E[V_{k+1}^Y(\phi)] = \gamma_{k+1}^Y(\phi). \tag{12}$$

Since the procedures δ_{k+1}^{XY} and δ_{k+1}^X specify the same first observation, any difference in the numbers of mistakes made in the two procedures depends on the remaining k observations. In turn, these numbers depend on whether $\phi(X) < \frac{1}{2}$ or $\phi(X) > \frac{1}{2}$ after the first observation on X has been taken. If $\phi(X) > \frac{1}{2}$, then at the second stage both the procedure δ_{k+1}^{XY} and the procedure δ_{k+1}^X specify that an observation should be taken on Y. Both procedures specify that an optimal procedure should be followed after the second stage, and therefore the two procedures agree throughout the entire process. On the other hand, if $\phi(X) < \frac{1}{2}$, then at the second stage the procedure δ_{k+1}^{XY} specifies an observation on Y, whereas the procedure δ_{k+1}^X specifies an observation on X because the optimal procedure should be adopted for the last k observations and, by our induction hypothesis, this optimal procedure is δ^*. Hence, if $\phi(X) > \frac{1}{2}$, then

$$E[V_{k+1}^X(\phi)|\phi(X)] = 0. \tag{13}$$

If $\phi(X) < \frac{1}{2}$, the appropriate relation is

$$E[V_{k+1}^X(\phi)|\phi(X)] = m_k^Y[\phi(X)] - m_k^X[\phi(X)] = -\gamma_k[\phi(X)]. \tag{14}$$

Similarly, it can be shown that if the first observation is taken on Y and if $\phi(Y) < \frac{1}{2}$, then

$$E[V_{k+1}^Y(\phi)|\phi(Y)] = 0. \tag{15}$$

If $\phi(Y) > \frac{1}{2}$, then

$$E[V_{k+1}^Y(\phi)|\phi(Y)] = m_k{}^X[\phi(Y)] - m_k{}^Y[\phi(Y)] = \gamma_k[\phi(Y)]. \tag{16}$$

By the induction hypothesis, $\gamma_k(\phi)$ is nondecreasing and $\gamma_k(\frac{1}{2}) = 0$. Hence, it follows from Eqs. (13) and (14) that $E[V_{k+1}^X(\phi)|\phi(X)]$ is a nonincreasing function of $\phi(X)$, and it follows from Eqs. (15) and (16) that $E[V_{k+1}^Y(\phi)|\phi(Y)]$ is a nondecreasing function of $\phi(Y)$.

Let $h[\phi(X)] = E[V_{k+1}^X(\phi)|\phi(X)]$. Then, by Eq. (11),

$$\gamma_{k+1}^X(\phi) = E\{h[\phi(X)]\}. \tag{17}$$

Since h is a nonincreasing function of $\phi(X)$, it now follows from Eq. (17) and Lemma 3 that $\gamma_{k+1}^X(\phi)$ is a nonincreasing function of ϕ [see, e.g., Lehmann (1959), pp. 73–74 and 112]. Similarly, it follows from Eq. (12) and Lemma 3 that $\gamma_{k+1}^Y(\phi)$ is a nondecreasing function of ϕ. Together with Eq. (10), these statements imply that $\gamma_{k+1}(\phi)$ is a nondecreasing function of ϕ. $\blacksquare$

A development which is essentially the same as that just presented can be used for the following more general problem. Suppose that X and Y are not necessarily Bernoulli random variables but are arbitrary random variables for which the g.p.d.f.'s are f_X and f_Y. Let g and h be two given g.p.d.f.'s, and suppose either that $f_X = g$ and $f_Y = h$ or else that $f_X = h$ and $f_Y = g$. At each of n stages, suppose that an observation can be taken either on X or on Y. Every time an observation is taken on the random variable for which the g.p.d.f. is g, it is assumed that the statistician has made a mistake. The statistician's problem is to determine a procedure which minimizes the expected number of mistakes. Let

$$\phi = \Pr\{f_X = g, f_Y = h\} = 1 - \Pr\{f_X = h, f_Y = g\}. \tag{18}$$

Then the optimal procedure is again to take the first observation on X if $\phi < \frac{1}{2}$ and to take the first observation on Y if $\phi > \frac{1}{2}$.

Further Remarks and References

The main references for Secs. 14.5 to 14.7 are Bradt, Johnson, and Karlin (1956) and, as mentioned earlier, Feldman (1962). Thompson (1933) was one of the first authors to consider this type of problem. The papers by Robbins (1952, 1956a) stimulated much of the further work on the subject. Some other papers are those by Bellman (1956), Isbell (1959), Vogel

(1960a,b), Furukawa (1964), Mallows and Robbins (1964), and Brown (1965). The paper by Quisel (1965) includes a summary of two-armed-bandit problems.

There are obvious generalizations of two-armed-bandit problems to k-armed-bandit problems $(k \geq 2)$. In the other direction, there are certain problems of optimal stopping which are called one-armed-bandit problems; see Chernoff (1967b).

14.8 INVENTORY PROBLEMS

We shall now consider some problems pertaining to optimal inventory and stock control. Suppose that the statistician orders a certain product from a supplier and then sells it to the public. Suppose further that he can order for his stock any number of units of the product at a cost of c dollars per unit and that he sells the product at the price of p dollars per unit, where $p > c > 0$. It is assumed, however, that the demand for the product is a random variable, and we shall suppose that the statistician loses the entire cost of any part of his stock that he does not subsequently sell. Hence, if the statistician orders too little of the product at any time, he will lose potential profit by being unable to meet the full demand for the product and having to forego some sales. On the other hand, if he orders too much of the product, he will lose the amount of money he has spent for unsold stock. A problem of this type with just a single stage was presented in Exercise 13 of Chap. 7. The problem of finding an optimal procedure for ordering amounts of the product in a sequential process is more complicated.

Specifically, we shall consider a process with a fixed number of stages n. We shall suppose that at the beginning of each stage, the statistician can order any number of units $t \geq 0$ of the product at a cost of ct dollars. The t units are added to any stock which may have remained unsold during the earlier stages of the process. We shall let y be the total number of units in stock after the t units have been added. Next, the demand x at the given stage is observed. If $x \leq y$, the statistician can meet the entire demand at that stage from his stock. Hence, he will sell x units. If $x > y$, the statistician will sell his entire stock of y units. In either case, we shall let s denote the number of units which the statistician sells at that stage. Then his income from these sales is ps dollars, exclusive of his costs, and $y - s$ is the number of units in stock, or the stock level, at the beginning of the next stage.

For simplicity, we shall assume that the product can be measured in arbitrarily small quantities and that any positive amount t can be ordered. Furthermore, we shall assume that the demand at each stage has an absolutely continuous distribution which can be represented by a p.d.f.

Let X_j denote the demand at the jth stage $(j = 1, \ldots, n)$. We shall assume that $X_1, \ldots, X_n$ are independent and identically distributed positive random variables and that each has the specified p.d.f. f and a finite mean. If T_j represents the number of units of the product which the statistician orders at the beginning of the jth stage $(j = 1, \ldots, n)$ and if S_j represents the number which he sells at that stage, then the problem is to find a sequential procedure that maximizes his expected profit. This expected profit can be written as follows:

$$E\left[\sum_{j=1}^{n} (pS_j - cT_j)\right].\tag{1}$$

Let y denote the number of units in stock before the statistician orders more units at the first stage. For any number $y \geq 0$ and any positive integer n, we shall let $V_n(y)$ denote the value of the expected profit (1) from an optimal procedure in a process with n stages. Suppose that he orders a number of units $t \geq 0$ at the first stage and that the demand is $X_1 = x$. If $x \leq y + t$, then his income from sales is px, his stock level is reduced to $y + t - x$, and his expected profit from following an optimal procedure over the remaining $n - 1$ stages is $V_{n-1}(y + t - x)$. If $x > y + t$, then his income from the sale of his entire stock is $p(y + t)$, his stock level is reduced to 0, and his expected profit over the remaining $n - 1$ stages is $V_{n-1}(0)$. When the cost ct of the t units is subtracted, we obtain the following relation:

$$V_n(y) = \sup_{t \geq 0} \left\{ \int_0^{y+t} [px + V_{n-1}(y + t - x)]f(x)\,dx \right.$$
$$\left. + [p(y + t) + V_{n-1}(0)] \int_{y+t}^{\infty} f(x)\,dx - ct \right\}.\tag{2}$$

We shall let $V_0(y) = 0$ for all values of $y \geq 0$. Also, we shall let F denote the d.f. corresponding to the p.d.f. f. When only one stage remains, i.e., when $n = 1$, it will be found, by differentiating the right side of Eq. (2), that the statistician should order exactly the number of units t that will bring his stock level $y + t$ to the value for which the following relation is satisfied:

$$F(y + t) = \frac{p - c}{p}.\tag{3}$$

If the initial stock level y already exceeds this value, then no additional stock should be ordered. The solutions of Eq. (2) for the functions $V_1, V_2, \ldots$ become successively more complicated. Instead of considering this equation further, we shall adopt a different approach.

Because of the nature of the problem, it can be seen that for any value $y \geq 0$ and for $n = 1, 2, \ldots$, the following equation must be satisfied:

$$V_n(y) = \sup_{t \geq 0} [V_n(y + t) - ct].\qquad(4)$$

Equation (4) expresses the fact that the statistician can, in effect, change his initial stock level y to any other level $y + t$ by ordering t units at a cost of ct. For each value of y, we shall let $t = t_n(y)$ be a value of t at which the supremum in Eq. (4) is attained. This supremum must actually be attained at some value of t because V_n is a continuous function, and very large values of t cannot be optimal and need not be considered.

Suppose that for some initial stock level $y = y_0$, the number $t_n(y_0) > 0$. It will then be optimal for the statistician to order a positive number of units to bring his stock level up to the value $y_n^* = y_0 + t_n(y_0)$. We shall now show that when n stages remain, the optimal procedure is as follows: If the initial stock level $y < y_n^*$, the statistician should order enough units at the first stage to bring the stock level up to the value y_n^*. If $y \geq y_n^*$, the optimal procedure at the first stage is to order nothing.

To demonstrate that the above procedure is optimal, we shall first consider an initial value y such that $y_0 \leq y \leq y_n^*$. Then, by letting $t = t' - (y - y_0)$, we can rewrite Eq. (4) as follows:

$$V_n(y) = \sup_{t' \geq y - y_0} [V_n(y_0 + t') - ct' + c(y - y_0)].\qquad(5)$$

It now follows from the definition of $t_n(y_0)$ that the supremum in Eq. (5) is attained when $t' = t_n(y_0)$ and, hence, when $t = t_n(y_0) - (y - y_0)$. In other words, $t_n(y) = y_n^* - y$, and the following result is obtained:

$$y + t_n(y) = y_n^* \qquad y_0 \leq y \leq y_n^*.\qquad(6)$$

Now we shall widen the interval of initial values y for which Eq. (6) is correct. We shall consider any value y such that $y_0 - t_n(y_0) \leq y \leq y_0$ or, if $y_0 - t_n(y_0) < 0$, such that $0 \leq y \leq y_0$. It is clear from the nature of the problem that since the optimal procedure is to order $t_n(y_0)$ units when the initial stock level is y_0, it is optimal to order at least $t_n(y_0)$ units when the initial stock level is $y < y_0$. In other words, $t_n(y) \geq t_n(y_0)$. More generally, $t_n(y)$ can be chosen as a nonincreasing function of y. Hence, Eq. (4) can be rewritten as follows:

$$\begin{aligned}V_n(y) &= \sup_{t \geq t_n(y_0)} [V_n(y + t) - ct] \\ &= \sup_{t' \geq y - [y_0 - t_n(y_0)]} [V_n(y_0 + t') - ct' - c(y_0 - y)].\end{aligned}\qquad(7)$$

The supremum in Eq. (7) is attained when $t' = t_n(y_0)$ or, equivalently, when $t = t_n(y_0) - (y - y_0)$. This result is the same as that previously obtained. Hence, we have established the fact that Eq. (6) is correct for all initial values y in the interval $y_0 - t_n(y_0) \leq y \leq y_n^*$.

We can now consider initial values y in the interval $y_0 - 2t_n(y_0) \leq y \leq y_0 - t_n(y_0)$, and we can proceed in this way until Eq. (6) has been established for all values of y in the interval $0 \leq y \leq y_n^*$.

It follows from the foregoing development that when n stages remain, there exists a stock level y_n^* such that the optimal procedure is as follows: If the stock level is below y_n^*, the statistician should order enough to bring the stock level up to y_n^*. If the stock level is not below y_n^*, he should order nothing. For $j = 1, \ldots, n$, let y_j^* be the corresponding optimal stock level when j stages remain. Then, by Eq. (3), y_1^* must satisfy the following equation:

$$F(y_1^*) = \frac{p - c}{p}. \tag{8}$$

Furthermore, it must be true that $y_1^* \leq y_2^* \leq \cdots \leq y_n^*$ for the following reason: The larger the number of stages which remain, the larger is the total number of units of the product which the statistician is likely to sell and, hence, the larger is the number of units which he can risk keeping in stock. These results are summarized in the next theorem.

Theorem 1 *Consider the n-stage inventory problem defined by Eq. (2). There exist numbers $0 < y_1^* \leq y_2^* \leq \cdots \leq y_n^*$ with the following property: If the stock level is y_j when j stages remain $(j = 1, \ldots, n)$, then an optimal procedure is to order $y_j^* - y_j$ units if $y_j < y_j^*$ and to order 0 units if $y_j \geq y_j^*$.*

We shall not consider any further the computation of the numbers y_j^* for $j = 2, \ldots, n$.

14.9 INVENTORY PROBLEMS WITH AN INFINITE NUMBER OF STAGES

We shall now modify the inventory problem as follows. Instead of a process with a fixed finite number of stages, we shall consider a process with an infinite number of stages. Also, we shall introduce a discount factor β. Thus both the cost of the units ordered at the nth stage and the income realized from sales at the nth stage will be multiplied by the factor β^{n-1}. The problem again is to find a sequential procedure that maximizes the expected profit. If we let T_j represent the number of units ordered by the statistician at the beginning of the jth stage $(j = 1, 2, \ldots)$ and S_j represent the number he sells at that stage, the expected profit can be expressed as follows:

$$E\left[\sum_{j=1}^{\infty} \beta^{j-1}(pS_j - cT_j) \right]. \tag{1}$$

Again, we shall let X_j be the demand at the jth stage ($j = 1, 2, \ldots$). Moreover, we shall assume that the random variables in the sequence $X_1, X_2, \ldots$ are independent and identically distributed and that the p.d.f. of each is f.

For any number $y \geq 0$, we shall let $V(y)$ denote the value of the expected profit (1) from an optimal procedure when the statistician's initial stock level is y. The function V will satisfy a functional equation similar to Eq. (2) of Sec. 14.8. Instead of changing the subscript for V on the right side of that equation, we must introduce the discount factor β. However, the basic reasoning remains the same, and we can obtain the following relation:

$$V(y) = \sup_{t \geq 0} \left\{ \int_0^{y+t} [px + \beta V(y + t - x)]f(x)\, dx \right.$$
$$\left. + [p(y + t) + \beta V(0)] \int_{y+t}^{\infty} f(x)\, dx - ct \right\}. \quad (2)$$

The solution of this problem has a simpler form than has the solution of the problem in which there are a fixed finite number of stages. Because of the stationary properties of the problem with an infinite number of stages, the optimal procedure simply specifies the number of units $t(y)$ which the statistician should order when his stock level is y, regardless of the stage of the process.

By the same reasoning which led to Eq. (4) of Sec. 14.8, we can obtain the following result:

$$V(y) = \sup_{t \geq 0} [V(y + t) - ct]. \quad (3)$$

Also, it follows from precisely the same argument as that given for V_n that it is optimal to order just enough units at each stage to bring the stock level up to a certain fixed value y^*. If the stock level exceeds y^* during some of the initial stages of the process, then no additional stock is ordered at those stages. We shall now determine the optimal value y^*.

For any value $y > 0$, we shall let $U(y)$ denote the expected profit if the initial stock level is y and the procedure adopted at each stage is to order just enough units to bring the stock level back up to its original value y. The values of $U(y)$ can be computed as follows.

Let $\Pi(y)$ be the expected profit from the sales at the first stage minus the discounted cost of bringing the stock level back up to y units at the beginning of the second stage. Then

$$\Pi(y) = \int_0^y (px - \beta cx)f(x)\, dx + (py - \beta cy) \int_y^{\infty} f(x)\, dx$$
$$= (p - \beta c) \left[\int_0^y xf(x)\, dx + y \int_y^{\infty} f(x)\, dx \right]. \quad (4)$$

Since the stock level at this time is again y, the expected discounted profit from the sales at the second stage and the subsequent ordering at the next stage is simply $\beta\Pi(y)$. By continuing in this way through the infinite sequence of stages, we can obtain the following result:

$$U(y) = \Pi(y)(1 + \beta + \beta^2 + \cdots) = \frac{\Pi(y)}{1 - \beta}. \tag{5}$$

We shall now suppose that the initial stock level is 0 and that the procedure adopted at each stage is to order just enough units to bring the stock level up to y, and we shall let $M(y)$ be the expected profit from this procedure. Since the cost of the units ordered at the beginning of the first stage is cy, it follows that

$$M(y) = U(y) - cy. \tag{6}$$

We know that the optimal procedure is of this form for some value of y. Hence, the optimal value y^* will be the value which maximizes $M(y)$. This value can be found by elementary differentiation. If F again denotes the d.f. corresponding to the p.d.f. f, then it follows from Eqs. (4) to (6) that

$$M'(y) = \frac{p - \beta c}{1 - \beta}[1 - F(y)] - c. \tag{7}$$

By solving the equation $M'(y) = 0$, we find that the required value y^* must satisfy the following relation:

$$F(y^*) = \frac{p - c}{p - \beta c}. \tag{8}$$

We can summarize these results in the following theorem.

Theorem 1 *Consider the infinite-stage inventory problem defined by Eq. (2). Let y^* be a number which satisfies Eq. (8). If y is the stock level before units are ordered at any stage, then an optimal procedure is to order $y^* - y$ units if $y < y^*$ and to order 0 units if $y \geq y^*$.*

If $\beta = 0$, then the problem is really a one-stage problem and, as we should expect, y^* and the value y_1^* found in Sec. 14.8 are equal. On the other hand, as $\beta \to 1$, future sales are not discounted seriously. Therefore, the statistician would suffer no great loss by having stock which remains unsold for a long period of time, provided that it is ultimately sold. For this reason, $y^* \to \infty$ as $\beta \to 1$.

Further Remarks and References

The literature on inventory problems has been strongly influenced by the papers of Arrow, Harris, and Marschak (1951), and Dvoretzky, Kiofor, and Wolfowitz (1952a,b). The books edited by Arrow, Karlin, and Scarf (1958, 1962) and Scarf, Gilford, and Shelly (1963) contain many developments in inventory theory. Bellman (1957a), chap. 5, treats the problems we have discussed, as well as several others. Some other books on the subject are Wagner (1962), Hadley and Whitin (1963), and parts of Hadley (1964). Veinott (1966) has published a survey article on inventory theory containing 118 references.

14.10 CONTROL PROBLEMS

The problems to be considered in the next few sections fall under the general heading of adaptive stochastic control theory. In each problem of this class, an optimal procedure must be determined for a sequential process which has a fixed finite number of stages and which involves a quadratic loss function at each stage. We shall see that these quadratic loss functions lead, in turn, to optimal decision procedures based on linear functions of the variables which the statistician can control. Because of this linearity, the optimal procedures can be determined explicitly. We shall begin the development by considering a relatively simple, one-dimensional control problem.

Let n be a fixed positive integer, and consider a finite sequence of $n + 1$ random variables $X_1, \ldots, X_{n+1}$. The values of these variables can be thought of as the states of a stochastic system at the various stages of an n-stage sequential process. Thus, X_1 denotes the initial state of the system, and $X_2, \ldots, X_{n+1}$ denote the states of the system at the n successive stages.

We shall suppose that at any given stage j $(j = 1, \ldots, n)$, the distribution of the next state X_{j+1} depends only on the current state X_j and on the value u_j of a real variable, called the *control variable*, which can be chosen by the statistician. Specifically, we shall assume that the process evolves according to the following system of equations:

$$X_{j+1} = \alpha_j X_j + \beta_j + u_j + Z_j \qquad j = 1, \ldots, n. \tag{1}$$

Here, α_j and β_j are given constants, with $\alpha_j \neq 0$, u_j is a value which can be chosen by the statistician after the value of X_j has been observed, and Z_j is a random variable having a normal distribution with mean 0 and variance γ_j^2. The variables $Z_1, \ldots, Z_n$ represent the random disturbances in the system, and they are assumed to be independent. Also, it is assumed that the initial state $X_1 = x_1$ is known.

Before we indicate the criteria by which the statistician chooses the values $u_1, \ldots, u_n$ of the control variable, we shall describe in more detail the evolution of the process governed by the system of equations (1). The initial state $X_1 = x_1$ of the particle will be known, and the statistician will choose a value u_1 of the control variable. Then the next state X_2 of the particle has a normal distribution with mean $\alpha_1 x_1 + \beta_1 + u_1$ and variance $\gamma_1{}^2$. After the position $X_2 = x_2$ has been observed, the statistician chooses a value u_2 of the control variable. The next state X_3 then has a normal distribution with mean $\alpha_2 x_2 + \beta_2 + u_2$ and variance $\gamma_2{}^2$. The process continues in this way until the terminal position X_{n+1} is observed.

We shall suppose that at each stage j $(j = 1, \ldots, n)$, the statistician must choose a value of the control variable that will keep the next state X_{j+1} of the system close to a specified target value t_j. We shall assume that the loss due to missing the target value at this stage is $q_j(X_{j+1} - t_j)^2$, where $q_j \geq 0$. It may be true that $q_j = 0$ for certain values of j. In fact, in many interesting problems, $q_j = 0$ for $j = 1, \ldots, n - 1$. These values for q_j are appropriate when the statistician is interested only in the distance between the terminal state X_{n+1} of the system and some terminal target value t_n. However, it is always assumed that $q_n > 0$ since the terminal state X_{n+1} would not otherwise be of interest and the final stage of the process could be eliminated from the problem.

If the statistician could choose the value u_j of the control variable without any constraint or any cost, his optimal choice at the jth stage would be the value of u_j which makes the expectation $\alpha_j x_j + \beta_j + u_j$ of the next state X_{j+1} equal to its target value t_j. In a practical situation, however, the cost of controlling the process may make it necessary for the statistician to find a proper balance between the cost of control and the loss from missing the target. We shall assume that the cost of choosing the value u_j of the control variable at the jth stage is $r_j u_j{}^2$, where $r_j \geq 0$.

It follows from the above remarks that the total loss λ_j at the jth stage $(j = 1, \ldots, n)$ can be represented as follows:

$$\lambda_j = q_j(X_{j+1} - t_j)^2 + r_j u_j{}^2. \tag{2}$$

The total loss over the entire process is the sum $\sum_{j=1}^{n} \lambda_j$. A sequence of values $u_1, \ldots, u_n$ of the control variable must be chosen which minimizes the expected value of this sum.

Since this control problem has only a finite number of stages, we can determine the optimal sequence of values of the control variable by backward induction. We shall consider a given stage j $(j = 1, \ldots, n)$, and we shall suppose that the position $X_j = x_j$ has just been observed and that the value u_j must now be chosen. The total loss over the remaining stages of the process will be $\sum_{i=j}^{n} \lambda_i$.

It can be seen from Eq. (1) that the optimal choice of u_j will depend only on the value x_j and will not be affected directly by the earlier states of the system or by the earlier values of the control variable. We shall let $L_j(x_j)$ denote the expected value of the sum $\sum_{i=j}^{n} \lambda_i$ when u_j and all subsequent values of the control variable are chosen in an optimal fashion. In particular, $L_1(x_1)$ is the minimum expected total loss over the entire process that can be attained when the initial state X_1 is x_1.

For $j = 1, \ldots, n$, we shall let E_j denote any expectation which is computed with respect to the conditional distribution of X_{j+1} when $X_j = x_j$ and the value of u_j is given. If we define the function $L_{n+1}(x_{n+1})$ to be identically 0, then the functions $L_1, \ldots, L_n$ must satisfy the following relation for $j = 1, \ldots, n$:

$$L_j(x_j) = \inf_{u_j} E_j[\lambda_j + L_{j+1}(X_{j+1})]. \tag{3}$$

We shall now prove by an induction argument that for $j = 1, \ldots, n$, the function L_j is a certain quadratic function of the following form:

$$L_j(x_j) = a_{j-1}(x_j - b_{j-1})^2 + c_{j-1}. \tag{4}$$

The values of a_{j-1}, b_{j-1}, and c_{j-1} can be computed explicitly. Furthermore, we shall show that the optimal value u_j of the control variable is

$$u_j = \frac{q_j t_j + a_j b_j - (q_j + a_j)(\alpha_j x_j + \beta_j)}{q_j + a_j + r_j}. \tag{5}$$

The right side of Eq. (5) is simply a linear function of x_j.

Since the function L_{n+1} is identically 0, it has the form specified in Eq. (4) with $a_n = b_n = c_n = 0$. We shall now suppose that for some value of j ($j = 1, \ldots, n$), the function L_{j+1} has the form specified in Eq. (4). We shall show that u_j must then have the form specified in Eq. (5) and that the function L_j must again have the form specified in Eq. (4).

It follows from Eq. (1) that for any constant δ,

$$E_j[(X_{j+1} - \delta)^2] = (\alpha_j x_j + \beta_j + u_j - \delta)^2 + \gamma_j^2. \tag{6}$$

Hence, by Eq. (2),

$$E_j(\lambda_j) = q_j(\alpha_j x_j + \beta_j + u_j - t_j)^2 + r_j u_j^2 + q_j \gamma_j^2. \tag{7}$$

Furthermore, by Eq. (4) and the induction hypothesis,

$$E_j[L_{j+1}(X_{j+1})] = a_j(\alpha_j x_j + \beta_j + u_j - b_j)^2 + a_j \gamma_j^2 + c_j. \tag{8}$$

It can be seen from Eq. (3) that the optimal value of u_j must minimize the sum of the values in Eqs. (7) and (8). Since this sum is simply a quadratic function of u_j, the optimal value of u_j can be found by elementary differentiation, and it can be verified that this optimal value is specified by Eq. (5).

When this optimal value of u_j is substituted into Eqs. (7) and (8), it can be seen that their sum $L_j(x_j)$ will be a certain quadratic function of x_j. Hence, L_j also has the form required by Eq. (4), and the induction argument has been completed. After some algebraic simplification, it can be found that the constants a_{j-1}, b_{j-1}, and c_{j-1} which appear in Eq. (4) will satisfy the following equations:

$$
\begin{aligned}
a_{j-1} &= \frac{\alpha_j^2 r_j (q_j + a_j)}{q_j + a_j + r_j}, \\
b_{j-1} &= \frac{1}{\alpha_j} \left(\frac{q_j t_j + a_j b_j}{q_j + a_j} - \beta_j \right), \\
c_{j-1} &= \frac{q_j a_j (t_j - b_j)^2}{q_j + a_j} + (q_j + a_j)\gamma_j^2 + c_j.
\end{aligned}
\tag{9}
$$

The entire sequence of values of a_{j-1}, b_{j-1}, and c_{j-1}, for $j = 1, \ldots, n$, can be determined in reverse order from Eq. (9) together with the given values $a_n = b_n = c_n = 0$. The optimal sequence of values $u_1, \ldots, u_n$ of the control variable can then be found from Eq. (5). Finally, the minimum expected total loss $L_1(x_1)$ can be found from Eq. (4) when $j = 1$.

We have already emphasized the fact that the optimal value of the control variable at each stage is a linear function of the state of the system at that stage. The following two properties of the solutions are also very important. First, the variances $\gamma_1^2, \ldots, \gamma_n^2$ do not appear in Eq. (5), and they do not appear in the expressions for a_{j-1} and b_{j-1} given in Eq. (9). Hence, these variances need not actually be known in order to specify the optimal sequence of values $u_1, \ldots, u_n$. Second, the normality of the distributions of the random disturbances $Z_1, \ldots, Z_n$ was never used in this example. Therefore, the optimal sequence of values we have derived here will be optimal in any process governed by Eq. (1), provided that the disturbances $Z_1, \ldots, Z_n$ are independent and that each has a distribution for which the mean is 0 and the variance is finite. These facts greatly increase the range of application of the results presented in this section.

14.11 OPTIMAL CONTROL WHEN THE PROCESS CANNOT BE OBSERVED WITHOUT ERROR

We shall now introduce a complication in the control problem considered in the preceding section by supposing that as the process evolves, it is not possible for the statistician to observe the states $X_1, \ldots, X_n$ without error. Specifically, we shall now assume that the value of the state X_j is not available to the statistician at the time at which he must choose a value u_j of the control variable. Instead, he must make this choice after observing the value of a random variable Y_j whose distribution typically depends on the state X_j and whose observed value therefore provides at least some information about the value of X_j. Formally, the observation process is described by the following system of equations:

$$Y_j = \varepsilon_j X_j + V_j \qquad j = 1, \ldots, n. \tag{1}$$

Here, ε_j is a given constant and V_j is a random variable having a normal distribution with mean 0 and variance σ_j^2. Furthermore, the random variables $V_1, \ldots, V_n$, the random variables $Z_1, \ldots, Z_n$ in Eq. (1) of Sec. 14.10, and the initial state X_1 are all assumed to be independent. It may be true that $\varepsilon_j = 0$ for some value of j. The observation Y_j at that stage provides no additional information about the state X_j. It may also be true that $\varepsilon_j \neq 0$ but that $\sigma_j^2 = 0$ for some value of j. The state X_j at that stage can be observed exactly.

We shall now consider the manner in which the process governed by Eq. (1) of Sec. 14.10 and Eq. (1) of this section will evolve. In our previous work, it was assumed that the initial state X_1 was a given number x_1. Here, however, we shall assume that the value of x_1 is unknown and that X_1 has a normal distribution with mean m_1 and variance g_1^2. At the beginning of the process, the statistician will observe the value of Y_1, and this observation will lead to a revised posterior distribution of the state X_1. On the basis of this distribution, the statistician will choose the control value u_1, and this choice will lead to a distribution for the next state X_2. Again, the value of this state cannot be observed directly. However, the statistician can observe the value of Y_2, and this observation will lead to a revised posterior distribution for X_2. He can then choose the control value u_2 and can continue in this way until the final control value u_n has been chosen.

As before, a sequence of values $u_1, \ldots, u_n$ of the control variable must be chosen which will minimize the expected value of the total loss $\sum_{j=1}^{n} \lambda_n$, as specified in Sec. 14.10. At each stage j $(j = 1, \ldots, n)$ at which the statistician must choose a control value u_j, his information can be represented by the posterior distribution of the state X_j after the value of Y_j has been observed. Since the prior distribution of the initial state X_1 is assumed to be a normal distribution and since each of the random variables $Z_1, \ldots, Z_n$ and $V_1, \ldots, V_n$ has a normal distribution, it can be seen that the posterior distribution of X_j at each stage will again be a normal distribution. We shall now consider the manner in which this distribution changes from one stage of the process to another.

We have already assumed that the prior distribution of X_1 is a normal distribution with mean m_1 and variance g_1^2. Moreover, for $j = 2, \ldots, n$, we shall assume that after the value of u_{j-1} has been chosen but before the value of Y_j has been observed, the distribution of X_j is a normal distribution with mean m_j and variance g_j^2.

For $j = 1, \ldots, n$, it follows from Eq. (1) that after the value $Y_j = y_j$ has been observed, the posterior distribution of X_j will be a normal distribution for which the mean m_j^* and the variance g_j^{*2} are as follows:

$$m_j^* = \frac{\sigma_j^2 m_j + \varepsilon_j g_j^2 y_j}{\sigma_j^2 + \varepsilon_j^2 g_j^2} \tag{2}$$

and

$$g_j^{*2} = \frac{g_j^2 \sigma_j^2}{\sigma_j^2 + \varepsilon_j^2 g_j^2} \tag{3}$$

In these control problems, we shall describe normal distributions in terms of their variances rather than their precisions. This notation is advantageous here because we must repeatedly compute the expected value of a quadratic function of X_j and the distributions of sums of independent random variables.

Suppose now, for a given value of j ($j = 1, \ldots, n$), that the posterior distribution of X_j after the value of Y_j has been observed is a normal distribution with mean m_j^* and variance g_j^{*2}. By Eq. (1) of Sec. 14.10, it can be seen that for any choice of the control value u_j, the next state X_{j+1} will have a normal distribution for which the mean m_{j+1} and the variance g_{j+1}^2 are

$$m_{j+1} = \alpha_j m_j^* + \beta_j + u_j \tag{4}$$

and

$$g_{j+1}^2 = \alpha_j^2 g_j^{*2} + \gamma_j^2. \tag{5}$$

Furthermore, when Y_j is regarded as a random variable, then the mean m_j^* specified by Eq. (2) becomes a random variable, and we can easily compute its distribution. By Eq. (1), the observation Y_j has a normal distribution with mean $\varepsilon_j m_j$ and variance $\varepsilon_j^2 g_j^2 + \sigma_j^2$. Therefore, by Eq. (2), the distribution of m_j^* will be a normal distribution for which the mean $E_{j-1}(m_j^*)$ and the variance $\mathrm{Var}_{j-1}(m_j^*)$ are

$$E_{j-1}(m_j^*) = m_j \tag{6}$$

and

$$\mathrm{Var}_{j-1}(m_j^*) = \frac{\varepsilon_j^2 g_j^4}{\sigma_j^2 + \varepsilon_j^2 g_j^2}. \tag{7}$$

The subscript $j - 1$ in Eqs. (6) and (7) is used to indicate that the mean and variance of m_j^* are computed after the value of u_{j-1} has been chosen but before the value of Y_j has been observed.

It can be seen from Eqs. (3) and (5) that the variance g_j^{*2} of the posterior distribution of X_j changes in a deterministic way from one stage to another because neither the values of the observations nor the values chosen for the controls affect the variance.

Equations (2) to (7) characterize the distributions which are of interest to the statistician during the control process. We can now determine the optimal values $u_1, \ldots, u_n$ of the control variable by the method of backward induction. At any stage j ($j = 1, \ldots, n$), the optimal choice of the control value u_j will depend only on the posterior distribution of X_j at that time. Since the variance g_j^{*2} of the posterior distribution changes in a fixed and predetermined way from one stage to another, the posterior distribution of X_j can be characterized simply by its mean m_j^*. For any given value of m_j^*, we shall let $L_j(m_j^*)$ denote the expected value of the sum $\sum_{i=j}^n \lambda_i$ when u_j and all subsequent values of the control variable are chosen in an optimal fashion. As before, it follows from Eqs. (1) and (2) of Sec. 14.10 that the functions $L_1, \ldots, L_n$ must satisfy the following relation for $j = 1, \ldots, n$:

$$L_j(m_j^*) = \inf_{u_j} E_j[q_j(X_{j+1} - t_j)^2 + r_j u_j^2 + L_{j+1}(m_{j+1}^*)]. \tag{8}$$

Again, we shall define the function L_{n+1} to be identically 0.

We shall now prove by induction that for $j = 1, \ldots, n$, the function L_j is a certain quadratic function of the following form:

$$L_j(m_j^*) = a_{j-1}(m_j^* - b_{j-1})^2 + c_{j-1}. \tag{9}$$

We shall also show that the optimal value u_j of the control variable is

$$u_j = \frac{q_j t_j + a_j b_j - (q_j + a_j)(\alpha_j m_j^* + \beta_j)}{q_j + a_j + r_j}. \tag{10}$$

Moreover, for $j = 1, \ldots, n$, the values of a_j and b_j will be the same as those found in Sec. 14.10 but the value of c_j will be different.

Since the function L_{n+1} is identically 0, it has the form specified in Eq. (9) with $a_n = b_n = c_n = 0$. We shall now suppose that for some value of j ($j = 1, \ldots, n$), the function L_{j+1} has the form specified in Eq. (9). We shall show that u_j must then have the form specified in Eq. (10) and that the function L_j must again have the form specified in Eq. (9).

For any choice of the control value u_j, the expected value on the right side of Eq. (8) will be

$$\begin{aligned}
E_j[q_j(X_{j+1} - t_j)^2 &+ r_j u_j^2 + a_j(m_{j+1}^* - b_j)^2 + c_j] \\
&= q_j[(m_{j+1} - t_j)^2 + g_{j+1}^2] + r_j u_j^2 \\
&\quad + a_j[(m_{j+1} - b_j)^2 + \mathrm{Var}_j(m_{j+1}^*)] + c_j. \tag{11}
\end{aligned}$$

As shown in Eq. (4), m_{j+1} is a simple linear combination of m_j^* and u_j. Furthermore, as shown in Eq. (7), $\text{Var}_j(m_{j+1}^*)$ is a constant which does not involve m_j^* or u_j. Hence, the right side of Eq. (11) is a quadratic function of u_j, and by an elementary computation, the minimizing value is found to be the value specified by Eq. (10).

When this optimum value of u_j is substituted into Eq. (11), it is found that $L_j(m_j^*)$ will be a certain quadratic function of the form required by Eq. (9). Furthermore, the values of a_{j-1} and b_{j-1} will again be specified by Eq. (9) of Sec. 14.10, together with the given values $a_n = b_n = 0$. The induction argument has now been completed. The optimal sequence of values $u_1, \ldots, u_n$ can be found from Eq. (10) by using the values $a_n = b_n = 0$ and the relation (9) of Sec. 14.10. The value of c_j is not needed for the computation of the optimal sequence of control values, and it is not given here.

The optimal value of u_j, as specified by Eq. (10), is completely analogous to the optimal value of u_j when the state X_j can be observed exactly at each stage, as specified by Eq. (5) of Sec. 14.10. Thus, if the actual state X_j is not known to the statistician at the time he must choose a value u_j for the control variable, then the optimal procedure is simply to replace the unknown exact value x_j in Eq. (5) of Sec. 14.10 by the expected value m_j^* of X_j at that time. This phenomenon occurs in various decision problems involving quadratic loss functions [see, e.g., Simon (1956) or Theil (1964)]. Furthermore, the minimum expected losses specified by Eq. (9) and by Eq. (4) of Sec. 14.10 are also of the same form, since the only difference is the replacement of x_j by m_j^*. However, although the constants a_{j-1} and b_{j-1} are defined the same way in both equations, the additional term c_{j-1} will be larger in Eq. (9) than in Eq. (4) of Sec. 14.10 because that term will reflect the extra loss due to the statistician's uncertainty about the actual states X_j.

14.12 MULTIDIMENSIONAL CONTROL PROBLEMS

We shall now consider a generalization of the problem in the preceding section by supposing that the state $\mathbf{X}_j$ is a k-dimensional vector ($k \geq 1$). Since the discussion will be completely analogous to that just presented, some details will be omitted here.

We shall assume that the process evolves according to the following system of equations:

$$\mathbf{X}_{j+1} = \alpha_j \mathbf{X}_j + \beta_j + \mathbf{u}_j + \mathbf{Z}_j \qquad j = 1, \ldots, n. \tag{1}$$

Here, $\mathbf{X}_j$ is a k-dimensional vector, α_j is a given nonsingular $k \times k$ matrix, β_j is a given k-dimensional vector, $\mathbf{u}_j$ is a k-dimensional control vector the value of which can be chosen by the statistician, $\mathbf{Z}_j$ is a k-dimensional random vector which has a multivariate normal distribution with mean vector $\mathbf{0}$ and $k \times k$ covariance matrix Γ_j, and the random vectors $\mathbf{Z}_1$, $\ldots$, $\mathbf{Z}_n$ and $\mathbf{X}_1$ are independent.

We shall assume that the observation process is governed by the following system of equations:

$$\mathbf{Y}_j = \varepsilon_j \mathbf{X}_j + \mathbf{V}_j \qquad j = 1, \ldots, n. \tag{2}$$

Here, $\mathbf{Y}_j$ is a p-dimensional vector, ε_j is a given $p \times k$ matrix for which p is not necessarily equal to k, $\mathbf{V}_j$ is a p-dimensional random vector which has a multivariate normal distribution with mean vector $\mathbf{0}$ and nonsingular $p \times p$ covariance matrix Σ_j, and the random vectors $\mathbf{Z}_1, \ldots, \mathbf{Z}_n$, $\mathbf{V}_1, \ldots, \mathbf{V}_n$, and $\mathbf{X}_1$ are independent. In any particular problem of the type now being considered, the dimension p of the observation vector $\mathbf{Y}_j$ may be less than, equal to, or greater than the dimension k of the state vector $\mathbf{X}_j$.

We shall suppose that a sequence of k-dimensional target vectors $\mathbf{t}_1, \ldots, \mathbf{t}_n$ is given. Also, we shall assume that the total loss λ_j at the jth stage $(j = 1, \ldots, n)$ is the sum of the loss due to the difference between the state $\mathbf{X}_{j+1}$ and the target vector $\mathbf{t}_j$ and the cost of using the value $\mathbf{u}_j$ for the control vector. Specifically, it is assumed that λ_j can be expressed as follows, for $j = 1, \ldots, n$:

$$\lambda_j = (\mathbf{X}_{j+1} - \mathbf{t}_j)'\mathbf{q}_j(\mathbf{X}_{j+1} - \mathbf{t}_j) + \mathbf{u}_j'\mathbf{r}_j\mathbf{u}_j. \tag{3}$$

Here, $\mathbf{q}_j$ and $\mathbf{r}_j$ are given symmetric $k \times k$ nonnegative definite matrices. The statistician must choose a sequence $\mathbf{u}_1, \ldots, \mathbf{u}_n$ that will minimize the expected value of the sum $\sum_{j=1}^{n} \lambda_j$.

Finally, we shall suppose that the prior distribution of the initial state $\mathbf{X}_1$ is a multivariate normal distribution with mean vector $\mathbf{m}_1$ and nonsingular $k \times k$ covariance matrix $\mathbf{G}_1$. Then, at each stage of the process, the distribution of $\mathbf{X}_j$ will also be a multivariate normal distribution. For $j = 2, \ldots, n$, we shall assume that after the value of $\mathbf{u}_{j-1}$ has been chosen but before the value of $\mathbf{Y}_j$ has been observed, the state $\mathbf{X}_j$ has a distribution for which the mean vector is $\mathbf{m}_j$ and the covariance matrix is $\mathbf{G}_j$.

Then, for $j = 1, \ldots, n$, after the value $\mathbf{Y}_j = \mathbf{y}_j$ has been observed, the state $\mathbf{X}_j$ will have a posterior distribution for which the mean vector $\mathbf{m}_j^*$ and the covariance matrix $\mathbf{G}_j^*$ are

$$\mathbf{m}_j^* = (\varepsilon_j'\Sigma_j^{-1}\varepsilon_j + \mathbf{G}_j^{-1})^{-1}(\varepsilon_j'\Sigma_j^{-1}\mathbf{y}_j + \mathbf{G}_j^{-1}\mathbf{m}_j) \tag{4}$$

and

$$\mathbf{G}_j^* = (\varepsilon_j'\Sigma_j^{-1}\varepsilon_j + \mathbf{G}_j^{-1})^{-1}. \tag{5}$$

Furthermore, for any choice of the control value u_j, the next state X_{j+1} will have a distribution for which the mean vector m_{j+1} and the covariance matrix G_{j+1} are

$$m_{j+1} = \alpha_j m_j^* + \beta_j + u_j \tag{6}$$

and

$$G_{j+1} = \alpha_j G_j^* \alpha_j' + \Gamma_j. \tag{7}$$

When Y_j is regarded as a random vector, then the mean vector m_j^* specified by Eq. (4) becomes a random vector whose mean vector $E_{j-1}(m_j^*)$ and covariance matrix $\text{Cov}_{j-1}(m_j^*)$ are

$$E_{j-1}(m_j^*) = m_j \tag{8}$$

and

$$\text{Cov}_{j-1}(m_j^*) = G_j - (\epsilon_j' \Sigma_j^{-1} \epsilon_j + G_j^{-1})^{-1}. \tag{9}$$

The subscript $j - 1$ in Eqs. (8) and (9) again indicates that these values are computed after the value of u_{j-1} has been chosen but before the value of Y_j has been observed.

For $j = 1, \ldots, n$, we shall let $L_j(m_j^*)$ denote the expected value of the sum $\sum_{i=j}^{n} \lambda_i$ when the posterior mean vector of X_j is m_j^* and when u_j and all subsequent values of the control variable are chosen in an optimal fashion. If we define the function L_{n+1} to be identically 0, then the following relation will be satisfied for $j = 1, \ldots, n$:

$$L_j(m_j^*) = \inf_{u_j} E_j[\lambda_j + L_{j+1}(m_{j+1}^*)]. \tag{10}$$

We shall now prove by an induction argument that for $j = 1, \ldots, n$, the function L_j must have the following form:

$$L_j(m_j^*) = (m_j^* - b_{j-1})' a_{j-1}(m_j^* - b_{j-1}) + c_{j-1}. \tag{11}$$

Here, a_{j-1} will be a specified symmetric $k \times k$ nonnegative definite matrix, b_{j-1} will be a specified k-dimensional vector, and c_{j-1} will be a specified number. Furthermore, we shall show that the optimal value u_j of the control vector is

$$u_j = (q_j + a_j + r_j)^{-1}[q_j t_j + a_j b_j - (q_j + a_j)(\alpha_j m_j^* + \beta_j)]. \tag{12}$$

We shall assume that the matrix $q_j + a_j$ is nonsingular for $j = 1, \ldots, n$ (see Exercise 9). Hence, the matrix $q_j + a_j + r_j$ will also be nonsingular.

Since the function L_{n+1} is identically 0, it has the form specified in Eq. (11) with $a_n = 0$, $b_n = 0$, and $c_n = 0$. Now suppose that for some value of j ($j = 1, \ldots, n$), the function L_{j+1} has the form specified in

Eq. (11). Then, by Eqs. (3) and (8) and Exercise 11 of Chap. 11,

$$E_j[\lambda_j + L_{j+1}(\mathbf{m}_{j+1}^*)] = (\mathbf{m}_{j+1} - \mathbf{t}_j)'\mathbf{q}_j(\mathbf{m}_{j+1} - \mathbf{t}_j) + \text{tr } (\mathbf{q}_j G_{j+1})$$
$$+ \mathbf{u}_j'\mathbf{r}_j\mathbf{u}_j + (\mathbf{m}_{j+1} - \mathbf{b}_j)'\mathbf{a}_j(\mathbf{m}_{j+1} - \mathbf{b}_j)$$
$$+ \text{tr } [\mathbf{a}_j \text{ Cov}_j(\mathbf{m}_{j+1}^*)] + c_j. \quad (13)$$

When $\mathbf{m}_{j+1}$ is replaced by the value given in Eq. (6), we find that the right side of Eq. (13) is a quadratic function of the vector $\mathbf{u}_j$. The minimizing value of $\mathbf{u}_j$ is then found to be as specified by Eq. (12).

When this optimum value of $\mathbf{u}_j$ is substituted into the right side of Eq. (13), the value of $L_j(\mathbf{m}_j^*)$ is obtained, and it can be seen that this value has the form required by Eq. (11). The induction argument is now complete. After some algebraic simplification, it can be found that the values $\mathbf{a}_{j-1}$ and $\mathbf{b}_{j-1}$ which appear in Eq. (11) will satisfy the following relations:

$$\mathbf{a}_{j-1} = \alpha_j'[(\mathbf{q}_j + \mathbf{a}_j) - (\mathbf{q}_j + \mathbf{a}_j)(\mathbf{q}_j + \mathbf{a}_j + \mathbf{r}_j)^{-1}(\mathbf{q}_j + \mathbf{a}_j)]\alpha_j,$$
$$\mathbf{b}_{j-1} = \alpha_j^{-1}[(\mathbf{q}_j + \mathbf{a}_j)^{-1}(\mathbf{q}_j\mathbf{t}_j + \mathbf{a}_j\mathbf{b}_j) - \beta_j]. \quad (14)$$

The value of c_{j-1} is not needed in Eq. (12) in order to be able to determine the optimal value $\mathbf{u}_j$ of the control vector, and therefore it is omitted here. Equations (12) and (14) can be checked against the corresponding one-dimensional results given by Eq. (10) of Sec. 14.11 and Eq. (9) of Sec. 14.10. Together with the values $\mathbf{a}_n = \mathbf{0}$ and $\mathbf{b}_n = \mathbf{0}$, Eqs. (12) and (14) determine the optimal sequence $\mathbf{u}_1, \ldots, \mathbf{u}_n$.

14.13 CONTROL PROBLEMS WITH ACTUATION ERRORS

In this section we shall consider again a one-dimensional control process in which the exact state of the system can be observed at each stage. However, we shall now consider a problem which incorporates the following realistic feature: By exercising control over the system at any stage, the statistician introduces into the system a random actuation error whose magnitude will depend on the magnitude of the control variable. This actuation error will affect the next state of the system. Hence, for $j = 1, \ldots, n$, the value u_j of the control variable will affect not only the mean of the state X_{j+1} at the next stage but also its variance.

Specifically, we shall assume that the process evolves according to the following system of equations:

$$X_{j+1} = \alpha_j X_j + \beta_j + u_j + H_j u_j + Z_j \quad j = 1, \ldots, n. \quad (1)$$

Here, the constants α_j and β_j, the control variable u_j, and the random disturbance Z_j have the same interpretation as in Sec. 14.10. The product of the random variable H_j and the control variable u_j in Eq. (1) represents the actuation error. It is assumed that $E(H_j) = 0$ and that

$\text{Var}(H_j) = \eta_j^2$. Furthermore, it is assumed that the initial position $X_1 = x_1$ is fixed and that the random variables $H_1, \ldots, H_n$ and $Z_1, \ldots, Z_n$ are all independent.

For any given state $X_j = x_j$ and any control value u_j, the mean of the next state X_{j+1} will again be $\alpha_j x_j + \beta_j + u_j$, as in the problem considered in Sec. 14.10, but its variance will now be $u_j^2 \eta_j^2 + \gamma_j^2$, whereas in the previous problem it was simply γ_j^2. However, if the total loss is again represented by the sum $\sum_{j=1}^n \lambda_j$, where λ_j is given by Eq. (2) of Sec. 14.10, it will be shown that the optimal control value u_j at each stage will again be a linear function of the state x_j at that stage. The present problem is a generalization of the problem in Sec. 14.10 since that problem can be obtained by assuming that the variance $\eta_j^2 = 0$ for $j = 1, \ldots, n$.

If the functions $L_j(x_j)$ for $j = 1, \ldots, n+1$ are defined as in Sec. 14.10, then these functions must again satisfy Eq. (3) of Sec. 14.10. It can again be shown by an induction argument that $L_j(x_j)$ must be a quadratic function of the form indicated in Eq. (4) of Sec. 14.10. In the present problem, $a_n = b_n = c_n = 0$, and for $j = 1, \ldots, n$, the following relation must be satisfied:

$$a_{j-1} = \frac{\alpha_j^2(q_j + a_j)[r_j + \eta_j^2(q_j + a_j)]}{(q_j + a_j)(1 + \eta_j^2) + r_j}. \tag{2}$$

The values of b_{j-1} and c_{j-1} in this problem are again specified by Eq. (9) of Sec. 14.10. Furthermore, it can be established by the induction argument that the optimal value u_j of the control variable, for $j = 1, \ldots, n$, is as follows:

$$u_j = \frac{q_j l_j + a_j b_j - (q_j + a_j)(\alpha_j x_j + \beta_j)}{(q_j + a_j)(1 + \eta_j^2) + r_j}. \tag{3}$$

The derivation of these results is required in Exercise 10.

As in Sec. 14.10, the optimal sequence $u_1, \ldots, u_n$ does not depend on the values of the variances $\gamma_1^2, \ldots, \gamma_n^2$. Hence, these values need not actually be known by the statistician. Furthermore, the values $u_1, \ldots, u_n$ specified by Eq. (3) will be optimal provided only that the random variables $H_1, \ldots, H_n$ and $Z_1, \ldots, Z_n$ are independent and that their first two moments satisfy the given requirements. No further assumptions about the distributions of these variables need be made.

Further Remarks and References

The book by Aoki (1967) discusses many problems of the types considered in Secs. 14.10 to 14.13 and their extensions. It also contains a large bibliography on control theory. Other books on this topic are those by Tou (1964), Fel'dbaum (1966), and Sawaragi, Sunahara, and Nakamizo

(1967). Kleindorfer and Kleindorfer (1967) have considered an extension of the problem described in Sec. 14.12 in which the loss functions can include both linear terms and quadratic terms. Vande Linde (1967) has considered an extension in which at each stage the statistician can choose from among several control variables having different characteristics and costs.

In many control problems, some of the coefficients defining the process are unknown and a prior distribution must be assigned to their values. In a problem of this type, the optimal procedure typically will not have one of the simple linear forms developed here. Some of these problems are considered in the books mentioned above.

14.14 SEARCH PROBLEMS

One of the interesting features of the search problems which will be studied in this section and the next one is that although it seems to be very difficult to solve the basic functional equations, the optimal procedures can nevertheless be found by other means, and they have very simple forms.

Suppose that an object is hidden in one of r possible locations ($r \geq 2$), and let p_i be the prior probability that the object is in location i. Here, $p_i > 0$ $(i = 1, \ldots, r)$ and $\sum_{i=1}^{r} p_i = 1$. We shall assume that the statistician must find the object but that he can search in only one location at a time. Hence, he must devise a sequential search procedure which specifies at each stage that a certain one of the r locations is to be searched.

We shall assume that even though the correct location is searched at some particular stage, there may be a positive probability that the object will be overlooked in this search. Specifically, we shall assume that even if the object is in location i $(i = 1, \ldots, r)$, there is probability α_i $(0 \leq \alpha_i < 1)$ that the object will not be found in a particular search of that location. Of course, if the object is not in location i, it certainly will not be found in any search of location i. The fact that the probability α_i of overlooking the object remains the same for every search of location i indicates that for any specified search procedure and any particular location of the object, the outcomes of all searches are independent. If $0 < \alpha_i < 1$ for some location i and if the object has not been found in several searches of that location, there remains only a small posterior probability that the object is actually in location i. Moreover, if $\alpha_i = 0$ for some location, then the statistician never has to make more than one search of that location.

The general process may be described as follows: Suppose that, at some stage, p_i is the probability that the object is in location i $(i = 1, \ldots, r)$. Also, suppose that when one of the locations, say location j, is searched, the object is not found. Then, for $i = 1, \ldots, r$, the posterior probability p_i^* that the object is in location i is

$$
\begin{cases}
p_j^* = \dfrac{p_j \alpha_j}{p_j \alpha_j + 1 - p_j}, \\[2mm]
p_i^* = \dfrac{p_i}{p_j \alpha_j + 1 - p_j} & \text{for } i \neq j.
\end{cases}
\tag{1}
$$

We shall assume that the statistician must continue searching until the object has been found and that the cost of each search of location i is $c_i > 0$ $(i = 1, \ldots, r)$. A procedure must be determined that will minimize the expected total cost of the searching process.

For any probabilities $p_1, \ldots, p_r$ such that $\Sigma_{i=1}^r p_i = 1$, we shall let $L(p_1, \ldots, p_r)$ denote the expected total cost of the searching process when p_i is the probability that the object is in location i $(i = 1, \ldots, r)$. The functional equation that must be satisfied by L can be derived as follows:

Suppose that the first search is made in a certain location, say location j. Then the probability is $p_j(1 - \alpha_j)$ that the object will be found in the first search, and the searching process will then be terminated. On the other hand, the probability is $p_j \alpha_j + (1 - p_j)$ that the object will not be found in the first search. If it is not found, the posterior probabilities $p_1^*, \ldots, p_r^*$ are as specified by Eq. (1), and the expected cost of the remainder of the searching process when an optimal procedure is adopted is $L(p_1^*, \ldots, p_r^*)$. After we add the cost c_j of the first search, the expected total cost of searching in location j first and then continuing with an optimal procedure will be $c_j + (p_j \alpha_j + 1 - p_j)L(p_1^*, \ldots, p_r^*)$. Since one of the r locations must be searched first, $L(p_1, \ldots, p_r)$ must satisfy the following equation:

$$
\begin{aligned}
L(p_1, &\ldots, p_r) \\
&= \min_{j=1,\ldots,r} \{c_j + (p_j \alpha_j + 1 - p_j)L(p_1^*, \ldots, p_r^*)\}.
\end{aligned}
\tag{2}
$$

It can be shown that $L(p_1, \ldots, p_r) < \infty$ for any probabilities $p_1, \ldots, p_r$. In fact, since $L(p_1, \ldots, p_r)$ is the expected total cost of an optimal procedure, its value cannot be greater than the expected cost of the procedure under which locations 1 through r are searched cyclically until the object has been found. We shall derive a bound for the expected cost of this cyclic procedure.

Suppose first that the object is in location i. Then $1/(1 - \alpha_i)$ is the expected number of searches in location i which will be needed to find the object. Since the cost of each cycle of searches of all r locations is $c_1 + \cdots + c_r$, it follows that the expected cost of finding the object will not be more than $(c_1 + \cdots + c_r)/(1 - \alpha_i)$. If we compute the expectation over the r possible locations of the object, we find that the expected total cost cannot be more than

$$(c_1 + \cdots + c_r) \sum_{i=1}^{r} \frac{p_i}{1 - \alpha_i}. \tag{3}$$

Hence, this value is an upper bound for the value of $L(p_1, \ldots, p_r)$.

As mentioned above, it seems to be very difficult to solve the functional equation (2). We shall therefore proceed along a different line of attack. For any sequential procedure and for $i = 1, \ldots, r$ and $j = 1, 2, \ldots$, we shall let Π_{ij} denote the probability under any procedure that the object will be found for the first time and the search cease during the jth search of location i. Then, regardless of the number of times other locations have been searched before the jth search of location i is made, the probability Π_{ij} is

$$\Pi_{ij} = p_i \alpha_i^{j-1}(1 - \alpha_i) \qquad i = 1, \ldots, r;$$
$$j = 1, 2, \ldots. \tag{4}$$

Every search procedure specifies, at each stage n ($n = 1, 2, \ldots$), that a certain one of the locations $1, \ldots, r$ should be searched if the object has not yet been found. For any given procedure δ, we shall let γ_n denote the cost of the search that is made at the nth stage, and we shall let λ_n denote the probability that the object will be found at the nth stage ($n = 1, 2, \ldots$). For instance, if δ specifies that location i should be searched for the jth time at the nth stage, then $\gamma_n = c_i$ and $\lambda_n = \Pi_{ij}$. As usual, we shall let N denote the total number of searches required to find the object. Then the expected total cost $\rho(\delta)$ of finding the object can be written as follows:

$$\rho(\delta) = \sum_{n=1}^{\infty} \left(\sum_{m=1}^{n} \gamma_m \right) \lambda_n = \sum_{m=1}^{\infty} \left(\sum_{n=m}^{\infty} \lambda_n \right) \gamma_m$$
$$= \sum_{m=1}^{\infty} \gamma_m \Pr(N \geq m)$$
$$= \gamma_1 + \gamma_2(1 - \lambda_1) + \gamma_3(1 - \lambda_1 - \lambda_2) + \cdots. \tag{5}$$

The final result in Eq. (5) has the advantage that it is appropriate even for a procedure δ under which there is positive probability of continuing indefinitely without ever finding the object. For such a procedure, the expected cost $\rho(\delta)$ will be infinite.

We shall now show that the optimal procedure specifies that the search should be performed so that the following relation is satisfied:

$$\frac{\lambda_1}{\gamma_1} \geq \frac{\lambda_2}{\gamma_2} \geq \cdots \geq \frac{\lambda_n}{\gamma_n} \geq \cdots. \tag{6}$$

In other words, if all values of the ratio Π_{ij}/c_i for all values of i and j are arranged in order of decreasing magnitude, then this ordering is the sequence in which the searches should be made. For instance, if a particular ratio Π_{ij}/c_i is the nth largest value in the ordering, then the jth search of location i should be made at the nth stage. Before we proceed with the proof of the optimality of this procedure, we shall verify the fact that the procedure is feasible in the sense that it does not specify, for example, making the fourth search of some location before it has specified making the third search of that location. It can be seen from Eq. (4) that the procedure is feasible because, for each location i, the probability Π_{ij} decreases as j increases and the cost c_i remains fixed. Indeed, this same fact guarantees that it is always possible to arrange all the values of Π_{ij}/c_i in a nonincreasing sequence.

Suppose now that δ is a procedure for which $\rho(\delta) < \infty$ and for which, for some value of n, the following relation is satisfied:

$$\frac{\lambda_n}{\gamma_n} < \frac{\lambda_{n+1}}{\gamma_{n+1}}. \tag{7}$$

Suppose also that δ specifies that location i should be searched at the nth stage and location k should be searched at the $(n + 1)$st stage. By the relation (7), we know that $i \neq k$. Let δ^* be the procedure which specifies that location k should be searched at the nth stage and location i should be searched at the $(n + 1)$st stage but which agrees with the procedure δ at all other stages. If γ_m^* and λ_m^* denote the cost and probability at the mth stage for the procedure δ^*, then the following relations must be satisfied:

$$\begin{aligned}
\gamma_n^* &= \gamma_{n+1} \quad \text{and} \quad \lambda_n^* = \lambda_{n+1}; \\
\gamma_{n+1}^* &= \gamma_n \quad \text{and} \quad \lambda_{n+1}^* = \lambda_n; \\
\gamma_m^* &= \gamma_m \quad \text{and} \quad \lambda_m^* = \lambda_m \quad \text{for } m \neq n \text{ and } m \neq n + 1.
\end{aligned} \tag{8}$$

We can now obtain the following result from Eqs. (5), (7), and (8):

$$\begin{aligned}
\rho(\delta) - \rho(\delta^*) = \;& \gamma_n(1 - \lambda_1 - \cdots - \lambda_{n-1}) \\
& + \gamma_{n+1}(1 - \lambda_1 - \cdots - \lambda_{n-1} - \lambda_n) \\
& - \gamma_{n+1}(1 - \lambda_1 - \cdots - \lambda_{n-1}) \\
& - \gamma_n(1 - \lambda_1 - \cdots - \lambda_{n-1} - \lambda_{n+1}) \\
= \;& \gamma_n\lambda_{n+1} - \gamma_{n+1}\lambda_n > 0.
\end{aligned} \tag{9}$$

Therefore, the expected total cost is smaller for the procedure δ^* than for the procedure δ.

We have just shown that if a search procedure does not satisfy the relation (6), then its expected total cost can be reduced. Therefore, the procedure cannot be optimal. We can now conclude that the procedure which specifies that the searching should be done in accordance with the relation (6) must be optimal. If, at any stage, two or more values of the ratio Π_{ij}/c_i are equal, these values may be ordered arbitrarily among themselves.

The method we have just used to derive the optimal procedure differs somewhat from the methods we have been using previously in this chapter. The dynamic point of view was almost completely abandoned because there was no updating of posterior distributions and all stages of the process were considered simultaneously rather than sequentially. However, it can be seen from Eqs. (1) and (4) that the optimal procedure we derived has the following simple and forceful dynamic interpretation. At any given stage of the process, let $p_1^*, \ldots, p_r^*$ be the current posterior probabilities that the object is in each of the r possible locations. Then the next search should be made in the location for which the value of $p_i^*(1 - \alpha_i)/c_i$ is a maximum. Since $p_i^*(1 - \alpha_i)$ is the probability of actually finding the object in a single search of location i, the optimal procedure at each stage is simply to search in the location for which there is the highest probability per unit search cost of finding the object in the next search.

Suppose now that the search costs c_i $(i = 1, \ldots, r)$ are equal. Under these conditions, the optimal procedure which has just been found will minimize the expected number of searches needed to find the object. Since the value of c_i is the same for each search, the optimal procedure is simply to arrange all the probabilities Π_{ij} determined by Eq. (4) in a decreasing sequence and at the nth stage to search in the location indicated by the nth term in the sequence. In the next section we shall show that this procedure also has other optimal properties.

14.15 SEARCH PROBLEMS WITH EQUAL COSTS

In this section we shall continue to assume that the search costs c_i $(i = 1, \ldots, r)$ are equal. Moreover, we shall now consider the possibility that the object is not in any of the r locations. Thus, if p_i again denotes the probability that the object is in location i $(i = 1, \ldots, r)$, then we shall now assume only that $\sum_{i=1}^{r} p_i \leq 1$. If there is positive probability that the object is not in any of the r locations, then $\sum_{i=1}^{r} p_i < 1$. However, if the object has not been found in a single search of location j, then the posterior probabilities $p_1^*, \ldots, p_r^*$ will again be specified by Eq. (1) of Sec. 14.14. Furthermore, the probability Π_{ij} of finding the object for the first time during the jth search of location i will again be specified by Eq. (4) of Sec. 14.14.

We shall now consider the problem of finding a search procedure which maximizes the probability of finding the object within a fixed number of searches n. Since Π_{ij} is the probability of finding the object on the jth search of location i, it follows that the optimal procedure is simply to make the n searches which correspond to the n largest values of Π_{ij}.

Since the searching process is carried out one stage at a time, we can return to the dynamic interpretation of the optimal procedure described near the end of Sec. 14.14. Here, the procedure is particularly simple, and the following discussion will show that its optimality is especially strong.

At any given stage of the process, let $p_1^*, \ldots, p_r^*$ be the current probabilities that the object is in each of the r possible locations. Then the next search should be made in the location for which the value of $p_i^*(1 - \alpha_i)$ is a maximum. Since this number is simply the probability of finding the object during the next search, if that search is made in location i, the optimal procedure is the following myopic procedure: At each stage, the search should be made in the location for which there is the highest probability that the object will be found at that stage. This procedure maximizes simultaneously for all positive integers n the probability that the object will be found in not more than n searches. If the object is actually in one of the r locations and the statistician continues searching until he finds the object, then, as was found in Sec. 14.14, this procedure minimizes the expected number of searches $E(N)$. In addition, we can now conclude that for any nondecreasing function g, this procedure also minimizes $E[g(N)]$.

We shall continue to assume that the object may not be in any of the r locations, and we shall now consider a problem in which the statistician is allowed to stop searching at any stage before the object has been found if he pays an additional cost. Specifically, we shall suppose that the cost c_i of each search of location i is 1 unit ($i = 1, \ldots, r$) and that the additional cost of terminating the search without having found the object is c ($c > 1$). A procedure must be determined which minimizes the expected total cost. In this problem, the statistician not only must choose a search procedure that specifies the location to be searched at each stage but also must choose a stopping rule that specifies when the searching process should be stopped if the object has not been found. As in Chap. 12, it is convenient to separate consideration of the search procedure from consideration of the stopping rule. Hence, we shall assume that any search procedure δ specifies the location to be searched at each stage of an infinite sequence. We shall now show that as long as the statistician continues searching, he should conform to the optimal procedure described in this section.

Suppose that the statistician decides to use a given search procedure δ and to stop the process if the object has not been found by the nth stage. Let N again denote the total number of searches which would be required to find the object when the procedure δ is followed without stopping, and let $N = \infty$ if the object would never be found. Then the expected total cost $\rho(\delta, n)$ can be written as follows:

$$\rho(\delta, n) = \sum_{m=1}^{n} m \Pr(N = m) + (n + c) \Pr(N > n)$$

$$= \sum_{m=1}^{n} \Pr(N \geq m) + c \Pr(N > n). \tag{1}$$

For any fixed value of n, the optimal procedure described earlier in this section simultaneously minimizes each of the probabilities $\Pr(N \geq m)$ for $m = 1, \ldots, n$ and $\Pr(N > n)$. Hence, it follows from Eq. (1) that regardless of the stopping rule used by the statistician, he should conform to that optimal search procedure as long as he continues searching. Thus, at each stage the search should be made in the location for which there is the highest probability that the object will be found at that stage.

It is somewhat more difficult to determine the optimal number of searches n after which the searching process should be terminated. This problem will not be discussed here, but it has been studied by Chew (1967).

Further Remarks and References

Search problems of the type we have discussed in this section and the preceding one have been studied by Staroverov (1963); by Matula (1964), who also describes the work of Blackwell on these problems; by Black (1965); by Chew (1967); and by Kadane (1968), on whose work much of our discussion is based. Similar problems have been considered by Bellman (1957a). Various other types of search problems have also been treated in the literature. Enslow (1966) has compiled a bibliography on this topic.

14.16 UNCERTAINTY FUNCTIONS AND STATISTICAL DECISION PROBLEMS

We shall now consider a statistical decision problem which is characterized by a parameter W whose values belong to the parameter space Ω, by a decision space D, and by a nonnegative loss function defined on the product space $\Omega \times D$. As in the earlier chapters, we shall let $\rho_0(\xi)$ denote the risk when the prior distribution of W is ξ and the statistician chooses the best decision in D, that is, the Bayes decision against ξ.

We shall suppose that before the statistician chooses a decision in D, he can gain information about the value of W by taking exactly n observations from some class of available random variables. The choices of the random variables to be observed at each of the n stages can be made sequentially. Specifically, we shall assume that there is a given class $\mathcal{F}$ of random variables X, each of which is characterized by a family of g.p.d.f.'s $\{f_X(\cdot \mid w), w \in \Omega\}$, where $f_X(\cdot \mid w)$ is the g.p.d.f. of X when $W = w$. At each of the n stages, the statistician selects a random variable from the class $\mathcal{F}$ and observes its value. As usual, we shall assume that the outcomes of the n observations are independent in the following sense: Suppose that the random variable X is chosen for observation at some given stage. Then, regardless of the choices and outcomes of the observations at all preceding stages, the conditional g.p.d.f. of the observation at the given stage, when $W = w$, is $f_X(\cdot \mid w)$. Furthermore, we shall assume that the statistician can take independent observations (in the above sense) at different stages on the same random variable $X \in \mathcal{F}$. Thus, if the statistician wishes to do so, he can take all n observations on a single random variable $X \in \mathcal{F}$.

In different but equivalent terms, each of the random variables $X \in \mathcal{F}$ represents an *experiment* which the statistician can perform, and any experiment in $\mathcal{F}$ can be replicated at different stages. Thus, the phrase "performing the experiment X" and the phrase "taking an observation on the random variable X" are two ways of expressing the same idea.

We shall let X_j denote the random variable selected from the class $\mathcal{F}$ for observation at the jth stage $(j = 1, 2, \ldots, n)$. Also, for any prior distribution ξ of W, we shall let $\xi(X_1, \ldots, X_n)$ denote the posterior distribution of W at the end of the nth stage. When the sequential procedure for selecting the random variables $X_1, \ldots, X_n$ has been specified but the values of these variables have not yet been observed, this posterior distribution should be regarded as a random distribution. The problem is to determine a sequential procedure that minimizes the expected terminal risk $E\{\rho_0[\xi(X_1, \ldots, X_n)]\}$.

In order to simplify the discussion, we shall assume throughout this section and the next one that the parameter W can have only a finite number of values. We shall suppose therefore that Ω contains just k points, that is, that $\Omega = \{w_1, \ldots, w_k\}$. Also, we shall let Ξ denote the set of all probability distributions ξ on Ω. Thus, Ξ is the $(k - 1)$-dimensional simplex of vectors $\xi = (\xi_1, \ldots, \xi_k)$ such that $\xi_i \geq 0$ $(i = 1, \ldots, k)$ and $\xi_1 + \cdots + \xi_k = 1$.

We know from Theorem 1 of Sec. 8.4 that the risk function ρ_0 is a nonnegative, concave function on the set Ξ. In other words, for any two distributions $\phi_1 \in \Xi$ and $\phi_2 \in \Xi$ and for any constant α $(0 \leq \alpha \leq 1)$, the following relation must be satisfied:

$$\rho_0[\alpha\phi_1 + (1 - \alpha)\phi_2] \geq \alpha\rho_0(\phi_1) + (1 - \alpha)\rho_0(\phi_2). \tag{1}$$

Since nonnegativity and concavity are the only properties of ρ_0 which will be used in this section and the next one, we can widen the scope of our discussion by introducing the following definition: Any nonnegative, concave function Υ which is defined on the set Ξ of all possible distributions of W is called an *uncertainty function*. The problem to be considered here can now be formulated as follows: For any given prior distribution ξ of W, for any class $\mathfrak{F}$ of possible observations, for any uncertainty function Υ defined on the set Ξ, and for any fixed number of observations n, the statistician must determine a procedure that minimizes the expected terminal uncertainty $E\{\Upsilon[\xi(X_1, \ldots, X_n)]\}$.

In a problem in which the decision space D and the loss function L are specified, the uncertainty function Υ will be simply the risk ρ_0. In some problems, however, the statistician can assign the uncertainty function Υ directly, without explicitly introducing the more detailed structure of a statistical decision problem. One useful and popular choice is the *entropy function*, which is defined for any distribution $\xi = (\xi_1, \ldots, \xi_k)$ as follows:

$$\Upsilon(\xi) = -\sum_{i=1}^{k} \xi_i \log \xi_i. \tag{2}$$

This function is basic to the mathematical theory of information and communication [see Shannon and Weaver (1949) and Khinchin (1957)]. Another important uncertainty function is defined by the equation $\Upsilon(\xi) = 1 - \max\{\xi_1, \ldots, \xi_k\}$.

Since the process has a fixed finite number of stages, an optimal procedure can easily be characterized by backward induction. Suppose that after the statistician has selected the random variables $X_1, \ldots, X_{n-1}$ and has observed their values $x_1, \ldots, x_{n-1}$, the posterior distribution of W is ξ_{n-1}. Then $\xi_{n-1}(X_n)$ will be the posterior distribution after all n observations have been taken, and an optimal choice of the final random variable X_n will be one for which

$$E\{\Upsilon[\xi_{n-1}(X_n)]\} = \inf_{X \in \mathfrak{F}} E\{\Upsilon[\xi_{n-1}(X)]\}. \tag{3}$$

We shall let $\Upsilon_0 = \Upsilon$ and shall define the function Υ_1 on the set Ξ as follows:

$$\Upsilon_1(\phi) = \inf_{X \in \mathfrak{F}} E\{\Upsilon_0[\phi(X)]\} \qquad \text{for } \phi \in \Xi. \tag{4}$$

Then, if ξ_{n-1} is the posterior distribution of W when just one observation remains to be taken, $\Upsilon_1(\xi_{n-1})$ will be the minimum value of the expected uncertainty which can be obtained.

In general, we shall let $\Upsilon_1, \Upsilon_2, \ldots, \Upsilon_n$ be functions on Ξ which are defined recursively by the following relation:

$$\Upsilon_{j+1}(\phi) = \inf_{X \in \mathfrak{F}} E\{\Upsilon_j[\phi(X)]\} \qquad \text{for } \phi \in \Xi. \tag{5}$$

Then, if ϕ is the posterior distribution of W when j observations remain to be taken, $\Upsilon_j(\phi)$ will be the minimum value of the expected terminal uncertainty. In particular, if ξ is the prior distribution of W, then $\Upsilon_n(\xi)$ will be the minimum value of $E\{\Upsilon[\xi(X_1, \ldots, X_n)]\}$ that can be obtained from any sequential procedure.

An optimal procedure at the first stage is to select a random variable $X_1 \in \mathfrak{F}$ for which

$$\Upsilon_n(\xi) = E\{\Upsilon_{n-1}[\xi(X_1)]\}. \tag{6}$$

At the $(j + 1)$st stage, after the statistician has observed the values $X_1 = x_1, \ldots, X_j = x_j$ and has computed the posterior distribution ξ_j, the optimal procedure is to select a random variable $X_{j+1} \in \mathfrak{F}$ for which

$$\Upsilon_{n-j}(\xi_j) = E\{\Upsilon_{n-j-1}[\xi_j(X_{j+1})]\}. \tag{7}$$

Although the optimal sequential procedure can be constructed in this way theoretically, the actual computation may be quite difficult. Bradt and Karlin (1956) have studied some special problems of this type. They give examples which show how complicated the optimal procedure can be, even in a problem having an extremely simple appearance (also, see Exercise 17). However, as we shall see in the next section, there are some problems in which the optimal procedure has a very simple form.

Further Remarks and References

Much of the material in this section and the next one is based on the papers by DeGroot (1962, 1966).

The theory of the sequential selection of experiments for large samples has been developed by Chernoff (1959, 1960, 1961b) and by Albert (1961), Bessler (1960), Abramson (1966), and Bohrer (1966). Anderson (1964a) and Whittle (1965) have also studied related problems.

A formal notion of the amount of information which can be obtained from an experiment can be introduced in terms of an uncertainty function. For any given uncertainty function Υ on the set Ξ, the *expected amount of information* $I(X, \Upsilon, \xi)$ in an experiment X, when the distribution of W is $\xi \in \Xi$, is defined to be the expected reduction in uncertainty which results from performing the experiment X. In other words, the information is defined by the following equation:

$$I(X, \Upsilon, \xi) = \Upsilon(\xi) - E\{\Upsilon[\xi(X)]\}. \tag{8}$$

This concept has been studied by Lindley (1956, 1957), when it is assumed that Υ is the entropy function defined by Eq. (2). Of course, there are a variety of definitions and uses of the concept of information in the theory of statistical inference, such as the Fisher information matrix defined by

Eq. (2) of Sec. 10.11. Various definitions have been discussed by Kullback (1959), Sakaguchi (1957, 1959, 1964, 1966), Mallows (1959), Rényi (1961, 1964, 1966, 1967), and Hackleman (1967). Other discussions are those by Kempthorne (1966) and Kullback (1967).

Economic aspects of gaining information have been studied by Marschak (1954, 1959, 1963b, 1964), by Marschak and Miyasawa (1968), and by Radner (1961, 1962).

14.17 SUFFICIENT EXPERIMENTS

We shall now consider again the optimal sequential procedure for selecting n experiments, as developed in Sec. 14.16. In the simplest possible form of this optimal procedure, all n observations are to be taken on the same random variable $X \in \mathfrak{F}$. In this section we shall consider the conditions under which the optimal procedure will have this special form.

Theorem 1 *Suppose that there exists a random variable $X^* \in \mathfrak{F}$ such that, for any distribution $\phi \in \Xi$ and any other random variable $X \in \mathfrak{F}$,*

$$E\{T[\phi(X^*)]\} \leq E\{T[\phi(X)]\}. \tag{1}$$

Then, for any prior distribution ξ, a sequential procedure that minimizes the expected terminal uncertainty $E\{T[\xi(X_1, \ldots, X_n)]\}$ is to take all n observations on the random variable X^*.

Proof It follows from the relation (1) and the discussion in regard to Eq. (3) of Sec. 14.16 that the nth observation should always be made on the random variable X^*, regardless of which random variables have been selected at the earlier stages and regardless of their observed values. Hence, for any distribution $\phi \in \Xi$,

$$T_1(\phi) = E\{T[\phi(X^*)]\}. \tag{2}$$

Suppose that for any distribution $\phi \in \Xi$ and any other random variable $X \in \mathfrak{F}$, it can be shown that

$$E\{T_1[\phi(X^*)]\} \leq E\{T_1[\phi(X)]\}. \tag{3}$$

It will then follow by similar reasoning that the $(n-1)$st observation should always be made on the random variable X^*. When the same argument is used at each stage, it will follow by induction that all n observations should be made on X^*. We shall now show that the relation (3) is satisfied.

For any given distribution $\phi \in \Xi$, we shall let $\phi_1^* = \phi(X^*)$ denote the posterior distribution after an observation has been taken on the random variable X^*. Then, for any random variable $X \in \mathfrak{F}$, the following relation can be obtained:

$$E\{T[\phi(X^*, X)]\} = E(E\{T[\phi(X^*, X)]|X^*\}) = E(E\{T[\phi_1^*(X)]|\phi_1^*\})$$
$$\geq E(E\{T[\phi_1^*(X^*)]|\phi_1^*\}) = E[T_1(\phi_1^*)] = E\{T_1[\phi(X^*)]\}. \quad (4)$$

Similarly, for any given distribution $\phi \in \Xi$ and any random variable $X \in \mathcal{F}$, we shall let $\phi_1 = \phi(X)$ denote the posterior distribution after an observation has been taken on X. We know that $\phi(X^*, X) = \phi(X, X^*)$, because the posterior distribution will be the same regardless of the order in which the observations are taken. Hence, we can obtain the following relation:

$$E\{T[\phi(X^*, X)]\} = E\{T[\phi(X, X^*)]\} = E(E\{T[\phi(X, X^*)]|X\})$$
$$= E(E\{T[\phi_1(X^*)]|\phi_1\}) = E[T_1(\phi_1)] = E\{T_1[\phi(X)]\}. \quad (5)$$

Together, the relations (4) and (5) yield the relation (3).■

We shall now consider a basic concept of experimentation that yields conditions under which the relation (1) will be satisfied. Let X and Y be random variables, or experiments, in the class $\mathcal{F}$ whose values are in the sample spaces S_X and S_Y, respectively. It is said that the experiment Y is *sufficient* for the experiment X if there exists a nonnegative function h on the product space $S_X \times S_Y$ for which the following three relations are satisfied:

$$f_X(x|w) = \int_{S_Y} h(x, y) f_Y(y|w) \, d\mu(y) \qquad \text{for } w \in \Omega \text{ and } x \in S_X, \quad (6)$$

$$\int_{S_X} h(x, y) \, d\mu(x) = 1 \qquad \text{for } y \in S_Y, \quad (7)$$

and

$$0 < \int_{S_Y} h(x, y) \, d\mu(y) < \infty \qquad \text{for } x \in S_X. \quad (8)$$

A nonnegative function h which satisfies Eq. (7) is called a *stochastic transformation* from Y to X. For each fixed value $y \in S_Y$, the function $h(\cdot, y)$ is a g.p.d.f. on S_X. Since this function does not involve the parameter W, a point $x \in S_X$ could be generated in accordance with this g.p.d.f. by means of an auxiliary randomization. Thus, the import of Eq. (6) is that Y is sufficient for X if, regardless of the value of the parameter W, an observation on Y and an auxiliary randomization make it possible to generate a random variable which has the same distribution as X. The integrability condition on h in the relation (8) is introduced as a technical convenience. These concepts were developed by Blackwell (1951, 1953).

It is intuitively clear that if Y is sufficient for X, then the statistician should never perform the experiment X when Y is available because performing X is equivalent to performing Y and then subjecting the outcome to a random transformation that can only obscure any information about the value of W which may have been contained in that outcome. The next theorem formalizes the statement that the experiment Y must be at least as informative as the experiment X. We shall first give a helpful lemma, and we shall introduce some convenient notation which will be used to expedite the proofs of the lemma and the theorem.

Let A be the set of all vectors $\mathbf{a} = (a_1, \ldots, a_k)$ such that $a_i \geq 0$ ($i = 1, \ldots, k$). Then Ξ is the subset of A containing all vectors $\mathbf{a}$ such that $\Sigma_{i=1}^{k} a_i = 1$. For any vectors $\mathbf{a} \in A$ and $\mathbf{b} \in A$, we shall let $\mathbf{a} \cdot \mathbf{b} = \Sigma_{i=1}^{k} a_i b_i$. Also, if $\mathbf{a} \cdot \mathbf{b} > 0$, we shall let $\mathbf{a} \otimes \mathbf{b}$ be the vector in Ξ which is specified by the following equation:

$$\mathbf{a} \otimes \mathbf{b} = \frac{1}{\mathbf{a} \cdot \mathbf{b}} (a_1 b_1, \ldots, a_k b_k). \tag{9}$$

If $\mathbf{a} \cdot \mathbf{b} = 0$, we can let $\mathbf{a} \otimes \mathbf{b}$ be defined as any arbitrary vector in Ξ.

Lemma 1 *Let Υ be an uncertainty function, and let $\phi \in \Xi$ be any fixed vector. Also, let the function v be defined on the set A by the following relation:*

$$v(\mathbf{a}) = (\phi \cdot \mathbf{a})\Upsilon(\phi \otimes \mathbf{a}) \qquad \text{for } \mathbf{a} \in A. \tag{10}$$

Then v is a concave function on the set A.

Proof Consider any vectors $\mathbf{a} \in A$ and $\mathbf{b} \in A$ and any positive constants α and β such that $\alpha + \beta = 1$. It must be shown that

$$v(\alpha\mathbf{a} + \beta\mathbf{b}) \geq \alpha v(\mathbf{a}) + \beta v(\mathbf{b}). \tag{11}$$

If either $\phi \cdot \mathbf{a} > 0$ or $\phi \cdot \mathbf{b} > 0$, then an elementary computation yields the following result:

$$v(\alpha\mathbf{a} + \beta\mathbf{b}) = [\alpha(\phi \cdot \mathbf{a}) + \beta(\phi \cdot \mathbf{b})]\Upsilon[\alpha^*(\phi \otimes \mathbf{a}) + \beta^*(\phi \otimes \mathbf{b})],$$

$$\tag{12}$$

where

$$\alpha^* = \frac{\alpha(\phi \cdot \mathbf{a})}{\alpha(\phi \cdot \mathbf{a}) + \beta(\phi \cdot \mathbf{b})} \qquad \text{and} \qquad \beta^* = 1 - \alpha^*. \tag{13}$$

Since Υ is concave, it can be seen from Eqs. (12) and (13) that the following relation must be satisfied:

$$\begin{aligned} v(\alpha\mathbf{a} + \beta\mathbf{b}) &\geq [\alpha(\phi \cdot \mathbf{a}) + \beta(\phi \cdot \mathbf{b})][\alpha^*\Upsilon(\phi \otimes \mathbf{a}) + \beta^*\Upsilon(\phi \otimes \mathbf{b})] \\ &= \alpha(\phi \cdot \mathbf{a})\Upsilon(\phi \otimes \mathbf{a}) + \beta(\phi \cdot \mathbf{b})\Upsilon(\phi \otimes \mathbf{b}) \\ &= \alpha v(\mathbf{a}) + \beta v(\mathbf{b}). \end{aligned} \tag{14}$$

Hence, the relation (11) is satisfied. Finally, if both $\phi \cdot \mathbf{a} = 0$ and $\phi \cdot \mathbf{b} = 0$, then each side of the relation (11) vanishes.∎

Theorem 2 *Suppose that the experiment $Y \epsilon \mathfrak{F}$ is sufficient for the experiment $X \epsilon \mathfrak{F}$. Then, for any uncertainty function Υ and any distribution $\phi \epsilon \Xi$,*

$$E\{\Upsilon[\phi(X)]\} \geq E\{\Upsilon[\phi(Y)]\}. \tag{15}$$

Proof Any nonnegative function $g(\cdot)$ defined on the parameter space Ω can be regarded as a vector $(g_1, \ldots, g_k)$ in the set A if we let $g_i = g(w_i)$ for $i = 1, \ldots, k$. We shall utilize this convention in the proof without further discussion of it.

Since Y is sufficient for X, there exists a nonnegative function h on the product space $S_X \times S_Y$ for which the relations (6) to (8) are satisfied. For each point $x \epsilon S_X$, we shall let $\psi(\cdot | x)$ be the function which is defined at each point in Ω as follows:

$$\psi(w_i | x) = \frac{\int_{S_Y} h(x, y) f_Y(y | w_i)\, d\mu(y)}{\int_{S_Y} h(x, y)\, d\mu(y)} \qquad \text{for } i = 1, \ldots, k. \tag{16}$$

From Eq. (10) and the usual formula for the posterior distribution $\phi(X)$, it can be shown that the following relation must be satisfied:

$$E\{\Upsilon[\phi(X)]\} = \int_{S_X} v[f_X(x | \cdot)]\, d\mu(x). \tag{17}$$

By Eqs. (6) and (16), this relation can be rewritten as follows:

$$E\{\Upsilon[\phi(X)]\} = \int_{S_X} v[\psi(\cdot | x)] \left[\int_{S_Y} h(x, y)\, d\mu(y) \right] d\mu(x). \tag{18}$$

In Eq. (18), we have utilized the fact that $v(\alpha \mathbf{a}) = \alpha v(\mathbf{a})$ for all vectors $\mathbf{a} \epsilon A$ and all constants $\alpha \geq 0$.

For each point $x \epsilon S_X$, let $f^*(\cdot | x)$ denote the g.p.d.f. on the sample space S_Y defined as follows:

$$f^*(y | x) = \frac{h(x, y)}{\int_{S_Y} h(x, y)\, d\mu(y)} \qquad \text{for } y \epsilon S_Y. \tag{19}$$

Then it can be seen from Eq. (16) that for $i = 1, \ldots, k$, the value of $\psi(w_i | x)$ is the expectation of the function $f_Y(\cdot | w_i)$ when Y has the g.p.d.f. $f^*(\cdot | x)$ specified by Eq. (19). By Lemma 1, v is a concave function on the set A. Therefore, by an application of Jensen's inequality for a concave function of a k-dimensional random vector [see, e.g., Ferguson (1967), p. 76], we can obtain the following relation:

$$v[\psi(\cdot | x)] \geq \int_{S_Y} v[f_Y(y | \cdot)] f^*(y | x)\, d\mu(y). \tag{20}$$

Hence, from Eqs. (18) to (20), it follows that

$$E\{\Upsilon[\phi(X)]\} \geq \int_{S_X} \int_{S_Y} v[f_Y(y | \cdot)] h(x, y)\, d\mu(y)\, d\mu(x). \tag{21}$$

Finally, by reversing the order of integration in the relation (21) and using Eq. (7), we can obtain the following relation:

$$E\{\Upsilon[\phi(X)]\} \geq \int_{S_Y} v[f_Y(y| \cdot)] \, d\mu(y) = E\{\Upsilon[\phi(Y)]\}. \tag{22}$$

The equality in the relation (22) simply expresses Eq. (17) for the experiment Y.∎

The final result, which states that performing n replications of a sufficient experiment must be an optimal procedure, now follows immediately from Theorems 1 and 2.

Corollary 1 *Suppose that there exists a random variable $X^* \in \mathfrak{F}$ which is sufficient for any other random variable $X \in \mathfrak{F}$. Then, for any uncertainty function Υ on Ξ, for any prior distribution $\xi \in \Xi$, and for any positive integer n, the sequential procedure that minimizes the expected terminal uncertainty $E\{\Upsilon[\xi(X_1, \ldots , X_n)]\}$ is to take all n observations on the random variable X^*.*

14.18 EXAMPLES OF SUFFICIENT EXPERIMENTS

In this section we shall illustrate the concept of sufficient experiments by considering three examples.

EXAMPLE 1 Suppose that the parameter W can have only two values w_1 and w_2, and consider two random variables X and Y which are defined as follows. The random variable X can have only the values 0 and 1, with the following probabilities:

$$\begin{array}{ll} \Pr(X = 0|W = w_1) = \tfrac{1}{4}, & \Pr(X = 1|W = w_1) = \tfrac{3}{4}; \\ \Pr(X = 0|W = w_2) = \tfrac{3}{4}, & \Pr(X = 1|W = w_2) = \tfrac{1}{4}. \end{array} \tag{1}$$

If $W = w_1$, the random variable Y has a uniform distribution on the interval $(0, 1)$, and if $W = w_2$, the random variable Y has a uniform distribution on the interval $(\tfrac{1}{2}, \tfrac{3}{2})$.

In this example, Y is sufficient for X. To demonstrate this fact, we shall define a random variable $Z = z(Y)$ as follows:

$$Z = \begin{cases} 0 & \text{if } Y \geq \tfrac{3}{4}, \\ 1 & \text{if } Y < \tfrac{3}{4}. \end{cases} \tag{2}$$

Then it can be verified directly that Z has the same distribution as X if $W = w_1$ and that Z also has the same distribution as X if $W = w_2$.

An alternative demonstration that Y is sufficient for X can be based on an explicit construction of a function h satisfying relations (6) to (8) of Sec. 14.17.

Consider now a statistical decision problem in which the statistician must choose a decision d from a given decision space and in which his loss will depend on the value of W and the decision d which he chooses. Suppose that the statistician can take a fixed total of n observations sequentially before he chooses a decision d and that at each of the n stages he may take either an observation on X or an observation on Y. Then the optimal procedure is to take all n observations on Y.

EXAMPLE 2 Suppose that the parameter W can have only two values w_1 and w_2. Let X and Y be two random variables each of which can have only the values 0 and 1, and suppose that the p.f.'s are as follows:

$$
\begin{aligned}
f_X(1|w_1) &= \tfrac{1}{3} = 1 - f_X(0|w_1), \\
f_X(1|w_2) &= \tfrac{1}{2} = 1 - f_X(0|w_2); \\
f_Y(1|w_1) &= \tfrac{2}{3} = 1 - f_Y(0|w_1), \\
f_Y(1|w_2) &= \tfrac{1}{5} = 1 - f_Y(0|w_2).
\end{aligned}
\tag{3}
$$

We shall illustrate another method for determining whether one experiment is sufficient for another by proving that Y is sufficient for X.

According to Eq. (6) of Sec. 14.17, we must show that there is a nonnegative function $h(x, y)$, where $x = 0, 1$ and $y = 0, 1$, such that the following equations are satisfied both for $x = 0$ and for $x = 1$:

$$
\begin{aligned}
f_X(x|w_1) &= h(x, 0)f_Y(0|w_1) + h(x, 1)f_Y(1|w_1), \\
f_X(x|w_2) &= h(x, 0)f_Y(0|w_2) + h(x, 1)f_Y(1|w_2).
\end{aligned}
\tag{4}
$$

Since x can have either of the two values $x = 0$ or $x = 1$, Eqs. (4) represent four equations which are to be solved for the four numbers $h(x, y)$. However, $f_X(0|w_i) + f_X(1|w_i) = f_Y(0|w_i) + f_Y(1|w_i) = 1$ for $i = 1, 2$. Also, by Eq. (7) of Sec. 14.17, $h(0, 0) + h(1, 0) = h(0, 1) + h(1, 1) = 1$. Hence, it is sufficient to solve the pair of equations obtained by letting $x = 1$ in the relations (4). When the numerical values in Eq. (3) are used, the following equations are obtained:

$$
\begin{aligned}
\tfrac{1}{3} &= \tfrac{1}{3}h(1, 0) + \tfrac{2}{3}h(1, 1), \\
\tfrac{1}{2} &= \tfrac{4}{5}h(1, 0) + \tfrac{1}{5}h(1, 1).
\end{aligned}
\tag{5}
$$

The unique values found by solving these equations are $h(1, 0) = \tfrac{7}{12}$ and $h(1, 1) = \tfrac{1}{6}$. It follows that $h(0, 0) = \tfrac{5}{12}$ and $h(0, 1) = \tfrac{5}{6}$. Since each of these values is in the interval $[0, 1]$, the function h is a stochastic transformation from Y to X. Hence, Y is sufficient for X.

General conditions under which one random variable will be sufficient in a simple dichotomous experiment of this type are given in Exercise 19.

EXAMPLE 3 Suppose that the parameter W can have any value on the real line, and let X and Y be random variables which have the following distributions: For any given value w of W, the random variable X has a normal distribution with mean w and variance 3 and the random variable Y has a normal distribution with mean w and variance 1. Obviously, because of its smaller variance, an observation on Y yields more information about the value of W than does an observation on X. We shall formalize this statement by showing that Y is sufficient for X.

Let Z be a random variable which is independent of Y and W and which has a normal distribution with mean 0 and variance 2. Then, for any given value of W, the random variable $Y + Z$ has the same distribution as the random variable X. Hence, Y is sufficient for X.

Further Remarks and References

Sufficient experiments are discussed in the books by Blackwell and Girshick (1954), chap. 12, and by Lehmann (1959), sec. 3.4, and in the lecture notes by Sakaguchi (1964, 1966), which also include some of the other problems discussed in this and the preceding chapters. This topic has also been studied at a more abstract level by LeCam (1964), Strassen (1965), Morse and Sacksteder (1966), and Sacksteder (1967).

EXERCISES

1. Consider a betting problem of the type discussed in Sec. 14.4. Suppose that the probability is $\frac{1}{3}$ that the statistician will win on any play of a certain game and that his utility function U, as a function of his fortune y, is as follows:

$$U(y) = \begin{cases} y & \text{for } 0 \le y \le 1, \\ 3y - 2 & \text{for } y > 1. \end{cases}$$

Suppose also that the statistician's initial fortune is Y_0 and that the game is to consist of exactly two plays. Show that an optimal procedure is not to bet at all if $Y_0 \le \frac{2}{3}$ or if $Y_0 \ge \frac{4}{3}$ and to bet his entire fortune Y_0 on either the first play or the second play, but not on both, if $\frac{2}{3} < Y_0 < \frac{4}{3}$.

2. Consider a betting problem of the type discussed in Sec. 14.4. Suppose that the statistician's probability of winning on any particular play is $p > \frac{1}{2}$ and that his utility function U is

$$U(y) = y^\alpha \qquad y \ge 0.$$

Here $0 < \alpha < 1$. Suppose also that the game is to consist of exactly n plays. Show that an optimal procedure is to bet the fixed proportion β of his current fortune on each play of the game, where β is the unique solution of the following equation:

$$\left(\frac{1 - \beta}{1 + \beta}\right)^{1-\alpha} = \frac{1 - p}{p}.$$

3. Consider a two-armed-bandit problem in which the random variables X and Y can have only the values 0 and 1 and the p.f.'s are as specified by Eq. (2) of Sec. 14.6. Suppose that W_1 can have only the two values $\frac{2}{3}$ and $\frac{1}{3}$ and that $\Pr(W_1 = \frac{2}{3}) = \xi = 1 - \Pr(W_1 = \frac{1}{3})$. Suppose also that $W_2 = \frac{1}{2}$. Show that if

exactly two observations are to be taken, the first one should be taken on X if $\xi > \frac{4}{9}$ and that the expected value $V_2(\xi)$ of the sum of the two observations is

$$V_2(\xi) = \begin{cases} 1 & \text{for } \xi < \frac{4}{9}, \\ \dfrac{\xi}{2} + \dfrac{7}{9} & \text{for } \frac{4}{9} \leq \xi \leq \frac{2}{3}, \\ \dfrac{2}{3}(\xi + 1) & \text{for } \xi > \frac{2}{3}. \end{cases}$$

4. Consider a two-armed-bandit problem in which, for any given values $W_1 = w_1$ and $W_2 = w_2$, the random variable X has a normal distribution with mean w_1 and precision 1 and the random variable Y has a normal distribution with mean w_2 and precision 1. Suppose that under the prior distribution, the parameters W_1 and W_2 are independent and that W_i has a normal distribution with mean μ_i and precision τ_i ($i = 1, 2$). Let $\alpha(\tau) = [\tau(\tau + 1)]^{\frac{1}{2}}$ for $\tau > 0$, and let the function Ψ be defined by Eq. (1) of Sec. 11.9. Show that if exactly two observations are to be taken, the first one should be taken on X when the following relation is satisfied:

$$\alpha(\tau_2)\Psi[\alpha(\tau_1)(\mu_2 - \mu_1)] > \alpha(\tau_1)\Psi[\alpha(\tau_2)(\mu_1 - \mu_2)].$$

When this inequality is reversed, the first observation should be taken on Y. Show that this criterion implies, in particular, that the first observation should be taken on X when $\mu_1 \geq \mu_2$ and $\tau_1 \leq \tau_2$.

5. In the inventory problem defined by Eq. (2) of Sec. 14.8, let

$$G(y) = \int_0^y [1 - F(x)] \, dx \qquad y \geq 0.$$

If y_1^* satisfies Eq. (8) of Sec. 14.8, show that

$$V_1(y) = \begin{cases} pG(y_1^*) + c(y - y_1^*) & \text{if } y < y_1^*, \\ pG(y) & \text{if } y \geq y_1^*. \end{cases}$$

6. Let $p = 2$ and $c = 1$ in the inventory problem defined by Eq. (2) of Sec. 14.8, and suppose that f is the p.d.f. of the uniform distribution on the interval $(0, 1)$. Show that when two stages remain, the optimal value y_2^*, as defined in Theorem 1 of Sec. 14.8, is

$$y_2^* = \frac{\sqrt{7} - 1}{2}.$$

7. Consider the control problem described in Sec. 14.10. Suppose that $\alpha_j = 1$ and $\beta_j = 0$ for $j = 1, \ldots, n$. Suppose also that $q_j = q$, $r_j = r$, and $t_j = t$ for $j = 1, \ldots, n$, where q, r, and t are given numbers such that $q > 0$ and $r > 0$. (a) Show that $b_{j-1} = t$ for $j = 1, \ldots, n$. (b) Let the number a^* be defined as follows:

$$a^* = \frac{(q^2 + 4rq)^{\frac{1}{2}} - q}{2}.$$

Show that $a_0 \to a^*$ as $n \to \infty$. Hence, show that in a problem with a large number of stages, the optimal value u_1 of the control variable at the first stage is approximately as follows:

$$u_1 = \frac{q + a^*}{q + a^* + r}(t - x_1).$$

8. Consider the control problem described in Sec. 14.10. Suppose that $\alpha_j = 1$ and $\beta_j = 0$ for $j = 1, \ldots, n$. Suppose also that $q_n = q > 0$ and that $q_j = 0$ for

$j = 1, \ldots, n - 1$. Finally, suppose that $r_j = r > 0$ and $t_j = t$ for $j = 1, \ldots, n$.
(a) Show that for $j = 1, \ldots, n$,

$$a_{n-j} = \left(\frac{j}{r} + \frac{1}{q}\right)^{-1}.$$

(b) Show that $b_{j-1} = t$ for $j = 1, \ldots, n$. (c) Show that for $j = 1, \ldots, n$, the optimal value u_j of the control variable is

$$u_j = \frac{q}{r + (n - j + 1)q} (t - x_j).$$

9. Consider the control problem described in Sec. 14.12. Suppose that the matrix q_n is nonsingular and that each of the matrices α_j and r_j is nonsingular for $j = 1, \ldots, n$. Show that the matrix a_{j-1} must also be nonsingular for $j = 1, \ldots, n$.

10. Consider the control problem described in Sec. 14.13. Prove that $L_j(x_j) = a_{j-1}(x_j - b_{j-1})^2 + c_{j-1}$ for $j = 1, \ldots, n$, where $a_n = b_n = c_n = 0$, a_{j-1} satisfies Eq. (2) of Sec. 14.13 for $j = 1, \ldots, n$, and b_{j-1} and c_{j-1} satisfy Eq. (9) of Sec. 14.10. Prove also that for $j = 1, \ldots, n$, the optimal value u_j of the control variable is specified by Eq. (3) of Sec. 14.13.

11. Consider the control problem described in Sec. 14.13. Suppose that $\alpha_j = 1$ and $\beta_j = 0$ for $j = 1, \ldots, n$. Suppose also that $q_j = q$, $r_j = r$, $t_j = t$, and $\eta_j^2 = \eta^2$ for $j = 1, \ldots, n$, where $q > 0$, $r > 0$, and $\eta > 0$. (a) Show that $b_{j-1} = t$ for $j = 1, \ldots, n$. (b) Let the number a^* be defined as follows:

$$a^* = \frac{[q^2(1 + \eta^2)^2 + 4qr]^{\frac{1}{2}} + q(\eta^2 - 1)}{2}.$$

Show that $a_0 \to a^*$ as $n \to \infty$. Hence, show that in a problem with a large number of stages, the optimal value u_1 of the control variable at the first stage is approximately as follows:

$$u_1 = \frac{q + a^*}{(q + a^*)(1 + \eta^2) + r} (t - x_1).$$

12. Consider the control problem described in Sec. 14.13. Suppose that $\alpha_j = 1$, $\beta_j = 0$, $r_j = 0$, and $\eta_j^2 = \eta^2$ for $j = 1, \ldots, n$, where $\eta^2 > 0$. Let $\zeta = \eta^2/(1 + \eta^2)$. (a) Show that for $j = 1, \ldots, n$,

$$a_{j-1} = \sum_{i=j}^{n} \zeta^{i+1-j} q_i$$

and

$$b_{j-1} = \frac{\sum_{i=j}^{n} \zeta^i q_i t_i}{\sum_{i=j}^{n} \zeta^i q_i}.$$

(b) Show that for $j = 1, \ldots, n$, the optimal value u_j of the control variable is

$$u_j = (1 - \zeta)(b_{j-1} - x_j).$$

(c) If, in addition, $t_j = t$ for $j = 1, \ldots, n$, show that $u_j = (1 - \zeta)(t - x_j)$ for $j = 1, \ldots, n$. (Note: This sequence $u_1, \ldots, u_n$ is optimal for any values of $q_1, \ldots, q_n$.)

13. Let $X_1, \ldots, X_r$ be r random variables, and let f and g be two given distinct p.d.f.'s on the real line. Suppose that the statistician knows that the p.d.f. of exactly one of the random variables $X_1, \ldots, X_r$ is f and the p.d.f. of each of the

other $r - 1$ random variables is g but that he does not know which variable has the p.d.f. f. At each stage of a sequential process, the statistician can take an observation on any one of the random variables $X_1, \ldots, X_r$. Also, each observation on X_i costs c_i $(i = 1, \ldots, r)$. At any stage he may stop sampling and decide that a certain one of the random variables has the p.d.f. f. For $i = 1, \ldots, r$ and $j = 1, \ldots, r$, let L_{ij} be the cost to the statistician if he decides that the p.d.f. of X_j is f when, in fact, the p.d.f. of X_i is f. For $i = 1, \ldots, r$, let p_i be the prior probability that the p.d.f. of X_i is f, and let $\rho(p_1, \ldots, p_r)$ be the total risk from the Bayes sequential decision procedure. Furthermore, use the following definitions:

$$\rho_0(p_1, \ldots, p_r) = \min_j \sum_{i=1}^{r} p_i L_{ij}.$$

For $j = 1, \ldots, r$ and $-\infty < x < \infty$,

$$p_j^*(x, j) = \frac{p_j f(x)}{p_j f(x) + (1 - p_j)g(x)}.$$

For $i \neq j$,

$$p_i^*(x, j) = \frac{p_i g(x)}{p_j f(x) + (1 - p_j)g(x)}.$$

Also,

$$\rho'(p_1, \ldots, p_r)$$
$$= \min_j \left\{ c_j + \int_{-\infty}^{\infty} \rho[p_1^*(x, j), \ldots, p_r^*(x, j)][p_j f(x) + (1 - p_j)g(x)] \, dx \right\}.$$

Show that ρ must satisfy the following functional equation:

$$\rho(p_1, \ldots, p_r) = \min \{\rho_0(p_1, \ldots, p_r), \rho'(p_1, \ldots, p_r)\}.$$

14. Consider the conditions described in Exercise 13, but suppose now that the statistician must take exactly one observation and then must decide that a certain one of the random variables $X_1, \ldots, X_r$ has the p.d.f. f. Suppose that the costs c_i $(i = 1, \ldots, r)$ are equal, that $L_{ii} = 0$ for $i = 1, \ldots, r$ and $L_{ij} = 1$ for $i \neq j$, and that $p_1 \geq p_2 \geq \cdots \geq p_r$. Let the subsets A and B of the real line be defined as follows:

$$A = \{x: p_1 f(x) \geq p_2 g(x)\},$$
$$B = \{x: p_1 g(x) \geq p_2 f(x)\}.$$

Prove that an optimal procedure may be described as follows: The statistician should observe X_1 if

$$\int_A p_1 f(x) \, dx + \int_{A^c} p_2 g(x) \, dx > \int_B p_1 g(x) \, dx + \int_{B^c} p_2 f(x) \, dx.$$

He should observe X_2 if this inequality is reversed, and he need not ever observe $X_3, \ldots, X_r$.

15. Suppose that an object must be in one of r locations at any time but can change its location at each stage of a sequential process. Suppose that the movements of the object form a Markov chain with stationary transition probabilities f_{ij} $(i, j = 1, \ldots, r)$. Thus, if it is known that the object is in location i at a given stage, f_{ij} is the probability that the object will be in location j at the next stage. At each stage, the statistician can search any one of the locations. However, there is the

probability α_i $(0 \leq \alpha_i < 1)$ that he will overlook the object in location i ($i = 1$, . . . , r), even when it is there. Suppose that each search of location i costs c_i.
For $i = 1$, . . . , r, let p_i be the probability that the object is in location i at the time at which the statistician must make his first search. After this search, the object moves in accordance with the transition probabilities given above, and the statistician then makes his second search. The process continues in this way until the statistician finds the object. Let $L(p_1, \ldots, p_r)$ be the expected total cost of the optimal search procedure. Show that $L(p_1, \ldots, p_r) < \infty$ and that L satisfies the following functional equation:

$$L(p_1, \ldots, p_k) = \min_j \{c_j + (p_j\alpha_j + 1 - p_j)L(p_1', \ldots, p_r')\},$$

where $p_k' = \Sigma_{i=1}^r p_i^* f_{ik}$, for $k = 1, \ldots, r$, and $p_1^*, \ldots, p_r^*$ are as specified by Eq. (1) of Sec. 14.14.

16. Suppose that an object can be in any one of three locations and that, as in Exercise 15, it changes its location in accordance with a Markov chain for which the transition probabilities f_{ij} are as follows:

$$f_{11} = f_{12} = 0, \qquad f_{13} = 1;$$
$$f_{21} = \tfrac{1}{7}, \qquad f_{22} = \tfrac{4}{7}, \qquad f_{23} = \tfrac{2}{7};$$
$$f_{31} = f_{32} = 0, \qquad f_{33} = 1.$$

Also, suppose that there is zero probability that the statistician will overlook the object when it is actually in the location being searched. Suppose, finally, that the probability of finding the object in not more than two searches must be maximized. Show that the optimal procedure is to make the first search in location 2, no matter how small may be the prior probability that the object will be in location 2.

17. Consider a problem involving a parameter W which can have only two values w_1 and w_2. Hence, the set Ξ contains all vectors $\xi = (\xi_1, \xi_2)$, where $\xi_i = \Pr(W = w_i)$ for $i = 1$, 2. Let the uncertainty function T on Ξ be specified by the equation $T(\xi) = \min \{\xi_1, \xi_2\}$. Suppose that the class $\mathfrak{F}$ of available experiments contains just two random variables X and Y each of which can have only the values 0 and 1 and that the conditional distributions are as follows:

$$f_X(1|w_1) = \tfrac{1}{2}, \qquad f_X(0|w_1) = \tfrac{1}{2},$$
$$f_X(1|w_2) = \tfrac{1}{4}, \qquad f_X(0|w_2) = \tfrac{3}{4};$$
$$f_Y(1|w_1) = \tfrac{3}{4}, \qquad f_Y(0|w_1) = \tfrac{1}{4},$$
$$f_Y(1|w_2) = \tfrac{1}{2}, \qquad f_Y(0|w_2) = \tfrac{1}{2}.$$

Suppose also that the statistician is permitted to take exactly two observations and must minimize the expected uncertainty of the posterior distribution of W.

(a) Show that for any given prior distribution ξ, an optimal procedure specifies that the first observation should be made on either X or Y if $0 < \xi_1 < \tfrac{1}{5}$, on X if $\tfrac{1}{5} < \xi_1 < \tfrac{2}{5}$, on Y if $\tfrac{2}{5} < \xi_1 < \tfrac{1}{2}$, on X if $\tfrac{1}{2} < \xi_1 < \tfrac{3}{5}$, on Y if $\tfrac{3}{5} < \xi_1 < \tfrac{4}{5}$, and on either X or Y if $\tfrac{4}{5} < \xi_1 < 1$.

(b) Show that the minimum expected terminal uncertainty, which is denoted by $T_2(\xi)$, satisfies the equation $T_2(\{\xi_1, \xi_2\}) = T_2(\{\xi_2, \xi_1\})$.

(c) Show that for $0 \leq \xi_1 \leq \tfrac{1}{2}$, the value of $T_2(\xi)$ is

$$T_2(\xi) = \begin{cases} \xi_1 & \text{for } 0 \leq \xi_1 \leq \tfrac{1}{5}, \\ \tfrac{3}{4}\xi_1 + \tfrac{1}{16}\xi_2 & \text{for } \tfrac{1}{5} < \xi_1 \leq \tfrac{1}{3}, \\ \tfrac{5}{8}\xi_1 + \tfrac{1}{8}\xi_2 & \text{for } \tfrac{1}{3} < \xi_1 \leq \tfrac{2}{5}, \\ \tfrac{7}{16}\xi_1 + \tfrac{1}{4}\xi_2 & \text{for } \tfrac{2}{5} < \xi_1 \leq \tfrac{1}{2}. \end{cases}$$

18. Let W be a parameter which can have only two values w_1 and w_2. Let X and Y be random variables for which the distributions are as follows: If $W = w_i$, X has a normal distribution with mean a_i and precision 1, and Y has a normal distribution with mean b_i and precision 1. Show that if $a_1 = b_2$ and $a_2 = b_1$, then X and Y are equivalent in the sense that each is sufficient for the other.

19. Let W be a parameter which can have only two values w_1 and w_2. Also, let X and Y be random variables each of which can have only the values 0 and 1. Suppose that their p.f.'s are

$$f_X(1|w_1) = a_1 = 1 - f_X(0|w_1),$$
$$f_X(1|w_2) = a_2 = 1 - f_X(0|w_2);$$
$$f_Y(1|w_1) = b_1 = 1 - f_Y(0|w_1),$$
$$f_Y(1|w_2) = b_2 = 1 - f_Y(0|w_2).$$

Assuming that $b_1 < b_2$, show that Y is sufficient for X if, and only if, either of the following relations is satisfied:

$$\frac{1 - b_2}{1 - b_1} \leq \frac{1 - a_2}{1 - a_1} \leq \frac{a_2}{a_1} \leq \frac{b_2}{b_1}$$

or

$$\frac{1 - b_2}{1 - b_1} \leq \frac{a_2}{a_1} \leq \frac{1 - a_2}{1 - a_1} \leq \frac{b_2}{b_1}.$$

20. Let W be a parameter which can have any positive value. Let X and Y be random variables for which the distributions are as follows: For any given positive value w of W, the random variable X/w has a χ^2 distribution with α_1 degrees of freedom and the random variable Y/w has a χ^2 distribution with α_2 degrees of freedom ($\alpha_1 < \alpha_2$). Show that Y is sufficient for X.

Hint: If Z has a beta distribution with parameters $\frac{1}{2}\alpha_1$ and $\frac{1}{2}\alpha_2 - \frac{1}{2}\alpha_1$ and if Z and Y are independent, then the random variable ZY has the same distribution as the random variable X for any given value of W.

21. [See Blackwell and Girshick (1954), chap. 12; Lehmann (1959), p. 76.] The individuals in a large population can be divided into four classes as follows: Members of one class possess both of two characteristics A and B; members of a second class possess characteristic A but not characteristic B; members of a third class possess characteristic B but not A; and members of the fourth class possess neither characteristic A nor B. It is known that the total proportion of individuals having the characteristic A is α and that the total proportion of individuals having the characteristic B is β, where α and β are given numbers such that $\alpha < \beta \leq \frac{1}{2}$. However, the proportion W of individuals in the population having both characteristic A and characteristic B is unknown.

In order to gain information about the value of W, the statistician plans to select a sequential sample of n individuals and to note which of the characteristics each individual possesses. Each of the n individuals can be selected at random from any one of the following five classes: (a) the subpopulation $\Pi(A)$ of all individuals possessing characteristic A, (b) the subpopulation $\Pi(B)$ of all individuals possessing characteristic B, (c) the subpopulation $\Pi(A^c)$ of all individuals who do not possess characteristic A, (d) the subpopulation $\Pi(B^c)$ of all individuals who do not possess characteristic B, and (e) the whole population Π_0.

Show that the experiment of selecting an individual from the subpopulation $\Pi(A)$ and noting whether or not he possesses characteristic B is sufficient for the experiment of selecting an individual from any one of the other four classes $\Pi(B)$, $\Pi(A^c)$, $\Pi(B^c)$, or Π_0. Prove, therefore, that the statistician should select all n individuals from the subpopulation $\Pi(A)$.

Hint: Let $X(A)$ be 1 if an individual selected from $\Pi(A)$ also possesses characteristic B, and let $X(A)$ be 0 if he does not possess characteristic B. Similarly, let $X(B)$ be 1 if an individual selected from $\Pi(B)$ also possesses characteristic A, and let $X(B)$ be 0 otherwise. Let $X(A^c)$ be 1 if an individual selected from $\Pi(A^c)$ possesses characteristic B, and let $X(A^c)$ be 0 otherwise. Let $X(B^c)$ be 1 if an individual selected from $\Pi(B^c)$ possesses characteristic A, and let $X(B^c)$ be 0 otherwise. Finally, let X_0 be 1 if an individual selected from Π_0 possesses both characteristic A and characteristic B, and let X_0 be 0 otherwise. Then the various probabilities will be as follows:

$$\Pr[X(A) \ = 1|W = w] = \frac{w}{\alpha};$$

$$\Pr[X(B) \ = 1|W = w] = \frac{w}{\beta};$$

$$\Pr[X(A^c) = 1|W = w] = \frac{\beta - w}{1 - \alpha};$$

$$\Pr[X(B^c) = 1|W = w] = \frac{\alpha - w}{1 - \beta};$$

$$\Pr[X_0 \ = 1|W = w] = w.$$

22. The durability T of the items in a certain manufactured lot has a normal distribution with unknown mean W and known variance $\sigma_1{}^2$. If an item is selected from the lot, its durability $T = t$ cannot be observed directly. However, it is possible to make measurements on the item which are independent and identically distributed, and each such measurement has a normal distribution with mean t and known variance $\sigma_2{}^2$. It is desired to learn about the value of W by selecting a random sample of items from the lot and making measurements on each of these items.

For any fixed positive integer n, show that the experiment in which n items are selected at random from the lot and one measurement is made on each of them is sufficient for the experiment in which one item is selected at random from the lot and n measurements are made on it.

Hint: Show first that the average of the n measurements which are made will be a sufficient statistic in each of the two experiments, and then compare the two experiments in which only these real-valued statistics are considered.

references

Abramson, L. R. (1966). Asymptotic sequential design of experiments with two random variables. *J. Roy. Statist. Soc.* (B)**28**:73–87.

Albert, A. E. (1961). The sequential design of experiments for infinitely many states of nature. *Ann. Math. Statist.* **32**:774–799.

Allais, M. (1953). Le comportement de l'homme rationnel devant le risque: Critique des postulats et axioms de l'école Americaine. *Econometrica* **21**:503–546.

Amster, S. J. (1963). A modified Bayes stopping rule. *Ann. Math. Statist.* **34**:1404–1413.

Anderson, T. W. (1958). *An Introduction to Multivariate Statistical Analysis.* John Wiley & Sons, Inc., New York.

———. (1960). A modification of the sequential probability ratio test to reduce the sample size. *Ann. Math. Statist.* **31**:165–197.

———. (1964a). On Bayes procedures for a problem with choice of observations. *Ann. Math. Statist.* **35**:1128–1135.

———. (1964b). Sequential analysis with delayed observations. *J. Am. Statist. Assoc.* **59**:1006–1015.

———, and Friedman, M. (1960). A limitation of the optimum property of the sequential probability ratio test. *Contributions to Probability and Statistics* (ed. by Olkin et al.), pp. 57–69. Stanford University Press, Stanford, Calif.

Ando, A., and Kaufman, G. M. (1965). Bayesian analysis of the independent multinormal process—neither mean nor precision known. *J. Am. Statist. Assoc.* **60**:347–358.

Andrews, F. C., and Blum, J. R. (1966). On expectation in sequential sampling. *Statistica Neerlandica* **20**:9–18.

Anscombe, F. J. (1963). Sequential medical trials. *J. Am. Statist. Assoc.* **58**:365–383.
———. (1964a). Some remarks on Bayesian statistics. *Human Judgments and Optimality* (ed. by Shelly and Bryan), pp. 155–177. John Wiley & Sons, Inc., New York.
———. (1964b). Normal likelihood functions. *Ann. Inst. Statist. Math.* **16**:1–91.
———, and Aumann, R. J. (1963). A definition of subjective probability. *Ann. Math. Statist.* **34**:199–205.
Aoki, M. (1967). *Optimization of Stochastic Systems.* Academic Press Inc., New York.
Arrow, K. J. (1962). Optimal capital adjustment. *Studies in Applied Probability and Management Science* (ed. by Arrow, Karlin, and Scarf), pp. 1–17. Stanford University Press, Stanford, Calif.
———. (1964). Optimal capital policy, the cost of capital, and myopic decision rules. *Ann. Inst. Statist. Math.* **16**:21–30.
———, Blackwell, D., and Girshick, M. A. (1949). Bayes and minimax solutions of sequential decision problems. *Econometrica* **17**:213–244.
———, Harris, T. E., and Marschak, J. (1951). Optimal inventory policy. *Econometrica* **19**:250–272.
———, Karlin, S., and Scarf, H. (1958). *Studies in the Mathematical Theory of Inventory and Production.* Stanford University Press, Stanford, Calif.
———, Karlin, S., and Scarf, H. (eds.). (1962). *Studies in Applied Probability and Management Science.* Stanford University Press, Stanford, Calif.
Aumann, R. J. (1964). Subjective programming. *Human Judgments and Optimality* (ed. by Shelly and Bryan), pp. 217–242. John Wiley & Sons, Inc., New York.
Bahadur, R. R. (1954). Sufficiency and statistical decision functions. *Ann. Math. Statist.* **25**:423–462.
———. (1958). A note on the fundamental identity of sequential analysis. *Ann. Math. Statist.* **29**:534–543.
Barnard, G. A. (1949). Statistical inference (with discussion). *J. Roy. Statist. Soc.* (B)**11**:115–149.
———. (1958). Thomas Bayes—a biographical note (with a reproduction of "An essay towards solving a problem in the doctrine of chances," by Thomas Bayes). *Biometrika* **45**:293–315.
———. (1962). Comments on Stein's "A Remark on the Likelihood Principle." *J. Roy. Statist. Soc.* (A)**125**:569–573.
———. (1967). The use of the likelihood function in statistical practice. *Proc. 5th Berkeley Symp. Math. Statist. Probability* **1**:27–40. University of California Press, Berkeley, Calif.
———, Jenkins, G. M., and Winsten, C. B. (1962). Likelihood inference and time series (with discussion). *J. Roy. Statist. Soc.* (A)**125**:321–372.
Bather, J. A. (1962). Bayes procedures for deciding the sign of a normal mean. *Proc. Cambridge Phil. Soc.* **58**:599–620.
Becker, G. M., DeGroot, M. H., and Marschak, J. (1963). Stochastic models of choice behavior. *Behavioral Sci.* **8**:41–55. Reprinted in *Decision Making* (ed. by Edwards and Tversky) (1967). Penguin Books, Inc., Baltimore.
———, ———, and ———. (1964). Measuring utility by a single-response sequential method. *Behavioral Sci.* **9**:226–232.
Bellman, R. (1956). A problem in the sequential design of experiments. *Sankhyā* **16**:221–229.
———. (1957a). *Dynamic Programming.* Princeton University Press, Princeton, N.J.

———. (1957b). On a generalization of the fundamental identity of Wald. *Proc. Cambridge Phil. Soc.* **53**:257–259.

———. (1961). *Adaptive Control Processes: A Guided Tour.* Princeton University Press, Princeton, N.J.

Berk, R. H. (1966). Limiting behavior of posterior distributions when the model is incorrect. *Ann. Math. Statist.* **37**:51–58.

Bessler, S. A. (1960). Theory and applications of the sequential design of experiments, k-actions and infinitely many experiments, parts I and II. Unpublished technical reports, Department of Statistics, Stanford University, Stanford, Calif.

Bickel, P. J., and Yahav, J. A. (1967). Asymptotically pointwise optimal procedures in sequential analysis. *Proc. 5th Berkeley Symp. Math. Statist. Probability* **1**:401–413. University of California Press, Berkeley, Calif.

Birnbaum, A. (1962). On the foundations of statistical inference (with discussion). *J. Am. Statist. Assoc.* **57**:269–326.

Black, W. L. (1965). Discrete sequential search. *Inform. Control* **8**:156–162.

Blackwell, D. (1951). Comparison of experiments. *Proc. 2nd Berkeley Symp. Math. Statist. Probability*, pp. 93–102. University of California Press, Berkeley, Calif.

———. (1953). Equivalent comparison of experiments. *Ann. Math. Statist.* **24**:265–272.

———. (1961). On the functional equation of dynamic programming. *J. Math. Anal. Appl.* **2**:273–276.

———. (1962). Discrete dynamic programming. *Ann. Math. Statist.* **33**:719–726.

———. (1964). Memoryless strategies in finite-stage dynamic programming. *Ann. Math. Statist.* **35**:863–865.

———. (1965). Discounted dynamic programming. *Ann. Math. Statist.* **36**:226–235.

———. (1967). Positive dynamic programming. *Proc. 5th Berkeley Symp. Math. Statist. Probability* **1**:415–418. University of California Press, Berkeley, Calif.

———, and Dubins, L. (1962). Merging of opinions with increasing information. *Ann. Math. Statist.* **33**:882–886.

———, and Girshick, M. A. (1954). *Theory of Games and Statistical Decisions.* John Wiley & Sons, Inc., New York.

Blasbalg, H. (1957). Transformation of the fundamental relationships in sequential analysis. *Ann. Math. Statist.* **28**:1024–1027.

Bloch, D. A., and Watson, G. S. (1967). A Bayesian study of the multinomial distribution. *Ann. Math. Statist.* **38**:1423–1435.

Bohrer, R. (1966). On Bayes sequential design with two random variables. *Biometrika* **53**:469–475.

Box, G. E. P., and Draper, N. R. (1965). The Bayesian estimation of common parameters from several processes. *Biometrika* **52**:355–365.

———, and Tiao, G. C. (1964). A Bayesian approach to the importance of assumptions applied to the comparison of variances. *Biometrika* **51**:153–167.

Bracken, J., and Schleifer, A., Jr. (1964). *Tables for Normal Sampling with Unknown Variance.* Division of Research, Graduate School of Business Administration, Harvard University, Boston.

Bradt, R. N., Johnson, S. M., and Karlin, S. (1956). On sequential designs for maximizing the sum of n observations. *Ann. Math. Statist.* **27**:1060–1074.

———, and Karlin, S. (1956). On the design and comparison of certain dichotomous experiments. *Ann. Math. Statist.* **27**:390–409.

Bramblett, J. E. (1965). Some approximations to optimal stopping procedures.

Unpublished doctoral dissertation, Department of Mathematical Statistics, Columbia University, New York.

Breakwell, J., and Chernoff, H. (1962). Sequential tests for the mean of a normal distribution II (large t). *Ann. Math. Statist.* **33**:162–173.

Breiman, L. (1961). Optimal gambling systems for favorable games. *Proc. 4th Berkeley Symp. Math. Statist. Probability* **1**:65–78. University of California Press, Berkeley, Calif.

——. (1964). Stopping-rule problems. *Applied Combinatorial Mathematics* (ed. by Beckenbach), pp. 284–319. John Wiley & Sons, Inc., New York.

Brewer, K. R. W. (1963). Decisions under uncertainty: Comment. *Quart. J. Econ.* **77**:159–161.

Brown, L. (1965). Optimal policies for a sequential decision process. *J. Soc. Ind. Appl. Math.* **13**:37–45.

Brunk, H. D. (1965). *An Introduction to Mathematical Statistics*, 2d ed. Blaisdell, New York.

Burkholder, D. L., and Wijsman, R. A. (1963). Optimum properties and admissibility of sequential tests. *Ann. Math. Statist.* **34**:1–17.

Carnap, R. (1962). *Logical Foundations of Probability.* The University of Chicago Press, Chicago.

Cayley, A. (1875). Mathematical questions and their solutions. *Educational Times* **22**:18–19.

Chanda, K. C. (1954). A note on the consistency and maxima of the roots of likelihood equations. *Biometrika* **41**:56–61.

Cheng, Ping. (1963). Bayes sequential procedures with K-decision actions for the exponential Pólya type III distributions. *Chinese Math.* **4**:131–147.

Chernoff, H. (1952). A measure of asymptotic efficiency for tests of a hypothesis based on the sum of observations. *Ann. Math. Statist.* **23**:493–507.

——. (1954). On the distribution of the likelihood ratio. *Ann. Math. Statist.* **25**:573–578.

——. (1956). Large-sample theory: Parametric case. *Ann. Math. Statist.* **27**:1–22.

——. (1959). Sequential design of experiments. *Ann. Math. Statist.* **30**:755–770.

——. (1960). Motivation for an approach to the sequential design of experiments. *Information and Decision Processes* (ed. by Machol), pp. 15–26. McGraw-Hill Book Company, New York.

——. (1961a). Sequential tests for the mean of a normal distribution. *Proc. 4th Berkeley Symp. Math. Statist. Probability* **1**:79–91. University of California Press, Berkeley, Calif.

——. (1961b). Sequential experimentation. *Bull. Intern. Statist. Inst.* **38**:3–9.

——. (1965a). Sequential tests for the mean of a normal distribution III (small t). *Ann. Math. Statist.* **36**:28–54.

——. (1965b). Sequential tests for the mean of a normal distribution IV (discrete case). *Ann. Math. Statist.* **36**:55–68.

——. (1967a). A note on risk and maximal regular generalized submartingales in stopping problems. *Ann. Math. Statist.* **38**:606–607.

——. (1967b). Sequential models for clinical trials. *Proc. 5th Berkeley Symp. Math. Statist. Probability* **4**:805–812. University of California Press, Berkeley, Calif.

——, and Moses, L. E. (1959). *Elementary Decision Theory.* John Wiley & Sons, Inc., New York.

——, and Ray, S. N. (1965). A Bayes sequential sampling inspection plan. *Ann. Math. Statist.* **36**:1387–1407.

Chew, M. C., Jr. (1967). A sequential search procedure. *Ann. Math. Statist.* **38**: 494–502.

Chow, Y. S. (1967). On the expected value of a stopped submartingale. *Ann. Math. Statist.* **38**:608–609.

——, Moriguti, S., Robbins, H., and Samuels, S. M. (1964). Optimal selection based on relative rank (the "Secretary Problem"). *Israel J. Math.* **2**:81–90.

——, and Robbins, H. (1961). A martingale system theorem and applications. *Proc. 4th Berkeley Symp. Math. Statist. Probability* **1**:93–104. University of California Press, Berkeley, Calif.

——, and ——. (1963). On optimal stopping rules. *Z. Wahrscheinlichkeitstheorie und Verw. Gebiete* **2**:33–49.

——, and ——. (1965). On optimal stopping rules for S_n/n. *Illinois J. Math.* **9**:444–454.

——, and ——. (1967a). A class of optimal stopping problems. *Proc. 5th Berkeley Symp. Math. Statist. Probability* **1**:419–426. University of California Press, Berkeley, Calif.

——, and ——. (1967b). On values associated with a stochastic sequence. *Proc. 5th Berkeley Symp. Math. Statist. Probability* **1**:427–440. University of California Press, Berkeley, Calif.

——, ——, and Teicher, H. (1965). Moments of randomly stopped sums. *Ann. Math. Statist.* **36**:789–799.

Chung, K. L. (1960). *Markov Chains with Stationary Transition Probabilities.* Springer-Verlag OHG, Berlin.

Cramér, H. (1946). *Mathematical Methods of Statistics.* Princeton University Press, Princeton, N.J.

Darmois, G. (1935). Sur les lois de probabilité à estimation exhaustive. *C. R. Acad. Sci. Paris* **260**:1265–1266.

Davidson, D., Suppes, P., and Siegel, S. (1957). *Decision Making: An Experimental Approach.* Stanford University Press, Stanford, Calif.

Debreu, G. (1960). Topological methods in cardinal utility theory. *Mathematical Methods in the Social Sciences* (ed. by Arrow, Karlin, and Suppes), pp. 16–26. Stanford University Press, Stanford, Calif.

de Finetti, B. (1937). Foresight: Its logical laws, its subjective sources. Translated and reprinted in *Studies in Subjective Probability* (ed. by Kyburg and Smokler) (1964), pp. 93–158. John Wiley & Sons, Inc., New York.

DeGroot, M. H. (1962). Uncertainty, information, and sequential experiments. *Ann. Math. Statist.* **33**:404–419.

——. (1966). Optimal allocation of observations. *Ann. Inst. Statist. Math.* **18**:13–28.

——. (1968). Some problems of optimal stopping. *J. Roy. Statist. Soc.* (B)**30**:108–122.

——, and Nadler, J. (1958). Some aspects of the use of the sequential probability ratio test. *J. Am. Statist. Assoc.* **53**:187–199.

——, and Rao, M. M. (1963). Bayes estimation with convex loss. *Ann. Math. Statist.* **34**:839–846.

——, and ——. (1966). Multidimensional information inequalities and prediction. *Multivariate Analysis* (ed. by Krishnaiah), pp. 287–313. Academic Press Inc., New York.

Derman, C. (1962). On sequential decisions and Markov chains. *Management Sci.* **9**:16–24.

——. (1963). Stable sequential rules and Markov chains. *J. Math. Anal. Appl.* **6**:257–265.

————. (1964). On sequential control processes. *Ann. Math. Statist.* **35**:341–350.

————. (1966). Denumerable state Markovian decision processes—average cost criterion. *Ann. Math. Statist.* **37**:1545–1553.

————, and Sacks, J. (1960). Replacement of periodically inspected equipment. *Naval Res. Logist. Quart.* **7**:597–607.

————, and Veinott, A. F., Jr. (1967). A solution to a countable system of equations arising in Markovian decision processes. *Ann. Math. Statist.* **38**:582–584.

Deutsch, R. (1965). *Estimation Theory.* Prentice-Hall, Inc., Englewood Cliffs, N.J.

Doob, J. L. (1953). *Stochastic Processes.* John Wiley & Sons, Inc., New York.

Doss, S. A. D. C. (1962). A note on consistency and asymptotic efficiency of maximum likelihood estimates in multi-parametric problems. *Calcutta Statist. Assoc. Bull.* **11**:85–93.

————. (1963). On consistency and asymptotic efficiency of maximum likelihood estimates. *J. Indian Soc. Agr. Statist.* **15**:232–241.

Dubins, L. E., and Freedman, D. A. (1966). On the expected value of a stopped martingale. *Ann. Math. Statist.* **37**:1505–1509.

————, and Savage, L. J. (1965). *How to Gamble If You Must: Inequalities for Stochastic Processes.* McGraw-Hill Book Company, New York.

————, and Teicher, H. (1967). Optimal stopping when the future is discounted. *Ann. Math. Statist.* **38**:601–605.

Duncan, D. B. (1965). A Bayesian approach to multiple comparisons. *Technometrics* **7**:171–222.

Dvoretzky, A. (1967). Existence and properties of certain optimal stopping rules. *Proc. 5th Berkeley Symp. Math. Statist. Probability* **1**:441–452. University of California Press, Berkeley, Calif.

————, Kiefer, J., and Wolfowitz, J. (1952a). The inventory problem: I. Case of known distributions of demand. *Econometrica* **20**:187–222.

————, ————, and ————. (1952b). The inventory problem: II. Case of unknown distributions of demand. *Econometrica* **20**:451–466.

————, ————, and ————. (1953). Sequential decision problems for processes with continuous time parameter. Testing hypotheses. *Ann. Math. Statist.* **24**:254–264.

Dynkin, E. B. (1961). Necessary and sufficient statistics for a family of probability distributions. *Selected Transl. Math. Statist. Probability* **1**:23–40. Institute of Mathematical Statistics and American Mathematical Society, Providence, R.I. (Originally published in Russian, 1951.)

————. (1965). *Markov Processes* (2 vols.). Academic Press Inc., New York.

Edwards, W., Lindman, H., and Savage, L. J. (1963). Bayesian statistical inference for psychological research. *Psychol. Rev.* **70**:193–242. Reprinted in *Readings in Mathematical Psychology*, vol. 2 (ed. by Luce, Bush, and Galanter) (1965). John Wiley & Sons, Inc., New York.

Eggleston, H. G. (1958). *Convexity.* Cambridge University Press, London.

Elfving, G. (1967). A persistency problem connected with a point process. *J. Appl. Probability* **4**:77–89.

Ellsberg, D. (1961). Risk, ambiguity, and the Savage axioms. *Quart. J. Econ.* **75**:643–669.

Enslow, P., Jr. (1966). A bibliography of search theory and reconnaissance theory literature. *Naval Res. Logist. Quart.* **13**:177–202.

Evans, I. G. (1964). Bayesian estimation of the variance of a normal distribution. *J. Roy. Statist. Soc.* (B)**26**:63–68.

————. (1965). Bayesian estimation of parameters of a multivariate normal distribution. *J. Roy. Statist. Soc.* (B)**27**:270–283.

Fabius, J. (1964). Asymptotic behavior of Bayes' estimates. *Ann. Math. Statist.* **35**:846-856.

Farrell, R. H. (1964a). Estimators of a location parameter in the absolutely continuous case. *Ann. Math. Statist.* **35**:949-998.

———. (1964b). Limit theorems for stopped random walks. *Ann. Math. Statist.* **35**:1332-1343.

———. (1966a). Limit theorems for stopped random walks, II. *Ann. Math. Statist.* **37**:860-865.

———. (1966b). Limit theorems for stopped random walks, III. *Ann. Math. Statist.* **37**:1510-1527.

Fel'dbaum, A. A. (1966). *Optimal Control Systems.* Academic Press Inc., New York.

Feldman, D. (1962). Contributions to the "two-armed bandit" problem. *Ann. Math. Statist.* **33**:847-856.

Feller, W. (1957). *An Introduction to Probability Theory and Its Applications*, vol. 1, 2d ed. John Wiley & Sons, Inc., New York.

———. (1966). *An Introduction to Probability Theory and Its Applications*, vol. 2. John Wiley & Sons, Inc., New York.

Fellner, W. (1961). Distortion of subjective probabilities as a reaction to uncertainty. *Quart. J. Econ.* **75**:670-689.

———. (1963). Slanted subjective probabilities and randomization: Reply to Howard Raiffa and K. R. W. Brewer. *Quart. J. Econ.* **77**:676-690.

———. (1965). *Probability and Profit.* Richard D. Irwin, Inc., Homewood, Ill.

Ferguson, T. S. (1965). Betting systems which minimize the probability of ruin. *J. Soc. Ind. Appl. Math.* **13**:795-818.

———. (1967). *Mathematical Statistics: A Decision Theoretic Approach.* Academic Press Inc., New York.

Fishburn, P. C. (1964). *Decision and Value Theory.* John Wiley & Sons, Inc., New York.

———. (1967a). Bounded expected utility. *Ann. Math. Statist.* **38**:1054-1060.

———. (1967b). Preference-based definitions of subjective probability. *Ann. Math. Statist.* **38**:1605-1617.

———. (1968). Utility theory. *Management Sci.* **14**:335-378.

Fisher, L. (1968). On recurrent denumerable decision processes. *Ann. Math. Statist.* **39**:424-434.

———, and Ross, S. M. (1968). An example in denumerable decision processes. *Ann. Math. Statist.* **39**:674-675.

Fisher, R. A. (1922). On the mathematical foundations of theoretical statistics. *Phil. Trans. Roy. Soc. London* (A)**222**:309-368. Reprinted in Fisher, R. A. (1950). *Contributions to Mathematical Statistics.* John Wiley & Sons, Inc., New York.

———. (1956). *Statistical Methods and Scientific Inference*, Oliver & Boyd Ltd., Edinburgh and London. (2d ed. Hafner Publishing Company, Inc., New York, 1959.)

Fisz, M. (1963). *Probability Theory and Mathematical Statistics*, 3d ed. John Wiley & Sons, Inc., New York.

Fraser, D. A. S. (1963). On sufficiency and the exponential family. *J. Roy. Statist. Soc.* (B)**25**:115-123.

Freedman, D. A. (1963). On the asymptotic behavior of Bayes estimates in the discrete case. *Ann. Math. Statist.* **34**:1386-1403.

———. (1965). On the asymptotic behavior of Bayes estimates in the discrete case, II. *Ann. Math. Statist.* **36**:454-456.

———. (1967). Timid play is optimal. *Ann. Math. Statist.* **38**:1281-1283.

————, and Purves, R. (1967). Timid play is optimal, II. *Ann. Math. Statist.* 38:1284–1285.

Freeman, H. (1963). *Introduction to Statistical Inference.* Addison-Wesley Publishing Company, Inc., Reading, Mass.

Friedman, M., and Savage, L. J. (1948). The utility analysis of choice involving risk. *J. Political Econ.* 56:279–304.

————, and ————. (1952). The expected utility hypotheses and the measurability of utility. *J. Political Econ.* 60:463–474.

Furukawa, N. (1964). On some properties of an optimal strategy in the "two-armed bandit" problem. *Mem. Fac. Sci. Kyushu Univ.* (A)18:74–88.

Geisser, S. (1964). Posterior odds for multivariate normal classifications. *J. Roy. Statist. Soc.* (B)26:69–76.

————. (1965a). Bayesian estimation in multivariate analysis. *Ann. Math. Statist.* 36:150–159.

————. (1965b). A Bayes approach for combining correlated estimates. *J. Am. Statist. Assoc.* 60:602–607.

————. (1966). Predictive discrimination. *Multivariate Analysis* (ed. by Krishnaiah), pp. 149–163. Academic Press Inc., New York.

————, and Cornfield, J. (1963). Posterior distributions for multivariate normal parameters. *J. Roy. Statist. Soc.* (B)25:368–376.

Ghosh, J. K. (1961). On the optimality of probability ratio tests in sequential and multiple sampling. *Calcutta Statist. Assoc. Bull.* 10:73–92.

————. (1964). Bayes solutions in sequential problems for two or more terminal decisions and related results. *Calcutta Statist. Assoc. Bull.* 13:101–122.

Gilbert, J. P., and Mosteller, F. (1966). Recognizing the maximum of a sequence. *J. Am. Statist. Assoc.* 61:35–73.

Gnedenko, B. V. (1962). *The Theory of Probability.* Chelsea Publishing Company, New York.

Good, I. J. (1950). *Probability and the Weighing of Evidence.* Hafner Publishing Company, Inc., New York.

————. (1965). *The Estimation of Probabilities.* The M.I.T. Press, Cambridge, Mass.

Grigelionis, B. I., and Shiryaev, A. N. (1965). Criteria of "truncation" for the optimal stopping time in sequential analysis. *Theory Probability Appl.* 10:541–552.

Grundy, P. M., Healy, M. J. R., and Rees, D. H. (1956). Economic choice of the amount of experimentation (with discussion). *J. Roy. Statist. Soc.* (B)18:32–55.

Gupta, S. S. (1963). Bibliography on the multivariate normal integrals and related topics. *Ann. Math. Statist.* 34:829–838.

Guttman, I. (1960). On a problem of L. Moser. *Can. Math. Bull.* 3:35–39.

Hackleman, R. P. (1967). Metric spaces of distribution functions and statistical information. Unpublished technical report, Department of Statistics, Carnegie-Mellon University, Pittsburgh, Pa.

Hadley, G. (1964). *Nonlinear and Dynamic Programming.* Addison-Wesley Publishing Company, Inc., Reading, Mass.

————. (1967). *Introduction to Probability and Statistical Decision Theory.* Holden-Day, Inc., San Francisco.

————, and Whitin, T. (1963). *Analysis of Inventory Systems.* Prentice-Hall, Inc., Englewood Cliffs, N.J.

Haggstrom, G. W. (1966). Optimal stopping and experimental design. *Ann. Math. Statist.* 37:7–29.

Halmos, P. R. (1950). *Measure Theory.* D. Van Nostrand Company, Inc., Princeton, N.J.

————, and Savage, L. J. (1949). Application of the Radon-Nikodym theorem to the theory of sufficient statistics. *Ann. Math. Statist.* 20:225-241.

Hannan, J. F. (1957). Approximations to Bayes risk in repeated play. *Contrib. Theory Games* 3:97-139.

————, and Robbins, H. (1955). Asymptotic solutions of the compound decision problem for two completely specified distributions. *Ann. Math. Statist.* 26:37-51.

————, and Van Ryzin, J. R. (1965). Rate of convergence in the compound decision problem for two completely specified distributions. *Ann. Math. Statist.* 36:1743-1752.

Harris, B. (1966). *Theory of Probability.* Addison-Wesley Publishing Company, Inc., Reading, Mass.

Hartigan, J. (1964). Invariant prior distributions. *Ann. Math. Statist.* 35:836-845.

Herstein, I. N., and Milnor, J. (1953). An axiomatic approach to measurable utility. *Econometrica* 21:291-297.

Hill, B. M. (1965). Inference about variance components in the one-way model. *J. Am. Statist. Assoc.* 60:806-825.

Hoeffding, W. (1960). Lower bounds for the expected sample size and the average risk of a sequential procedure. *Ann. Math. Statist.* 31:352-368.

Hogg, R. V., and Craig, A. T. (1965). *Introduction to Mathematical Statistics,* 2d ed. The Macmillan Company, New York.

Holland, J. D. (1962). The Reverend Thomas Bayes, F.R.S. (1702-61). *J. Roy. Statist. Soc.* (A)125:451-461.

Howard, R. (1960). *Dynamic Programming and Markov Processes.* The M.I.T. Press, Cambridge, Mass., and John Wiley & Sons, Inc., New York.

Huzurbazar, V. S. (1948). The likelihood equation, consistency, and maxima of the likelihood function. *Ann. Eugenics (London)* 14:185-200.

Isbell, J. R. (1959). On a problem of Robbins. *Ann. Math. Statist.* 30:606-610.

Jackson, J. E. (1960). Bibliography on sequential analysis. *J. Am. Statist. Assoc.* 55:561-580.

James, W., and Stein, C. (1961). Estimation with quadratic loss. *Proc. 4th Berkeley Symp. Math. Statist. Probability* 1:361-380. University of California Press, Berkeley, Calif.

Jeffreys, H. (1961). *Theory of Probability,* 3d ed. Oxford University Press, London.

Johns, M. V., Jr. (1957). Non-parametric empirical Bayes procedures. *Ann. Math. Statist.* 28:649-669.

————. (1961). An empirical Bayes approach to non-parametric two-way classification. *Studies in Item Analysis and Prediction* (ed. by Solomon), pp. 221-232. Stanford University Press, Stanford, Calif.

Johnson, N. L. (1959). A proof of Wald's theorem on cumulative sums. *Ann. Math. Statist.* 30:1245-1247. [Correction note. 32:1344 (1961).]

Johnson, R. A. (1967). An asymptotic expansion for posterior distributions. *Ann. Math. Statist.* 38:1899-1906.

Kadane, J. B. (1968). Discrete search and the Neyman-Pearson lemma. *J. Math. Anal. Appl.* 22:156-171.

Karlin, S. (1955a). The structure of dynamic programming problems. *Naval Res. Logist. Quart.* 2:285-294.

————. (1955b). Decision theory of Pólya type distributions. Case of two actions, I. *Proc. 3rd Berkeley Symp. Math. Statist. Probability* 1:115-128. University of California Press, Berkeley, Calif.

——. (1957a). Pólya type distributions, II. *Ann. Math. Statist.* **28**:281-308.

——. (1957b). Pólya type distributions, III. Admissibility for multiaction problems. *Ann. Math. Statist.* **28**:839-860.

——. (1958). Pólya type distributions, IV. Some principles of selecting a single procedure from a complete class. *Ann. Math. Statist.* **29**:1-21.

——, and Rubin, H. (1956a). The theory of decision procedures for distributions with monotone likelihood ratio. *Ann. Math. Statist.* **27**:272-299.

——, and ——. (1956b). Distributions possessing a monotone likelihood ratio. *J. Am. Statist. Assoc.* **51**:637-643.

Kelly, J. L., Jr. (1956). A new interpretation of information rate. *Bell System Tech. J.* **35**:917-926.

Kemeny, J. G., and Snell, J. L. (1958). Semimartingales of Markov chains. *Ann. Math. Statist.* **29**:143-152.

——, and ——. (1960). *Finite Markov Chains.* D. Van Nostrand Company, Inc., Princeton, N.J.

——, ——, and Knapp, A. W. (1966). *Denumerable Markov Chains.* D. Van Nostrand Company, Inc., Princeton, N.J.

Kempthorne, O. (1966). Some aspects of experimental inference. *J. Am. Statist. Assoc.* **61**:11-34.

Keynes, J. M. (1921). *A Treatise on Probability.* Macmillan & Co., Ltd., London. Reprinted (1962). Harper Torchbooks, Harper & Row, Publishers, New York.

Khinchin, A. I. (1957). *Mathematical Foundations of Information Theory.* Dover Publications, Inc., New York.

Kiefer, J., and Sacks, J. (1963). Asymptotically optimum sequential inference and design. *Ann. Math. Statist.* **34**:705-750.

——, and Weiss, L. (1957). Some properties of generalized sequential probability ratio tests. *Ann. Math. Statist.* **28**:57-75.

Kleindorfer, G. B., and Kleindorfer, P. R. (1967). Quadratic performance criteria with linear terms in discrete-time control. *IEEE Trans. Autom. Control* (AC)**12**:320-321.

Kolmogorov, A. N., and Fomin, S. V. (1961). *Measure, Lebesgue Integrals, and Hilbert Space.* Academic Press Inc., New York.

Koopman, B. O. (1936). On distributions admitting a sufficient statistic. *Trans. Am. Math. Soc.* **39**:399-409.

——. (1940). The bases of probability. *Bull. Am. Math. Soc.* **46**:763-774. Reprinted in *Studies in Subjective Probability* (ed. by Kyburg and Smokler) (1964). John Wiley & Sons, Inc., New York.

Koopmans, T. C., Diamond, P. A., and Williamson, R. E. (1964). Stationary utility and time perspective. *Econometrica* **32**:82-100.

Kraft, C. H., Pratt, J. W., and Seidenberg, A. (1959). Intuitive probability on finite sets. *Ann. Math. Statist.* **30**:408-419.

Krickeberg, K. (1965). *Probability Theory.* Addison-Wesley Publishing Company, Inc., Reading, Mass.

Kullback, S. (1959). *Information Theory and Statistics.* John Wiley & Sons, Inc., New York.

——. (1967). The two concepts of information. *J. Am. Statist. Assoc.* **62**:685-686.

Kulldorff, G. (1957). On the conditions for consistency and asymptotic efficiency of maximum likelihood estimates. *Skand. Aktuarietidskr.* **40**:129-144.

Kyburg, H. E., Jr., and Smokler, H. E. (eds.). (1964). *Studies in Subjective Probability.* John Wiley & Sons, Inc., New York.

Le Cam, L. (1953). On some asymptotic properties of maximum likelihood estimates and related Bayes' estimates. *Univ. Calif. Publ. Statist.* 1·277-329

———. (1955). An extension of Wald's theory of statistical decision functions. *Ann. Math. Statist.* **26**:69–81.

———. (1956). On the asymptotic theory of estimation and testing hypotheses. *Proc. 3rd Berkeley Symp. Math. Statist. Probability* **1**:129–156. University of California Press, Berkeley, Calif.

———. (1964). Sufficiency and approximate sufficiency. *Ann. Math. Statist.* **35**:1419–1456.

———. (1966). Likelihood functions for large numbers of independent observations. *Research Papers in Statistics* (ed. by David), pp. 167–187. John Wiley & Sons, Inc., New York.

Lechner, J. A. (1962). Optimum decision procedures for a Poisson process parameter. *Ann. Math. Statist.* **33**:1384–1402.

Lehmann, E. L. (1959). *Testing Statistical Hypotheses.* John Wiley & Sons, Inc., New York.

———, and Scheffé, H. (1950). Completeness, similar regions, and unbiased estimation. *Sankhyā* **10**:305–340.

Lévy, P. (1937). *Théorie de l'Addition des Variables Aléatoires.* Gauthier-Villars, Paris.

Lindgren, B. W. (1962). *Statistical Theory.* The Macmillan Company, New York.

Lindley, D. V. (1956). On the measure of the information provided by an experiment. *Ann. Math. Statist.* **27**:986–1005.

———. (1957). Binomial sampling schemes and the concept of information. *Biometrika* **44**:179–186.

———. (1961a). Dynamic programming and decision theory. *Appl. Statist.* **10**:39–51.

———. (1961b). The use of prior probability distributions in statistical inference and decisions. *Proc. 4th Berkeley Symp. Math. Statist. Probability* **1**:453–468. University of California Press, Berkeley, Calif.

———. (1964). The Bayesian analysis of contingency tables. *Ann. Math. Statist.* **35**:1622–1643.

———. (1965). *Introduction to Probability and Statistics from a Bayesian Viewpoint; Part 1, Probability; Part 2, Inference.* Cambridge University Press, London.

———, and Barnett, B. N. (1965). Sequential sampling: Two decision problems with linear losses for binomial and normal variables. *Biometrika* **52**:507–532.

Linnik, Yu. V. (1963). On the Behrens-Fisher problem. *Bull. Inst. Intern. Statist.* **40**(2):833–841.

Loève, M. (1963). *Probability Theory,* 3d ed. D. Van Nostrand Company, Inc., Princeton, N.J.

Lorden, G. (1967). Integrated risk of asymptotically Bayes sequential tests. *Ann. Math. Statist.* **38**:1399–1422.

Luce, R. D. (1959). *Individual Choice Behavior.* John Wiley & Sons, Inc., New York.

———, and Raiffa, H. (1957). *Games and Decisions.* John Wiley & Sons, Inc., New York.

———, and Suppes, P. (1965). Preference, utility, and subjective probability. *Handbook of Mathematical Psychology* (ed. by Luce, Bush, and Galanter), vol. 3, pp. 249–410. John Wiley & Sons, Inc., New York.

MacQueen, J. B. (1961). A problem in survival. *Ann. Math. Statist.* **32**:605–610.

———. (1964a). Optimal policies for a class of search and evaluation problems. *Management Sci.* **10**:746–759.

———. (1964b). A problem in making resources last. *Management Sci.* **11**:341–347.

———. (1966). A modified dynamic programming method for Markovian decision problems. *J. Math. Anal. Appl.* **14**:38–43.

————, and Miller, R. G., Jr. (1960). Optimal persistence policies. *Operations Res.* **8**:362–380.

Maitra, A. (1965). Dynamic programming for countable state systems. *Sankhyā* (A)**27**:241–248.

————. (1966). A note on undiscounted dynamic programming. *Ann. Math. Statist.* **37**:1042–1044.

Mallows, C. L. (1959). The information in an experiment. *J. Roy. Statist. Soc.* (B)**21**:67–72.

————, and Robbins, H. (1964). Some problems of optimal sampling strategy. *J. Math. Anal. Appl.* **8**:90–103.

Mandelbrot, B. (1959). A note on a class of skew distribution functions: Analysis and critique of a paper by H. A. Simon. *Inform. Control* **2**:90.

————. (1961a). Final note on a class of skew distribution functions: Analysis and critique of a model due to H. A. Simon. *Inform. Control* **4**:198–216.

————. (1961b). Post scriptum to "Final Note." *Inform. Control* **4**:300–304.

————. (1965). A class of long-tailed probability distributions and the empirical distribution of city sizes. *Mathematical Explorations in Behavioral Science* (ed. by Massarik and Ratoosh), pp. 322–332. Richard D. Irwin, Inc., Homewood, Ill.

Marschak, J. (1950). Rational behavior, uncertain prospects, and measurable utility. *Econometrica* **18**:111–141.

————. (1954). Towards an economic theory of organization and information. *Decision Processes* (ed. by Thrall, Coombs, and Davis), pp. 187–220. John Wiley & Sons, Inc., New York.

————. (1959). Remarks on the economics of information. *Contributions to Scientific Research in Management*, pp. 79–100. Western Data Processing Center, University of California, Los Angeles.

————. (1963a). On adaptive programming. *Management Sci.* **9**:517–526.

————. (1963b). The payoff-relevant description of states and acts. *Econometrica* **31**:719–725.

————. (1964). Problems in information economics. *Management Controls: New Directions in Basic Research* (ed. by Bonini, Jaedicke, and Wagner), pp. 38–74. McGraw-Hill Book Company, New York.

————, and Miyasawa, K. (1968). Economic comparability of information systems. *Intern. Econ. Rev.* **9**:137–174.

Martin, J. J. (1967). *Bayesian Decision Problems and Markov Chains.* John Wiley & Sons, Inc., New York.

Matthes, T. K. (1963). On the optimality of sequential probability ratio tests. *Ann. Math. Statist.* **34**:18–21.

Matula, D. (1964). A periodic optimal search. *Am. Math. Monthly* **71**:15–21.

McCall, J. J. (1965). The economics of information and optimal stopping rules. *J. Business* **38**:300–317.

McCord, J. R., III, and Moroney, R. M., Jr. (1964). *Introduction to Probability Theory.* The Macmillan Company, New York.

Meyer, P. A. (1966). *Probability and Potentials.* Blaisdell Publishing Co., Waltham, Mass.

Mikhalevich, V. S. (1956). Sequential Bayes solutions and optimal methods of statistical acceptance control. *Theory Probability Appl.* **1**:395–421.

Miller, H. D. (1961). A generalization of Wald's identity with applications to random walks. *Ann. Math. Statist.* **32**:549–560.

Molenaar, W., and van der Velde, E. A. (1967). How to survive a fixed number of fair bets. *Ann. Math. Statist.* **38**:1278–1280.

Mood, A. M., and Graybill, F. A. (1963). *Introduction to the Theory of Statistics*, 2d ed. McGraw-Hill Book Company, New York.

Moriguti, S., and Robbins, H. (1962). A Bayes test of "$p \leq \frac{1}{2}$" versus "$p > \frac{1}{2}$." *Rept. Statist. Appl. Res. Union Japan. Scientists Engrs.* **9**:39–60.

Morse, N., and Sacksteder, R. (1966). Statistical isomorphism. *Ann. Math. Statist.* **37**:203–213.

Moser, L. (1956). On a problem of Cayley. *Scripta Math.* **22**:289–292.

Mosteller, F., and Nogee, P. (1951). An experimental measurement of utility. *J. Political Econ.* **59**:371–404.

———, and Wallace, D. L. (1964). *Inference and Disputed Authorship: The Federalist.* Addison-Wesley Publishing Company, Inc., Reading, Mass.

Nagel, E. (1937). Principles of the theory of probability. *International Encyclopedia of Unified Science* (1939), vol. 1, no. 6. The University of Chicago Press, Chicago.

Neveu, J. (1965). *Mathematical Foundations of the Calculus of Probability.* Holden-Day, Inc., San Francisco.

Neyman, J. (1929). Contribution to the theory of certain test criteria. *Bull. Inst. Intern. Statist.* 3–48. Reprinted in *A Selection of Early Statistical Papers of J. Neyman* (1967). University of California Press, Berkeley, Calif.

———. (1949). Contribution to the theory of the χ^2 test. *Proc. Berkeley Symp. Math. Statist. Probability* 239–273. University of California Press, Berkeley, Calif. Reprinted in *A Selection of Early Statistical Papers of J. Neyman* (1967). University of California Press, Berkeley, Calif.

——— (1967). R. A. Fisher (1890–1962): An appreciation. *Science* **156**:1456–1460.

Papoulis, A. (1965). *Probability, Random Variables, and Stochastic Processes.* McGraw-Hill Book Company, New York.

Parzen, E. (1960). *Modern Probability Theory and Its Applications.* John Wiley & Sons, Inc., New York.

———. (1962). *Stochastic Processes.* Holden-Day, Inc., San Francisco.

Pfeiffer, P. (1965). *Concepts of Probability Theory.* McGraw-Hill Book Company, New York.

Pitman, E. J. G. (1936). Sufficient statistics and intrinsic accuracy. *Proc. Cambridge Phil. Soc.* **32**:567–579.

Pratt, J. W. (1964). Risk aversion in the small and in the large. *Econometrica* **32**:122–136.

———. (1966). The outer needle of some Bayes sequential continuation regions. *Biometrika* **53**:455–467.

———, Raiffa, H., and Schlaifer, R. (1964). The foundations of decision under uncertainty: An elementary exposition. *J. Am. Statist. Assoc.* **59**:353–375.

———, ———, and ———. (1965). *Introduction to Statistical Decision Theory*, prelim. ed. McGraw-Hill Book Company, New York.

Prescott, E. C. (1967). Adaptive decision rules for macro economic planning. Unpublished Ph.D. dissertation, Graduate School of Industrial Administration, Carnegie-Mellon University, Pittsburgh, Pa.

Quisel, K. (1965). Extensions of the two-armed bandit and related processes with on-line experimentation. Unpublished technical report, Institute for Mathematical Studies in the Social Sciences, Stanford University, Stanford, Calif.

Radner, R. (1961). The evaluation of information in organizations. *Proc. 4th Berkeley Symp. Math. Statist. Probability* **1**:491–530. University of California Press, Berkeley, Calif.

———. (1962). Team decision problems. *Ann. Math. Statist.* **33**:857–881.

———. (1964). Mathematical specification of goals for decision problems. *Human*

Judgments and Optimality (ed. by Shelly and Bryan), pp. 178–216. John Wiley & Sons, Inc., New York.

Raiffa, H. (1961). Risk, ambiguity, and the Savage axioms: Comment. *Quart. J. Econ.* **75**:690–694.

———, and Schlaifer, R. (1961). *Applied Statistical Decision Theory.* Division of Research, Graduate School of Business Administration, Harvard University, Boston.

Ramsey, F. P. (1926). Truth and probability. Reprinted in *The Foundations of Mathematics and Other Logical Essays* (ed. by Braithwaite) (1950), Humanities Press, New York, and in *Studies in Subjective Probability* (ed. by Kyburg and Smokler) (1964), John Wiley & Sons, Inc., New York.

Rao, C. R. (1965). *Linear Statistical Inference and Its Applications.* John Wiley & Sons, Inc., New York.

Ray, S. N. (1963). Some sequential Bayes procedures for comparing two binomial parameters when observations are taken in pairs. Department of Statistics, University of North Carolina, Chapel Hill, N.C.

———. (1965). Bounds on the maximum sample size of a Bayes sequential procedure. *Ann. Math. Statist.* **36**:859–878.

Reichenbach, H. (1949). *The Theory of Probability.* University of California Press, Berkeley, Calif.

Rényi, A. (1961). On measures of entropy and information. *Proc. 4th Berkeley Symp. Math. Statist. Probability* **1**:547–561. University of California Press, Berkeley, Calif.

———. (1964). On the amount of information concerning an unknown parameter in a sequence of observations. *Publ. Math. Inst. Hung. Acad. Sci.* **9**:617–624.

———. (1966). On the amount of missing information and the Neyman-Pearson lemma. *Research Papers in Statistics* (ed. by David), pp. 281–288. John Wiley & Sons, Inc., New York.

———. (1967). On some basic problems of statistics from the point of view of information theory. *Proc. 5th Berkeley Symp. Math. Statist. Probability* **1**:531–543. University of California Press, Berkeley, Calif.

Robbins, H. E. (1952). Some aspects of the sequential design of experiments. *Bull. Am. Math. Soc.* **55**:527–535.

———. (1956a). A sequential design with finite memory. *Proc. Natl. Acad. Sci.* **42**:920–923.

———. (1956b). An empirical Bayes approach to statistics. *Proc. 3rd Berkeley Symp. Math. Statist. Probability* **1**:157–164. University of California Press, Berkeley, Calif.

———. (1964). The empirical Bayes approach to statistical decision problems. *Ann. Math. Statist.* **35**:1–20.

Roberts, H. V. (1963). Risk, ambiguity, and the Savage axioms: Comment (with reply by D. Ellsberg). *Quart. J. Econ.* **77**:327–342.

Ross, S. M. (1968). Non-discounted denumerable Markovian decision models. *Ann. Math. Statist.* **39**:412–423.

Ruben, H. (1959). A theorem on the cumulative product of independent random variables. *Proc. Cambridge Phil. Soc.* **55**:333–337.

Sacks, J. (1963). Generalized Bayes solutions in estimation problems. *Ann. Math. Statist.* **34**:751–768.

Sacksteder, R. (1967). A note on statistical equivalence. *Ann. Math. Statist.* **38**:787–794.

Sakaguchi, M. (1957). Notes on statistical applications of information theory, III. *Rept. Statist. Appl. Res. Union Japan. Scientists Engrs.* **5**:9–16.

———. (1959). Notes on statistical applications of information theory, IV. *Rept. Statist. Appl. Res. Union Japan. Scientists Engrs.* **6**:54–57.

———. (1961). Dynamic programming of some sequential sampling design. *J. Math. Anal. Appl.* **2**:446–466.

———. (1964). *Information Theory and Decision Making.* Unpublished lecture notes, Statistics Department, The George Washington University, Washington, D.C.

———. (1966). Topics in information and decision processes. Unpublished technical report, Statistics Department, The George Washington University, Washington, D.C.

Samuel, E. (1963a). Asymptotic solutions of the sequential compound decision problem. *Ann. Math. Statist.* **34**:1079–1094.

———. (1963b). An empirical Bayes approach to the testing of certain parametric hypotheses. *Ann. Math. Statist.* **34**:1370–1385.

———. (1964). Convergence of the losses of certain decision rules for the sequential compound decision problem. *Ann. Math. Statist.* **35**:1606–1621.

———. (1965a). Sequential compound estimators. *Ann. Math. Statist.* **36**:879–889.

———. (1965b). On simple rules for the compound decision problem. *J. Roy. Statist. Soc.* (B)**27**:238–244.

Savage, L. J. (1954). *The Foundations of Statistics.* John Wiley & Sons, Inc., New York.

———. (1957). When different pairs of hypotheses have the same family of likelihood-ratio test regions. *Ann. Math. Statist.* **28**:1028–1032.

———. (1961). *The Subjective Basis of Statistical Practice.* Unpublished report, Department of Mathematics, University of Michigan, Ann Arbor, Mich.

———, et al. (1962). *The Foundations of Statistical Inference.* Methuen & Co., Ltd., London.

Sawaragi, Y., Sunahara, Y., and Nakamizo, T. (1967). *Statistical Decision Theory in Adaptive Control Systems.* Academic Press Inc., New York.

Scarf, H., Gilford, D., and Shelly, M. (eds.). (1963). *Multistage Inventory Models and Techniques.* Stanford University Press, Stanford, Calif.

Scheffé, H. (1959). *Analysis of Variance.* John Wiley & Sons, Inc., New York.

Schlaifer, R. (1961). *Introduction to Statistics for Business Decisions.* McGraw-Hill Book Company, New York.

Schwartz, L. (1965). On Bayes procedures. *Z. Wahrscheinlichkeitstheorie und Verw. Gebiete* **4**:10–26.

Schwarz, G. (1962). Asymptotic shapes of Bayes sequential testing regions. *Ann. Math. Statist.* **33**:224–236.

Scott, D. (1964). Measurement structures and linear inequalities. *J. Math. Psychol.* **1**:233–247.

Shannon, C. E., and Weaver, W. (1949). *The Mathematical Theory of Communication.* The University of Illinois Press, Urbana, Ill. (Paperback edition, 1963.)

Siegmund, D. O. (1967). Some problems in the theory of optimal stopping rules. *Ann. Math. Statist.* **38**:1627–1640.

Simon, H. A. (1955). On a class of skew distribution functions. *Biometrika* **42**:425–440.

———. (1956). Dynamic programming under uncertainty with a quadratic criterion function. *Econometrica* **24**:74–81.

———. (1960). Some further notes on a class of skew distribution functions. *Inform. Control* **3**:80.

———. (1961a). Reply to "Final Note" by Benoit Mandelbrot. *Inform. Control* **4**:217–223.

————. (1961b). Reply to Dr. Mandelbrot's Post Scriptum. *Inform. Control* **4**:305–308.

Snell, J. L. (1952). Applications of martingale system theorems. *Trans. Am. Math. Soc.* **73**:293–312.

Sobel, M. (1953). An essentially complete class of decision functions for certain standard sequential problems. *Ann. Math. Statist.* **24**:319–337.

————, and Wald, A. (1949). A sequential decision procedure for choosing one of three hypotheses concerning the unknown mean of a normal distribution. *Ann. Math. Statist.* **20**:502–522.

Staroverov, O. V. (1963). On a searching problem. *Theory Probability Appl.* **8**:184–187.

Stein, C. (1946). A note on cumulative sums. *Ann. Math. Statist.* **17**:498–499.

————. (1956). Inadmissibility of the usual estimator for the mean of a multivariate normal distribution. *Proc. 3rd Berkeley Symp. Math. Statist. Probability* **1**:197–206. University of California Press, Berkeley, Calif.

————. (1962a). Confidence sets for the mean of a multivariate normal distribution (with discussion). *J. Roy. Statist. Soc.* (B)**24**:265–296.

————. (1962b). A remark on the likelihood principle. *J. Roy. Statist. Soc.* (A)**125**: 565–568.

————. (1964). Inadmissibility of the usual estimator for the variance of a normal distribution with unknown mean. *Ann. Inst. Statist. Math.* **16**:155–160.

————. (1965). Approximation of improper prior measures by prior probability measures. *Bernoulli, Bayes, Laplace Anniversary Volume* (ed. by Neyman and Le Cam), pp. 217–240. Springer-Verlag OHG, Berlin.

Stone, M. (1963). The posterior *t* distribution. *Ann. Math. Statist.* **34**:568–573.

————, and Springer, B. G. F. (1965). A paradox involving quasi prior distributions. *Biometrika* **52**:623–627.

Strassen, V. (1965). The existence of probability measures with given marginals. *Ann. Math. Statist.* **36**:423–439.

Strauch, R. E. (1966). Negative dynamic programming. *Ann. Math. Statist.* **37**:871–890.

Suppes, P., and Walsh, K. (1959). A nonlinear model for the experimental measurement of utility. *Behavioral Sci.* **4**:204–211.

Teicher, H. (1966). Higher moments of randomly stopped sums. *Teoriya Veroyatnostei i ee Primeneniya* **11**:179–185.

————, and Wolfowitz, J. (1966). Existence of optimal stopping rules for linear and quadratic rewards. *Z. Wahrscheinlichkeitstheorie und Verw. Gebiete* **5**:361–368.

Theil, H. (1964). *Optimal Decision Rules for Government and Industry.* Rand McNally & Company, Chicago.

Thompson, W. R. (1933). On the likelihood that one unknown probability exceeds another in view of the evidence of two samples. *Biometrika* **25**:285–294.

Thorp, E. O. (1961). A favorable strategy for twenty-one. *Proc. Natl. Acad. Sci.* **47**:110–112.

————. (1962). *Beat the Dealer.* Blaisdell, New York.

Thrall, R. M., Coombs, C. H., and Davis, R. L. (eds.). (1954). *Decision Processes.* John Wiley & Sons, Inc., New York.

Tiao, G. C., and Tan, W. Y. (1965). Bayesian analysis of random-effect models in the analysis of variance. I. Posterior distribution of variance-components. *Biometrika* **52**:37–53.

————, and Zellner, A. (1964a). Bayes' theorem and the use of prior knowledge in regression analysis. *Biometrika* **51**:219–230.

———, and ———. (1964b). On the Bayesian estimation of multivariate regression. *J. Roy. Statist. Soc.* (B)**26**:277–285.

Tou, J. T. (1964). *Modern Control Theory.* McGraw-Hill Book Company, New York.

Tucker, H. G. (1962). *An Introduction to Probability and Mathematical Statistics.* Academic Press Inc., New York.

Tweedie, M. C. K. (1960). Generalizations of Wald's fundamental identity of sequential analysis to Markov chains. *Proc. Cambridge Phil. Soc.* **56**:205–214.

Vande Linde, V. D. (1967). Optimal observation policies in linear stochastic systems. Unpublished Ph.D. dissertation, Department of Electrical Engineering, Carnegie-Mellon University, Pittsburgh, Pa.

Van Ryzin, J. (1966a). The sequential compound decision problem with $m \times n$ finite loss matrix. *Ann. Math. Statist.* **37**:954–975.

———. (1966b). Repetitive play in finite statistical games with unknown distributions. *Ann. Math. Statist.* **37**:976–994.

Veinott, A. F., Jr. (1966). The status of mathematical inventory theory. *Management Sci.* **12**:745–777.

Ville, J. (1939). *Étude Critique de la Notion de Collectif.* Gauthier-Villars, Paris.

Villegas, C. (1964). On qualitative probability σ-algebras. *Ann. Math. Statist.* **35**:1787–1796.

Vogel, W. (1960a). A sequential design for the two-armed bandit. *Ann. Math. Statist.* **31**:430–443.

———. (1960b). An asymptotic minimax theorem for the two-armed bandit problem. *Ann. Math. Statist.* **31**:444–451.

von Mises, R. (1957). *Probability, Statistics, and Truth.* The Macmillan Company, New York.

———. (1964). *Mathematical Theory of Probability and Statistics.* Academic Press Inc., New York.

Von Neumann, J., and Morgenstern, O. (1947). *Theory of Games and Economic Behavior,* 2d ed. Princeton University Press, Princeton, N.J.

Wagner, H. (1962). *Statistical Management of Inventory Systems.* John Wiley & Sons, Inc., New York.

Wald, A. (1947). *Sequential Analysis.* John Wiley & Sons, Inc., New York.

———. (1949). Note on the consistency of the maximum likelihood estimate. *Ann. Math. Statist.* **20**:595–601.

———. (1950). *Statistical Decision Functions.* John Wiley & Sons, Inc., New York.

———, and Wolfowitz, J. (1948). Optimum character of the sequential probability ratio test. *Ann. Math. Statist.* **19**:326–339.

———, and ———. (1950). Bayes solutions of sequential decision problems. *Ann. Math. Statist.* **21**:82–99.

Watson, G. S. (1966). Some Bayesian methods related to χ^2. *Bull. Inst. Intern. Statist.* **41**:64–76.

Weiss, L. (1961). *Statistical Decision Theory.* McGraw-Hill Book Company, New York.

———. (1963). The relative maxima of the likelihood function. *Skand. Aktuarietidskr.* **46**:162–166.

———. (1964). Sequential Bayes procedures which never observe more than a bounded number of observations. *Ann. Inst. Statist. Math.* **15**:177–185.

Wetherill, G. B. (1961). Bayesian sequential analysis. *Biometrika* **48**:281–292.

———. (1966). *Sequential Methods in Statistics.* John Wiley & Sons, Inc., New York.

Whittle, P. (1964). Some general results in sequential analysis. *Biometrika* **51**:123–141.

———. (1965). Some general results in sequential design (with discussion). *J. Roy. Statist. Soc.* (B)**27**:371–394.

Wilder, R. L. (1965). *Introduction to the Foundations of Mathematics*, 2d ed. John Wiley & Sons, Inc., New York.

Wilks, S. S. (1962). *Mathematical Statistics*. John Wiley & Sons, Inc., New York.

Yahav, J. A. (1966). On optimal stopping. *Ann. Math. Statist.* **37**:30–35.

Yao, Y. (1965). An approximate degrees of freedom solution to the multivariate Behrens-Fisher problem. *Biometrika* **52**:139–147.

Zellner, A., and Chetty, V. K. (1965). Prediction and decision problems in regression models from the Bayesian point of view. *J. Am. Statist. Assoc.* **60**:608 616.

———, and Tiao, G. C. (1964). Bayesian analysis of the regression model with autocorrelated errors. *J. Am. Statist. Assoc.* **59**:763–778.

supplementary bibliography

Aggarwal, O. P. (1959). Bayes and minimax procedures in sampling from finite and infinite populations—I. *Ann. Math. Statist.* **30**:206–218.

Aitchison, J. (1964). Bayesian tolerance regions. *J. Roy. Statist. Soc.* (B)**26**:161–175.

Albert, G. E. (1954). On the computation of the sampling characteristics of a general class of sequential decision problems. *Ann. Math. Statist.* **25**:340–356.

———. (1956). Accurate sequential tests on the mean of an exponential distribution. *Ann. Math. Statist.* **27**:460–470.

Anscombe, F. J. (1958). Rectifying inspection of a continuous output. *J. Am. Statist. Assoc.* **53**:702–719.

———. (1960). Notes on sequential sampling plans. *J. Roy. Statist. Soc.* (A)**123**:297–306.

———. (1961). Rectifying inspection of lots. *J. Am. Statist. Assoc.* **56**:807–823.

———. (1963). Tests of goodness of fit. *J. Roy. Statist. Soc.* (B)**25**:81–94.

Antelman, G. R. (1965). Insensitivity to non-optimal design in Bayesian decision theory. *J. Am. Statist. Assoc.* **60**:584–601.

Armitage, P. (1950). Sequential analysis with more than two alternative hypotheses, and its relation to discriminant function analysis. *J. Roy. Statist. Soc.* (B)**12**:137–144.

———. (1963). Sequential medical trials: Some comments on F. J. Anscombe's paper. *J. Am. Statist. Assoc.* **58**:384–387.

Ash, M., and Jones, W. (1964). Optimal strategies for maximum-number games. *J. Math. Anal. Appl.* **9**:138–140.

Barankin, E. W., and Maitra, A. P. (1963). Generalizations of the Fisher-Darmois-Koopman-Pitman theorem on sufficient statistics. *Sankhyā* (A)**25**:217–244.

Barnard, G. A. (1954). Sampling inspection and statistical decisions (with discussion). *J. Roy. Statist. Soc.* (B)16:151–174.

Bartholomew, D. J. (1965). A comparison of some Bayesian and frequentist inferences. *Biometrika* 52:19–35.

Bather, J. A. (1963). Control charts and the minimization of costs (with discussion). *J. Roy. Statist. Soc.* (B)25:49–80.

———. (1965). Invariant conditional distributions. *Ann. Math. Statist.* 36:829–846.

———. (1967). On a quickest detection problem. *Ann. Math. Statist.* 38:711–724.

Bechhofer, R. (1960). A note on the limiting relative efficiency of the Wald sequential probability ratio test. *J. Am. Statist. Assoc.* 55:660–663.

Beightler, C. S., and Mitten, L. G. (1964). Design of an optimal sequence of interrelated sampling plans. *J. Am. Statist. Assoc.* 59:96–104.

Bellman, R. (1961). A mathematical formulation of variational processes of adaptive type. *Proc. 4th Berkeley Symp. Math. Statist. Probability* 1:37–48. University of California Press, Berkeley, Calif.

———. (1965). Functional equations. *Handbook of Mathematical Psychology* (ed. by Luce, Bush, and Galanter), vol. 3, pp. 487–513. John Wiley & Sons, Inc., New York.

Berkson, J. (1930). Bayes theorem. *Ann. Math. Statist.* 1:42–56.

Bhat, B. R. (1964). Bayes solution of sequential decision problem for Markov dependent observations. *Ann. Math. Statist.* 35:1656–1662.

Bhattacharya, S. K. (1967). Bayesian approach to life testing and reliability estimation. *J. Am. Statist. Assoc.* 62:48–62.

Bickel, P. J., and Blackwell, D. (1967). A note on Bayes estimates. *Ann. Math. Statist.* 38:1907–1911.

———, and Yahav, J. A. (1968). Asymptotically optimal Bayes and minimax procedures in sequential estimation. *Ann. Math. Statist.* 39:442–456.

Birnbaum, A. (1961). On the foundations of statistical inference: Binary experiments. *Ann. Math. Statist.* 32:414–432.

Borch, K. (1967). The theory of risk (with discussion). *J. Roy. Statist. Soc.* (B)29:432–467.

Box, G. E. P., and Cox, D. R. (1964). An analysis of transformations (with discussion). *J. Roy. Statist. Soc.* (B)26:211–252.

———, and Jenkins, G. M. (1962). Some statistical aspects of adaptive optimization and control (with discussion). *J. Roy. Statist. Soc.* (B)24:297–343.

———, and Tiao, G. C. (1962). A further look at robustness via Bayes' theorem. *Biometrika* 49:419–433.

———, and ———. (1965). A change in level of a non-stationary time series. *Biometrika* 52:181–192.

———, and ———. (1967). Bayesian analysis of a three-component hierarchical design model. *Biometrika* 54:109–125.

———, and ———. (1968). Bayesian estimation of means for the random effect model. *J. Am. Statist. Assoc.* 63:174–181.

Bracken, J. (1966). Percentage points of the beta distribution for use in Bayesian analysis of Bernoulli processes. *Technometrics* 8:687–694.

Braga-Illa, A. (1964). A simple approach to the Bayes choice criterion: The method of extreme probabilities. *J. Am. Statist. Assoc.* 59:1227–1230.

Bulinskaya, E. V. (1967). Optimum inventory policies with a convex ordering cost function. *Theory Probability Appl.* 12:9–21.

Chernoff, H., and Zacks, S. (1964). Estimating the current mean of a normal distribution which is subject to changes in time. *Ann. Math. Statist.* 35:999–1018.

Chipman, J. S. (1960). The foundations of utility. *Econometrica* 28:193–224.

Churchman, C. W. (1956). Problems of value measurement for a theory of induction and decisions. *Proc. 3rd Berkeley Symp. Math. Statist. Probability* 5:53–59. University of California Press, Berkeley, Calif.

Cohen, A. (1965). Estimates of linear combinations of the parameters in the mean vector of a multivariate distribution. *Ann. Math. Statist.* 36:78–87.

Colton, T. (1963). A model for selecting one of two medical treatments. *J. Am. Statist. Assoc.* 58:388–400.

———. (1965). A two-stage model for selecting one of two treatments. *Biometrics* 21:169–180.

Copeland, A. H., Sr. (1956). Probabilities, observations and predictions. *Proc. 3rd Berkeley Symp. Math. Statist. Probability* 2:41–48. University of California Press, Berkeley, Calif.

Cox, D. R. (1958). Some problems connected with statistical inference. *Ann. Math. Statist.* 29:357–372.

———. (1960). Serial sampling acceptance schemes derived from Bayes' theorem. *Technometrics* 2:353–360.

de Finetti, B. (1961). The Bayesian approach to the rejection of outliers. *Proc. 4th Berkeley Symp. Math. Statist. Probability* 1:199–210. University of California Press, Berkeley, Calif.

Dickey, J. M. (1968). Smoothed estimates for multinomial cell probabilities. *Ann. Math. Statist.* 39:561–566.

Draper, N. R., and Hunter, W. G. (1966). Design of experiments for parameter estimation in multiresponse situations. *Biometrika* 53:525–533.

———, and ———. (1967a). The use of prior distributions in the design of experiments for parameter estimation in non-linear situations. *Biometrika* 54:147–153.

———, and ———. (1967b). The use of prior distributions in the design of experiments for parameter estimation in non-linear situations: Multiresponse case. *Biometrika* 54:662–665.

Duncan, D. B. (1961). Bayes rules for a common multiple comparisons problem and related Student-t problems. *Ann. Math. Statist.* 32:1013–1033.

Dunnett, C. W. (1960). On selecting the largest of k normal population means (with discussion). *J. Roy. Statist. Soc.* (B)22:1–40.

Dunsmore, I. R. (1966). A Bayesian approach to classification. *J. Roy. Statist. Soc.* (B)28:568–577.

Edwards, W., and Tversky, A. (eds.). (1967). *Decision Making.* Penguin Books, Inc., Baltimore.

Ehrenfeld, S., and Zacks, S. (1963). Optimal strategies in factorial experiments. *Ann. Math. Statist.* 34:780–791.

Eisenberg, E., and Gale, D. (1959). Consensus of subjective probabilities: The parimutuel method. *Ann. Math. Statist.* 30:165–168.

Ellison, B. E. (1962). A classification problem in which information about alternative distributions is based on samples. *Ann. Math. Statist.* 33:213–223.

Ericson, W. A. (1965). Optimal stratified sampling using prior information. *J. Am. Statist. Assoc.* 60:750–771.

———. (1967a). On the economic choice of experiment sizes for decision regarding certain linear combinations. *J. Roy. Statist. Soc.* (B)29:503–512.

———. (1967b). Optimal sample design with nonresponse. *J. Am. Statist. Assoc.* 62:63–78.

Farrell, R. H. (1968). Towards a theory of generalized Bayes tests. *Ann. Math. Statist.* 39:1–22.

Fraser, D. A. S. (1963). On the sufficiency and likelihood principles. *J. Am. Statist. Assoc.* **58**:641–647.

——. (1965). On information in statistics. *Ann. Math. Statist.* **36**:890–896.

Gart, J. J. (1959). An extension of the Cramér-Rao inequality. *Ann. Math. Statist.* **30**:367–380.

Ghosh, M. N. (1960). Bounds for the expected sample size in a sequential probability ratio test. *J. Roy. Statist. Soc.* (B)**22**:360–367.

Ghurye, S. G., and Wallace, D. L. (1959). A convolutive class of monotone likelihood ratio families. *Ann. Math. Statist.* **30**:1158–1164.

Girshick, M. A. (1958). An extension of the optimum property of the sequential probability ratio test. *Ann. Math. Statist.* **29**:288–290.

——, Karlin, S., and Royden, H. L. (1957). Multistage statistical decision pro cedures. *Ann. Math. Statist.* **28**:111–125.

——, and Rubin, H. (1952). A Bayes approach to a quality control model. *Ann. Math. Statist.* **23**:114–125.

——, and Savage, L. J. (1951). Bayes and minimax estimates for quadratic loss functions. *Proc. 2nd Berkeley Symp. Math. Statist. Probability*, pp. 53–74. University of California Press, Berkeley, Calif.

Good, I. J. (1952). Rational decisions. *J. Roy. Statist. Soc.* (B)**14**:107–114.

——. (1967). A Bayesian significance test for multinomial distributions (with discussion). *J. Roy. Statist. Soc.* (B)**29**:399–431.

Grettenberg, T. L. (1964). The ordering of finite experiments. *Trans. 3rd Prague Conf. Inform. Theory, Statist. Decision Functions, Random Processes*, pp. 193–206. Publishing House of the Czechoslovak Academy of Sciences, Prague.

Guthrie, D., Jr., and Johns, M. V., Jr. (1959). Bayes acceptance sampling procedures for large lots. *Ann. Math. Statist.* **30**:896–925.

Guttman, I., and Tiao, G. C. (1964). A Bayesian approach to some best population problems. *Ann. Math. Statist.* **35**:825–835.

Haggstrom, G. W. (1967). Optimal sequential procedures when more than one stop is required. *Ann. Math. Statist.* **38**:1618–1626.

Hald, A. (1960). The compound hypergeometric distribution and a system of single sampling inspection plans based on prior distributions and costs. *Technometrics* **2**:275–340.

——. (1967). Asymptotic properties of Bayesian single sampling plans. *J. Roy. Statist. Soc.* (B)**29**:162–173 and 586.

Hall, W. J., Wijsman, R. A., and Ghosh, J. K. (1965). The relationship between sufficiency and invariance with applications in sequential analysis. *Ann. Math. Statist.* **36**:575–614.

Hartigan, J. A. (1967). The likelihood and invariance principles. *J. Roy. Statist. Soc.* (B)**29**:533–539.

Hildreth, C. (1963). Bayesian statisticians and remote clients. *Econometrica* **31**:422–438.

Hill, B. M. (1963a). The three-parameter lognormal distribution and Bayesian analysis of a point-source epidemic. *J. Am. Statist. Assoc.* **58**:72–84.

——. (1963b). Information for estimating the proportions in mixtures of exponential and normal distributions. *J. Am. Statist. Assoc.* **58**:918–932.

Hodges, J. L., Jr., and Lehmann, E. L. (1952). The use of previous experience in reaching statistical decisions. *Ann. Math. Statist.* **23**:396–407.

Hoeffding, W. (1956). The role of assumptions in statistical decisions. *Proc. 3rd Berkeley Symp. Math. Statist. Probability* **1**:105–114. University of California Press, Berkeley, Calif.

Hoel, P. G., and Peterson, R. P. (1949). A solution to the problem of optimum classification. *Ann. Math. Statist.* **20**:433-438.

Karlin, S. (1958). Admissibility for estimation with quadratic loss. *Ann. Math. Statist.* **29**:406-436.

———, and Truax, D. (1960). Slippage problems. *Ann. Math. Statist.* **31**:296-324.

Kemp, K. W. (1958). Formulae for calculating the operating characteristic and the average sample number of some sequential tests. *J. Roy. Statist. Soc.* (B)**20**:379-386.

Kerridge, D. F. (1961). Inaccuracy and inference. *J. Roy. Statist. Soc.* (B)**23**:184-194.

———. (1963). Bounds for the frequency of misleading Bayes inferences. *Ann. Math. Statist.* **34**:1109-1110.

Kiefer, J., and Schwartz, R. (1965). Admissible Bayes character of T^2-, R^2-, and other fully invariant tests for classical multivariate normal problems. *Ann. Math. Statist.* **36**:747-770.

Kraft, C. H., and van Eeden, C. (1964). Bayesian bio-assay. *Ann. Math. Statist.* **35**:886-890.

Le Cam, L. (1954). Note on a theorem of Lionel Weiss. *Ann. Math. Statist.* **25**:791-793.

Lechner, J. A. (1964). Optimality and the OC curve for the Wald SPRT. *J. Am. Statist. Assoc.* **59**:464-468.

Lehmann, E. L. (1957a). A theory of some multiple decision problems. *Ann. Math. Statist.* **28**:1-25.

———. (1957b). A theory of some multiple decision problems, II. *Ann. Math. Statist.* **28**:547-572.

Lindley, D. V. (1953). Statistical inference (with discussion). *J. Roy. Statist. Soc.* (B)**15**:30-76.

Luce, R. D. (1968). On the numerical representation of qualitative conditional probability. *Ann. Math. Statist.* **39**:481-491.

Manne, A. (1960). Linear programming and sequential decisions. *Management Sci.* **6**:259-267.

Marks, B. L. (1962). Some optimal sequential schemes for estimating the mean of a cumulative normal quantal response curve. *J. Roy. Statist. Soc.* (B)**24**:393-400.

Messick, S., and Brayfield, A. H. (eds.). (1964). *Decision and Choice: Contributions of Sidney Siegel.* McGraw-Hill Book Company, New York.

Novick, M. R., and Grizzle, J. E. (1965). A Bayesian approach to the analysis of data from clinical trials. *J. Am. Statist. Assoc.* **60**:81-96.

Orford, R. J. (1963). Optimal stochastic control systems. *J. Math. Anal. Appl.* **6**:419-429.

Page, E. S. (1954). An improvement to Wald's approximations for some properties of sequential tests. *J. Roy. Statist. Soc.* (B)**16**:136-139.

Pfanzagl, J. (1963). Sampling procedures based on prior distributions and costs. *Technometrics* **5**:47-61.

Plackett, R. L. (1966). Current trends in statistical inference. *J. Roy. Statist. Soc.* (A)**129**:249-267.

Pratt, J. W. (1965). Bayesian interpretation of standard inference statements (with discussion). *J. Roy. Statist. Soc.* (B)**27**:169-203.

Roberts, C. D. (1963). An asymptotically optimal sequential design for comparing several experimental categories with a control. *Ann. Math. Statist.* **34**:1486-1493.

Roberts, H. V. (1965). Probabilistic prediction. *J. Am. Statist. Assoc.* **60**:50-62.

――――. (1967). Informative stopping rules and inferences about population size. *J. Am. Statist. Assoc.* **62**:763–775.

Scarf, H. (1959). Bayes solutions of the statistical inventory problem. *Ann. Math. Statist.* **30**:490–508.

Shaw, L. G. (1965). Optimum stochastic control. *Disciplines and Techniques of Systems Control* (ed. by Peschon), pp. 123–185. Blaisdell, New York.

Shelly, M. W., II, and Bryan, G. L. (1964). *Human Judgments and Optimality.* John Wiley & Sons, Inc., New York.

Shiryaev, A. N. (1964). On Markov sufficient statistics in non-additive Bayes problems of sequential analysis. *Theory Probability Appl.* **9**:604–618.

Siegmund, D. O. (1967). Some one-sided stopping rules. *Ann. Math. Statist.* **38**:1641–1646.

Skibinsky, M. (1960). Some properties of a class of Bayes two-stage tests. *Ann. Math. Statist.* **31**:332–351.

Smith, C. A. B. (1961). Consistency in statistical inference and decision (with discussion). *J. Roy. Statist. Soc.* (B)**23**:1–37 [and **28**(1966):252].

――――. (1965). Personal probability and statistical analysis (with discussion). *J. Roy. Statist. Soc.* (A)**128**:469–499.

Snell, J. L. (1965). Stochastic processes. *Handbook of Mathematical Psychology* (ed. by Luce, Bush, and Galanter), vol. 3, pp. 411–485. John Wiley & Sons, Inc., New York.

Springer, M. D., and Thompson, W. E. (1966). Bayesian confidence limits for the product of N binomial parameters. *Biometrika* **53**:611–613.

Steinhaus, H. (1957). The problem of estimation. *Ann. Math. Statist.* **28**:633–648.

Stone, M. (1959). Application of a measure of information to the design and comparison of regression experiments. *Ann. Math. Statist.* **30**:55–70.

――――. (1961). The opinion pool. *Ann. Math. Statist.* **32**:1339–1342.

――――. (1963). Robustness of non-ideal decision procedures. *J. Am. Statist. Assoc.* **58**:480–486.

――――. (1964). Comments on a posterior distribution of Geisser and Cornfield. *J. Roy. Statist. Soc.* (B)**26**:274–276.

Suppes, P. (1956). The role of subjective probability and utility in decision-making. *Proc. 3rd Berkeley Symp. Math. Statist. Probability* **5**:61–73. University of California Press, Berkeley, Calif.

――――. (1960). Some open problems in the foundations of subjective probability. *Information and Decision Processes* (ed. by Machol), pp. 162–170. McGraw-Hill Book Company, New York.

――――. (1961). Behavioristic foundations of utility. *Econometrica* **29**:186–202.

Suzuki, Y. (1960). On sampling inspection plans. *Ann. Inst. Statist. Math.* **11**:71–79.

Thatcher, A. R. (1964). Relationships between Bayesian and confidence limits for predictions (with discussion). *J. Roy. Statist. Soc.* (B)**26**:176–210.

Theil, H. (1963). On the use of incomplete prior information in regression analysis. *J. Am. Statist. Assoc.* **58**:401–414.

Thompson, J. W. (1964). A property of some symmetric two-stage sequential procedures. *Ann. Math. Statist.* **35**:755–761.

Tiao, G. C., and Tan, W. Y. (1966). Bayesian analysis of random-effect models in the analysis of variance. II. Effect of autocorrelated errors. *Biometrika* **53**:477–495.

Tingey, F. H., and Merrill, J. A. (1959). Minimum risk specification limits. *J. Am. Statist. Assoc.* **54**:260–274.

Vagholkar, M. K., and Wetherill, G. B. (1960). The most economical binomial sequential probability ratio test. *Biometrika* **47**:103–109.

Villegas, C. (1967). On qualitative probability. *Am. Math. Monthly* **74**:661-669.

Weiler, H. (1965). The use of incomplete beta functions for prior distributions in binomial sampling. *Technometrics* **7**:335-347.

Weiss, L. (1956). On the uniqueness of Wald sequential tests. *Ann. Math. Statist.* **27**:1178-1181.

———. (1963). On estimating scale and location parameters. *J. Am. Statist. Assoc.* **58**:658-659.

Wetherill, G. B. (1959). The most economical sequential sampling scheme for inspection by variables. *J. Roy. Statist. Soc.* (B)**21**:400-408.

———. (1960). Some remarks on the Bayesian solution of the single sample inspection scheme. *Technometrics* **2**:341-352.

———, and Campling, G. E. G. (1966). The decision theory approach to sampling inspection (with discussion). *J. Roy. Statist. Soc.* (B)**28**:381-416.

Wijsman, R. A. (1960). A monotonicity property of the sequential probability ratio test. *Ann. Math. Statist.* **31**:677-684.

———. (1963). Existence, uniqueness and monotonicity of sequential probability ratio tests. *Ann. Math. Statist.* **34**:1541-1548.

Winkler, R. L. (1967a). The assessment of prior distributions in Bayesian analysis. *J. Am. Statist. Assoc.* **62**:776-800.

———. (1967b). The quantification of judgment: Some methodological suggestions. *J. Am. Statist. Assoc.* **62**:1105-1120.

indexes

name index

subject index